Joint IMEKO TC1-TC7-TC13 Symposium: Measurement Science Challenges in Natural and Social Sciences

Journal of Physics: Conference Series Volume 1044

Rio de Janeiro, Brazil
31 July - 3 August 2017

Editor:

Rodrigo P.B. Costa-Felix

ISBN: 978-1-5108-6494-8
ISSN: 1742-6588

Printed from e-media with permission by:

Curran Associates, Inc.
57 Morehouse Lane
Red Hook, NY 12571

CURRAN ASSOCIATES INC. proceedings.com

Some format issues inherent in the e-media version may also appear in this print version.

Printed by Curran Associates, Inc. (2018)

For permission requests, please contact the Institute of Physics
at the address below.

Institute of Physics
Dirac House, Temple Back
Bristol BS1 6BE UK

Phone: 44 1 17 929 7481
Fax: 44 1 17 920 0979

techtracking@iop.org

Additional copies of this publication are available from:

Curran Associates, Inc.
57 Morehouse Lane
Red Hook, NY 12571 USA
Phone: 845-758-0400
Fax: 845-758-2633
Email: curran@proceedings.com
Web: www.proceedings.com

Table of contents

Volume 1044

2017 Joint IMEKO TC1-TC7-TC13 Symposium: Measurement Science Challenges in Natural and Social Sciences

31 July to 3 August 2017, Rio de Janeiro, Brazil

Accepted papers received: 29 May 2018
Published online: 18 June 2018

Preface

Papers

IOP Conf. Series: Journal of Physics: Conf. Series **1044** (2018) 011001 doi:10.1088/1742-6596/1044/1/011001

2017 Joint IMEKO TC1-TC7-TC13 Symposium: "Measurement Science Challenges in Natural and Social Sciences"

PREFACE

The 2017 Joint IMEKO TC1-TC7-TC13 Symposium: "Measurement Science Challenges in Natural and Social Sciences" was held in Rio de Janeiro, Brazil, from July 31st to August 3rd, 2017. The Joint Symposium was organized by the Pontifical Catholic University of Rio de Janeiro (PUC-Rio), the Brazilian Society of Metrology (SBM), and the International Measurement Confederation (IMEKO). The present volume reports the work of the 2017 Joint IMEKO TC1-TC7-TC13 Symposium.

The scope of the Joint Symposium covered a broad and diverse range of topics that are relevant to the Sponsoring Technical Committees of IMEKO: TC1 – Education and Training in Measurement and Instrumentation; TC7 – Measurement Science; TC13 – Measurements in Biology and Medicine.

Fostering a mature reflection about the relevance of developing a framework of research activities able to ensure reliable and globally comparable measurement results in all fields of science, the symposium positively inspired participants to share their knowledge and experience in challenging new areas, particularly in the advances of measuring more complex phenomena such as those from the Biomedical, Behavioural and Social Sciences. The keynote lectures promoted intensive scientific discussions and exchange of information regarding these quite advanced topics in biomeasurement; biostandards; metrological traceability of biological quantities and Rasch measurement model for Behavioral and Social Sciences.

The highly interactive sessions of the Joint Symposium effectively promoted fruitful debates, approaching scholars, researchers, students, regulators, and professionals engaged in the art and science of measurements, from 16 countries, including Mexico, Ukraine, Australia, and Malaysia.

This volume includes seventy-three papers organized according to the presentation sessions. Each article was independently peer-reviewed by at least two reviewers from the Symposium Steering and International Programme Committees.

The Symposium, the first edition of the Joint IMEKO TC1-TC7-TC13 event to take place in a Latin American country, was held in Rio de Janeiro, at the Convention Centre of Rio Othon Palace Hotel. Undoubtedly, the venue offered a right balance between relevant factors: the quality of the infrastructure and the beauty of the location, by the sea, benefiting from the emblematic Copacabana beach, surrounded by the mountains of the Atlantic Forest. Altogether, these features certainly contributed to providing an inspiring atmosphere for the highly multi-inter-and-transdisciplinary nature of the scientific debate that has enthusiastically emerged from the Symposium.

The organizers express their most profound gratitude and appreciation to the prominent participants of the Symposium and all the members of the Symposium Steering and International Program Committees, many of whom contributed as reviewers providing the evaluations and recommendations of the papers included in this volume. We would also like to express our special thanks to our sponsors, partners, and notably, to the event's team at Brazilian Society of Metrology for their hard work in the organization.

We hope that the present collection of diversified scientific contributions provides a breeding ground to inspire the next steps towards new advances, innovations and overcoming of the challenges of measurement science.

Elisabeth Costa Monteiro (PUC-Rio) – Conference chair

Rodrigo P.B. Costa-Felix (Inmetro) – Editor-in-chief

Carlos Roberto Hall Barbosa (PUC-Rio) – General secretary

IOP Conf. Series: Journal of Physics: Conf. Series **1044** (2018) 011002 doi:10.1088/1742-6596/1044/1/011002

Peer review statement

All papers published in this volume of *Journal of Physics: Conference Series* have been peer reviewed through processes administered by the proceedings Editors. Reviews were conducted by expert referees to the professional and scientific standards expected of a proceedings journal published by IOP Publishing.

A contactless ammeter based on GMR magnetometers

C Schuina Neves[1], D P Magalhães[2], C R Hall Barbosa[1], E C Oliveira[1,3]

[1] Postgraduate Program in Metrology, [2] Department of Electrical Engineerig, Pontifical Catholic University of Rio de Janeiro, Rio de Janeiro, 22541-900, Brazil
[3] Petrobras Transporte S.A., Technology Management, Rio de Janeiro, RJ, 20091-060, Brazil
E-mail: camilaschuina@aluno.puc-rio.br

Abstract. This work presents a contactless ammeter prototype based on GMR magnetometers, able to measure direct and alternating electrical currents. Two techniques were used to solve the inverse problem, which is, given the output voltages of the GMR sensors, to estimate the electric current flowing through the conductor - nonlinear curve fitting and artificial neural networks. The performance of both techniques was compared based on the standard uncertainty of the data calculated by each one. The results obtained with the neural networks are significantly better.

1. Introduction

Ammeters, essential for various applications in the industry, are instruments whose purpose is the measurement of electric currents in conductors. They can be of several types depending on the physical principle used for their implementation and on the need to interrupt the circuit for the current to be measured [1] [2] [3]. It is expected that the developed ammeter prototype will perform measurements of direct and alternating currents with high resolution, in relation to the present clamp ammeters based on coils and Hall effect sensors. The proposed system, initially presented in [2] [3], was developed with the purpose of also estimating the distance between the sensor and the conductor.

The main purpose in this work is to compare the performance of the technique used to solve the inverse problem, that is needed to estimate the current, previously presented in [3],nonlinear curve fitting, with artificial neural networks (ANN).

This paper is divided into: Section 2 presents the operating principle of giant magnetoresistance (GMR) magnetometers and how to use them to build an ammeter. Section 3 shows step by step how the data were acquired, pre-processed and selected. The computational results are described in Section 4. At last, Section 5 presents the conclusions.

2. Concepts

The following sections explain the operating principle of GMR magnetometers and how they are applied to the measurement of electric currents.

2.1. Giant magnetoresistance (GMR)

The basic operating principle of GMR is the variation of the resistance of a material as a function of an external magnetic field [4]. This variation, typically between 10 and 20 %, if compared to the maximum sensitivity of other magnetic sensors, is very large, hence the name giant magnetoresistance. Magnetoresistance can be found in classical semiconductors, particularly in magnetic semiconductors and, like the Hall effect, has its origin in the Lorentz force [4].

Although a single resistance can be used as a sensor element, the Wheatstone bridge configuration is a good recommendation because, in this topology, it is possible to obtain a differential voltage output as a function of the resistance variation [4] with lower influence of external factors such as temperature. In the development of this project it was used a commercial GMR, an integrated circuit, model AA005-02, manufactured by NVE Corporation [5]. The topology of the transducer is a Wheatstone half-bridge, having two magnetically shielded GMRs and two GMRs that are sensitive to the magnetic field. Without the presence of an external magnetic field the bridge is balanced and the output of the bridge is zero. In the presence of an external magnetic field, the field-sensitive GMRs change their values, unbalancing the bridge and generating a differential voltage output. The output signal of the bridge is directly proportional to the supply voltage/current, with a typical sensitivity of 0.45 mV/V/Oe.

2.2. *GRM based ammeter*

An electric current passing through a wire generates a circular magnetic field around it that varies with the intensity of the current and with the distance from the conductor to the measuring point [1]. By applying the Biot-Savart law to a straight conductor it is observed that the magnetic field H, measured at a distance r_1 from the conductor, is given by

$$H = \frac{I}{500 r_1}, \tag{1}$$

where I is the electric current in amperes, r_1 is the distance between the conductor and the sensor in meters and H is the magnetic field in oersteds (1 Oe = 1000/4π A/m).

Analyzing Eq. (1), it is noticed that the use of only one GMR sensor is not sufficient, since it is necessary to know the distance r_1 to estimate the electric current I. Thus, a configuration has been previously proposed with two GMR sensors [2] [3], separated by a fixed distance (D = 3 cm), in addition to a permanent magnet that generates the magnetic field of 15 Oe (so that the sensor operates in its linear range), as shown in Figure 1.

By measuring the voltage generated by both GMR sensors it is possible to estimate the current I, regardless of the distance r_1 of the conductor from the sensors, and also estimate the distance r_1 between the sensor GMR_1 and the conductor.

The electronic conditioning and reading circuit shown in Figure 2 was designed and implemented. The electronic circuit has the functions of feeding the GMR sensors with a DC current of approximately 3 mA (current source based on the LM318 operational amplifier), as well as reading the differential output voltage of the bridge and amplifying it by 50 times (instrumentation amplifier INA129, with gain defined by the 1 kΩ resistor). The offset voltage level of -5.0625 V allows a zero output voltage, V_{out}, to be obtained when H = 15 Oe (polarization field). Thus, for a current of 0 A in the wire, there will be an output of 0 V.

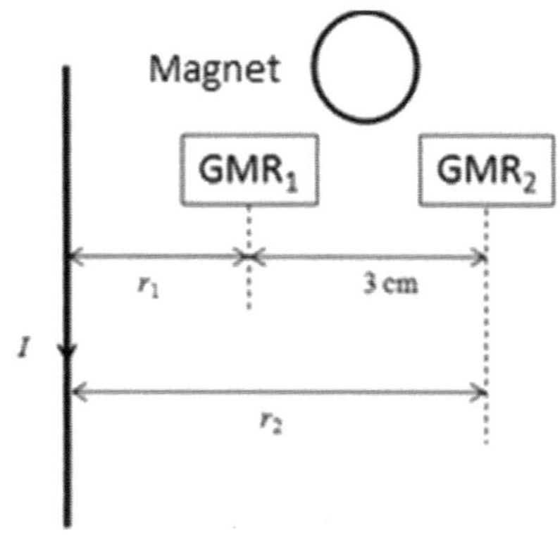

Figure 1. Schematic diagram of the ammeter.

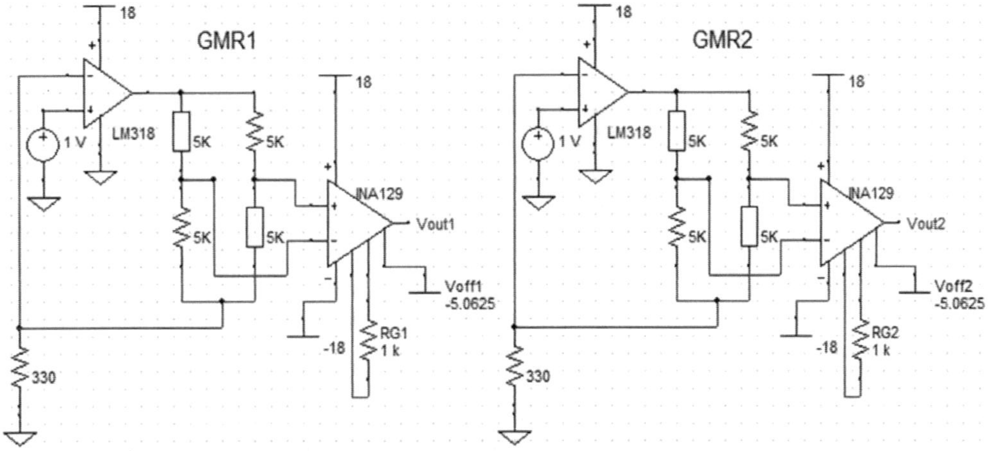

Figure 2. Schematic diagram of the electronic conditioning circuit of GMR magnetometers.

3. Acquisition, pre-processing and selection of data

To observe the effect of the hysteresis and any other fluctuations, ten tests with currents varying between -3 A and 3 A were performed, in steps of approximately 0.2 A. The result of the ten test cycles can be seen in Figure 3, containing 602 electrical current values and the respective output voltages.

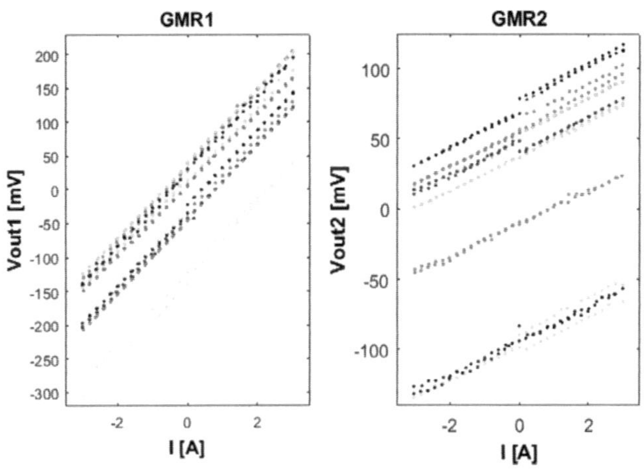

Figure 3. Results of the ten test cycles measured by the 2 GMR sensors. Each color corresponds to one test cycle.

To minimize sensor errors and improve the reliability of the measurements, the data were acquired using an A/D converter, model NI-USB 6001, with 14 bits resolution. In eight tests, for each sensor, an acquisition frequency of 10 kHz was used, with measurements taken every 2 s and, in two tests, with a frequency of acquisition of 6 kHz, and measurements taken every 3 s.

Analyzing Figure 3, it is possible to observe a variation on the signal offset, probably caused by the temperature effect or other external interferences. To solve this problem, and to remove the signal offset, the linear coefficient of the estimated trend lines for each test was excluded. Figure 4 shows the final data set and the overall trend lines, which show small linear coefficients and large determination coefficients (R^2).

IOP Conf. Series: Journal of Physics: Conf. Series **1044** (2018) 012001 doi:10.1088/1742-6596/1044/1/012001

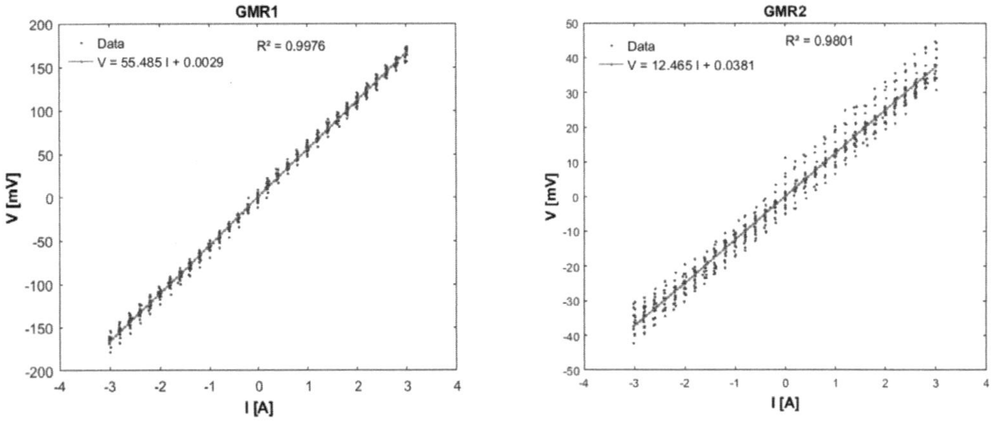

Figure 4. Final experimental data after removal of individual offsets.

4. Inverse Problem

As the objective is the development of an ammeter, it is necessary to solve the inverse problem, which is, given the output voltages of the GMR sensors, measured by the data acquisition system, to estimate the electric current flowing through the conductors. For that, two techniques were used: nonlinear curve fitting and artificial neural networks. The following subsections present the details of each of the techniques.

4.1. Artificial Neural Networks

A two-layer feed-forward network, with sigmoidal transfer function in the hidden layer and linear transfer function in the output layer was created using Matlab's Neural Network Fitting Tool.

The final data set described in section 3 was randomly divided into training, validation and testing, respectively 70 %, 20 % and 10 % of the data. The network was trained with Levenberg-Marquardt backpropagation algorithm. Training process automatically stops when generalization stops improving, as indicated by an increase in the mean squared error (MSE) of the validation samples. The training process continued until the validation error failed to decrease for six iterations [6].

In order to determine the optimal number of neurons in the hidden layer, 5000 networks were trained, with the number of hidden neurons varying from one to fifty. For each number of neurons, a hundred networks were trained and it was calculated an average Root Mean Squared Error (RMSE) for training and validation data. After repeating these tests five times, it has been obtained an average RMSE around 68-78 mA and a maximum improvement with the number of neurons around 8 mA.

Considering the results obtained previously, a small number of neurons in the hidden layer already presents satisfactory results to solve the proposed problem. Thus, the tests were performed using 5 and 10 neurons in the hidden layer. For each number of neurons, the neural networks were trained 50 times, and the one with best validation RMSE was chosen for testing. The following steps show the performance results for both networks.

The regression analysis in Figure 5 and Figure 6 shows that the data estimated by the neural networks are strongly correlated with the experimental data, with values of R (correlation coefficient) higher than 0.999 for validation, training and testing steps, for both network configurations.

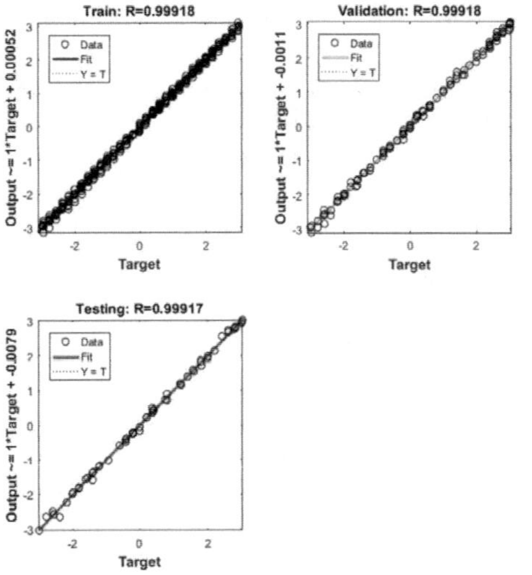

Figure 5. Regression analysis for a network with 5 neurons in the hidden layer.

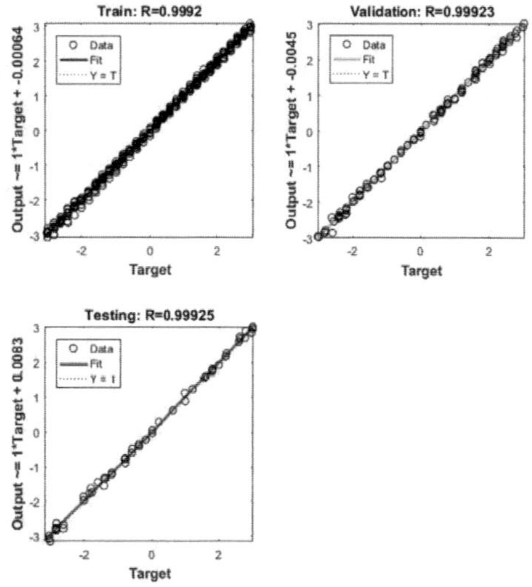

Figure 6. Regression analysis for a network with 10 neurons in the hidden layer.

At last, the test RMSE was calculated (Table 1) for each network. The network can predict the electrical current value, given the GMRs voltage values, with high accuracy, with errors around 71 – 76 mA.

Table 1. Test RMSE vs Hidden layer neurons

Hidden layer neurons	RMSE values [mA]
5	71.76
10	76.72

4.2. Nonlinear curve fitting

To solve the nonlinear curve fitting problem, described in Eqs. (2) to (5), it was used Matlab's function "lsqcurvefit", which adjusts the parameters of the problem, minimizing the error between the theoretical and experimental data. Given the arrangement of the system, the output voltages of the circuit can be deduced from the electric current and several parameters. For a specific sensor n, Eqs. (2) to (5) present a step-by-step calculation of the output voltage in the circuit (V_n) as a function of current (I), sensor sensitivity (K_n), the magnetic field related to the permanent magnet (H_{0n}), the offset voltage in the sensor due to internal factors (V_{0n}), the INA129 gain (G_n) and the offset voltage (V_{offn}) of the INA129. H_n represents the magnetic field at which the sensor is located and V_n represents the output voltage of the circuit, which is measured by the data acquisition system.

$$H_n = \frac{I}{2\pi r} + H_{0_n} \tag{2}$$

$$V_{n_{GMR}} = K_n H_n + V_{0_n} \tag{3}$$

$$V_n = V_{n_{GMR}} G_n + V_{off_n} \tag{4}$$

$$V_n = \left[K_n\left(\frac{I}{2\pi r} + H_{0_n}\right) + V_{0_n}\right] G_n + V_{off_n} \tag{5}$$

By means of the data and the analytical equation, the "lsqcurvefit" function of Matlab, the values of the parameters that best fit the problem are estimated, in order to generate a function that follows the phenomenon and is better approximated to the experimental data, in terms of the mean squared error.

For this, it is necessary to choose initial values for the parameters to be optimized, as well as limits (upper and lower) for the parameters, as indicated in Table 2.

Table 2. Parameters for the nonlinear curve fitting algorithm.

Parameter	Initial value	Lower Limit	Upper limit
K_1 and K_2	6.75 mV/Oe	6.75 mV/Oe	9.75 mV/Oe
r_1	0.01 m	0.008 m	0.012 m
r_2	0.04 m	0.038 m	0.042 m
H_{01} and H_{02}	15 Oe	13.5 Oe	16.5 Oe
V_{01} and V_{02}	0 V	-100 mV	100 mV
G_1 and G_2	50	49	51
V_{off1} and V_{off2}	-5.0625 V	-6 V	+ ∞ V

The data were processed and the optimal parameters found, as indicated in Table 3, and used to calculate Eqs. (6) and (7) of the output voltage in the circuits as a function of the current in the conductor and the distance r_1 to the sensor 1.

Table 3. Optimal parameters found by the curve fitting algorithm.

Parameter	Optimal value	Parameter	Optimal value
K_1	6.7500 mV/Oe	V_{01}	6.5 mV
K_2	6.7500 mV/Oe	V_{02}	35.6 mV
r_1	1.1929 cm	G_1	49.0273
r_2	4.200 cm	G_2	49.0000
H_{01}	15.0470 Oe	V_{off1}	-4.9799 V
H_{02}	14.5058 Oe	V_{off2}	-4.7995 V

Finally, in order to verify the reliability of the system, the respective current values were calculated for the different output voltage values of the GMRs obtained in section 3. Figure 7 shows the graph of the measured points, where the x-axis represents the nominal current values and the y-axis represents the current values calculated by means of Eqs. (6) and (7).

$$V_1 = 0.6619.\frac{I}{r_1} + 0.0029 \quad (6)$$

$$V_2 = 0.6615.\frac{I}{r_1 + 0,03} + 0.0087 \quad (7)$$

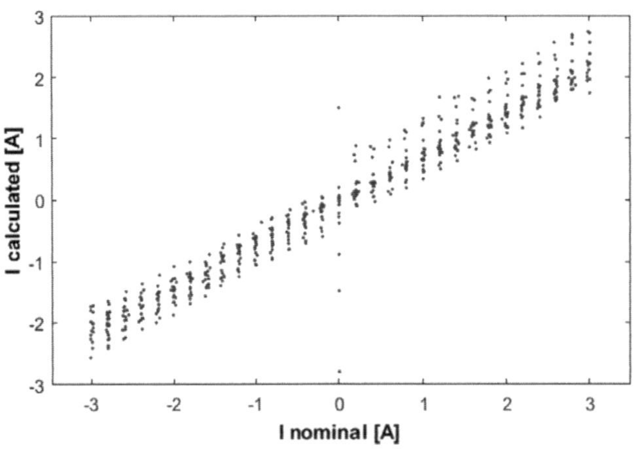

Figure 7. Nominal vs calculated current values.

5. Discussion and conclusions

The performance of both techniques was compared based on the standard uncertainty of data calculated by each one, as given by Table 4.

Table 4. Uncertainty of the inverse problem solution methods.

	Uncertainty [A]
Nonlinear regression	0.54006
ANN with 5 neurons	0.07133
ANN with 10 neurons	0.07052

The results show that the artificial neural networks technique fits much better the solution of this problem. In future works we intend to study the effects of temperature and to magnetically isolate the sensor, as well as improvements in the conditioning circuit, in order to eliminate the offset variations and interferences observed and obtain more consistent data, so as to improve the obtained uncertainty.

Acknowledgments
The authors would like to thank the Brazilian agencies CNPq, FAPERJ and FINEP for the financial support in the development of this research.

References

[1] P. Ripka, "Eletric current sensors: A review," *Measurement Science an Technology,* p. 23, 2010.

[2] C. L. S. Filho and V. B. Barbeta, "Amperímetro DC não invasivo utilizando sensor GMR," *Revista Pesquisa & Tecnologia FEI,* vol. 21, p. 5, 2001.

[3] C. R. H. Barbosa, M. C. Carvalho, C. Schuina, E. C. Silva and L. A. P. Gusmão, *Amperímetro sem contato baseado em sensores GMR,* Bento Gonçalves, RS, 2015.

[4] R. Càndid, C. María-Dolores and R. M. Diego, "Magnetic field sensors based on Giant Magnetoresistance (GMR) technology: Applications in electrical current sensing," *Sensors,* vol. 9, p. 24, 2010.

[5] N. Corporation. [Online]. Available: http://www.nve.com/Downloads/analog_catalog.pdf. [Accessed 20 11 2015].

[6] MATLAB, "MathWorks," [Online]. Available: https://www.mathworks.com/help/nnet/gs/fit-data-with-a-neural-network.html . [Accessed 15 01 2017].

A noncontact instrument based on ultrasound for the evaluation of asynchronous thoracoabdominal movement in respiratory diseases

C F Costa[1], N V Oliveira[1], E P Silva Júnior[1] and P L Melo[1,2,*]
[1]Biomedical Instrumentation Laboratory, Institute of Biology Roberto Alcantara Gomes and Faculty of Engineering, State University of Rio de Janeiro, Brazil
[2]Laboratory of Clinical and Experimental Research in Vascular Biology, State University of Rio de Janeiro, Brazil

E-mail: plopes@uerj.br

Abstract. Several respiratory diseases, as well as surgical procedures, may introduce changes in the thoracoabdominal movement. Non-intrusive techniques have the potential to simplify the exam of these patients. This work describes the design of a noncontact system able to evaluate changes in the thoracic and abdominal movement. Two ultrasonic transducers were used, providing analogue signals to a virtual instrument. In vitro evaluations using a flat surface in motion showed small errors in the evaluation of the asynchrony ($\Delta\varphi$=6.8O), and the relative contributions of the abdominal (CRAb) and thoracic (CRTx) compartments (CRAb=1.3%; CRTx=-1.3%). In vivo tests in normal subjects simulating abnormal conditions resulted in higher values of asynchrony ($p<0.005$), and differences in the CRAb ($p<0,02$) and CRTx ($p<0.005$). We concluded that the system is able to detect both, normal breathing and modifications associated with simulated respiratory disease.

1. Introduction

The respiratory process causes synchronized movements in the chest wall and abdomen. However, various respiratory diseases as well as surgical procedures cause changes in the timing of these movements. This asynchrony is usually classified into: (1) a delay between the movement of the chest (Ch) and the abdomen (Ab) during expansion or contraction and (2); a paradoxical movement, consisting of opposite movements between the chest and the abdomen, also known as full asynchrony [1, 2]. The diseases that cause changes in thoracoabdominal movement include asthma and chronic obstructive pulmonary disease (COPD).

In a previous study on healthy individuals and patients with COPD [3] we observed that in some individuals, the straps used to monitor respiratory movements could cause small changes in the normal ventilation pattern due to tactile stimulation as well as restriction due to the positioning of the straps. These transducers are obtrusive because they need to be adapted to the patient's body and may limit the application of the system in patients with respiratory diseases in an advanced state.

Using optical principles, Uhlig et al. [4] obtained promising results in the development of a monitor suitable for the evaluation of spontaneous breathing in adults and of a mechanical model that

* Corresponding author.

simulated high-frequency breathing. Other studies have shown promising results using systems based on optical principles [5, 6], including 3D movement analysis [7] and radar-based systems [8-10].

A promising alternative for the development of non-intrusive systems involves the use of ultrasound-based methods [11, 12]. The use of these transducers is suitable because the signals are easily generated and detected, the wavelength is relatively small compared with the movement of the chest and abdomen, and the low energy level does not negatively affect the monitored individuals.

This study describes the development of an entirely non-intrusive system based on ultrasonic transducers.

2. Materials and Methods

2.1. System development

2.1.1. Hardware design. Figure 1 depicts a simplified diagram of the instrument and transducers positioning. Two ultrasonic transducers (model 18-11117, Sick Automation Inc., Minneapolis, USA) were used [13]. The signals coming from the transducers were associated with a DC level generated from the positioning of the participant. Therefore, the first processing block consisted of a high-pass filter (23 mHz, first order). After that, we observed the first amplification stage with a gain of 2. The resulting signal was adapted to a second high-pass filter (23 mHz, first order) and an amplifier (gain of 455). The signal was further processed by a low-pass filter (Butterworth, fourth order, 1 Hz) to remove external noise and prevent aliasing. The two position signals were adapted to a data acquisition module (NI 6008, National Instruments, Austin, Texas) with a resolution of 12 bits, four channels, and a maximum sampling frequency (Fs) of 10 kHz. The Fs value used in this study was 196 Hz [3].

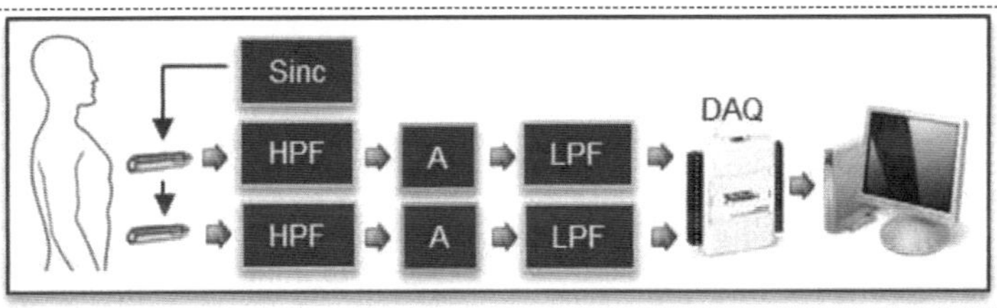

Figure 1: Simplified block diagram of the instrument. HPF = high-pass filter; A = amplifier and LPF = low-pass filter.

2.1.2. Software design. The asynchrony between the respiratory movements of the chest and abdomen is generally evaluated by analysing the phase difference (Δφ) between the signals obtained in these two compartments [14, 15]. This phase difference reflects the delay between excursions of the chest and abdomen. When the two compartments move in perfect synchrony, $\Delta\varphi=0^{\circ}$. With increasing asynchrony, Δφ increases to a maximum of 180°, which corresponds to the point at which the compartments are completely out of phase. Analysis of the Lissajous curves is the method traditionally used for obtaining Δφ [14]. However, this approach assumes that the excursions of the chest and abdomen are sine waves. In practice, these excursions are irregular in healthy individuals and tend to be more irregular in patients, which may introduce problems with these measurements [14]. A method that uses cross-correlation [15, 16] has been suggested for the analysis of thoracoabdominal asynchrony. In this particular application, this method presents significant advantages since it is independent of the waveforms and is robust in the presence of noise. In this study, this method was implemented using the subroutines available in LabVIEW 8.2 environment (National Instruments, Austin, TX).

2.2. System evaluation

2.2.1. In vitro tests. Before using the system in humans, its proper functioning was evaluated using a mechanical system that realistically simulated normal thoracoabdominal movement. To this end, we used a vibrating platform, which consisted of an oscillator connected to a DC-coupled amplifier, which in turn was connected to a loudspeaker (Arlen, 12 inches, Subcompact) that produced the desired movements. The system was tested over a moving range of approximately 2 mm using a sinusoidal signal of 0.3 Hz.

2.2.2. In vivo tests in healthy individuals during normal spontaneous respiration and simulated asynchronous conditions. These studies were conducted in 12 healthy subjects with no history of respiratory disease or smoking (mean age 31.0 ± 9.0 years, mean height 1.70 ± 0.09 m; mean weight 72.3 ± 13.6 kg). The volunteers were asked to perform 20 s analyses under four conditions: (1) normal breathing; (2) breathing primarily via the abdomen; (3) breathing primarily via the chest; and (4) in simulated asynchrony. The results are presented as the mean ± standard deviation. The results were compared using the Wilcoxon test. The protocol followed the guidelines of the Helsinki Declaration and Brazilian National Health Council Resolution No. 196/96 and was approved by the Research Ethics Committee of Pedro Ernesto University Hospital. Before the examinations, all subjects were fully informed about the content of the tests and signed a consent form.

3. Results

3.1. System development

Figure 2 illustrates the system and its use. The program consisted of three main modules, which were presented in a tab control structure. The first tab (Figure 3A) was dedicated to patient data (name, age, height, and weight) and the visual analysis of the signals of the respiratory movements measured during the test.

Figure 2: Photo illustrating the developed system and its use.

The second tab (Figure 3B) automatically calculated the mean peak and valley values of the abdominal movements (Ab) and thoracic movements (Tx) and the relative contribution of the abdomen (RCAb) and thorax (RCTx) to the total movement (TM = Ab + Tx) of the respiratory system (RCAb = 100% Ab/TM and RCTx = 100%Tx/TM). A graphic representation of the cross-correlation between the signals of the abdominal and thoracic movements and the $\Delta\varphi$ value are shown in the second tab.

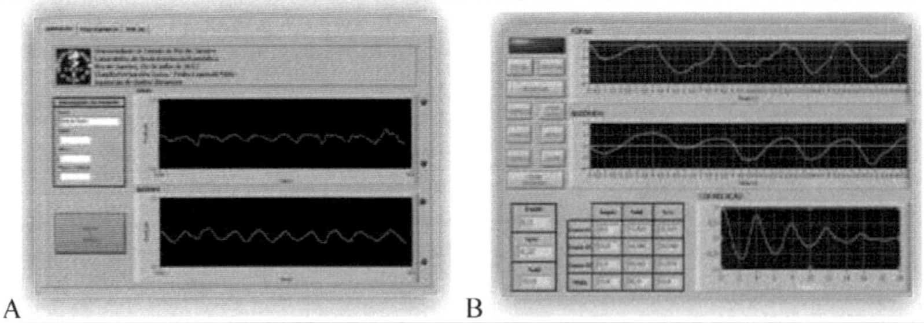

A B

Figure 3: Frontal panels of the software modules associated with data acquisition (A) and cross-correlation and asynchrony evaluation (B).

The final test result was obtained by averaging the three partial tests. After each partial test, the instrument allowed the researcher to save the data to an ASCII file for later visualization. The program allows the averaging of the measurements obtained in each partial test and saving the final data in an ASCII file.

3.2. *System evaluation*

3.2.1. In vitro tests. Figure 4 shows the in vitro test results. The asynchrony observed was small ($\Delta\varphi$ = 6.8°). Similarly, the error (ε) in the assessment of the contribution of each compartment was small (εRCTx = −1.3% and εRCAb = 1.3%).

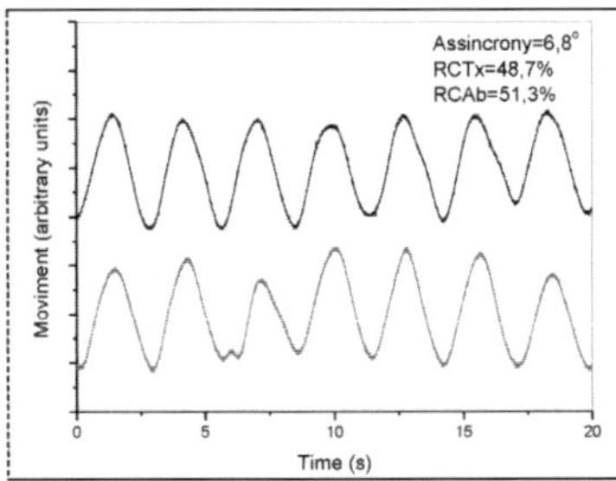

Figure 4: Signals obtained in vitro testing and results of asynchrony analyzes and the relative contribution of abdominal compartments (RCAB) and thoracic (RCTx).

3.2.2. In vivo tests in healthy individuals during normal spontaneous respiration and simulated asynchronous conditions. Figure 5 shows the typical plots of the movements measured in the chest and abdomen of a healthy individual during spontaneous breathing and during abnormal movement simulation. Figure 6 describes the analysis of the abdominal (A) and thoracic (B) relative contributions during spontaneous thoracic and abdominal ventilation. Note the phase difference observed during the simulation of asynchrony (C).

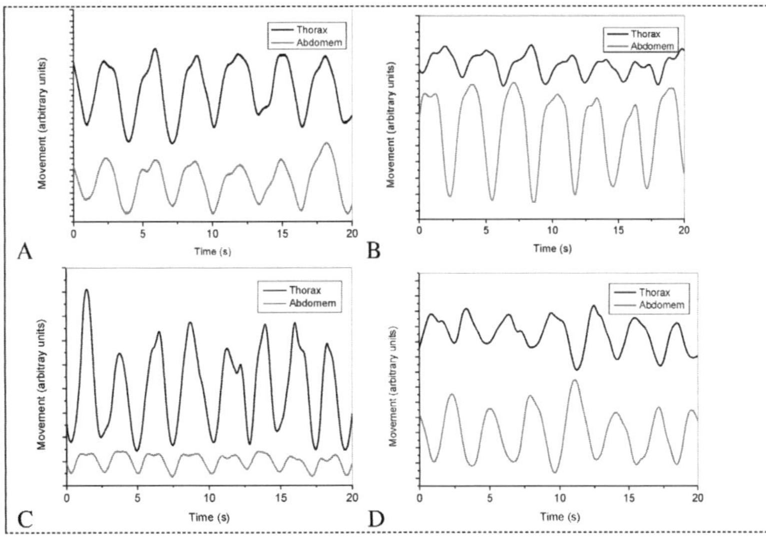

Figure 5: Typical movement waveforms in normal subjects under conditions of spontaneous ventilation (A) and during simulation of ventilation predominantly through the abdomen (B), through the thorax (C) and simulating asynchrony (D).

4. Discussion

The assessment of asynchronous respiratory movements is usually performed using straps with transducers around the chest and abdomen. However, these measurements can affect the pattern of the chest and abdominal movement [17, 18]. The system proposed in this study eliminated the need to adapt transducers to the patient.

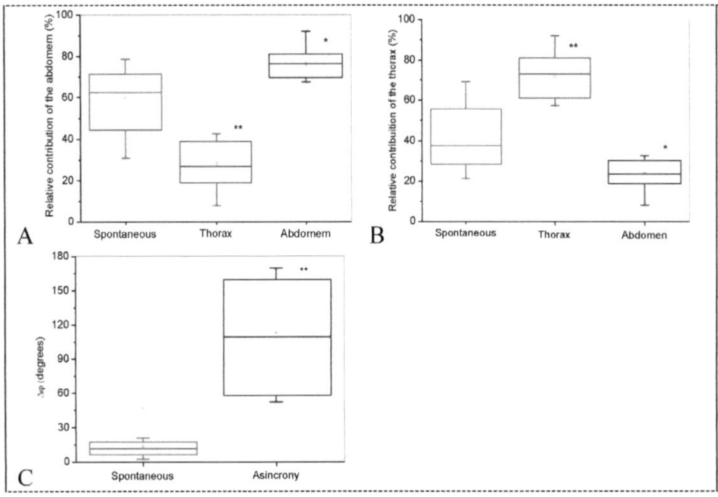

Figure 6: Relative contributions of the abdominal (A) and thoracic (B) compartments during spontaneous, thoracic and abdominal ventilation. Phase difference in the simulation of asynchrony (C). * and ** represent p<0.02 and p<0.005 compared to spontaneous ventilation, respectively.

During the initial phase of *in vitro* testing (Figure 4), errors smaller than 7° were observed. The theoretical values of RCTx and RCAb on *in vitro* testing were 50%. In practice, these tests presented errors smaller than 2.0% (Figure 4). Therefore, the cited values are appropriate for the target application.

The low $\Delta\varphi$ values observed in healthy individuals indicate the presence of synchronous movements between the chest and abdomen (Figure 5C). These results, as well as the RCAb and RCTx values found in healthy individuals (Figure 5), are consistent with the results of a previous study from our group using piezoelectric straps [3] and of other studies [2, 19, 20]. The relative contribution of simulated predominantly abdominal breathing (Figure 6A) was higher than under conditions of spontaneous breathing, indicating that the system properly identified this movement. Moreover, the system suitably identified the movement that simulated predominantly chest breathing (Figure 6B). The $\Delta\varphi$ values obtained by simulating asynchronous breathing ($\Delta\varphi$ = 113.9 ± 47.1°, Figure 6C) were higher than the values obtained during spontaneous breathing ($\Delta\varphi$ = 13.9 ± 12.5°; $p < 0.001$), indicating that the system properly identified this movement. Therefore, the description of the changes in the respiratory movements provided by our system are in close agreement with the physiological foundations involved in breathing, [19, 20] and the simulated abnormal respiratory movements were properly identified by the proposed system.

Conclusion & future work

A new instrument that allows a non-intrusive analysis of the respiratory system movement was developed. The applications of the system in a mechanical model and normal individuals simulating respiratory movement abnormalities showed results in close agreement with a literature, as well as with the physiological principles involved. The system has high potential to contribute to the improvement of the analysis of alterations due to respiratory diseases or associated to rehabilitation procedures. The next phases of this project include its use in the evaluation of patients with different respiratory diseases attended at our University Hospital and the development of a three dimensional version of the instrument to produce images associated with respiratory movements.

Acknowledgments

This study was supported by the Brazilian Council for Scientific and Technological Development (CNPq) and Rio de Janeiro State Research Supporting Foundation (FAPERJ).

References

[1] Sackner M A, Gonzalez H, Rodriguez M, Belsito A, Sackner D R and Grenvik S 1984 Assessment of asynchronous and paradoxic motion between rib cage and abdomen in normal subjects and in patients with chronic obstructive pulmonary disease *The American review of respiratory disease* **130** 588-93

[2] Tobin M J, Perez W, Guenther S M, Lodato R F and Dantzker D R 1987 Does rib cage-abdominal paradox signify respiratory muscle fatigue? *Journal of applied physiology* **63** 851-60

[3] da Silva Junior E P, Esteves G P, Dames K K and Melo P L 2011 A telemedicine instrument for Internet-based home monitoring of thoracoabdominal motion in patients with respiratory diseases *The Review of scientific instruments* **82** 014301

[4] Uhlig T, Kondo T and Sly P D 1997 Measurements of PEEP-induced changes in lung volume: evaluation of a laser monitor *Chest* **112** 107-12

[5] Cala S J, Kenyon C M, Ferrigno G, Carnevali P, Aliverti A, Pedotti A, Macklem P T and Rochester D F 1996 Chest wall and lung volume estimation by optical reflectance motion analysis *Journal of applied physiology* **81** 2680-9

[6] Kondo T, Uhlig T, Pemberton P and Sly P D 1997 Laser monitoring of chest wall displacement *The European respiratory journal* **10** 1865-9

[7] Kaneko H 2013 Estimating breathing movements of the chest and abdominal wall using a simple, newly developed breathing movement measuring device *Respiratory care*

[8] Bechet P, Mitran R and Munteanu M 2013 A non-contact method based on multiple signal classification algorithm to reduce the measurement time for accurately heart rate detection *The Review of scientific instruments* **84** 084707

[9] Hu W, Zhao Z, Wang Y, Zhang H and Lin F 2013 Noncontact Accurate Measurement of Cardiopulmonary Activity Using a Compact Quadrature Doppler Radar Sensor *IEEE transactions on bio-medical engineering*

[10] Dei D, Grazzini G, Luzi G, Pieraccini M, Atzeni C, Boncinelli S, Camiciottoli G, Castellani W, Marsili M and Dico J L 2009 Non-contact detection of breathing using a microwave sensor *Sensors* **9** 2574-85

[11] Min S D, Yoon D J, Yoon S W, Yun Y H and Lee M 2007 A study on a non-contacting respiration signal monitoring system using Doppler ultrasound *Medical & biological engineering & computing* **45** 1113-9

[12] O. Baltag, C. M. Rau and Banarescu A 2009 Non invasive monitoring of the cardio respiratory activity. In: *World Congress on Medical Physics and Biomedical Engineering,* (Munich, Germany: Springer) pp 529-31

[13] AG. S 2013 UM18-11117 - Ultrasonic sensors. In: *Datasheet,*

[14] Beydon N, Davis S D, Lombardi E, Allen J L, Arets H G, Aurora P, Bisgaard H, Davis G M, Ducharme F M, Eigen H, Gappa M, Gaultier C, Gustafsson P M, Hall G L, Hantos Z, Healy M J, Jones M H, Klug B, Lodrup Carlsen K C, McKenzie S A, Marchal F, Mayer O H, Merkus P J, Morris M G, Oostveen E, Pillow J J, Seddon P C, Silverman M, Sly P D, Stocks J, Tepper R S, Vilozni D, Wilson N M, American Thoracic Society/European Respiratory Society Working Group on I and Young Children Pulmonary Function T 2007 An official American Thoracic Society/European Respiratory Society statement: pulmonary function testing in preschool children *American journal of respiratory and critical care medicine* **175** 1304-45

[15] Black A M and Millard R K 2001 Assessing thoracoabdominal asynchrony *Clinical physiology* **21** 383-5

[16] Prisk G K, Hammer J and Newth C J 2002 Techniques for measurement of thoracoabdominal asynchrony *Pediatric pulmonology* **34** 462-72

[17] de Groote A, Verbandt Y, Paiva M and Mathys P 2000 Measurement of thoracoabdominal asynchrony: importance of sensor sensitivity to cross section deformations *Journal of applied physiology* **88** 1295-302

[18] Hershenson M B, Colin A A, Wohl M E and Stark A R 1990 Changes in the contribution of the rib cage to tidal breathing during infancy *The American review of respiratory disease* **141** 922-5

[19] Sharp J T, Goldberg N B, Druz W S, Fishman H C and Danon J 1977 Thoracoabdominal motion in chronic obstructive pulmonary disease *The American review of respiratory disease* **115** 47-56

[20] Konno K and Mead J 1967 Measurement of the separate volume changes of rib cage and abdomen during breathing *Journal of applied physiology: respiratory, environmental and exercise physiology* **22** 407-22

A nonintrusive system for monitoring the temperature distribution in the bed with applications in the analysis of positional snoring and the prevention of skin ulcerations

D G Oliveira[1], R Wajnsztajn[1], J A Silva[1], J A Mesquita Júnior[1] and P L Melo[1,2,*]
[1]Biomedical Instrumentation Laboratory, Institute of Biology and Faculty of Engineering, State University of Rio de Janeiro, Brasil
[2]Laboratory of Clinical and Experimental Research in Vascular Biology, State University of Rio de Janeiro, BR

E-mail: plopes@uerj.br

Abstract. The monitoring of patient position in bed is important in the diagnosis of positional snoring and in the prevention of ulcerations of the skin associated with long confinement to a bed. The aim of this work is to describe a system based in the measurement of the temperature distribution in bed able to perform these tasks. The device consists of a PC adapted to modules for measurements of temperature and sound level. The software is developed in LabVIEW environment and manages the reception, analysis and storage of sampled signals. In vivo studies showed adequate performance, confirming the high scientific and clinical potential of this system.

1. Introduction

Approximately 3% to 7 % of the adult population presents respiratory disorders during sleep, such as apnea and snoring [1]. The central phenomenon for the emergence of snoring is the increase muscle flaccidness of throat. The tone of the throat muscles is reduced during sleep, which progressively leads to contact between the walls that generates the characteristic noise. This can happen at any age, being more common among men from 45 years. Furthermore, anatomical factors such as obesity, small chin, narrow bite, roof of the mouth (palate) in warhead format, increased tonsils and adenoids may predispose to snoring. In short, all factors contributing to close the air passage and facilitate contact between the throat walls propitiate snoring.

Positional snoring is snoring that disappears with the change of posture. Although it can occur at various positions, the most common is in the decubitus dorsal position. This position can make the tongue project back due to gravitational effects and cause snoring. This disorder is easily treatable provided it is properly diagnosed the position that causes the problem.

This disturbance is usually diagnosed using a video camera with infrared, which records the patient throughout the night. However, for many patients it implies in loss of privacy. Additionally, this method is expensive and extremely laborious in terms of interpretation, which contributes to inhibit the clinical use of such exams. Thus, the diagnostic of this disease is usually performed by interview with the mate of the patient, an unreliable method. It is also worth noting that, although the effect of

* To whom any correspondence should be addressed.

position on sleep apnea is well known, objective evidences concerning the effect of position on snoring have not yet been presented [2]. A recent review pointed out that there is room for technical improvement of the devices to reduce discomfort and consequent disruption of sleep architecture as to improve compliance [3].

Long periods of confinement to bed may lead to local circulatory disorders causing ulcerations of the skin [4]. When pressure is maintained for long periods capillary endothelium is damaged, resulting in thrombosis of small vessels and consequent tissue necrosis and ulceration. The patient's body movement allows distributing the forces potentially causing ischemia, so that the monitoring of the mobility is an important factor in the prevention of ulceration.

This paper describes a system based on temperature sensors, analog and digital signal processing systems, integrated into a personal microcomputer, to identify the position of the patient in bed. The instrument has great potential for use in the applications described previously, reduced cost and allows patient privacy maintenance.

2. Materials and Methods

2.1. System development

Hardware design. Figure 1 describes a simplified block diagram of the instrument. The system evaluates the patient's position using a series of 14 thermistors distributed in equal distances on the bed by means of a band. This apparatus allows evaluating the temperature profile of the bed, providing an indirect assessment of the position of the patient. The thermistors (350 Ω at 25 °C, NTC) were excited by low-current sources (715 μA) in order to avoid self-heating [5]. The resulting signals are amplified and filtered (low-pass, 0.5Hz, first order) to reduce noise and interference. The operation principle is based on the temperature difference between the ambient (usually 25 °C) and the body temperature (usually around 37 °C). Thus, the increase in ambient temperature can introduce unacceptable errors. To avoid this inconvenience, the room temperature is constantly monitored by a thermistor positioned in a different point, separated from the band used to evaluate the patient's position. Experimental studies were performed in order to obtain the resistance versus temperature characteristic curve of the thermistors used in the position sensing subsystem, as well as in the thermistor used in monitoring the room temperature [5]. Based on these studies were prepared subroutines converting the measured voltage to temperature. Thus, the system allows visualizing the behaviour of the temperature of the 14 thermistors placed in the bed as well as the resulting temperature profile, which was obtained from a 4th order polynomial adjustment.

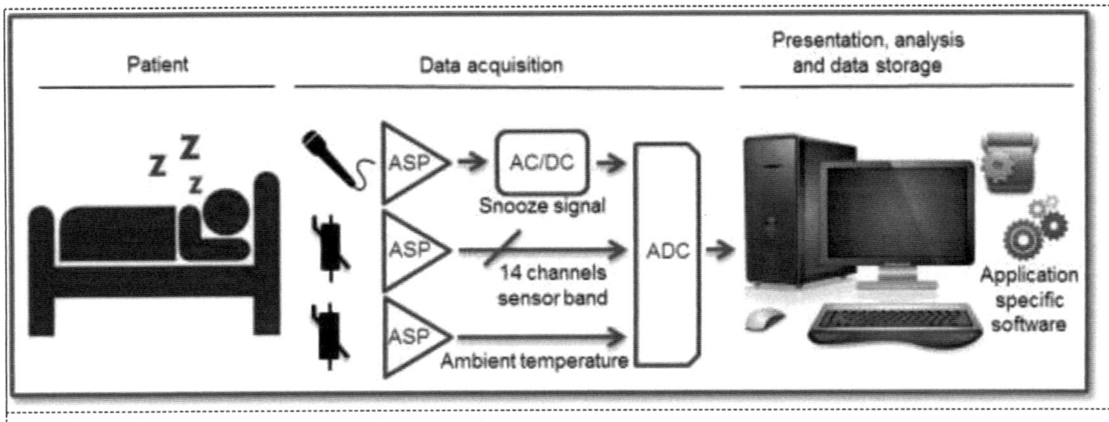

Figure 1: Simplified block diagram of the instrument. ASP: analog signal processing, AC/DC: alternate to continuous converter, ADC: analog to digital converter.

The snoring signal is measured by an electret microphone (CZN-15E), whose signal is amplified and applied to a full wave rectifier (AC/DC converter) [6]. The data acquisition system consists of a virtual instrument (VI) developed in LabVIEW environment (National Instruments, Austin, TX). The system also includes a data acquisition board presenting 12 bit of resolution (PCI6024E, National Instruments, Austin, TX).

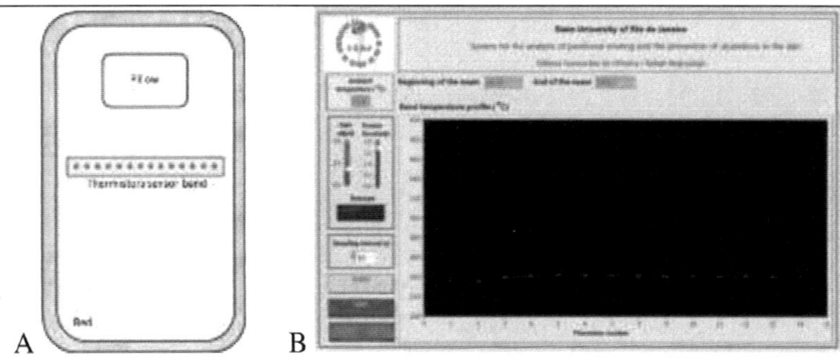

Figure 2: Superior view describing the positioning of the sensor band in the bed during the exams and the arrangement of the thermistors along the band (A). Frontal panel of the module associated with data acquisition (B).

2.1.1. Software design. The software was developed in LabVIEW environment (National Instruments, Austin, TX). The front panel of the VI is shown in figure 2. The system allows the user to set the time interval between measurements. In each measurement, it is performed a scan where the snoring signals, the bed temperature profile and the environment temperature are evaluated. Thus, the program can be used in both applications envisaged, for sleep studies as well as to monitor the movement of patients in bed preventing ulceration.

Used in the specific case of sleep studies, the snoring monitoring subsystem offers a gain adjustment in the amplifier circuit, an indicator on the front panel of the virtual instrument and a threshold adjustment that allows minimizing the effect of ambient noise. This optimizes the detection of the snoring signal. To this end, before use of the instrument, the level selector is set slightly above the ambient noise. The user simulates the patient snoring and the amplifier gain is adjusted so that, in the presence of snoring, the voltage at the AC/DC converter output is greater than that produced by the background environmental noise, indicating the presence of snoring. It was observed that, in practical exams, transient noises would be erroneously interpreted as snoring. To overcome this inconvenient, the system includes a discrimination routine, where acoustical noise signals with duration < 1s are eliminated. At the end of each exam the instrument allows researchers to save the data in an ASCII file so that they can be analysed later. In this file are stored the environmental temperature data, the temperature in each of 14 thermistors that composes the measurement band and the profile obtained from the adjusted polynomial. The file also includes data on the presence (1) or absence (0) of snoring, as well the scans times and the beginning and end of the exam.

2.2. System evaluation

2.2.1. In vivo simulated tests in healthy individuals. The ability of the system to identify the position of the patient in the bed and the presence of positional snoring was initially evaluated through a comparative analysis of the results obtained in three healthy individuals. The volunteers do not present history of respiratory diseases or snoring. The band was fixed on the bed at the level of the patient's waist height (figure 2A). The system performance in the identification of the patient´s position was evaluated in a simulated study conducted asking the volunteer to be positioned in frontal decubitus, dorsal, right side and left side.

IOP Conf. Series: Journal of Physics: Conf. Series **1044** (2018) 012003 doi:10.1088/1742-6596/1044/1/012003

The snoring detection system was tested by asking the volunteer to simulate snoring in the mentioned positions. The microphone was adapted to the bedside, as described in figure 3 Three snoring events were simulated in each position for each volunteer, resulting in 36 snoring events. The time (hh:mm:ss) when these events happened was manually recorded. These results were used as a reference in comparison with that obtained in the proposed system.

2.2.2. In vivo test in a snorer individual. This analysis was performed in a typical volunteer with clinical diagnosis of snoring. The volunteer was allowed to sleep freely during the afternoon (nap test). This exam followed a similar procedure to that used in the normal volunteers, using the manually recorded position and presence (or not) of snoring as a reference.

3. Results

3.1. System development

Before initiating the exams, the snoring monitoring subsystem positioned as described in figure 3A, and the gain and threshold adjusted in the front panel of the virtual instrument (figure 2B) to minimize the effect of ambient noise. Then, the exams initiate pressing the button "Start", having the duration controlled controlled by pressing the button "Exit". At the end of the exam, the system offers the possibility to save the results along the measurement time in txt format ("Save" option in figure 2B). The system allows the visualization of the temperature profilles obtained along the examination, as well as the presence of snoring during the examination. An indicative changes the colour (brown for red) during snoring events (figure 2B.

3.2. System evaluation

Figures 3 illustrate the typical temperature profiles obtained in the four positions studied. Clearly discriminative signatures were observed when the volunteer was positioned in dorsal, right side and left side decubitus (figure 3, A, B and C, respectively). Ventral decubitus (figure 3D), however, presented a similar signature to that presented in dorsal decubitus (figure 3A).

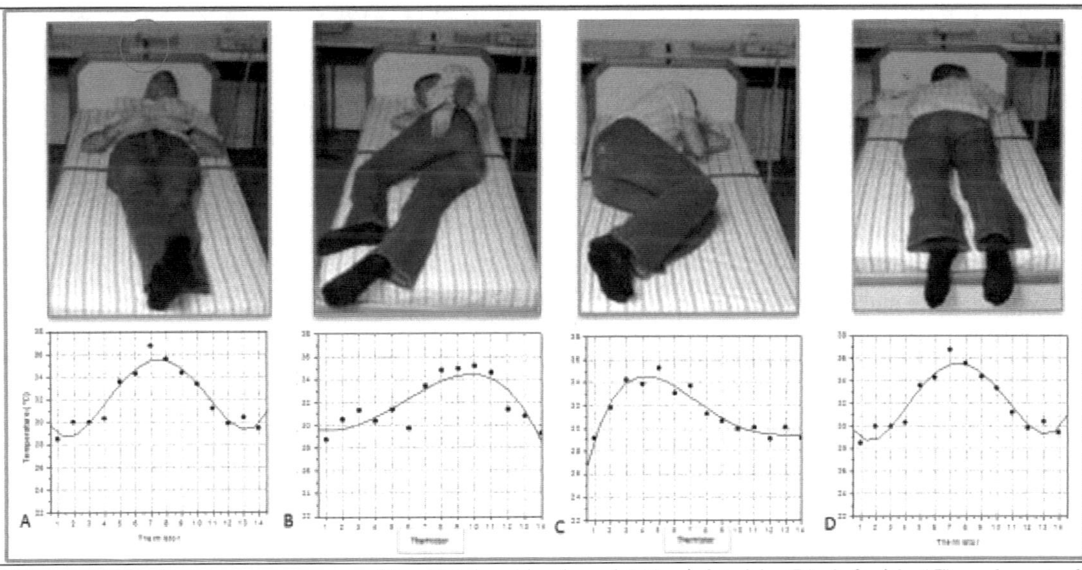

Figure 3: Typical temperature profiles obtained in dorsal (A), right side (B), left side (C) and ventral decubitus (D). Positioning of the microphone used in the snoring monitoring subsystem (A).

Typical results observed in simulation of snoring are described in figure 4. The system was able to correctly identify the presence of snoring in all of the 36 simulated events. The exam performed in the snorer lasted approximately one hour, and showed adequate identification of all of the events produced during of exam. Additionally, it was also noted that transient noises, such as the fall of glasses and cough, did not introduced erroneous measures.

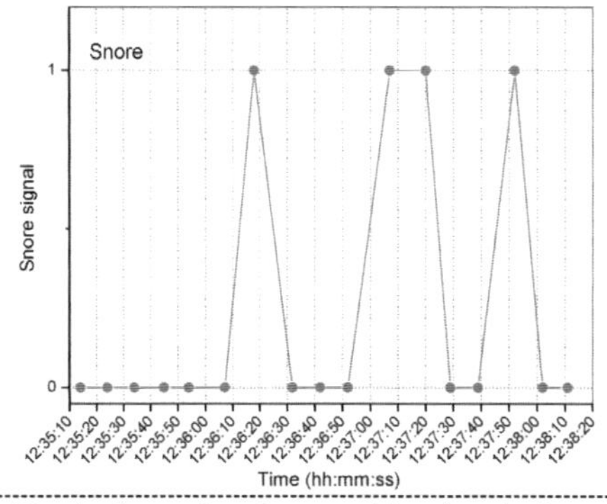

Figure 4: Typical results observed in simulation of snoring.

4. Discussion

The monitoring of the patient's position in bed has great practical interest. The diagnosis and treatment of positional apnea depend on the accurate identification of the position in which the individual snore during sleep. Another important application relates to patient monitoring in long periods of confinement to bed in order to avoid ulcerations on the skin. To this end, various methods of assessing the position in bed can be used, including methods based on pressure transducers [7], motion picture recording [8], as well as integrated temperature sensor circuits [8]. Methods based on pressure transducers have a high cost, while the image-based methods are intrusive, require large memory and complex and/or time-consuming analysis [8]. Solutions based on tilt sensors [9], three-axis accelerometers [10], nasal pressure monitoring [11] and wireless tracheal acoustic sensors [12] demands intrusive adaptations of the instrument to the patient´s body, which may reduce patient compliance to the exam. This study employs an completely nonintrusive alternative based on low-cost temperature transducers [8, 13], which allows monitoring the temperature profile of the bed. As can be seen in typical results described in figure 3, the individual's body positioning causes the temperature to rise on specific thermistors, approaching body temperature. Thus, the temperature profile obtained by the analysis of the 14 thermistors evaluates the position of the individual, whereas a temperature pattern is produced depending on the type of decubitus.

Similar to other instruments, the limitations of our system need to be recognized. The patterns produced by the right and left lateral decubitus are easily identified when compared with each other and the patterns achieved in the dorsal and frontal decubitus. However, the patterns produced by the dorsal and front decubitus are identical. Although this is not a problem in monitoring aimed at preventing ulcerations, this characteristic introduces limitations in the study of positional snoring. A possible solution to this problem would be to use a small emitter of infrared signal, which could be tailored to the patient's pajama and positioning a detector in the ceiling of the room where the test is performed. In dorsal decubitus position the signal would be picked up by the detector, while in ventral decubitus this signal would be blockaded, allowing the appropriate identification between the quoted decubitus. This solution, however, would introduce a small intrusive characteristic in the system. Another possible solution would use of a thermographic camera, which would be less intrusive.

Another possible limitation is the production of erroneous results due to the increase of the exam room temperature. We performed tests simulating the case of an unexpected room temperature increase due to the fault of the air conditioning system. The room temperature monitoring circuit adequately avoided the production of erroneous results. The presence of this fault, as well as the time-table in which the fault took place, are clearly described in the archives produced at the end of the exam and easily recognized by the user.

Tests simulating the effect of external transient noises such as coughing and falling objects in the identification of snoring showed an adequate discrimination, so that these transient signals were not interpreted as being associated with snoring. This was achieved adjusting a discrimination routine to eliminate acoustical noise signals with duration < 1s.

5. Conclusion & future work

A new instrument for the analysis of the position of patients in the bed was developed. The system allows a completely nonintrusive analysis, without the necessity of coupling any device to patient. Preliminary results in normal subjects and a snorer showed adequate results, providing evidence that the system may contribute to improve the diagnosis and treatment of positional snore, as well as to avoid ulceration in the skin. The use in the diagnosis of positional snore, however, depends on the introduction of a subsystem that allows the discrimination between the dorsal and ventral decubitus.

Future plans include to simplify the interpretation of the exams. Considering a sampling period of 10 s, a whole nigth exam would produce approximately 2880 measurements. Thus, an automatic interpretation of the temperature pattern produced by the instrument in terms of decubitus and the presence (or not) of snoring would simplify the clinical use of this system. We plan to develop this clinical diagnostic support system using machine learning algorithms [14].

References

[1] Punjabi N M 2008 The epidemiology of adult obstructive sleep apnea *Proceedings of the American Thoracic Society* **5** 136-43

[2] Nakano H, Ikeda T, Hayashi M, Ohshima E and Onizuja A 2003 Effects of body position on snoring in apneic an nonapneic snorers.: Sleep pp 169-72

[3] Ravesloot M J, van Maanen J P, Dun L and de Vries N 2013 The undervalued potential of positional therapy in position-dependent snoring and obstructive sleep apnea-a review of the literature *Sleep & breathing* **17** 39-49

[4] J.C Babenel, Fergunson-Pell M W and Beale A Q 1985 Monitoring the mobility of patintes in bed. Med & Biol Eng & Comput **23** pp 466-8

[5] Melo P L 2006 Thermistors. Encyclopedia of Medical Devices and Instrumentation, John Wiley & Sons.

[6] Tietze U and C. S 1978 Advanced Eletronic Circuits. Springer- Uerlag Berlin HeldelBerg)

[7] Miyamoto Y, Yonezawa Y, Maki H, Ogawa H, Hahn AW, Caldwell WM 2002 A system for monitoring heart pulse, respiration and posture in bed.Biomed Sci Instrum **38** pp 135-8

[8] Lu L, Tamura T and Togawa T 1999 Detection of body 'movements during sleep by monitoring of bed temperature.: physiol Meas pp 137-48

[9] Bignold J J, Mercer J D, Antic N A, McEvoy R D and Catcheside P G 2011 Accurate position monitoring and improved supine-dependent obstructive sleep apnea with a new position recording and supine avoidance device *Journal of clinical sleep medicine* **7** 376-83

[10] Levendowski D J, Veljkovic B, Seagraves S and Westbrook P R 2015 Capability of a neck worn device to measure sleep/wake, airway position, and differentiate benign snoring from obstructive sleep apnea *Journal of clinical monitoring and computing* **29** 53-64

[11] Mesquita Junior J A and Melo P L 2004 Respiratory monitoring system based on pressure measurement for the analysis of sleep breathing disorders: static and dynamic erros reduction and comparisons with thermistor and pneumotachograph.: Rev Sci Instrum **73** pp 760-7

[12] Młynczak M, Migacz E, M. M and W. K 2016 Detecting breathing and snoring episodes using a wireless tracheal sensor - a feasibility study *Journal of Biomedical and Health Informatics* **In press.**

[13] Tamaura T, Miyasako S, Fugimoto T, Ogawa M and Togawa T 1999 Assesment of bed temperature monitorin for detecting body movement during sleep: comparison with simultaneous video image recording and actigraph.: Med Eng Phys **21** pp 1-8

[14] Amaral J L, Lopes A J, Faria A C and Melo P L 2015 Machine learning algorithms and forced oscillation measurements to categorise the airway obstruction severity in chronic obstructive pulmonary disease *Computer methods and programs in biomedicine* **118** 186-97

A telemedicine system based on the forced oscillation technique for remote monitoring of patients with chronic respiratory diseases

N V Oliveira[1], M H N Santos[1], L M Aligleri[1] and P L Melo[1,2]
[1]Biomedical Instrumentation Laboratory, Institute of Biology Roberto Alcantara Gomes and Laboratory of Clinical and Experimental Research in Vascular Biology, State University of Rio de Janeiro, BR

E-mail: plopes@uerj.br

Abstract. Home telemonitoring is of great interest in respiratory medicine where a large number of people have long term conditions. We developed a telemedicine instrument for home monitoring of subjects with chronic respiratory diseases. The instrument measures respiratory impedance and connects to the Internet through a notebook. The development of the hardware and software used in the instrument is described, as well as initial in vivo results obtained in a normal and an asthmatic subject. These results were in close agreement with the physiology, confirming the high scientific and clinical potential of this system. Internet transmission tests showed adequate results. We concluded that the developed system could be a useful tool for the implementation of telemedicine services, contributing to reduce the costs of the assistance offered to patients with chronic respiratory diseases.

1. Introduction

The elevated prevalence of respiratory diseases has been leading to a growing demand of medical assistance services in the last years. This increase also results in a progressive elevation in the admission of patients, which is particularly important in case of chronic respiratory diseases, whose incidence will increase in a near future [1, 2]. The respiratory function exams are usually performed in specialized centers, which very often are distant from the patient's houses and workplaces.

In the last decades continuous advancements were observed in the telecommunication and computer science areas, allowing an important cost reduction and providing a significant opportunity for improving the efficiency of the health care. Thus, health services were developed using the concept of home care and telemedicine. However, the quality of these services highly depends on the use of portable instruments, which projects a growing demand for new instruments that allow a detailed analysis of respiratory parameters out of the hospital environment. Aiming to contribute in this direction, our research group have been developing portable systems for monitoring of respiratory parameters via the Internet [3-5].

The Forced Oscillation Technique (FOT) allows a detailed approach to investigate biomechanical characteristics of the respiratory system. This method is based on the application of sinusoidal excitation during normal respiration using an external pressure generator, allowing the measurement

[2] To whom any correspondence should be addressed.

of respiratory system impedance. For the patient, the main advantage of the FOT is the simplicity of the test administration, which requires little cooperation. This analysis provides parameters describing the resistive and reactive properties of the respiratory system that are complementary to traditional pulmonary assessment methods [6, 7]. These characteristics make FOT a potentially useful method in telemedicine.

In this context, the present work describes the development of a system for monitoring the respiratory impedance via Internet through the forced oscillation technique.

2. Materials and Methods

2.1. System development

2.1.1. Hardware design. Figure 1 describes a simplified block diagram of the instrument. The first stage consists of a sinusoidal oscillator with a 6 Hz frequency and approximately 1Vpp based on the PIC16F84A. The oscillator is controlled by the digital output of the data acquisition module. The resultant sinusoidal signal is connected to a DC power amplifier with voltage gain of 47 implemented with the IC LM3886 (Texas Instruments) coupled to a heatsink with thermal resistance of 1.72 °C/W (HS12643, Wasting HS). The output of the amplifier, in its turn, is adapted to a loudspeaker of 8" (Selenium, 140W) what converts the electric signals in pressure. The pressure signal produced by the loudspeaker excites the measurement systems of flow and pressure. The flow measurement system is composed by a pneumothacometer coupled to a differential pressure transducer (Honeywell 176PC), whose signal is amplified by an instrumentation amplifier (AD620, Analog Devices) and subsequently processed by a low-pass filter (Butterworth, 4th order, 6 Hz) and connected to one of the data acquisition system inputs. The pressure measurement system is composed by an identic pressure transducer, which is attached straightly to the tube that drives the oscillations to the patient´s mouth. The signal processing is identic to that used one in the flow measurement system. The A/D conversion system (NI USB-6009, National Instruments, Austin, Texas, USA) presents a resolution of 14 bits and allows a maximum sampling frequency (fs) up to 48 kHz. In the present project the used fs was 1024 Hz. A touch screen notebook (Dell Inspiron series 137000, Intel, Core i5, 4G of RAM, hard disk of 456 GB) was employed as a hardware platform.

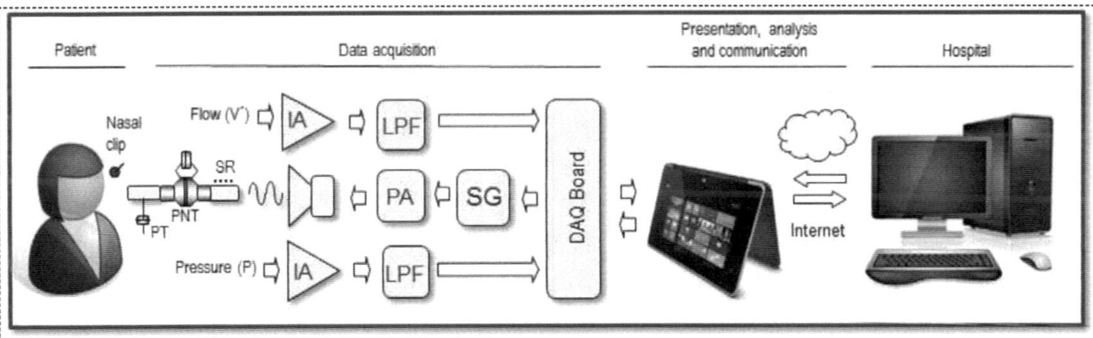

Figure 1: Simplified block diagram of the instrument. PT: pressure transducer, PNT: pneumothacometer, SR: shunt resistance, IA: instrumentation amplifier, LPF: low pass filter, PA: power amplifier, SG: sinusoidal generator.

2.1.2. Software design. The software was developed in LabVIEW 2012 environment (National Instruments, Austin, TX). It is composed by two main modules, the first is devoted to the acquisition, processing, presentation, quality control, saving the data and sending it to a server, aiming at remote use.

The signal processing is similar to the one previously described in [6]. Basically, the signals are separated in blocks of 1s, including blocks with 50 % of overlapping, multiplied by a Hanning window and processed through the Fourier Transformer. It also allows the estimate of the respiratory resistance and reactance in 6 Hz (R6 and X6, respectively). The program also allows the measurement of the average values of R6 and X6 relative to the inspiratory and expiratory phases. The impedances of the biological filter and of the mouthpiece used between the equipment and the volunteers during the measurement are compensated. The system also allows to automatically identify incorrect values [R6 or X6 greater or less than the average ± 5 SD (standard deviation)] resulting from deglutition, speech or cough during the test, as well as errors due to the saturation of the input of the A/D conversion module and negative values of resistance. The system calibration was performed using a reference mechanical load (7100R5, 5.0 $cmH_2O/L/s$, Hans Rudolph Inc., Coleparkway, Kansas, USA).

2.2. System evaluation

2.2.1. In vitro tests. Before using the system in humans, its appropriate functioning was valued using mechanical loads able to realistically simulate normal conditions (Rn=2.70cmH2O/L/s, In=12.82mcmH2O/L/s2) and respiratory obstruction (Robs=6.42cmH2O/L/s; Ipac=5.49mcmH2O/L/s2) [8].

2.2.2. In vivo tests in healthy and asthmatic individuals. The ability of the system to identify alterations associated with respiratory diseases was evaluated through a comparative analysis of the results obtained in a healthy individual that do not present history of respiratory diseases or tobacco use (male, 53 years, 171cm, 70 kg, percentile of the forced expiratory volume in the first second - VEF_1 = 104.4) and an asthmatic patient with slight respiratory obstruction (VEF_1 % = 65.6 %, male, 44 years, 65 kg, 171 cm). During the exams, the two individuals remained seated, with the head in a neutral position. The studies using the FOT were approved by the Ethics Committee of the Pedro Ernesto University Hospital. The volunteer given a signed permission before the realization of the tests.

2.2.3. Statistical analysis. The results are presented as mean±SD. The comparison between the results was effectuated through the Wilcoxon test [9] using ORIGIN 8.0. Values of $p<0.05$ were considered significant.

2.2.4. Tests of data transmission/reception. The remote monitoring and information exchange was tested with typical measurements, realistically simulating clinical conditions. The system was used during 17 days, performing one measure a day in the morning (8:00-9:00hs) and sending/receiving simulated medical recommendations and information. This resulted in fifty-one typical archives. The volunteer was in his house in Rio de Janeiro. These test files were sent/received to/from our laboratory in the Rio de Janeiro State University, which is associated with our University Hospital.

3. Results

3.1. System development

Figure 2 illustrates the developed system, showing in (A) the front panel of the program used in the patient module, based on touch screen controls, and in (B) the case containing the circuits, transducers and pneumatic subsystems. The elements used during the measures (biological filter, mouth and the nasal clip) are also described.

The exams are initialized by pressing the button "Iniciar exame", and it is completed 35 seconds latter. In order to identify measurement errors, associated with deglutition, speech or cough during the test, the pressure and flow signals can be visualized in the superior indicators of Figure 2A. The quality of the examination is evaluated through the coherence function (γ^2) (Figure 2, indicator in the middle part of the panel), which analyses the interference of the respiratory signal in the signals of pressure and flow, and consequently, in the result of the analyses. Examinations without interference results in γ^2=1, while more reduced values reflects higher interferences of the respiratory signal. In the present project, results obtained from procedures returning γ^2 values bellow 0.9 are discarded. At the end of the exam, the system offers the possibility to save the pressure and flow along the measurement time in txt format ("Salvar" option in Figure 2). The system allows the visualization of the γ^2 obtained along the examination, as well as the minimal value of γ^2 during the examination. An indicator changes its colour from green to red if the minimal γ^2 is below the stipulated threshold. Additionally, a message is shown for the patient asking that he/she remakes the examination. If this criteria is fulfilled, the program automatically send the results for a server, making the results available for the examination of the health care professional.

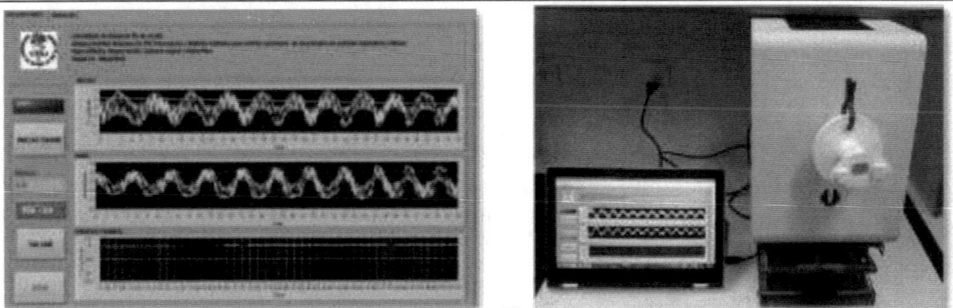

A B

Figure 2: Frontal panel of the module associated with data acquisition (A) and picture of the portable instrument for monitoring respiratory patients by forced oscillations (B).

The second module was designed for the medical environment. It allows the registration of the patient (name, sex, register, telephone, date of birth etc.), as well as the storage of specific observations on the patient (disease, time from diagnostic etc.). This module contains subroutines of receiving the examination data and of sending answers back to the patient using TCP/IP.

3.2. System evaluation

3.2.1. In vitro tests. Table 1 describes these tests. The observed errors were smaller than 7 %.

Table 1: Percentage error considering the impedance measurement in mechanical loads simulating normal and diseased volunteers.

	Reference	Measured	Error (%)
Rnormal ($cmH_2O/L/s$)	2.70	2.77	2.6
Inormal ($mcmH_2O/L/s^2$)	12.82	13.53	5.5
Rpatientes ($cmH_2O/L/s$)	6.42	6.17	-3.9
Ipatientes ($mcmH_2O/L/s^2$)	5.49	5.83	6.2

3.2.2. In vivo tests in healthy and asthmatic individuals. Figure 3A shows typical results of flow, respiratory resistance and reactance measured in the normal and the asthmatic individuals. Figure 4 presents the results obtained in the normal individual and in the asthmatic patient. Twenty one respiratory cycles were evaluated in the normal individual and twenty cycles in the asthmatic. The Rexp was significantly more elevated than the Rinsp in the normal individual ($p<0.05$), while the values of Xinsp and Xexp were not significantly different. In the asthmatic, The Rexp was not significantly different from the Rinsp, as well as the values of Xinsp and Xexp. The asthmatic presented values more elevated than the normal volunteer in terms of Rinsp ($p<0.0001$) and Rexp

IOP Conf. Series: Journal of Physics: Conf. Series **1044** (2018) 012004 doi:10.1088/1742-6596/1044/1/012004

($p<0.05$). On the other hand, the Rexp-Rinsp was more elevated in the normal individual ($p<0.01$). The Xinsp was more negative in the asthmatic ($p<0.02$), while Xexp and Xexp-Xinsp were not different from that observed in the normal individual.

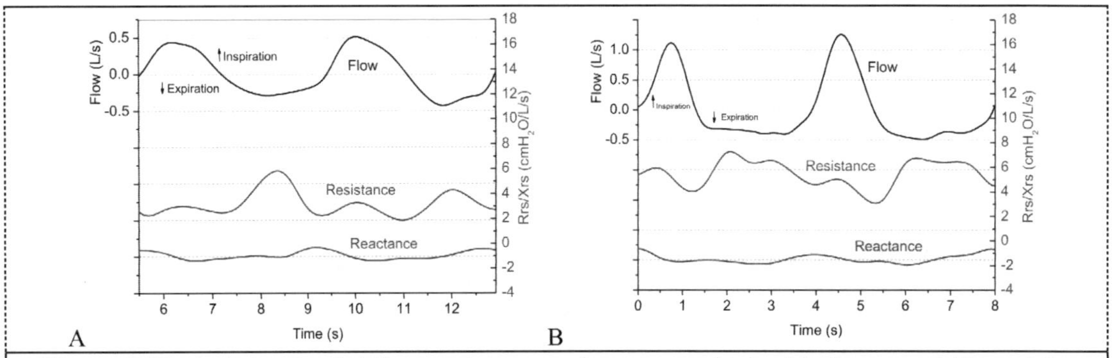

Figure 3: Typical signals of airflow, resistance and reactance obtained in the normal (A) and the asthmatic individual (B).

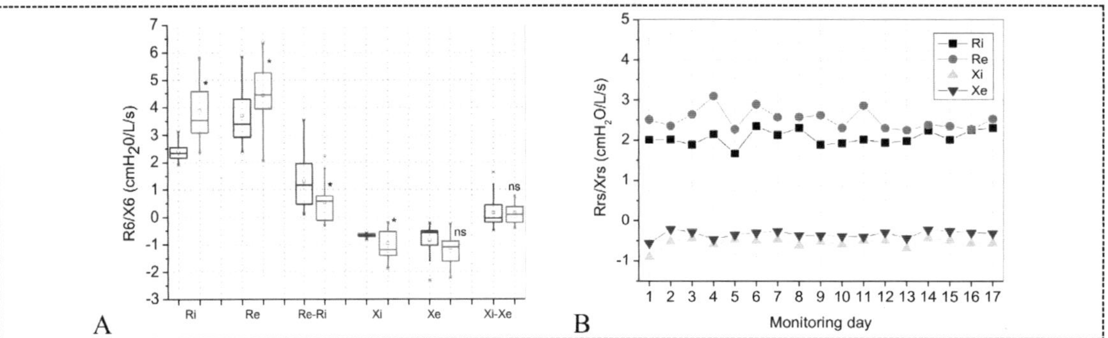

Figure 4: (A) Box-plot description of the values obtained in the normal individual (left and blue) and in the asthmatic patient (right and in red). The top and bottom of the box represent the percentages of 25% and 75% of the sample, while the circle represents the mean value. The bar represents the value of 50%, while the markers out of the box represent the percentages of 10% and 90%. * $P<0.05$; ns = not significant. (B) Impedance values obtained during seventeen days of remote monitoring in the normal individual.

4. Discussion

In vitro tests. The observed errors considering mechanical loads simulating normal conditions, as well as respiratory obstruction, were small (Table 1). They may be considered appropriated to the intended application.

In vivo tests. The observation of Re more elevated than the Ri in the normal individual, as well as the presence of similar values of Xi and Xe (Figure 4A), is consistent with previous works [10]. The pathophysiologic characteristics of asthma includes inflammation of the airways and bronchial hyperactivity, which results in air flow obstruction, and consequent alteration of the resistive and reactive properties of the respiratory system [11]. These alterations are coherent with the results described in the Figure 4A, which are also in line with works published previously [10]. It is important to emphasize that the studied asthmatic patient presents light asthma and is under adequate use of medication. This provides evidence of the high sensitivity of the impedance measurement method in the identification of the alterations in the respiratory mechanics of these patients.

Transmission / reception tests. During the tests it was observed an error in the transmission of the files (impedance: 3.1 MB; recommendations ≈ 1.0 kB) in only one of the examinations. It happened due to problems with the Internet connection. The measurements archive, however, was saved in the HD of the netbook. This problem was subsequently easily solved by creating a to-do list, responsible for sending all files that were not properly transmitted to the server at the beginning of each exam.

5. Conclusion & future work

A new portable instrument was developed for acquisition and transmission of the exams in patients with respiratory diseases. The system allows sending the results to a server, as well as the information exchange between the health professional and the patients, contributing to improve the assistance offered to patients with respiratory diseases. Future plans include to integrate the FOT home monitoring instrumentation described in the present work with machine learning algorithms [12, 13] contributing to improve the understanding and management of respiratory diseases and its exacerbations.

Acknowledgments

This study was supported by the Brazilian Council for Scientific and Technological Development (CNPq) and Rio de Janeiro State Research Supporting Foundation (FAPERJ).

References

[1] Anonymous 2013 The juggernaut of respiratory diseases gains momentum *The Lancet Respiratory Medicine* **1**

[2] GOLD 2016 Global Initiative For Chronic Obstructive Lung Disease – UPDATE (2016). "Global Strategy for the Diagnosis, Management, and prevention of Chronic Obstrutive Pulmonary Disease." (http://www.goldcopd.com: NHLBI/WHO)

[3] da Silva Junior E P, Esteves G P, Dames K K and Melo P L 2011 A telemedicine instrument for Internet-based home monitoring of thoracoabdominal motion in patients with respiratory diseases *The Review of scientific instruments* **82** 014301

[4] Silva Junior E P, Esteves G P, Faria A C and Melo P L 2010 An internet-based system for home monitoring of respiratory muscle disorders *Conference proceedings : ... Annual International Conference of the IEEE Engineering in Medicine and Biology Society. IEEE Engineering in Medicine and Biology Society. Conference* **2010** 5492-5

[5] Esteves G P, Silva Junior E P, Nunes L G, Greco C S and Melo P L 2010 Configurable portable/ambulatory instrument for the analysis of the coordination between respiration and swallowing *Conference proceedings : ... Annual International Conference of the IEEE Engineering in Medicine and Biology Society. IEEE Engineering in Medicine and Biology Society. Conference* **2010** 90-3

[6] Miranda I A, Faria A C D, Lopes A J, Jansen J M and de Melo P L 2013 On the Respiratory Mechanics Measured by Forced Oscillation Technique in Patients with Systemic Sclerosis *Plos One* **8**

[7] Oostveen E, MacLeod D, Lorino H, Farre R, Hantos Z, Desager K, Marchal F and Measurements E R S T F o R I 2003 The forced oscillation technique in clinical practice: methodology, recommendations and future developments *The European respiratory journal* **22** 1026-41

[8] Lima A N, Faria A C, Lopes A J, Jansen J M and Melo P L 2015 Forced oscillations and respiratory system modeling in adults with cystic fibrosis *Biomedical engineering online* **14** 11

[9] Dawson B and Trapp R G 2004 *Basic & clinical biostatistics* (New York: Lange Medical Books-McGraw-Hill, Medical Pub. Division)

[10] Paredi P, Goldman M, Alamen A, Ausin P, Usmani O S, Pride N B and Barnes P J 2010 Comparison of inspiratory and expiratory resistance and reactance in patients with asthma and chronic obstructive pulmonary disease *Thorax* **65** 263-7

[11] World Health Organization W H O 2015 GINA – Global Initiative for Asthma.

[12] Amaral J L, Lopes A J, Faria A C and Melo P L 2015 Machine learning algorithms and forced oscillation measurements to categorise the airway obstruction severity in chronic obstructive pulmonary disease *Computer methods and programs in biomedicine* **118** 186-97

[13] Amaral J L, Lopes A J, Veiga J, Faria A C and Melo P L 2017 High-accuracy Detection of Airway Obstruction in Asthma Using Machine Learning Algorithms and Forced Oscillation Measurements *Computer methods and programs in biomedicine*. In Press.

IOP Conf. Series: Journal of Physics: Conf. Series **1044** (2018) 012005 doi:10.1088/1742-6596/1044/1/012005

A Two-layer Semi Empirical Model of Nonlinear Bending of the Cantilevered Beam

I U Zulkarnay[1], T T Kaverzneva[2], D A Tarkhov[2,3,4], V A Tereshin[2], T V Vinokhodov[2] and D R Kapitsin[2]

[1]Bashkir State University, Ufa, 450076, ul.Z.Validi, 32, Russian Federation
[2]Peter the Great St. Petersburg Polytechnic University, 195251, St.Petersburg, ul.Polytechnicheskaya, 29, Russian Federation

[3]E-mail: dtarkhov@gmail.com

Abstract. This paper suggests a semi empirical model of nonlinear bending of the cantilevered beam. The model learns by using the results of measurement of nonlinear bending of the cantilevered metal pipe loaded by sinkers on the end. These studies are of interest for long-term forecasting of the behavior of building beams, various structural elements of load-lifting machines and mechanisms.

1.Introduction

Modeling complex technical objects is often hampered by insufficient knowledge of the processes occurring in them. The model of such an object in the form of a differential equation and boundary conditions can cause false ideas about the accuracy of modeling, since the structure and coefficients of the equations are not known accurately, even less accurately the boundary conditions are usually known. All these parameters are usually changing during the operation of the object, which makes the prediction of its state on the basis of the original model even more problematic.

To identify the state of an object, one can involve the results of observations of it, but the problem of identifying equations and boundary conditions from these data (the inverse problem) is usually much more complicated than the direct problem of solving a differential equation with boundary conditions. Previously, we solved such problems using our methodology for constructing the neural network model of the object by differential equations and additional data [1-8]. However, the training of neural networks requires a fairly large computational cost. To solve this problem, a new class of multi-layer models [9] was developed, with the help of which it is possible to do it without a complicated training procedure. This approach creates additional opportunities for combining classical and new methods. The essence of the approach is to apply known numerical methods for integrating differential equations to an interval with a variable upper limit. As a result, an approximate solution is obtained not in the form of a set of numerical values, but as a function of this upper limit. The method [9] can be used together with neural networks, but in this paper we managed without using them.

In this paper, this approach is tested on the bend of a cantilevered metal tube. The developed methods can be applied for long-term forecasting of the behavior of building beams, various structural elements of load-lifting machines and mechanisms, taking into account the real picture of wear, aging and corrosion of metal. In [10] the necessity of an estimation of influence of deterioration of the

[4] Author to whom any correspondence should be addressed

building equipment and the hand tool on working conditions of workers of building industry as the most traumatic kind of activity is revealed and justified. From this point of view, solving inverse problems in modeling the state of loaded elements will increase the reliability of forecasting results.

2.Material and methods

The measurements were carried out at the next experimental setup. A straight metal tube 1060 mm long was taken. And a mass of 116 g of circular cross-section with an outer diameter of 1 sm and a wall thickness of 2 mm, onto which labels were applied after 50 mm. One end of the tube, at a distance of 60 mm from the left end, was clamped in the vice, and loads with a weight of 100 g to 1300 g were attached alternately to the right with the help of a thread. A screen with a millimeter grid was located behind the experimental setup at a distance of 100 mm in a vertical plane. The horizontal and vertical lines of the grid were controlled by a bubble level with an error of no more than 3 mm per 1000 mm in length. Before the experimental installation at a distance of 1500 mm there was a camera, connected with a computer. In the center of the image was the middle of a tube with no load. To convert the coordinates from the pixel space into the space of a millimeter grid with optical distortion compensation, two calibration functions were constructed from two variables over a sufficiently large number of points. The resulting measurement error should not exceed 5 mm.

As a mathematical model we use the equation of a large static deflection of a thin homogeneous physically linear elastic rod under the action of distributed q and concentrated p forces [11].

$$\frac{d^2\theta}{dz^2} - \frac{mgL^2}{D}\left(\frac{m_i}{m} + z\right)\cos\theta = 0 \qquad (1)$$

where D and L are the constant flexural rigidity and length of the rod; θ - is the angle of inclination of the tangent; $z = 1 - s/L$ - s - the natural coordinate of the curved axis of the rod, measured from the seal, m - the mass of the rod, m_i - the mass of the load. In the experiment performed, the distributed and concentrated forces were the weights of the rod and the load at the end.

The boundary conditions have the form:

$$\left.\frac{d\theta}{dz}\right|_{z=0} = 0; \ \left.\theta\right|_{z=1} = \theta_0$$

The angle θ is related to the coordinates of the points on the rod by the equalities:

$$\frac{dx}{ds} = Cos(\theta); \frac{dy}{ds} = Sin(\theta); \qquad (2)$$

It should be pointed out that this equation describes the object in question inaccurately - the tube has a creep zone near the seal and is not a rod. This situation is typical for many applied problems, when processes in a real object are described by an approximate model in the form of differential equations, which is usually solved numerically. In addition, usually there are measurement data obtained during the process of observing the object. By standard methods, the model of an object that takes into account these data is difficult to construct. By our methods, an approximate model is constructed by the equation (in the problem under consideration, these are equations (1-2)), the parameters of which are refined from the measurement data.

We rewrite equation (1) in the form:

$$\frac{d^2\theta}{dz^2} = a\left(\mu_i + z\right)\cos\theta, \qquad (3)$$

where $a = \dfrac{mgL^2}{D}$, $\mu_i = \dfrac{m_i}{m}$.

As indicated earlier, equation (3) describes the process of bending a rod with a large error. This statement was confirmed by numerical experiments. To construct a more adequate model, we move from system (2)-(3) to its approximate parametric solution $x(s,\theta_0,a)$ and $y(s,\theta_0,a)$. The parameters θ_0, a are found by the method of least squares by minimizing the expression:

$$\sum_{j=1}^{N}(x(s_i,\theta_0,a)-x_i)^2+(y(s_i,\theta_0,a)-y_i)^2 \tag{4}$$

Here N is the number of points at which measurements were taken, $\{x_i, y_i\}$ - the measured coordinates of the points corresponding to the marks on the rod at a distance s_i from the embankment. To build more accurate models, we applied the approach developed in [9]. Its essence with respect to equation (3) is that the known formulas for the numerical solution of differential equations should be applied not to the interval $[0,1]$, but to an interval with a variable upper limit $[0,z]$. In this case, instead of a table of numbers, we get a function $\theta(z,\theta_0,a)$, and the parameters of the problem θ_0, a are among its arguments.

From $\theta(z,\theta_0,a)$ we pass to the original Cartesian coordinates, integrating (2) according to the Simpson formula for a variable-length interval:

$$x(s)=\frac{s}{6M}\left(1+\cos\left[\theta(1-s/l)\right]+4\sum_{i=1}^{M}\cos\left[\theta\left[\frac{1-s/l}{2M}(2i-1)\right]\right]+2\sum_{i=1}^{M-1}\cos\left[\theta\left[\frac{1-s/l}{M}i\right]\right]\right) \tag{5}$$

$$y(s)=\frac{s}{6M}\left(\sin\left[\theta(1-s/l)\right]+4\sum_{i=1}^{M}\sin\left[\theta\left[\frac{1-s/l}{2M}(2i-1)\right]\right]+2\sum_{i=1}^{M-1}\sin\left[\theta\left[\frac{1-s/l}{M}i\right]\right]\right) \tag{6}$$

In numerical experiments, we used this formula for $M=10$. As a result, we get the dependences $x(s,\theta_0,a)$ and $y(s,\theta_0,a)$. The parameters θ_0, a, as mentioned above, are calculated by minimizing expression (4).

We present the results of calculations for two variants of this approach. The first variant consists in applying to (3) the above-mentioned modification [9] of Störmer's method [12]. As a result, we obtain an approximate solution of the form:

$$\theta_1(z)\cong\theta_0+\frac{z^2}{4}a(\mu_i+z)\cos(\theta_0+\frac{z^2}{16}a(\mu_i+z)) \tag{7}$$

where $\theta_0=\theta(0)$ is the angle of the rod at its end.

Substituting (7) into (5) instead of θ, we obtain Simpson's formulas for $x_1(s,\theta_0,a)$ and $y_1(s,\theta_0,a)$. Parameters θ_0 and a are not formally constrained. In fact, the accuracy of this solution is higher, the smaller the parameter a, but in connection with the approximate character of equation (3) we are interested not in the smallness of the error of the solution of this equation, but in the accuracy of the correspondence to the measurement data.

The second variant consists in applying for the approximate solution of equation (3) an implicit Euler method [12] with two steps instead of the Störmer method. As a result, we obtain a system of two equations:

$$\begin{cases} \tilde{\theta}(z) \cong \theta_0 + \dfrac{z^2}{4} a(\mu_i + z)\cos(\tilde{\theta}(z)), \\ \theta(z) \cong 2\tilde{\theta}(z) - \theta_0 + \dfrac{z^2}{4} a(\mu_i + z)\cos(\theta(z)) \end{cases}$$

Whence we obtain an approximate solution:

$$\theta_2(z) \cong 2\frac{2\tilde{\theta}_2(z) - \theta_0 + 0.25z^2 a(\mu_i + z)}{\sqrt{1 + 0.5z^2 a(\mu_i + z)(2\tilde{\theta}_2(z) - \theta_0 + 0.25z^2 a(\mu_i + z))} + 1} \quad (8)$$

where:

$$\tilde{\theta}_2(z) = 2\frac{\theta_0 + 0.25z^2 a(\mu_i + z)}{\sqrt{1 + 0.5z^2 a(\mu_i + z)(\theta_0 + 0.25z^2 a(\mu_i + z))} + 1}$$

Substituting (8) into (5) instead of θ, we obtain Simpson's formulas for $x_2(s, \theta_0, a)$ and $y_2(s, \theta_0, a)$.

3. Calculation

Let's give the results of calculations for three values of the mass of the cargo $m_2 = 300$ g and $m_3 = 700$ g. Figure 1 compares the measurement data and the results of calculations using formulas (7)-(8) for $m_1 = 0$ g. For x_1 the standard deviation of the measurement results from the theoretical curve is 1.71, the ratio of the residual variance to the variance of the sample is 0.0033. For the standard deviation of the measurement results from the theoretical curve is 0.22, the ratio of the residual variance to the sample variance is 0.0037.

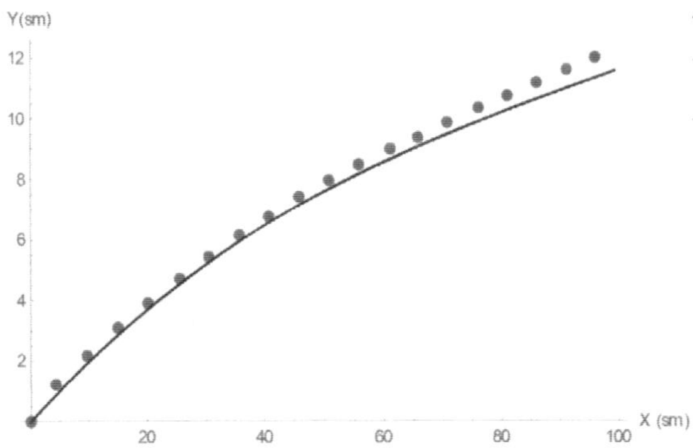

a)

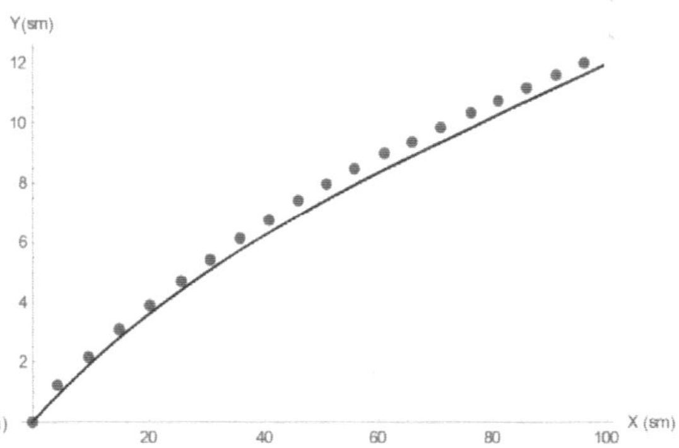

b)

Figure 1. Theoretical (curve) and experimental (points) rod deflection obtained using formula 7 (the left figure (a)) and formula 8 (the right figure (b)) for the mass of the cargo $m_1 = 0$ g. Here *Y* and *X* are Cartesian coordinates in *sm*.

The standard deviation of the measurement results from the theoretical curve $x_1(s,\theta_0,a)$ is 1.71, the ratio of the residual variance to the variance of the sample is 0.0033. For the standard deviation of the measurement results from the theoretical curve $y_1(s,\theta_0,a)$ is 0.22, the ratio of the residual variance to the sample variance is 0.0037.

The standard deviation of the measurement results from the theoretical curve $x_2(s,\theta_0,a)$ is 1.72, the ratio of the residual variance to the variance of the sample is 0.0033. The standard deviation of the measurement results from the theoretical curve $y_2(s,\theta_0,a)$ is 0.27, the ratio of the residual variance to the variance of the sample is 0.0056.

The discrepancy for the variable y is within the measurement error. The discrepancy with respect to the variable x is greater than the measurement error, which may indicate both the error of the model and the systematic error in the initial data. In the problem under consideration, we are interested in the shape of the rod bending, to which the measurement errors with respect to the variable x do not influence as much as the errors in the variable y.

Figure 2 shows the discrepancies between the theoretical and experimental data for the variable y, depending on the number of the measurement point, starting from the end of the rod.

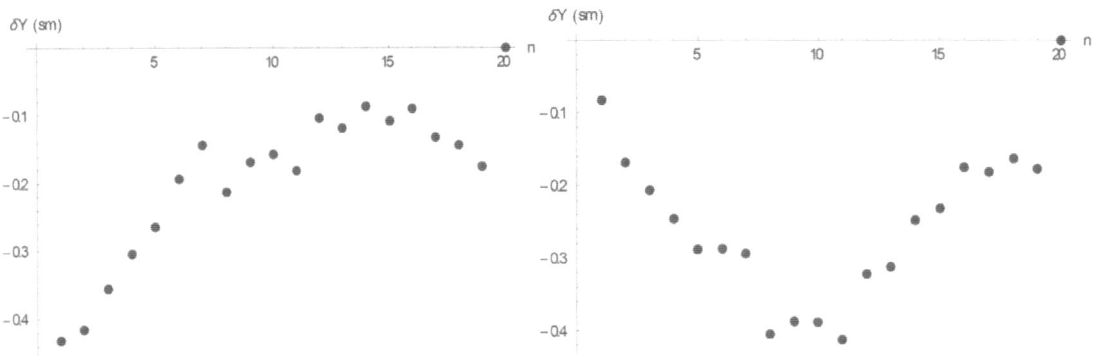

a) b)

Figure 2. Theoretical and experimental rod deflection curves obtained using formula 7 (the left figure (a)) and formula 8 (the right figure (b)). Here are values of the deflection on Cartesian coordinates Y and X in *sm*.

This figure shows that the deviations are not accidental. In figure 2a the error is maximal at the right end of the rod and decreases with approach to the left end. In figure 2b the error is maximal in the middle of the rod and decreases to its edges. This indicates the possibility of refinement of the model. Some methods for this refinement are discussed in the Results and Discussion section. For this task, this refinement was not carried out, since the problem is model and serves only to demonstrate methods. A similar conclusion can be drawn from the error graphs in the remaining cases. The corresponding graphs are not given in the article due to the limited space.

Figure 3 compares the measurement data and computational results using formulas (7)-(8) for $m_2 = 300$ g. For $x_1(s,\theta_0,a)$ the standard deviation of the measurement results from the theoretical curve is 1.20, the ratio of the residual variance to the sample variance is 0.0016. For $y_1(s,\theta_0,a)$ the standard deviation of the measurement results from the theoretical curve is 0.29, the ratio of the residual variance to the sample variance is 0.046.

For $x_2(s,\theta_0,a)$ the standard deviation of the measurement results from the theoretical curve is 1.13, the ratio of the residual variance to the variance of the sample is 0.0014. For $y_2(s,\theta_0,a)$ the standard deviation of the measurement results from the theoretical curve is 0.14, the ratio of the residual variance to the variance of the sample is 0.0105.

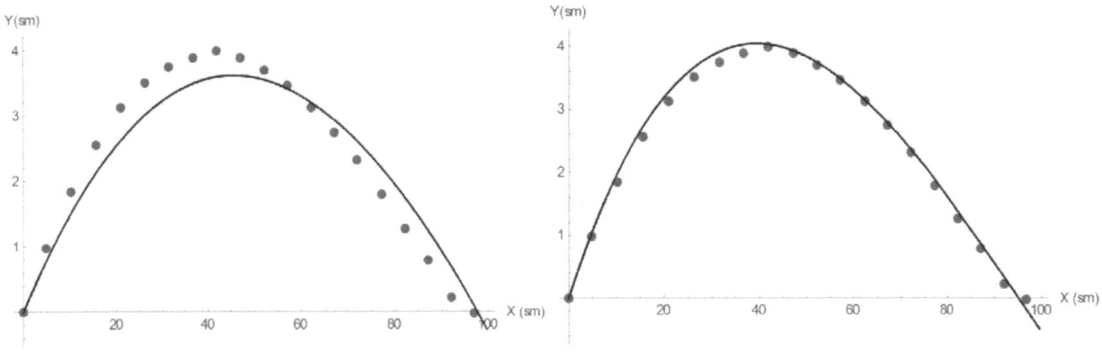

a) b)

Figure 3. Theoretical (curve) and experimental (points) rod deflection obtained using the formula 7 (the left figure (a)) and formula 8 (the right figure (b)) for the mass of the cargo $m_2 = 300$ g. Here *Y* and *X* are Cartesian coordinates in *sm*.

Figure 4 compares the measurement data and calculation results by formulas (7)-(8) for $m_3 = 700$ g.

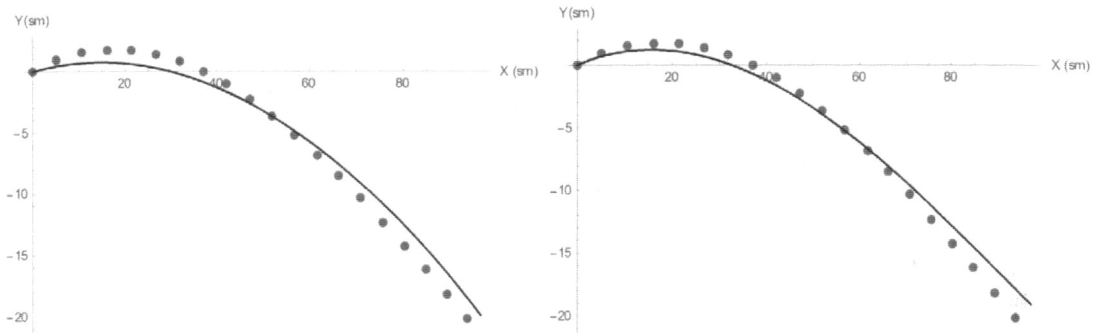

a) b)

Figure 4. Theoretical (curve) and experimental (points) rod deflection obtained using formulas 7 (the left figure (a)) and formula 8 (the right formula (b)) for the mass of the cargo $m_3 = 700$ g. Here *Y* and *X* are Cartesian coordinates in sm.

The conclusions that can be made on the basis of the error data are the same as in the case of a rod without a load. We note that similar results were obtained for other values of the mass of the cargo. For $x_1(s,\theta_0,a)$ the standard deviation of the measurement results from the theoretical curve is 1.28, the ratio of the residual variance to the variance of the sample is 0.0019. For $y_1(s,\theta_0,a)$ the standard deviation of the measurement results from the theoretical curve is 0.69, the ratio of the residual variance to the variance of the sample is 0.0089. For $x_2(s,\theta_0,a)$ the standard deviation of the measurement results from the theoretical curve is 1.29, the ratio of the residual variance to the sample variance is 0.0019. For $y_2(s,\theta_0,a)$ the standard deviation of the measurement results from the theoretical curve is 0.56, the ratio of the residual variance to the sample variance is 0.0058. We note that similar results were obtained for other values of the mass of the cargo.

4. Results and Discussion

A typical situation for practice is the situation when experiments conducted with a real object contradict the mathematical model in the form of differential equations obtained on the basis of application of general physical principles. Usually, in such a situation, one of two approaches is used.

The first approach is to try to refine the physical model of the object and obtain differential equations that reflect the processes occurring in it more accurately. A simpler case leads to refinement of model parameters, which is mathematically expressed in coefficients of inverse problems. To solve such problems, there are a number of approaches, one of which is the use of neural networks [4, 7]. A variant (to which the problem considered in this article belongs) is also possible, when no choice of parameters allows us to reflect experimental data with reasonable accuracy. In this case, it is necessary to change the structure of the model, which can lead to a sharp complication of differential equations and does not always end in success in a reasonable time.

The second approach consists in refusing to use differential equations. In this case, the model of the object is constructed empirically by interpolation from the experimental data. Interpolation uses the selected system of functions, for example polynomials or neural networks.

The methods developed here allow us to apply the third approach. It consists in obtaining approximate semiempirical formulas on the basis of an inaccurate differential model and measurement results, which may also have a large error. Known theorems on the error of numerical methods [12] allow us to assert that we can obtain an arbitrarily exact approximation to the solution of a differential equation by using a partition into a sufficiently large number of intervals (which leads to more complicated formulas). This complication is justified only if the model in the form of differential equations accurately describes the real object. It should be emphasized that the standard numerical methods [12] allow us to obtain only tables of numbers. Unlike them, our approach allows us to obtain formulas without the use of interpolation, which can be refined from the experimental data.

Such an approach is convenient in practically interesting problems for constructing semiempirical models in a situation where the accuracy of describing an object by differential equations is small or unknown.

References

[1] Kainov N U, Tarkhov D A and Shemyakina T A 2014 Application of neural network modeling to identification and prediction of problems in ecology data analysis for metallurgy and welding industry *Nonlinear Phenomena in Complex Systems* **17 1** 5763

[2] Vasilyev A and Tarkhov D 2014 Mathematical Models of Complex Systems on the Basis of Artificial Neural Networks *Nonlinear Phenomena in Complex Systems* **17 2** 327335

[3] Lazovskaya T V and Tarkhov D A 2015 Fresh approaches to the construction of parameterized neural network solutions of a stiff differential equation *St. Petersburg Polytechnical University Journal: Physics and Mathematics*

[4] Gorbachenko V I, Lazovskaya T V, Tarkhov D A, Vasilyev A N and Zhukov M V 2016 Neural Network Technique in Some Inverse Problems of Mathematical Physics *Springer International Publishing* (Switzerland) ed L Cheng pp 310–316

[5] Shemyakina T A, Tarkhov D A and Vasilyev A N 2016 Neural Network Technique for Processes Modeling in Porous Catalyst and Chemical Reactor *Springer International Publishing* (Switzerland) ed L Cheng pp 547–554

[6] Kaverzneva T, Lazovskaya T, Tarkhov D and Vasilyev A 2016 Neural network modeling of air pollution in tunnels according to indirect measurements *Journal of Physics: Conference Series* **772**

[7] Lazovskaya TV, Tarkhov D A and Vasilyev A N 2017 Parametric Neural Network Modeling in Engineering *Recent Patents on Engineering* **11 1** pp 10-15

[8] Lozhkin V, Tarkhov D, Timofeev V, Lozhkina O and Vasilyev A 2016 Differential neural network approach in information process for prediction of roadside air pollution by peat fire *IOP Conf. Series: Materials Science and Engineering* **158**

[9] Lazovskaya T and Tarkhov D 2016 Multilayer neural network models based on grid methods *IOP Conf. Series: Materials Science and Engineering* **158**

[10] Kaverzneva T T and Smirnova O V 2013 Wear-out Effect of Construction Equipment and Hand Tools on Workers' Labor Conditions *Bezopasnost′ v tekhnosfere (Safety in technosphere)* **3(42)** pp 14-18 (in Russian).

[11] Artyukhin Y P 2013 Arbitary Bending of a Cantilever Beam by a Conservative Force *Sc.Notes of Kazan Univ.Physic-Math sc* **2** pp 144-157

[12] Hairer E, Norsett S and Wanner G 1987 Solving Ordinary Differential Equations I: Nonstiff Problem (Berlin: Springer-Verlag)

ABSOLUTE STANDARDIZATION OF ^{22}Na, ^{65}Zn and ^{60}Co BY SUM-PEAK METHOD AND CRITICAL ANALYSIS OF INSTRUMENTATION

Journal of Physics: Conference Series

Araújo M T F[1, 2], Poledna R[1], da Silva R L[1], Delgado J U[1], de Almeida M C M[1], Lopes R T[2], Alfredo L F F[1, 2], Veras E V[1], Rangel J[1], Cruz P A L[1], da Silva C J[1].

[1]Laboratório Nacional de Metrologia das Radiações Ionizantes (LNMRI/IRD/CNEN), Av. Salvador Allende, s/n, Recreio, CEP 22780-160, Rio de Janeiro, Brazil, (miriamtaina@hotmail.com)
[2]Laboratório de Instrumentação Nuclear (LIN/PEN/COPPE/UFRJ), Caixa Postal 68509, CEP 21945-970, Rio de Janeiro, Brazil, (miriamtaina@hotmail.com)

Abstract. The sum-peak method was used to obtain the absolute activity value for the ^{60}Co, ^{65}Zn e ^{22}Na radioisotopes. Three HPGe detectors were used, being a planar, a coaxial and well type. The values of the activity obtained by sum-peak method are consistent, with values of uncertainty less 0.5% to for all radioisotopes considering k = 1. LNMRI has introduced a new Calibration System that uses a HPGE type well detector. Thus, a comparative study was made between the electronics available at the time of conception of the method and the one currently used.

1. Introduction

The application of the sum-peak method to obtain the absolute value of activity has been reported in several publications made by LNMRI. The method can be employed in emitters that decay by electronic capture and β^+ emitters that emitting a subsequent γ-ray. Here was considered three radioisotopes of interest that meet the requirements for the application of the method: ^{22}Na, ^{65}Zn and ^{60}Co.
The increasing use of radiopharmaceuticals composed of positron-emitting radionuclides in nuclear medicine has opened up recurrent discussions about their applications in Positron Emission Tomography (PET) imaging tests. In order to maintain the accuracy during the calibration procedure of the activity meters (activimeters) it is necessary to have a standardized source for this procedure to be performed. However, many difficulties end up appearing throughout the absolute standardization procedure of the sources due to the fact that most of the radionuclides used have short half-lives where many of them do not exceed 110 min [1]. The ^{22}Na reveals itself as an important target for study since it has one of the longest half-lives (2.6029 years) between the postiron emitters. The ^{22}Na presents no major difficulties for

the absolute calibration by the sum peak method because it has an interference-free spectrum. Its main characteristic is given by the decay β + which emits a subsequent gamma.

^{65}Zn decays by electron capture and around 2.5% by β^+ through two branches with intensities of ε_1 = 0,5075 and ε_2 = 0,4779, respectively. The electron capture branches are followed by γ-ray emission mainly of 1115.5 keV (0.5075 intensity) and two others of negligible intensity (less than 0.001). This fact makes this radionuclide practically a monoenergetic γ-ray emitter and applied as a reference source in γ-ray spectrometry [2]. For ^{65}Zn the sum peak method is easily applied since the planar type detector has a thin beryllium window which allows detection of the X-ray of interest of 8 keV.

^{60}Co is considered a perhaps the simplest nuclide exhibiting β-γ cascade decay, and can be easily standardized by sum-peak method. Regarding the instrumentation used by Brinkman and his collaborators who proposed the methodology in 1962, some aspects need to be carefully observed. These aspects of greater importance reflect the current days where more powerful and modern instrumentation is available.

2. Experimental

Three type's detectors of HPGe were used: a coaxial, a planar and a well-type with 20, 40 and 70% relative efficiency respectively. The well-type detector was submitted to an implantation and implementation procedure in the LNMRI and its performance regarding the accuracy and precision of results was expressed in comparison to the values obtained from the other two HPGe detectors of the laboratory. Each detector applied in the measurements has an appropriate electronics composed basically of the elements: High voltage supply, signal amplifier and a multichannel analyzer.

Point sources were prepared with the aid of a pycnometer, depositing drops of radionuclide solution in a polystyrene film, with a thickness of 0.05 mm, set in one acrylic ring. The ring has an external diameter of 15 mm, inner diameter of 4 mm and a thickness of 1 mm. Once dried, the sources were covered with the same polystyrene film [3]. The energies of 1173 keV and 1332 keV were used for a ^{60}Co measurement. The 8 keV X-ray and the 1115 keV gamma ray were selected to standardize the ^{65}Zn and ^{22}Na were selected as energies of 511 keV and 1274 keV. The measurement of sources was provided on various positions to applied sum-peak method. The counting time of each sample varied by about 24 h for each radionuclide and the dead-time was below 4% for all measurements. The control of radionuclidic impurities was also done and the presence of other radioisotopes in the samples analyzed was not verified.

3. Methodology

3.1 Sum-peak method

For all isotopes, the situation is simple, since there are only two lines for analysis. In this case we used two lines corresponding to X-ray and γ-ray or γ-γ interaction. The activity was obtained by the classical formula given by Brinkman et al. (1963).

$$No = N_T + \frac{N_1 N_2}{N_{12}} \quad (1)$$

Where, N_T are the counts under the total spectrum, N_1 is the x-ray count (K_α+K_β) or the first γ , N_2 is the less peak count and N_{12} is the sum-peak count.
The sum-peaks are characterized for each radioisotope as being: ^{60}Co: (1173 + 1232 = 2405 keV); ^{22}Na: (511 + 511 = 1022 keV; γ-ray = 1274 keV; 511 + 1274 = 1785 keV; 1022 + 1274 = 2296 keV); 65Zn: (1115 + 8 = 1123keV).

3.2 Efficiency Curve

The Efficiency Curve method consists of a methodology to obtain the photopeak efficiency for spectrometry system. The efficiency of photopeak is defined as the ratio between the rate of radiation detected by the system and the rate emitted by the source. With the use of calibrated standard sources of known energies and intensities, using the same detection system and the same counting geometry, an energy efficiency curve is obtained through the relation:

$$\varepsilon = \frac{S}{T.A.K.P_\gamma}$$

Where *S* is energy photopeak area of a standard source at (s^{-1}), *T* is the count time (s), *A* is activity of standard source at (s^{-1}), *K* is decay factor of the radionuclide standard source and *Pγ* is a probability of emission for the gamma radiation of photopeak in (s^{-1})
The efficiency curve was obtained using various radionuclidic standardized sources such as: ^{166m}Ho, ^{152}Eu, ^{133}Ba, ^{137}Cs, ^{134}Cs, ^{241}Am, ^{54}Mn, ^{65}Zn totaling 59 energy points and the graphic can be observed in Figure 1. The HPGe spectrometric system for detector 40% has been calibrated in efficiency through the use of standard point sources. The range of energy was established between 48 keV and 1427 keV originally. Then a cut was made considering only the energies above 100 keV since the low energy region was not necessary for the purpose of the calibration.

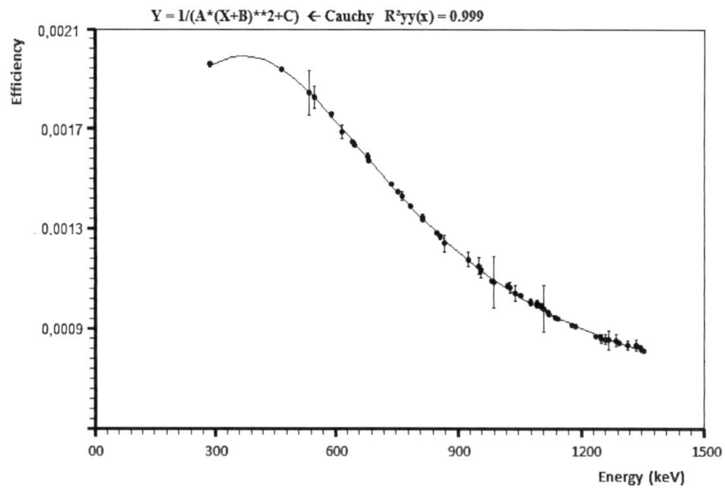

Figure 1. Efficiency curve obtained (linear – long scale) for the spectrometry system using the radionuclidic standards in point source geometry: *^{166m}Ho, ^{152}Eu, ^{133}Ba, ^{137}Cs, ^{134}Cs, ^{241}Am ^{54}Mn, ^{65}Zn.*

IOP Conf. Series: Journal of Physics: Conf. Series **1044** (2018) 012006 doi:10.1088/1742-6596/1044/1/012006

4. Results and discussion

The results obtained for the three radioisotopes are shown to be coherent and with reduced uncertainty values and compatible with those obtained by the 4πβ-γ system. Regarding instrumentation, it should be noted that at the time the standardization methodology known as sum-peak method was developed, the idealizers had a multichannel analyzer with 100-channel RIDL to measure N_1, N_2 and N_{12} while N_T was obtained by the intermediary of counter. In their study, there is no approach of the dead time parameter of the system since they did not have this function. The modern spectrometer with 16.000 channels has a dead-time management system and simultaneously provides N_T, N_1, N_2, and N_{12} values, as well as its uncertainties. Table 1 shows the mean activity values for the three radioisotopes of interest and Table 2 presents an example of uncertainty values for ^{60}Co, which is also valid for the other isotopes.

Table 1. Activity values for ^{22}Na (Sum-peak for detector wheel-type)

^{22}Na source	Activity (kBq)	Uncertainties (k = 1)	
		U_a%	U_b%
06S16	1.188	0.03	0.23
07S16	2.418	0.02	0.23
08S16	1.470	0.02	0.23
09S16	3.364	0.02	0.23
10S16	1.951	0.26	0.34

Table 2. Activity values for ^{22}Na (Efficiency Curve for detector 40%)

^{22}Na source	Activity (kBq)	Uncertainties (k = 1)	
		U_a%	U_b%
06S16	1.317	0.16	0.3
07S16	2.563	0.05	0.3
08S16	1.546	0.15	0.3
09S16	3.373	0.10	0.3
10S16	1.838	0.13	0.3

Table 3. Activity values for ^{60}Co (Sum-peak for detector wheel-type)

^{60}Co source	Activity (kBq)	Uncertainties (k = 1)	
		U_a%	U_b%
44S16	4.421	0.03	0.23
45S16	3.578	0.04	0.23
46S16	2.975	0.04	0.23

Table 4. Activity values for ^{60}Co (Sum-peak for detector 20%)

^{60}Co source	Activity (kBq)	Uncertainties (k = 1)	
		U_a%	U_b%
44S16	4.343	0.03	0.23
45S16	3.539	0.04	0.23
46S16	2.929	0.04	0.23

Table 5. Activity values for ^{65}Zn (Sum-peak for detector 20%)

^{65}Zn source	Activity (kBq)	Uncertainties (k = 1)	
		U_a%	U_b%
22S16	3.186	0.25	0.34
23S16	2.959	0.26	0.34
24S16	3.572	0.26	0.34
25S16	2.675	0.17	0.34
26S16	2.570	0.74	0.34

Table 6. Activity values for ^{60}Zn (Efficiency curve for detector 40%)

^{65}Zn source	Activity (kBq)	Uncertainties (k = 1)	
		U_a%	U_b%
22S16	2.887	0.27	0.7
23S16	2.315	0.46	0.7
24S16	3.723	0.18	0.7
25S16	*	*	*
26S16	2.320	0.31	0.7

IOP Conf. Series: Journal of Physics: Conf. Series **1044** (2018) 012006 doi:10.1088/1742-6596/1044/1/012006

Table 7. ^{60}Co uncertainty of activity measured by sum peak method.

Uncertainty Components	A (%)	B (%)
Live time	-	0.01
Weight	-	0.20
Half life	-	0.10
Background	-	0.02
Decay	-	0.01
Counting statistic	0.03	
Combined uncertainty (k = 1)	0.23	

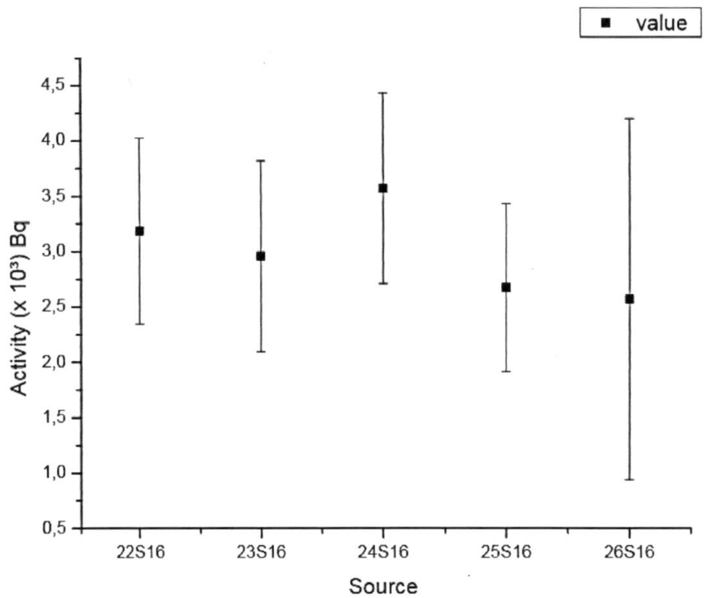

5. Conclusion

The objective of this work was to do an exercise in the use of the well-type HPGe detector that was implanted in the LNMRI recently. The sum-peak method made it possible to obtain values of activity with uncertainties of 0.23% for ^{60}Co, 0.23% for ^{22}Na and 0.26% for ^{65}Zn. In addition to being fast, this method made it possible to obtain good values so much for the activity as to the precise determination of emission probabilities for the radionuclides analyzed.

6. References

IOP Conf. Series: Journal of Physics: Conf. Series **1044** (2018) 012006 doi:10.1088/1742-6596/1044/1/012006

[1] Bernardes. E.M.O., Delgado. J.U., Tauhata, L., da Silva, C.J., Iwahara, A., Poledna, R., Paschoa, A.S., 2002. 166Ho: a multi-γ standard for the calibration of the Ge spectrometers. Appl. Radiat and Isotopes. 56, 157-161.

[2] Araújo M. T. F., Poledna R. Delgado J. U., Silva R. L., Iwahara A., da Silva C. J., Tauhata L., Oliveira A. E., de Almeida M. C. M., Lopes R. T., 2016. Absolute Standardization of the impurity 121Te associated to the production of the radiopharmaceutical 123I. International Journal of Appl. Radiat. and Isotopes, 109, 389-392.

[3] Brinkman, G. A., Aten, A. H. W., 1963. Absolute Standardization with a NaI (Tl) Crystal. International Journal of Appl. Radiat. and Isotopes, 14, 153-157.

[4] Brinkman, G. A.; Lindner, L.; Veenboer, J. TH., 1977. Sum-peak Calibration of 123I. International Journal of Appl. Radiat. and Isotopes, 28, 271-275.

Absolute standardization of the ^{133}Ba by sum-peak method and comparison with other methods

R L da Silva[1], J U Delgado [1], M C M de Almeida [1,2], R Poledna [1], A F F Lopes [1], M T F Araújo [1,3], O L Trindade [1], E V Veras [1], J Rangel [1], P A L Cruz [1] and C J Silva [1]
[1]LNMRI-Instituto de Radioproteção e Dosimetria, Rio de Janeiro, Brazil
[2]DINOR-Comissão Nacional de Energia Nuclear, Rio de Janeiro, Brazil
[3]COPPE-Universidade Federal do Rio de Janeiro, Rio de Janeiro, Brazil

E-mails: candida@cnen.gov.br, ronaldo@ird.gov.br

Abstract. This paper presents the absolute standardization of ^{133}Ba by sum-peak method. ^{133}Ba owns a complex decay and was used the sum coincidences between the average energy of characteristic X-rays (kα_1 and kα_2) and gamma-rays. The activity results presented uncertainty values below 0.3 % and were compared with those of other absolute activity measurement techniques. The results obtained are compared with other methods and confirm that this is an alternative, fast and efficient method for ^{133}Ba standardization.

1. Introduction

The sum- peak coincidence absolute method was introduced in the 1960s by Brinkman and his collaborators for standardization of radionuclides. This simple and challenging method of radiation measurement enables the direct standardization of radionuclides emitting two photons in coincidence using gamma spectrometry. It uses the peak counts of gamma and X-rays and their possible sums, plus the total spectrum count plus the extrapolation for energy equal zero. ^{133}Ba decays by electron capture emitting X-rays and gamma radiation in coincidence [1-4].

Recently ^{125}I activity was determined in 2011 using two sodium iodide detectors and compared with that obtained with an HPGe detector [5]. In 2007 de Almeida determined ^{51}Cr activity using an HPGe detector [2] . All of these authors used the sum-peak method. Many laboratories in recent years have published papers of standardization of simple decay radionuclides by sum-peak method [1, 2, 5].

Here ^{133}Ba standardization was performed, considering a radionuclide complex to use the sum-peak coincidence absolute method. It was then established the absolute sum-peak method is an efficient, fast and without many corrections for the analysis of this radionuclide when compared with other [6].

2. Experimental

Six sources of ^{133}Ba with activity varying from 1 kBq to 4 kBq were prepared in geometry 10R glass vial and were used in standardization in the absolute anti-coincidence and Ciemat/Nist methods. A glass ampoule with 2 cm was used for measurement by ionization chamber. An analytical balance model AX205 was used.

However six point sources, also prepared, were measured with a Canberra HPGe gamma-X planar detector, with 20 % relative efficiency and a power range of 3 to 300 keV. Applying a specific voltage divider, inserted between the preamplifier and the amplifier, as Figure 1, it was possible to increase up to 1100 keV the detector operation range [7].

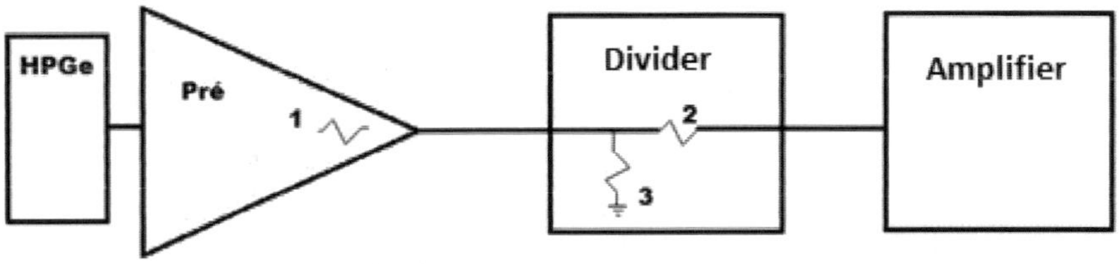

Figure 1. Voltage divider device.

The energies at 53, 276 and 356 keV from the electronic capture (85.4 %) were used for ^{133}Ba measurements. Other branch at 302 keV from the electronic capture (14.4 %) was also used. These energies are in agreement with the average of the X-rays energies ($k\alpha_1$ = 30.6 keV and $k\alpha_2$ = 30.9 keV).

The counting time for sources was 24 hours and 48 hours to the background radiation.

The photopeak areas were determined with the aid of deconvolution software that makes the separation of the double or triple photopeaks [8].

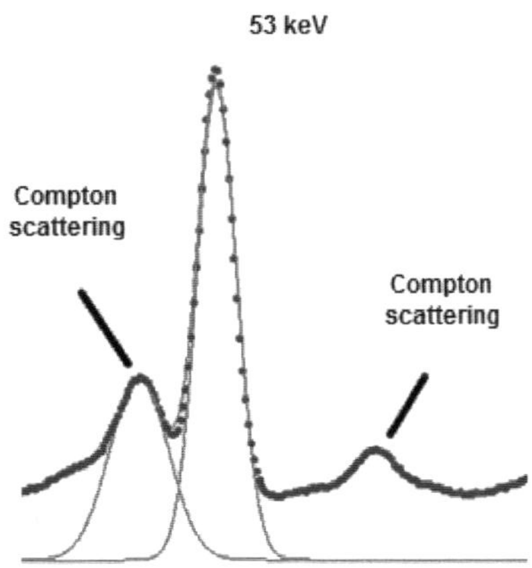

Figure 2. The 53 keV energy deconvolution COLEGRAM software.

IOP Conf. Series: Journal of Physics: Conf. Series **1044** (2018) 012007 doi:10.1088/1742-6596/1044/1/012007

3. Methodology

Figure 3 shows the decay of ^{133}Ba in a different form than conventional [9]. Level 6 represents the ground state of ^{133}Ba, and from level 5 to level 1 represents the various energy levels of the excited state of the ^{133}Cs. The decay EC1 represents the branch of electronic capture with 86.2 %. In it are the X-rays kα and kβ with the energies 30.8 and 35 keV, respectively. Similarly, EC2 represents the electron capture branch with 13.7 %, where the energies of the X-rays kα and kβ are respectively 30.8 keV and 35 keV.

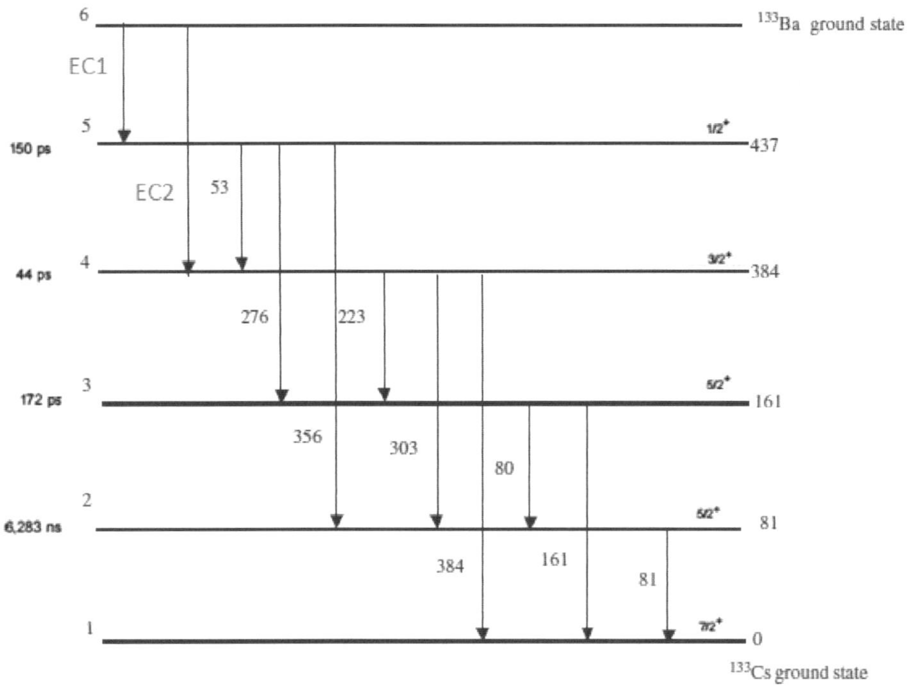

Figure 3. ^{133}Ba scheme decay modified.

Figure 4 shows the germanium detector spectrum in the energy range from 270 to 450 keV where the gamma energies and sums with the kα and kβ X-rays caused by the coincidence sum effect appear. In this figure some photopeaks are coming from possible escape peaks. The simple sums of the gamma rays with X-rays from the coincidence effect are well evidenced.

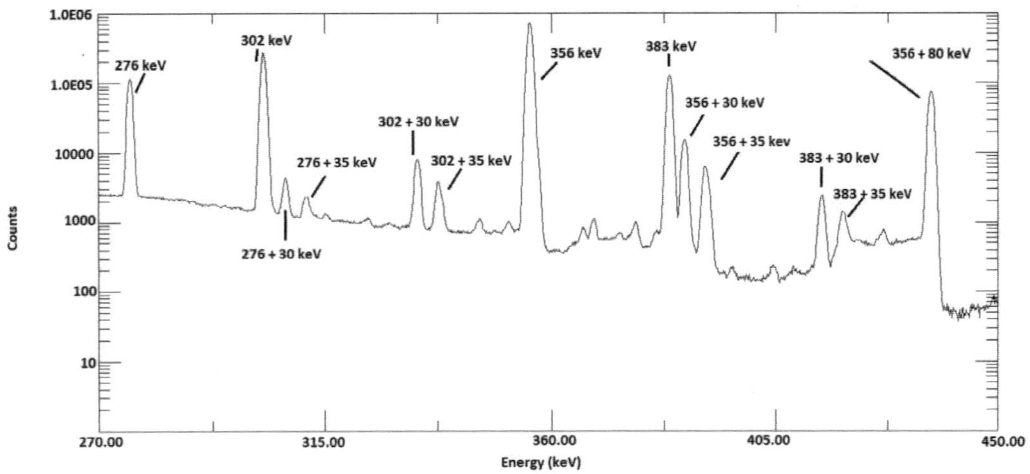

Figure 4. Energy spectrum from 270 keV to 450 keV of ^{133}Ba decay and its sums with the source on detector.

Equation 1 is used by the absolute sum-peak method for the different energies (x) [5].

$$N_0 = T + \frac{N\alpha N\gamma}{N\alpha\gamma} - \frac{N_\gamma^2 \, N\alpha\alpha}{N2_{\alpha\gamma}} \qquad (1)$$

where:

N_α ,N_γ, $N_{\alpha\alpha}$, $N_{\alpha\gamma}$, = keV for X- rays kα for gamma and sum-peak energy peak counts (E);

N_0 = total activity of the source;

T = the total counts of the entire spectrum.

4. Results and discussion

Point sources of ^{133}Ba with low activity were used, aiming at canceling the pile-up effects. This phenomenon occurs within the processing time of the electronic signal generated by the radiation mainly for sources with high activity. It was verified that the pile-up effect was below the detection limit of the system. The dead time in relation to the most active source was less than 5 %, this is the difference between the actual time and the counting time of the radiation.

In the standardization of the ^{133}Ba activity the photopeaks average $k\alpha_{1,2}$ $k_{\beta 1,2}$ energies (30.6 and 35 keV) were used in coincidence with the photopeaks of gamma energies (53, 276, 356 and 302 keV). For each source measured, four different activity values were obtained. The other γ energies were not analyzed due to interferences caused by other sums or because they are low probability energies.

The peak areas of different energies and their coincidences were determined using the de-convolution software.

Table 1 presents the uncertainty components considered for the determination of the activity.

Table 1. Uncertainty of the ^{133}Ba activity measured by sum- peak method.

Uncertainty Components	A (%)	B (%)
Dilution factor		0.05
Live time		< 0.01
Weight		0.20
Half-life		0.01
Background		< 0.01
Decay		< 0.01
Counting statistic	0.11	
Pile up	0.06	
Combined uncertainty (k = 1)		0.24

The final activity by sum-peak method was determined by the average of the results found by the observed coincidences and compared with the other methods. These results are shown in Table 2.

Table 2. Comparison of ^{133}Ba activity standardization between sum-peak and other methods.

Methods	Activity (Bq/g)	Uncertainties (k=1)
Sum-peak	35.338	0.24
Ionization chamber	35.454	0.29
Ciemat/Nist	34.959	0.48
Anti-coincidence	35.254	0.66

5. Conclusion

The final result of ^{133}Ba activity obtained by the sum-peak method is suitable for absolute standardization. This result, compared to those obtained by traditional absolute standardization methods, such as anti-coincidence, Ciemat/Nist and BIPM traceable ionization chamber, presented consistent value and coherent uncertainty, below 0.3 %. Thus, the sum-peak coincidence method can be used quickly and effectively to standardize in activity ^{133}Ba a radionuclide with complex decay.

References

[1] Brinkman G A and Aten Jr A H W 1963a Absolute standardization with a NaI (Tl) crystal-III. *Int. J. Appl. Radiat. Isot.* **14**, 503-510.

[2] de Almeida M C M, Iwahara A, Poledna R, da Silva C J and Delgado J U 2007 Absolute disintegration rate and 320 keV gamma ray emission probability of ^{51}Cr. *Nucl. Instrum. Meth. Phys. Res.* **A580**, 165–168.

[3] Araújo M T F, Poledna R, Delgado J U, Silva R L, Iwahara A, da Silva C J, Tauhata L, Oliveira A E, de Almeida M C M and Lopes R T 2013 Análise de Impurezas Radionuclídicas para o Radiofármaco (^{123}I) NaI utilizando a Espectrometria *Gama 7º Congresso Brasileiro de Metrologia* ISSN/ISBN 978-85-86768-07-1.

[4] *Practical Gamma-ray Spectrometry* - Handbooks, manuals, GILMORE GR, Nuclear Training Services Ltd, Warrington,UK, 2008.

[5] *Standardization of I-125 Sum-Peak Coincidence Counting* - Informes Técnicos Ciemat, Carles, A. G., Malonda, A. G., Espana, 2011.

[6] da Silva R L, de Almeida M C M, Delgado J U, Poledna R, Santos A, Veras E V, Rangel J and Trindade O L 2016 Metrological activity determination of ^{133}Ba by sum-peak absolute method *Journal of Physics: Conference series,* **733 (1).**

[7] da Silva R L, Estrada J J S, Delgado J U, Poledna R and de Almeida M C M 2007. Alteração na faixa de detecção e calibração de um sistema de espectrometria gama. *Revista Militar de Ciência e tecnologia.*

[8] Lépy M C Presentation of the COLEGRAM software, Version 4.0, Note technique LNHB/04/26.

[9] Novkovic D , Durasevic M , Kandic A, Vukanac I, Milosevic Z and Nadderd L 2007 Coincidence summing of X-and gamma rays of ^{133}Ba , *Nucl. Instrum. Meth. Phys. Res.* **A 582**, 592-602.

IOP Conf. Series: Journal of Physics: Conf. Series **1044** (2018) 012008 doi:10.1088/1742-6596/1044/1/012008

An Electromagnetic Force Balance Experiment for Measurement Science Education

L M Faller and H Zangl

Institute of Smart System Technologies, Sensors and Actuators, Alpen-Adria-Universitaet Klagenfurt, Universitaetsstrasse 65-67, 9020 Klagenfurt, Austria

E-mail: `lisa-marie.faller@aau.at`

Abstract. In the design and conception of all kinds of systems, measurement equipment and the respective measurement results play a crucial role. They provide the means to monitor the environment as well as the proper functioning of the respective systems. It is therefore mandatory to raise awareness of practitioners and engineers for important concepts such as measurement uncertainty, parameter uncertainty, their traceability and related consequences already in during education. In this paper, we present a teaching concept based on an electromagnetic force balance experiment towards the learning objectives such as electromagnetic force generation in actuators, utilization of the law of conservation of energy, experience with laboratory equipment and the SI system with aspects such as realizations of base units, traceability and in particular measurement uncertainty.

1. Introduction

In our modern society, measurement devices and sensors, the respective electronic evaluation and signal processing chains, are indispensable. Measurement science is employed to fulfil manifold tasks in consumer goods such as mobile phones and personal computers or tablets as well as in cars, (automated) production lines or safety and security systems. Virtually all electronic devices our society relies on in everyday private as well as working life are equipped with sensors and measurements electronics. When educating engineers to become the future designers of such measurement chains, critical analysis of sensor systems with respect to measurement uncertainty and traceability of results in context of standards, definitions ([1], [2]) and recent recommendations as well as research (e.g. [3] and [4]) thus seems mandatory. It is necessary to enhance the awareness with respect to measurement uncertainty, and uncertainty in modeling and parameter estimation in such systems and teach proper usage and adoption of the available methodology (- as in the Guide to the expression of Uncertainty in Measurement GUM [5, 6, 7] -) to prevent undesired system states, note at least in the context of safety critical systems. This is even more important when society more and more relies on electronic measurement systems, where in, e.g., automated production chains with human - robot interaction and collaboration, or autonomous in autonomous driving, lives depend on them.

In order to provide a practical experience in adoption, usage and especially limitations of the methods that are used to cope with these challenges, we provide a lab course for our Master students. Most of them are enrolled in the Master study program Information and Communications Engineering with a specialization in autonomous systems and robotics. It is assumed that the students are well familiar with the concept of measurement uncertainty (compare [8, 9, 10]), probability theory, estimation and detection theory [11] as well as electrical measurements and sensors.

In the previous lectures and courses, the methodologies are presented theoretically or in simulation. The concept of this work is now to provide practical experience. We also aim at giving an understanding of the concepts behind the SI system and the definition of basic units as well as their traceability.

2. Electromagnetic Force Balance

Balances are well-known experimental setups, where the most established is the Watt balance (e.g. [12]). Other setups rely on electrostatic force (e.g. [13]) or electromagnetic force (e.g. [14]). The Watt balance is the experiment suggested to determine the Planck's constant h with uncertainties of 1 part in 10^8 and is also suggested for the realization of the kg [12]. It is developed to measure at the lowest feasible uncertainty, i.e. highest possible accuracy. Other mentioned force balances as presented in, e.g. [13] and [14], are designed for traceability and, as well, for high accuracy.

The principle of the experiment we consider, although the setup is by far not as accurate, is similar to those mentioned before. It relies on the principle of energy conservation and the balance of electromagnetic energy E_{magn} and mechanical work W_{mech} when no external energy is introduced. We can state the following relations

$$dW_{mech} = Fds \tag{1}$$

$$E_{magn} = \frac{1}{2}LI^2 \tag{2}$$

$$dW_{mech} = -dE_{magn} \tag{3}$$

F is the force, L is the inductance, I is the current and s is the path or distance. The d-prefix indicates changes in the respective quantities. We can further apply Faraday's law of induction

$$U = N\frac{d\Phi}{dt} = L\frac{di}{dt} \tag{4}$$

where U is the voltage in a coil with N windings, induced by a change in the magnetic flux Φ. The current through the coil is denoted by i for the dynamic (AC) and I for the static (DC) quantity. Since the apparatus is used at equilibrium, i.e. no external energy is introduced, we know that voltage across the clamps needs to be zero (no electrical power), which further implies that the magnetic flux Φ needs to be constant, and we can thus modify (4) as

$$N\Phi = Li \tag{5}$$

and reformulate as

$$i = \frac{N\Phi}{L} \tag{6}$$

using (1) and (2) and the assumption of constant magnetic flux Φ, we get

$$F = \frac{dW_{mech}}{ds} = -\left.\frac{dE_{magn}}{ds}\right|_{\Phi=const} = -\frac{1}{2L^2}(N\Phi)^2\frac{dL}{ds} = -\frac{1}{2}\left(\frac{N\Phi}{L}\right)^2\frac{dL}{ds} \tag{7}$$

and using (6) we find

$$F = -\frac{1}{2}I^2\frac{dL}{ds} \tag{8}$$

3. Measurements and Evaluations

Using the setup as shown in Fig. 1, the coil can be moved above the ferromagnetic metal piece placed on the scale. The gradient of the inductance with respect to the path (distance between coil and feromagnetic metal piece), i.e. dL/ds, is measured (either with an LCR bridge or scope/multimeter and signal generator). The distance can be measured using either the ruler mounted at the height adjustment facility or using a caliper. The force is to be determined using the Eq.s as given above and is to be compared with the reference measurements of the electronic scale (see Fig. 1). To determine the inductance or change of inductance of the coil

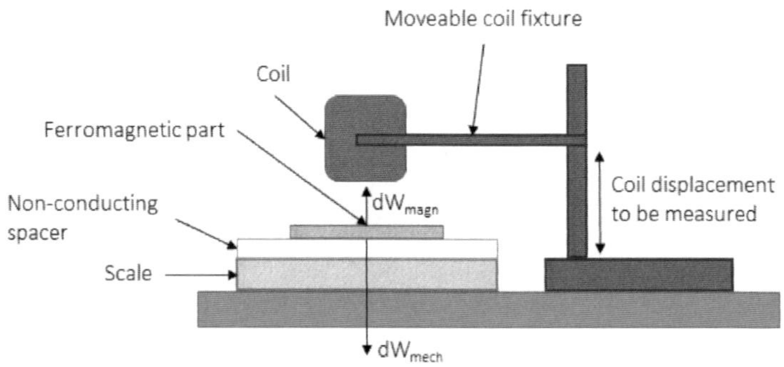

Figure 1. Left: photograph of the used laboratory setup. Right: schematic illustration of the experimental setup.

using the multimeter, it is necessary to record the change of the current i at fixed voltage u and frequency f. Using these values, it is possible to determine the coil inductance via impedance, using

$$Z = \frac{u}{i} = \sqrt{R^2 + X_L^2} \tag{9}$$

(10)

where the complex inductive resistance is

$$X_L = j\omega L \tag{11}$$

and where ω is the angular frequency and can be rewritten as $\omega = 2\pi f$. The inductance L can then be calculated e.g. as

$$L = \frac{\sqrt{Z^2 - R^2}}{j\omega} \tag{12}$$

The measured current i as function of the distance d at a constant voltage is given in Fig. 2 in the upper left. The respective inductance value is plotted in Fig. 2 upper right. The force as function of the current at a fixed distance is illustrated in Fig. 2 lower right. Bringing conductive material close to an energized coil, we can observe two different phenomena:

(i) In conductive material which is not magnetic (copper, aluminum etc.), the magnetic field of the approaching coil causes eddy currents. These eddy currents then produce a magnetic field which counteracts the primary field of the coil. This effect is used in, e.g., proximity sensors or non destructive testing of metallic parts.

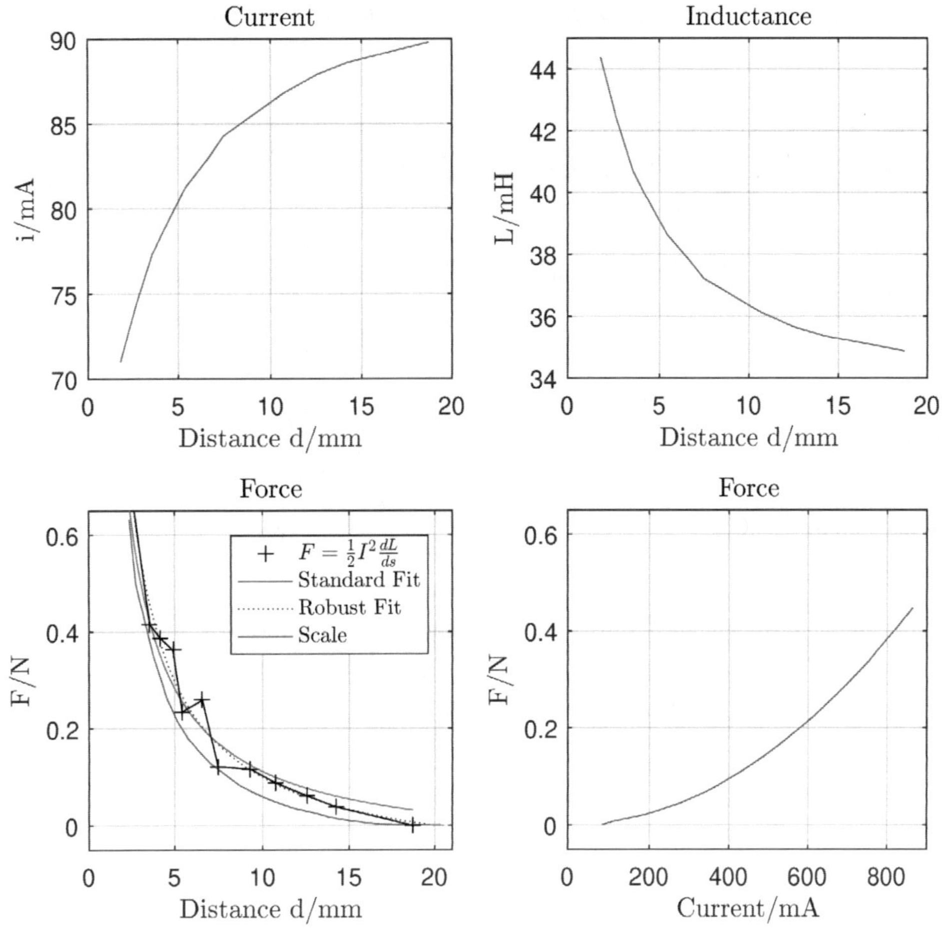

Figure 2. Measurement results and least squares fit for the force derived via inductance.

(ii) For ferromagnetic material, the prevalent effect is an increase in inductance due to the increased permeability μ_r (magnetic material instead of air). The magnetic field is enforced through the ferromagnetic material and its magnetization. This principle is used in transformers with ferromagnetic cores. The generation of eddy currents occurs in ferromagnetic material as well. To avoid this unwanted effect in, e.g., transformer cores, they are commonly made of electrically isolated sheet plates.

The lower left plot in Fig. 2 gives the force as dependent on the distance, once measured using the scale (blue) and once derived via inductance (black). Obviously, the curve for the derived force is noisier than the others. Here we see uncertainties introduced from the used measurement devices (e.g.: resolution, accuracy, settling time) as well as a low number of measurement points. Two types of fits are determined to describe the derived force as dependent on the distance using a semi-analytic model: *standard* and *robust*. Here, the choice of suitable model is crucial, in this case, we use a model of the form

$$F = \beta_1 e^{\frac{1}{-\beta_2 d}} + \frac{1}{\beta_3 d} \tag{13}$$

where the vector $\boldsymbol{\beta}$ holds the model coefficients determined through the fitting routine.
Using the *standard* fit, the model coefficients are determined using a simple least squares approach

$$\hat{\boldsymbol{\beta}} = (X^T X)^{-1} X^T y \tag{14}$$

where $\hat{\beta}$ are the estimated coefficients, X is the measurement matrix, and y is the response. Often, these measured quantities used to determine the response are also called predictors. In the weighted least squares, points are given more weights which are close to the fitted line, thus the influence of outliers (and large variation in measurement) can be reduced

$$\hat{\boldsymbol{\beta}} = (X^T W X)^{-1} X^T W y \tag{15}$$

where W is the weight matrix. In the case of the derived force as dependent on the distance, this *robust* fit can reduce the discrepancy between measured and derived dependence. Other approaches include the determination of outliers (e.g. violating the 1.5σ bound) and remove them before fitting.

4. Traceability

Traceability is defined according to the VIM [2] as "A property of a measurement result whereby the result can be related to a reference through a documented unbroken chain of calibrations, each contributing to the measurement uncertainty."
Traceability, as the name already implies, defines a concept which enables the experimenter to state the measurement uncertainty as dependent on the used equipment and setup. To be able to do so, it is necessary that all the used equipment is calibrated and the remaining uncertainty in the measurement is stated. The experimental setup used in this work is not traceable: the measurement equipment (ruler or caliper) of the distance between coil and ferromagnetic metal piece is not unique: using the ruler, there is no calibration routine performed, using the caliper no unique measuring position as well as procedure has been defined. The same is the case for the height adjustment facility.

5. General Teaching Concept

The respective course where this concept is applied is a laboratory course under the title of sensors and actuators. Since this course is suggested as successor to the Measurement and Signal Processing (MSP) course. In MSP, advanced stochastic concepts are covered to yield, in a certain sense, optimal estimators for measurements subject to various sources of noise. Thus, students are assumed familiar with probability theory and statistics. Hence, mathematical background to deal with measurement uncertainty is assumed known, as well as parameter estimation principles and regression.

The students are advised to form groups. In each unit, a group performs one of the four specifically designed experiments. Each experiment is elaborated to gain experience with laboratory equipment as well as the respective physics and theory.

In the considered experimental design, the students are supposed to

- Measure inductance as a function of the distance between a coil and a ferromagnetic part with signal generator and scope/multimeter and with LCR-bridge
- Perform regression: get a fitted analytic function for the inductance as function of the distance
- Determine the force as a function of the current
- Validate the results using the electronic scale

- Perform an uncertainty evaluation: is our result for the force *traceable*

The preliminary theory and equipment basics are introduced and explained at the beginning of each unit. At the end, the experimentally achieved results are discussed and have to be gathered in a report. Additionally, students are encouraged to give suggestions for improvement of the considered setup.

6. Suggestions for Improvement

Suggestions for improvement are given in terms of the accuracy as well as resolution. Contributors such as measurement instruments, environmental conditions and the measurement process are addressed. A few results are gathered in the following:

- The measurement of inductance using the LCR-meter is preferred over the calculation via impedance.
- Improved mounting of the coil (slack and instable)
- Height adjustment resolution is too small
- Place ruler (for distance measurement) differently: e.g., closer to the coil

7. Conclusion

The presented experiment is a simple to realize laboratory setup, based on which students can be familiarized with basic principles of sensors and actuators, measurement ncertainty, system modelling, the SI system and traceability. The experiment allows to demonstrate the influence of the uncertainties of the contributing equipment on the system. The principles of curve fitting and modelling are demonstrated by determining a semi-analytical model for dependence of the change of inductance dL on the change of distance ds. Furthermore, the difficulty in finding a suitable fit and influence of uncertainty in measurement are demonstrated.

Acknowledgments
Authors would like to thank the students of the Sensors and Actuators laboratory for their assistance in gathering the ideas for improvement of the described setup.

References

[1] BIPM JCGM 102:2011: Evaluation of measurement data - supplement 2 to the "guide to the expression of uncertainty in measurement" - extension to any number of output quantities
[2] BIPM JCGM 200:2012 : International Vocabulary of Metrology - Basic and General Concepts and Associated Terms (VIM 3rd edition)
[3] da Silva Hack P and ten Caten C S 2012 *IEEE Transactions on Instrumentation and Measurement* **61** 2116–2124
[4] Ferrero A and Salicone S 2012 *IEEE Transactions on Instrumentation and Measurement* **61** 2167–2178
[5] BIPM JCGM 100:2008: Guide to the expression of uncertainty in measurement
[6] BIPM JCGM 101:2008: Evaluation of measurement data - supplement 1 to the "guide to the expression of uncertainty in measurement" - propagation of distributions using a monte carlo method evaluaton of measurement data
[7] BIPM JCGM 102:2011: Evaluation of measurement data - supplement 2 to the "guide to the expression of uncertainty in measurement" - extension to any number of output quantities
[8] Zangl H and Hoermaier K 2017 *Measurement* **101** 257–264
[9] Zangl H, Zine-Zine M and Hoermaier K 2014 Utilization of software tools for uncertainty calculation in measurement science education *IMEKO joint Symposium 2014 TC1 - TC7 - TC13 Measurement Science Behind Safety and Security* (Funchal, Madeira Island)
[10] Faller L M, Zangl H and Leitzke J P 2016 Fault Injection Software Tools and Robust Design Principles for Reliability and Safety in Measurement Science Education *IMEKO Joint Symposium TC1-TC7-TC13*
[11] Kay S 1993 *Fundamentals of Statistical Signal Processing: Estimation Theory* Prentice Hall Signal Processing Series (New Jersey: Prentice Hall)

[12] Picard A, Fang H and Stock M 2007 Progress on the BIPM Watt Balance *IMEKO 20th TC3, 3rd TC16 and 1st TC22 International Conference*
[13] Pratt J R and Kramar J A 2006 SI Realization of Small Forces Using an Electrostatic Force Balance *XVIII IMEKO Worl Congress*
[14] Diethold C, Hilbrunner F, Frhlich T and Manske E 2012 Nanopositioning SSystem with Combined Force Measurement Based on Electromagnetic Force Compensated Balances *XX IMEKO World Congress*

IOP Conf. Series: Journal of Physics: Conf. Series **1044** (2018) 012009 doi:10.1088/1742-6596/1044/1/012009

An improved visual description of the VIM

Omar-Jair Purata-Sifuentes[1], **Anahí-Yridian Belman-Hernández** and **Andrea-Guadalupe Fernández-Torres**

Departamento de Estudios Organizacionales, Universidad of Guanajuato, Fraccionamiento I, El Establo, Campus Guanajuato, C.P. 36250, Guanajuato, México

Email: opurata@ugto.mx

Abstract. All the sciences and several disciplines have their own vocabulary, and metrology is not the exception. The international vocabulary of metrology (VIM) includes an annex with concept diagrams which represent most of the relations among the concepts defined in the main text of the vocabulary. Even when these diagrams follow international standards, they are usually confusing and sometimes hard to understand. OSSAD (Office System Support Analysis and Design) is a visual language that has been used to document management systems and has its own structure, syntax and grammar. Visual representation through OSSAD of originally blurred textual procedures has proved its usefulness in translating text into visual diagrams easy to understand for inexperienced users. The adaptation of OSSAD to visually description of the VIM is showed in this paper. A comparative analysis between the current concept diagrams and the proposed ones was done. The visual language has allowed the mapping of complex relationships between VIM concepts in a more natural way, and the results of a questionnaire suggest that comprehension of metrology terminology in classroom has been improved with the OSSAD visual description of the VIM.

1. Introduction

One of the functions of terminology is the representation of concept systems using concept diagrams. A concept diagram could be defined as [1] "a graphic representation of a set of units of knowledge, created by a unique combination of abstractions of a property of anything perceivable or conceivable, and structured according to the relations among them", i.e., the graphic representation of a concept system.

Even though there are many ISO (International Organization for Standardization) standards which contain a section –usually an informative annex– with concept diagrams, they are described with more detail in [2]. However, the concept diagrams included in ISO documents are not the friendly way to know such a specialized science or discipline (e.g., metrology) specially for the novice student. Even though an optional graphic representation has been already explored by ISO through the Unified Modeling Language (UML) [3], the scope of that representation is more oriented towards software development, rather than towards the understanding of the concept system.

OSSAD is a visual language developed initially for the support of office systems analysis and design [4]. However, several adaptations had been made for using it in documenting quality management systems [5], and as a tool for integration of management systems [6]. Its versatility has been proved, and is a friendly scheme for documenting management systems. This project aims to show how using the symbols and grammar rules of OSSAD could make improvements in the level of comprehension of the VIM [7]. It is important to note that the use of OSSAD is not meant to become a replacement for traditional concept diagrams, but is intended to be an alternative or complementary notation.

2. Mapping of terminology elements with OSSAD

OSSAD original methodology uses three levels of representation: an abstract model for systems, a descriptive model for processes, and a prescriptive model for instructions. Even when here the aim is not towards processes mapping, it is possible to apply a kind of "process approach" to the documentation of the several concept diagrams usually included in the terminology or vocabulary of ISO standards. In that sense, the whole vocabulary would be the "conceptual system" and the different concept diagrams included in a specific vocabulary would be the different "processes" which, of course, would be interrelated and interacting as a set aiming to a specific target. The table 1 describes the concepts in [1,2] used in the VIM, along with the corresponding proposed OSSAD modelling elements.

Table 1. Equivalence between the concepts and symbols proposed in [1,2] and their representation in OSSAD visual language.

[1, 2]	OSSAD symbol	Description
concept	concept	A rectangle with the name of the terminological concept. Lines could be single, double, dotted or heavy (see below).
	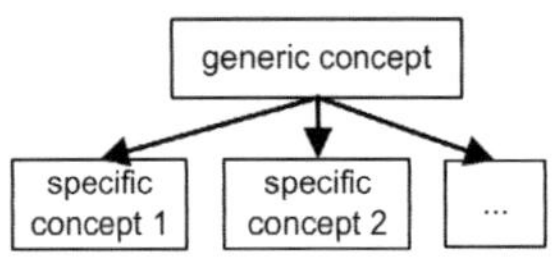 generic concept; specific concept 1; specific concept 2; ...	The first type of generic relation. Heavy arrows pointing towards every specific concept included. As in [7], when one or more specific concepts exist but are not represented, a rectangle with three dots is included.
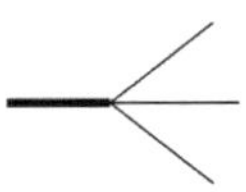	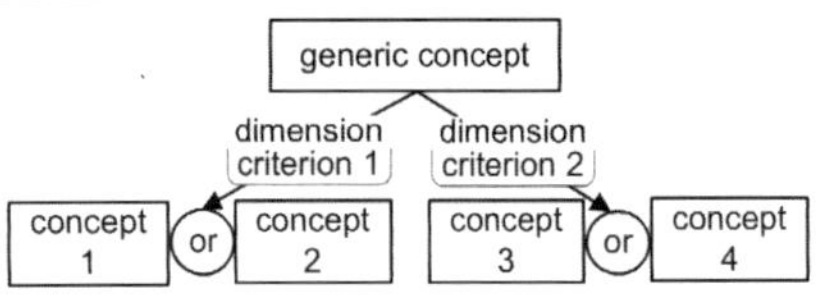 generic concept; dimension criterion 1; dimension criterion 2; concept 1; or; concept 2; concept 3; or; concept 4	The second type of generic relation is represented by an arrow with a tray which contains the dimension criterion. Specific concepts are linked with Boolean "OR" connectors.
	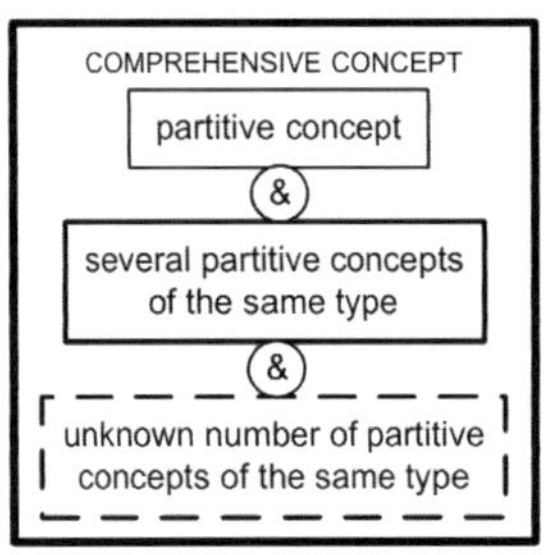 COMPREHENSIVE CONCEPT; partitive concept; &; several partitive concepts of the same type; &; unknown number of partitive concepts of the same type	A partitive relation is represented by a macro-rectangle with heavy line which title is the comprehensive concept. Every single partitive concept is represented in a rectangle. Several partitive concepts of a given type are represented by a heavy line rectangle, and if such plurality is uncertain a dotted line rectangle is used. All the concepts in the macro-rectangle are linked with Boolean "AND" (&) connectors.
(concept)	concept no defined	A concept not defined in the VIM is represented by double-line rectangles.
	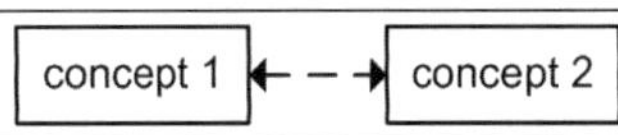 concept 1; concept 2	The associative relation is represented by a double-headed dotted-line arrow.
(see Fig. A#)	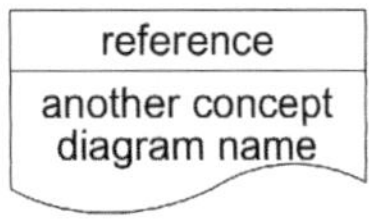 reference; another concept diagram name	A parenthetic reference to another concept diagram is indicated by a document symbol with two compartments used to divide the concept diagram reference and its name.

3. Examples of mapping of VIM concept diagrams with OSSAD

Application of the equivalences defined in table 1 on a couple of concept diagrams of the VIM, figures 1 and 2, has been made. As can be seen, is a more intuitive graphical description of the metrological concepts than the conceptual diagrams on figures A.1 and A.2 in the VIM.

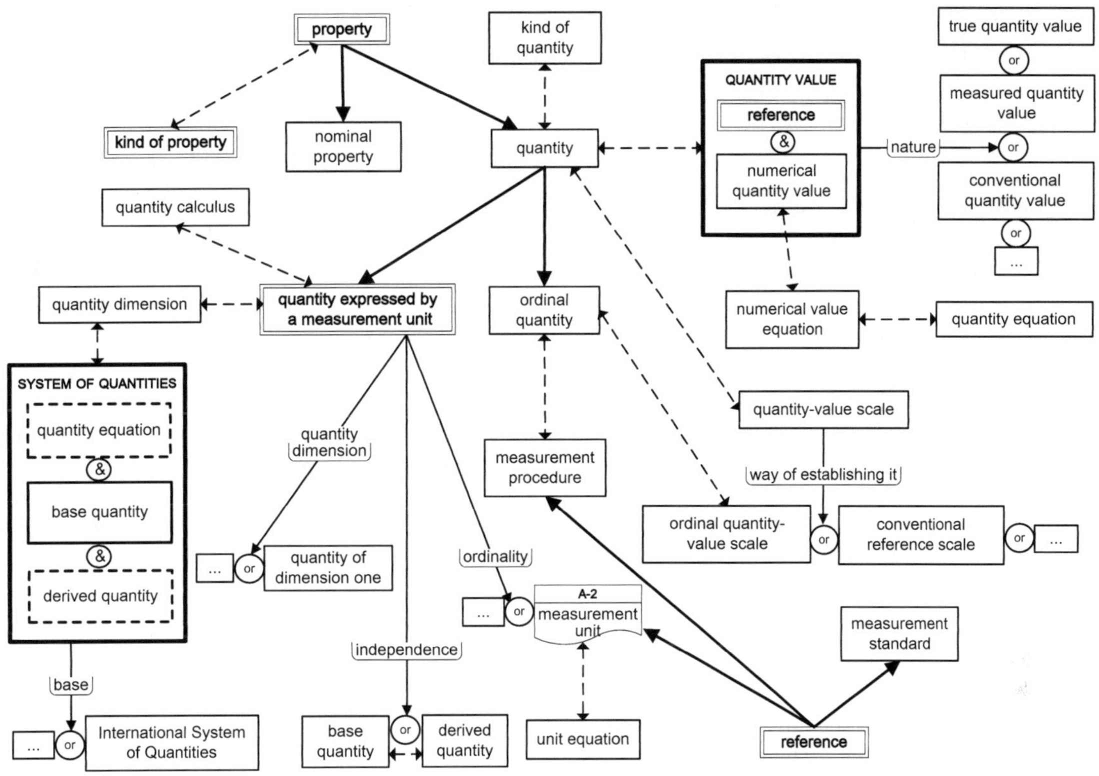

Figure 1. Equivalent OSSAD graphic description of the concept diagram for "quantity" in VIM.

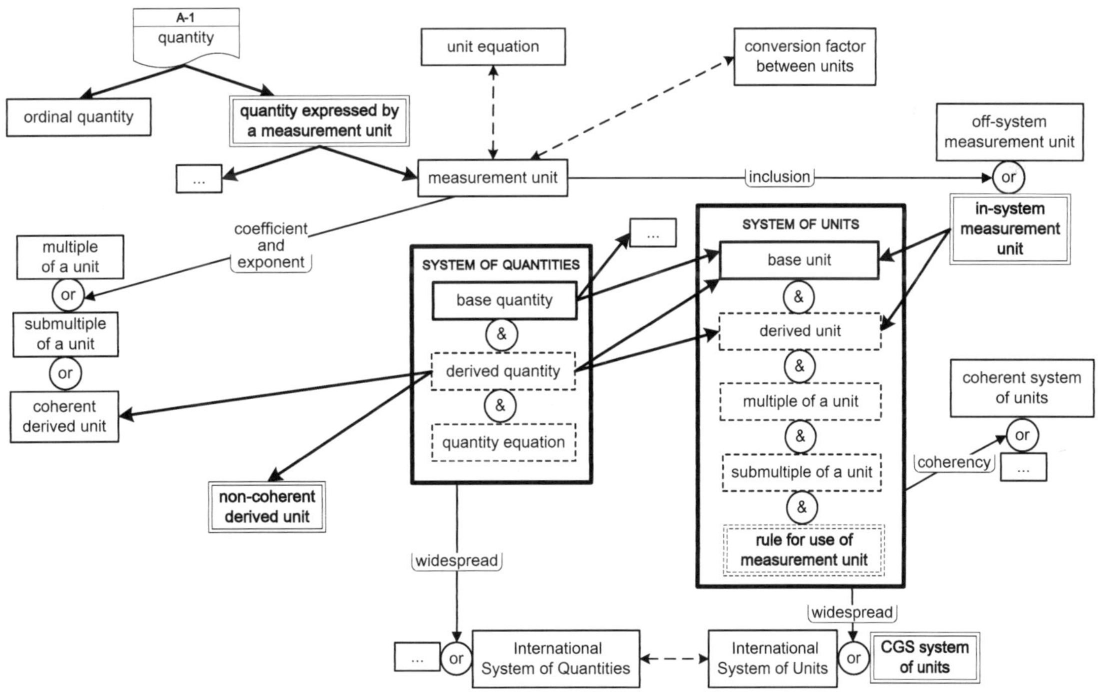

Figure 2. Equivalent OSSAD graphic description of the concept diagram for "measurement unit".

Among the most easily noted differences are: there is no crossing lines on the OSSAD diagrams, the partitive relations stand out notoriously due to the "concept macro-boxes", the document icon makes easy the connection between several diagrams, and the Boolean connectors made easier to understand the relationships between the concepts. However, these apparent advantages need to be tested, to be able to get some conclusions about its understanding for the novice metrologist.

4. Testing OSSAD diagrams vs VIM concept diagrams

A questionnaire (table 2) was developed and applied to under-graduated metrology course students. The students were divided into two groups with eleven students each one. Students in Group A were asked to study the VIM concept diagrams explanation (pp. 65-67 in [7]) for ten minutes, then the "quantity" concept diagram (p. 68 in [7]) was provided, along with another five minutes allowed for its analysis. Finally, the VIM concept diagrams explanation was withdrawn and the questionnaire was given with no time limit to respond it. On the other hand, students in the Group B were asked to study the OSSAD concept diagrams explanation which was based on the second and third columns of table 1, but with a very similar appearance to pp. 65-67 in [7]. After that, the "quantity" OSSAD concept diagram (figure 1) was provided with additional five minutes allowed for its analysis. As with the Group A, the same questionnaire was applied after the OSSAD concept diagrams explanation was withdrawn, no time limit to respond it. In both groups, the questionnaire was anonymous.

Table 2. Questionnaire to test the comprehension level of concept diagram representations.

#	Question or statement	Multiple choice answers
1	What kind of relation exists between the concepts '*ordinal quantity*' and '*measurement procedure*'?	a) partitive b) non-hierarchical c) generic
2	The definitions of the concepts '*reference*' and '*property*' are included in the VIM.	a) true b) false
3	The concepts '*quantity equation*', '*base quantity*' and '*derived quantity*' are ______________ the concept '*system of quantities*'.	a) associated concepts to b) partitive concepts of c) specific concepts of
4	The concepts '*ordinal quantity*' and '*quantity expressed by a measurement unit*' are ____________ the concept '*quantity*'.	a) associated concepts to b) partitive concepts of c) specific concepts of
5	According to "*quantity*" concept diagram, it is known how many partial concepts of the concept ____________ there are.	a) '*quantity equation*' b) '*base quantity*' c) '*derived quantity*'
6	According to "*quantity*" concept diagram, there are several partial concepts of the concept '*numeric quantity value*'.	a) true b) false
7	What kind of relation is there among the concept '*quantity value*' and the concepts '*true quantity value*', '*measured quantity value*', and '*conventional quantity value*'?	a) generic b) partitive c) associative

5. Results and discussion

The results of the quick comprehension questionnaire are showed in table 3. As can be seen, the Group A who worked with the original VIM concept diagrams had the same performance level than the Group B who worked with the proposed OSSAD concept diagrams. However, it is worth to note some remarkable issues:

i. The results of Group A are more balanced, almost all responses retain 5/6 or 4/7 ratios between right / wrong answers. On the other side, Group B has three questions with totally polarized answers, two questions were correctly answered and the other one was incorrectly answered.

ii. Group B reach 100 % comprehension with question #2, and more than 90 % with question #1.

iii. Group B failed with question #4. It will be argued below.

Table 3. Questionnaire results from undergraduate students attending metrology fundamentals course for the first time.

Question	Group A (VIM concept diagrams)		Group B (OSSAD concept diagrams)	
number	Right answers	Wrong answers	Right answers	Wrong answers
1	7	4	10	1
2	8	3	11	0
3	6	5	5	6
4	4	7	1	10
5	7	4	6	5
6	5	6	5	6
7	6	5	7	4
Total	**43**	**34**	**45**	**32**

Answers in Group A balanced around right/wrong could be an indication that the graphical description system of VIM depends a lot on the visual ability of the student, and her capability to work with naked words or phrases and different classes of lines only without other geometrical forms around.

Instead, Group B had almost half of their answers fully loaded to one side or another, and this is a clear indication of what are the strengths and the weaknesses for the OSSAD conceptual diagrams system. Concepts relationship asked in questions #1 and #2 were correctly answered by practically all the students. Question #4 appears to be a weakness, but a possible explanation had been traced to the fact that the Spanish version of figure 1 used in the questionnaire did not show the two arrows in bold that go from the concept 'quantity' to the concepts 'ordinal quantity' and 'quantity expressed by a measurement unit' starting from the same location point, as exemplified in the central column of the fourth row of table 1 (see also figure 3). With respect to the other questions, similar behavior to the VIM concept diagrams was obtained.

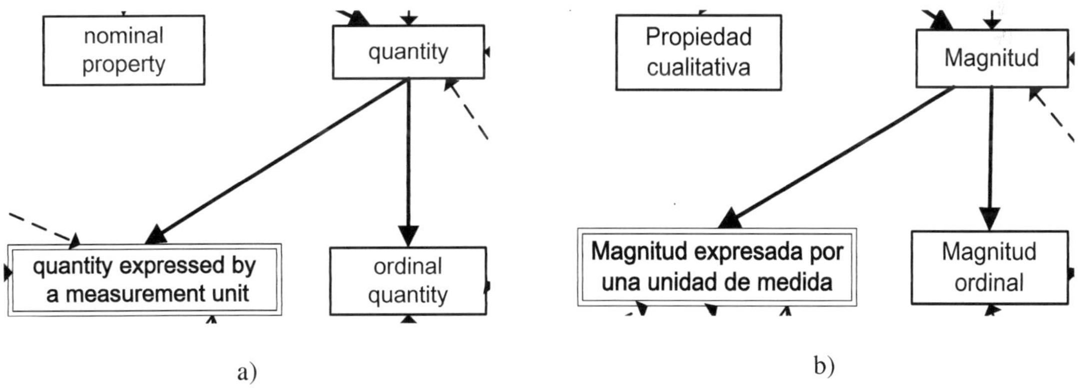

Figure 3. Little but striking difference in generic relation diagram: a) English, b) Spanish.

UML language is another option to visualize the VIM concept system, as showed in [3]. However, as stated in [8] UML is more oriented towards documentation of software-intensive system, including the object-oriented modeling for language programming. Nevertheless, it is another option which could be evaluated together with OSSAD and the current graphical representation of the VIM in a future research.

Finally, it is important to highlight that the scope of the concept diagrams of the VIM is more directed to verification of the systematic concepts definitions than to visual representations, so the visual expressive power of the system employed is weaker than OSSAD, which was born as a visual language and here is proposed an adaptation from its original intention, that is, the workflow process mapping.

6. Conclusions

OSSAD had been applied onto the graphical description of VIM concepts. It has been shown that the terminological concepts defined and used in the VIM can be represented by a limited set of symbols in the OSSAD language. A couple of VIM concept diagrams have been translated to the proposed OSSAD approach, and one of them has been used to test the relative level of understanding of the VIM's concepts relationships between twenty novice students of metrology. According to the results of a questionnaire (in Spanish) applied to them, the proposed representation system has a better visual expressive power. However, it is important to emphasize that the use of OSSAD is not meant to become a replacement for traditional concept diagrams, but is intended to be an alternative and complementary notation.

Acknowledgments

Belman-Hernández acknowledes the financial support of EDUCAFIN and Presidencia Municipal de Romita, Guanajuato, México. Fernández-Torres acknowledes the financial support of EDUCAFIN and Presidencia Municipal de Silao, Guanajuato, México.

References

[1] ISO 1087-1:2000 Terminology work – Vocabulary – Part 1: Theory and application (Geneva: International Organization for Standardization)

[2] ISO 704:2009 Terminology work – Principles and methods (Geneva: International Organization for Standardization)

[3] ISO 24156-1:2014 Graphic notations for concept modelling in terminology work and its relationship with UML – Part 1: Guidelines for using UML notation in terminology work (Geneva: International Organization for Standardization)

[4] Beslmüller E and Conrath D W 1989 *Esprit '89 – Proc. 6th Ann. ESPRIT Conf. (Brussels, Belgium, Nov 27th-Dec 1st)* (Netherlands: Springer) pp 865-77

[5] Berger C and Guillard S 2000 *La Rédaction Graphique des Procédures* (Paris: AFNOR)

[6] Purata O J 2015 *Proc. 18th QMOD-ICQSS Int. Conf. on Quality and Service Sciences (Seoul, Republic of Korea, 12-14 October 2015)* eds S M Dahlgaard-Park and J J Dahlgaard (Lund: Lund University Library Press) pp 950-63

[7] JCGM 200:2012 International vocabulary of metrology – Basic and general concepts and associated terms (VIM) 3rd ed. (Joint Committee for Guides in Metrology)

[8] ISO/IEC 19501:2005 Information technology – Open Distributd Processing – Unified Modelling Language (UML) Version 1.4.2 (Geneva: International Organization for Standardization)

IOP Conf. Series: Journal of Physics: Conf. Series **1044** (2018) 012010 doi:10.1088/1742-6596/1044/1/012010

Analyses of different methodologies employed for correcting systematic effect

N A Guimarães[1], W J L Garção[1, 2], A A Isnard[1], R G Santana[1]
[1] Teacher, *Instituto Federal do Rio de Janeiro - IFRJ Campus Volta Redonda,*
Rua Antônio Barreiros, 212, Volta Redonda, Rio de Janeiro, BRA
[2] Master Student, *Pontifícia Universidade Católica do Rio de Janeiro,*
Rua Marquês de São Vicente, 225 - Gávea, Rio de Janeiro, BRA

E-mail: nilmara.guimaraes@ifrj.edu.br

Abstract. Correcting indications, or compensating for systematic errors, is an indispensable factor in ensuring the reliability of measurement results in routine calibration laboratories. Different methodologies for the application of this correction can be adopted, since the guide for the expression of uncertainty in measurement ISO GUM is not sufficiently clear about whether: a) the correction procedure should be executed on each measure indication individually, or b) the correction procedure should be appliedtothe average of the measure indications. The main objective of the present work is to discuss and analyze the different methodologies that can be applied to the correction of indications through simulations and the various ways for defining the systematic effect itself. The expected result is to show that Type A uncertainty estimation depends directly on the method utilized for correcting indications. The main consequence is that measurement results always depend on the correction method chosen.

1. Introduction

Metrology is a multidisciplinary science that covers several areas of knowledge since it is not possible to monitor, control or even investigate [1] the effect of a given measurement without taking measurements. The International Vocabulary of Metrology [2] states that measurement isthe process of experimentally obtaining one or more quantity values that can reasonably be attributed to a quantity. Thus, because it is an experimental process, there are a number of factors that can be considered when performing a measurement.

ISO GUM [3] states that although systematic error, like random error, cannot be totally eliminated, it can often be reduced. If a systematic error arises from a recognized effect of an influence's quantity on the result of a measurand, hereafter termed a systematic effect, the effect can be quantified and, if it is significant in size relative to the required accuracy of the measurement, a correction or correction factor can be applied to compensate for this effect. It is assumed that, after correction, the expectation or expected value of the error arising from a systematic effect is zero.

1.1. Measurement uncertainty

An uncertainty is a "*non-negative parameter characterizing the dispersion of the quantity values being attributed to a measurand, based on the information used*" [2]. It is necessary to evaluate the uncertainties related to frequency distribution and probability-related experience or to other information to carry out an evaluation correctly. Measurement uncertainty evaluation consists of Type A and Type B uncertainty.

The components related to the evaluation of Type A uncertainty are generally those that most impact a measurement result, since they evaluate the variability of the measurements, both of the measurand, and the standard. There is a method for possible evaluation of the effect of the contribution of the random error, characterized by the standard deviation, according to equation 1 [3].

$$\mu = s_A/\sqrt{n} \tag{1}$$

Where:
μ corresponds to Type A uncertainty;
s_A corresponds to the sample standard deviation;
n is the number of observations.

The standard deviation is the positive square root of the variance, as shown in equation 2.

$$S_A = \sqrt{\left(\frac{1}{n-1}\sum_{j=1}^{n}(xi - \bar{X})^2\right)} \tag{2}$$

Where:
s_A corresponds to the sample standard deviation;
xi corresponds to the indication;
$\bar{X}$ corresponds to the mean of the measurand or standard;
n is the number of observations.

It is necessary to define measurement error and measurement uncertainty as different concepts. However, in the present work, the analysis concerns how much the method of correction of the systematic effect, known as a trend in the cases of finite measurements, can change the value of the estimated Type A uncertainty.

2. Methodology

At the present work was applied the methodology for defining systematic effect (bias) and the methodology for systematic effect correction.

2.1. Methodology for defining systematic effect (bias)

Bias can be defined by three different methods: through instrumental bias; through linear regression; and through interpolation with data of a calibration certificate [4].

Table 1 shows an example of data presented in the calibration certificate of a temperature sensor.

Table 1. Calibration certificate of a temperature sensor.

Standard (°C)	Measurand (°C)	Bias (°C)	Measurement uncertainty (°C)	k
20.000	20.01	0.010	0.05	2.10
40.000	40.07	0.070	0.05	2.10
100.000	100.20	0.200	0.05	2.10

In Method 1, the instrumental bias method, the value used is the one closest to the read value identified in the calibration certificate of the measuring instrument. For example, if the value indicated by the instrument was 20.08 °C, according to Table 1, the value set for the systematic effect would be +0.010 °C.

In Method 2, the linear regression method, it is necessary to determine the equation that correlates the value read with its respective trend, considering that the behavior of the instrument is linear. Figure 1 shows the linear correlation of the worked example.

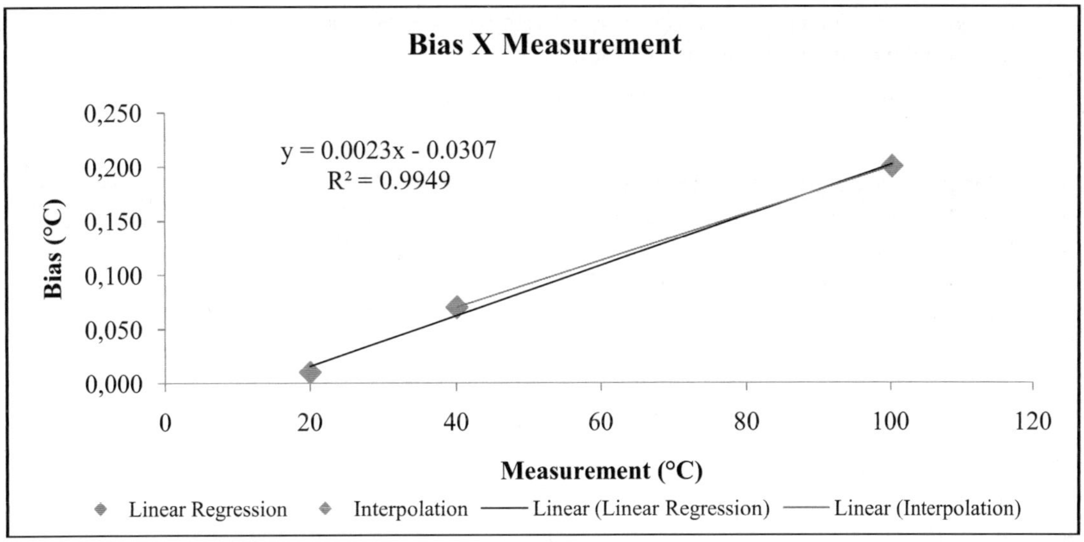

Figure 1 - Linear regression of the bias.

In this method, to define the bias value, it is necessary to replace the value of x in equation 3 with the value read by the instrument.

$$y = 0.0023x - 0.0307 \tag{3}$$

Where:
y corresponds to the estimated value of bias;
x corresponds to the indication of the measurand,

For example, if the value indicated by the instrument was 20.08 °C, according to equation 3, the value set for the systematic effect would be approximately 0.015 °C. This method produces an uncertainty because the value calculated by the equation is different from the actual value entered for generating it, known as the uncertainty of the regression adjustment.

In Method 3 - interpolation method, the instrumental bias is determined from the consideration that it varies linearly between the calibration points neighboring the reading, according to equation 4:

$$T = \left(\frac{L - L_1}{L_2 - L_1}\right)(T_2 - T_1) + T_1 \tag{4}$$

Where:
T corresponds to the estimated value of instrumental bias;
L is the indication of the measurand;
T_1 corresponds to the bias value in the calibration certificate of the point below the read value;
T_2 corresponds to the trend value in the calibration certificate of the point above the read value;
L_1 corresponds to the value of the measurand in the calibration certificate of the point below the read value;
L_2 corresponds to the value of the measurand in the calibration certificate of the point below the read value.

For example, if the value indicated by the instrument was 20.08 °C, according to equation 4, the value set for the systematic effect would be + 0.010209 °C, approximately 0.010 °C.

2.2. Methodology for systematic effect correction

Given that the correction is compensation for the systematic effect, that is, the bias with the signal exchanged, the following methods have been defined for carrying out the present work: individually for each reading presented by the measuring instrument and applied only to the averages of the measurements.

In Method A - systematic effect correction is performed for each reading, before determining the sample standard deviation, according to equation 5:

$$L_c = (L_i - T_i) \tag{5}$$

Where:
L_c corresponds to the corrected indication of the individual measurand;
L_i corresponds to each indication of the measurand;
T_i corresponds to the estimated value of the instrumental bias obtained using the methodology to define the systematic effect individually.

In Method B - a correction of the systematic effect is made in the mean value of the measurements, after determining the sample standard deviation, utilizing equation 6:

$$\bar{X}_c = (\bar{X} - T) \tag{6}$$

Where:
$\bar{X}_c$ corresponds to the corrected mean of the indications of the measurand;
$\bar{X}$ corresponds to the average of the indications of the measurand;
T corresponds to the estimated value of the instrumental bias obtained through the systematic effect definition methodology.

3. Results

Figure 2 presents a hypothetical chart of "bias T_i versus readings X_i" obtained through the calibration of a specific measurement instrument. In the same figure the point adjustment line (black line) is also presented that is calculated through the use of the least squares method.

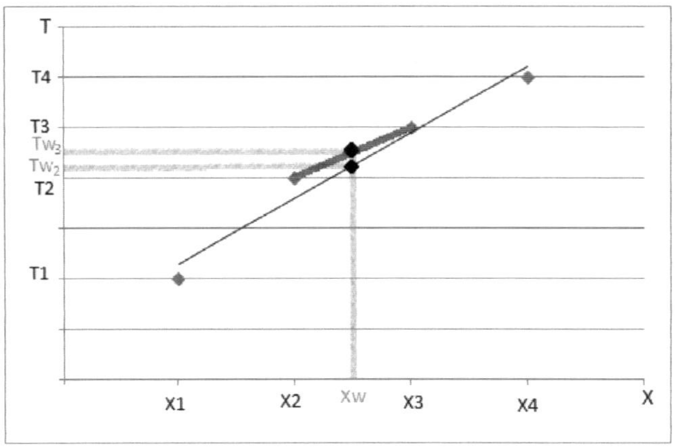

Figure 2 - Linear regression of the bias.

In order to find the value of the bias T_w corresponding to the reading X_w, such that $X_2 < X_w < X_3$, there are three different methodologies for reaching this objective, as previously described in the present work.

In Method 1, the bias value (T_2 or T_3) corresponding to the closest reading to X_w (X_2or X_3) would be adopted from the calibration certificate of the instrument. Adopting this methodology, the quality of theT_{w1} bias value will depend on how close the X_w reading is to X_2 or X_3. Considering that the X_w value is not chosen, but obtained through measurement, this means that this method presents a considerable level of unpredictability in the quality of the result obtained.

Through Method 2, the bias value adopted would be the value T_{w2} (shown in the figure) obtained through the equation of the adjustment line.

In Method 3, the bias value is obtained by linear interpolation from the points (X_2 , T_2) and (X_3, T_3), ie the point (X_w, T_{w3}) would be situated somewhere on the line (red line) connecting the points (X_2

, T_2) and (X_3, T_3). It is possible to infer that in the single case in which the adjustment line used in Method 2 coincides with the line connecting the points (X_2, T_2) and (X_3, T_3), the result of bias T_w obtained will be the same through the two methods. In any other case, where these lines do not coincide, linear interpolation always provides a T_w result more coherent with the initial hypothesis that can be assumed as linear the behavior of the function between (X_2 , T_2) and (X_3, T_3). Thus, for instruments of predominantly linear behavior, Method 3 is recommended among the 3 presented, since it does not give the level of unpredictability present in Method 1 and, in addition, provides the most coherent trend value, considering the initial hypothesis of function linearization and the points presented in the instrument calibration certificate.

Simulations were carried out to obtain uncertainty (Type A) values by utilizing data from the calibration certificate presented in Table 1 as well as data from Figure 1. Methods A and B (see section 2.2), for systematic effect correction, were carried out in investigating the three methodologies (methods 1, 2 and 3) for calculating x systematic effect (bias). The results are presented in Tables 2 and 3.

Table 2. Simulation results obtained with Method A for the correction of systematic effect. Methods 1, 2 and 3, for calculating systematic effect (bias) were tested.

MethodA	Method 1		Method 2		Method 3	
Nominal Value (°C)	Measure (°C)	Uncertainty Type A (°C)	Measure (°C)	Uncertainty Type A (°C)	Measure (°C)	Uncertainty Type A (°C)
40	42.2633333	0.3333333	42.2666667	**0.3325667**	42.2565636	0.3323363
60	67.2633333	0.3333333	67.2091667	**0.3325667**	67.2043905	0.3326127
80	86.4666667	0.8819171	86.4980333	**0.8798887**	86.4959255	0.8800104

Table 3. Simulation results obtained with Method B for the correction of systematic effect. Methods 1, 2 and 3, for calculating systematic effect (bias) were tested.

Method B	Method 1		Method 2		Method 3	
Nominal Value (°C)	Measure (°C)	Uncertainty Type A (°C)	Measure (°C)	Uncertainty Type A (°C)	Measure (°C)	Uncertainty Type A (°C)
40	42.26333333	0.3333333	42.2666667	0.3333333	42.2565636	0.3333333
60	67.26333333	0.3333333	67.2091667	0.3333333	67.2043905	0.3333333
80	86.46666667	0.8819171	86.4980333	0.8819171	86.4959255	0.8819171

Based on Tables 2 and 3, it can be concluded that Method 1, is indifferent for correcting mean value or values individually, since this does not generate change in the uncertainty of Type A. In Methods 2 and 3, correcting the values individually generates a contribution to Type A uncertainty smaller than correcting it in the mean value. Considering that the bias behavior of the simulated instrument is linear in most of the simulated results, it can be observed that Method 2 associated with Method A (in bold on the table) generated smaller contributions to Type A uncertainty when compared to the other method combinations. It should be noted that Method 2 also adds an uncertainty due to the approximation of the correction by a regression, which is not being considered in this paper, nor is it being taken into account some additional uncertainty contribution of Method 3.

4. Conclusion

The choice of a reading correction method should be based on prioritizing the practicality, and ease of operation, that it provides the user. However, the results presented may influence this choice, since reducing uncertainty is often the main objective. From this perspective, it can be concluded that using reading correction Method A (correcting the values individually before the statistical treatment) gives slightly lower uncertainty values in methods 2 and 3.

Method 1, on the other hand, has the advantage of being quite simple and direct, and it does not influence a decision to make corrections by Method A or B, since the uncertainty of this method is the same. However, it is not indicated when the measured value is too far from the points where performance is known. The results show that, regardless of the correction method employed (1, 2 or 3), the uncertainty values of Type A do not change, and only the change in the mean value is corrected. This value, in many cases, is the central value of the measurement result, which leads us to a more detailed analysis on which of the methods presents results that we can consider "more correct".

Furthermore, Method 2 (of linear regression) is probably the proper choice when a large volume of data is required and, combined with individual reading corrections (correction Method A), has generated results with the lowest Type A uncertainties. Future work is recommended to analyze the contribution of regression uncertainty, which is aggregated by this method, in expanded uncertainty.

On the other hand, through the use of individual reading correction (Method A) by interpolation (Method 3) we also obtained uncertainty values due to the dispersion of values significantly better than in Method 1, which means that the option for this correction strategy should be considered, for example in cases where the volume of data processed is lower, but especially in the case of instruments where the number of calibrated points is low, which leads to generally bad regression equations.

During the course of this work we observed several nuances that should be considered, such as the effect of instrument linearity in the choice of a correction strategy, the impact of the corrections on expanded uncertainty (not only in Type A), as well as methodology standardization for estimating the uncertainty arising from interpolation, among others. However, as these points distance us from the main objective of this paper, which is to compare the methods of correction A and B, we leave these observations as a suggestion for future work.

References

[1] Albertazzi, A; Sousa, a. *Metrologia: científica e industrial*. Editora Manole: 2008.

[2] BIPM, IEC, IFCC, ILAC, IUPAC, IUPAP, ISO, OIML (2012) *The international vocabulary of metrology - basic and general concepts and associated terms (VIM)*, 3rd edn. JCGM 200:2012. http://www.bipm.org/vim.

[3] BIPM, IEC, IFCC, ILAC, IUPAC, IUPAP, ISO, OIML (2008) *Evaluation of measurement data-guide for the expression of uncertainty in measurement*. JCGM 100:2008. http://www.bipm.org/en/publications/guides/gum.html.

[4] Santana, R G and Frota, M N. *Correção da tendência instrumental nas medições de massa na calibração de vidrarias volumétricas pelo método gravimétrico*. II CIMMEC, 2011.

IOP Conf. Series: Journal of Physics: Conf. Series **1044** (2018) 012011 doi:10.1088/1742-6596/1044/1/012011

Analysis of the airflow in the TA-2 subsonic wind tunnel

M L C C Reis[1,2], **M S Souza**[1], **M C C Araújo**[2]

[1]Instituto de Aeronáutica e Espaço, Pr. Mal. Eduardo Gomes, 50, CEP 12228904, Brasil.
[2]Instituto Tecnológico de Aeronáutica, Pr. Mal. Eduardo Gomes, 50, CEP 12228901, Brasil.

E-mail: marialuisamlccr@iae.cta.br

Abstract. Tests are being carried out at the TA-2 wind tunnel to obtain the velocity and pressure distributions at the test section. The airflow conditions must be known to fulfil the required accuracy of tests. A rake composed of Pitot tubes was used to measure local static and total pressures of the flow. The test section region considered here is related to the transverse central area. The estimated parameters are velocity and pressure coefficient. Uncertainties associated with these parameters are evaluated according to metrological standardisation. Calibration curves are supplied using the least squares fitting. Results revealed a gradient in airflow parameters.

1. Introduction

Wind tunnels are aerodynamic laboratories used to simulate the condition encountered by a body immersed in the flow. They are built in a great variety of configurations in terms of size and airflow regimes according to industrial and research needs [1].

The TA-2 is a subsonic facility located at the Aerodynamics Division of the Institute of Aeronautics and Space. It has a closed circuit and the test section is 2.10 m high, 3.30 m wide and 3.20 m long. The TA-2 wind tunnel has contributed to design optimization of aeronautical vehicles developed by the Brazilian industry and to the analysis of the flow around structures subject to wind loads such as offshore platforms, buildings and automotive vehicles.

Experimental aerodynamicists are primarily concerned about the uniformity of the airflow at the wind tunnel test section. Ideally, a test section should be free of gradients and turbulence. As these conditions are difficult to achieve in practice, tests are carried out periodically to calibrate the flow.

According to the international aerodynamic community, the calibration procedure of a wind tunnel depends on the type of tests conducted in the facility [2]. Reference [3] classifies four calibration categories: 1) calibration to correct the parameters which are measured in one location and are transferred to another point of the circuit; 2) analysis of experimental data repeatability employing standard models of the type wing/body/tail; 3) surveys to determine flow parameter uniformity; and 4) evaluation of the impact caused by changes in the circuit, hardware or test technique.

This study is concerned with the third category. Although the whole campaign encompasses also the evaluation of the boundary layer thickness and the turbulence level of the airflow, the results presented in this paper are related to the analysis of the velocity, V, and the pressure coefficient, C_p, distributions at the inner area of the TA-2 test section. A rake consisting of 15 Pitot tubes was

[1] To whom any correspondence should be addressed.

employed. The uncertainties associated with the estimated parameters were evaluated by using the law of propagation of uncertainty. Least squares fitting is used to supply calibration curves.

The knowledge of the distribution of the airflow field parameters helps to plan the test of aerodynamic models at the wind tunnel regarding accuracy, precision and uncertainty requirements. Information supplied by this campaign can be used to circuit improvements and represent a basis for the development of methodology for wall interference corrections.

2. Methodology

The test campaign analyses the distributions of the flow velocity and the flow pressure, measured at 25 positions of the TA-2 test section (figure 1a). The x axis represents the wind direction and is positive downstream of the flow. The y axis is the horizontal position and the positive direction is from left to right, when the observer is facing the airflow. The height from the floor is the z axis. The origin of the coordinate system $(x, y, z) = (0, 0, 0)$ is located at the center of the test section floor.

The velocity regimes considered in the tests are 30, 43, 60, 75, 86, 98 and 108 m/s, which corresponds to free stream dynamic pressures approximately equal to 50, 100, 200, 300, 400, 500 and 600 mmH_2O. The measured parameters are static temperature, T, and static, p, total, p_0, and dynamic, q, pressures.

The tests employ a rake composed of 15 Pitot tubes (figures 1b and 1c) to measure the local static pressure and the local total pressure. The Pitot tubes are located at 1982, 1851, 1720, 1590, 1459, 1329, 1198, 1067, 937, 806, 676, 545, 415, 283 and 153 mm from the test section floor. Free stream parameters of the airflow are measured by a Pitot tube and a thermometer located at the entrance of the test section. Pressure instrumentation is connected to the Pitot tubes to measure pressure signals. The mean and standard deviation of the temporal pressure signals obtained during the wind tunnel runs were estimated and considered in the uncertainty evaluation. The results presented in this paper are related to the central vertical plan, where $x = 0$ and $y = -120$, -60, 0, 60 and 120 mm (positions P11, P12, P13, P14 and P15, respectively).

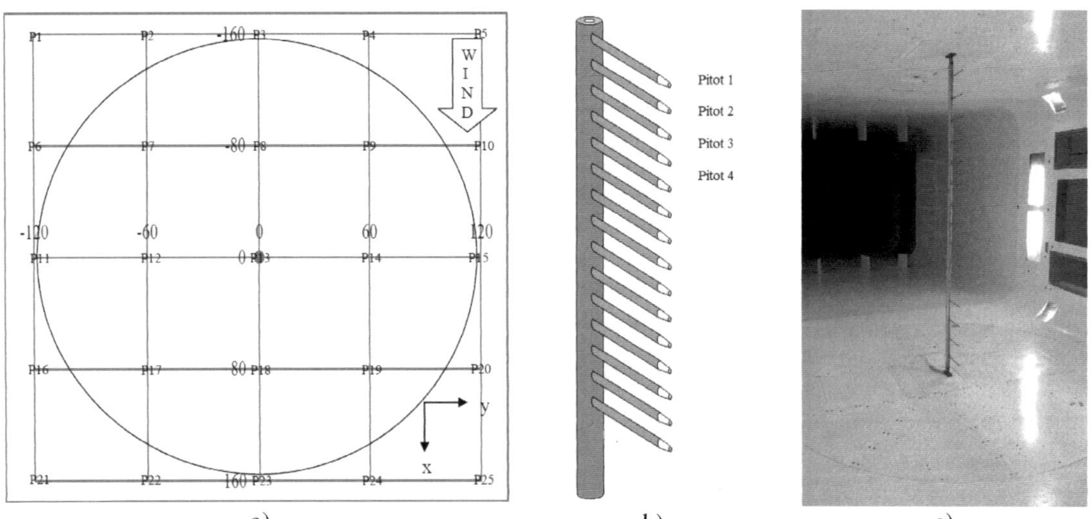

Figure 1. a) Scheme of the TA-2 test section floor where the rake is positioned. Unit: millimeter. The transverse central section corresponds to the positions P11, P12, P13, P14 e P15. b) Rake composed of 15 Pitot tubes. c) Rake positioned at P13 of the TA-2 test section.

Pressure instrumentation was calibrated prior to the tests. Excel® sheets, MatLab® and LabView® codes were implemented for data acquisition and data reduction.

2.1. *Mathematical modeling*

The Bernoulli equation for the incompressible flow is used to calculate the local airspeed, V [4]:

$$p_0 = p + \frac{1}{2}\rho V^2 \tag{1}$$

which for our study results in:

$$V = \left[\frac{2(p_{0local} - p_{local})}{\rho}\right]^{\frac{1}{2}} \tag{2}$$

In equation, p_{0local} is the total pressure and p_{local} is the static pressure measured at the rake tubes. The airflow density is calculated by:

$$\rho = \frac{p_\infty}{RT} \tag{3}$$

where:
p_∞ is the free stream static pressure measured at the entrance of the test section;
R is the gas constant equal to 287 J/(kg)(K); and
T is the temperature expressed in kelvin.

The pressure coefficient is defined by:

$$C_p \equiv \frac{p - p_\infty}{q_\infty} \tag{4}$$

where p is the local static pressure and p_∞ and q_∞ are the free stream static and dynamic pressures, respectively. For the present study, the numerator of equation (4) is measured by a differential sensor and the expression used to calculate C_p is:

$$C_p = \frac{\Delta p}{q_\infty} \tag{5}$$

2.2. *Uncertainty Evaluation*

The uncertainties associated with the pressure coefficient, u_{Cp}, and velocity, u_V, are evaluated by using the law of propagation of uncertainty [5]. For the case of independent variables, the combined standard uncertainty is the positive square root of:

$$u_y^2 = \sum_{i=1}^{N}\left(\frac{\partial y}{\partial x_i}\right)^2 u^2(x_i) \tag{6}$$

where y is the output quantity, the measurand [6], and x_i are the input quantities.

Applying equation (6) to equations (2), (3) and (5) results in:

$$u_V^2 = \left(\frac{1}{\rho} \left[\frac{2(p_{0local} - p_{local})}{\rho} \right]^{-\frac{1}{2}} \right)^2 u_{p_{0local}}^2 + \left(\frac{-1}{\rho} \left[\frac{2(p_{0local} - p_{local})}{\rho} \right]^{-\frac{1}{2}} \right)^2 u_{p_{local}}^2 + \left(\frac{(p_{local} - p_{0local})}{\rho^2} \left[\frac{2(p_{0local} - p_{local})}{\rho} \right]^{-\frac{1}{2}} \right)^2 u_\rho^2 \tag{7}$$

$$u_\rho^2 = \left(\frac{1}{RT} \right)^2 u_{p_\infty}^2 + \left(\frac{-p_\infty}{RT^2} \right)^2 u_T^2 \tag{8}$$

$$u_{C_p}^2 = \left(\frac{1}{q_\infty} \right)^2 u_{\Delta p}^2 + \left(\frac{-\Delta p}{q_\infty^2} \right)^2 u_{q_\infty}^2 \tag{9}$$

2.3. *The curve fitting*

The least squares method was applied to the local pressure coefficients, C_p, and local velocities, V, to supply calibration curves that represent the airflow behavior [7].

First order polynomials were fitted to the estimated parameters, *i. e.*, the calibration curves are of the kind:

$$y = a_0 + a_1 x \tag{10}$$

The law of propagation of uncertainty was also applied to equation (10) in order to obtain the uncertainty associated with the fitted curve. As the polynomial parameters a_0 and a_1 are correlated, the uncertainty of the output quantity y is written as [5]:

$$u_y^2 = (1)^2 u_{a_0}^2 + (x)^2 u_{a_1}^2 + 2u(a_0, a_1) \tag{11}$$

where the uncertainty of the input quantity x, u_x, was considered negligible and $u(a_0, a_1)$ is the covariance between a_0 and a_1.

The quality of the fit was evaluated by the chi-square quantity [8]:

$$\chi^2 = \sum_{j=1}^{n} \left(\frac{y_{measured} - y_{fit}}{u_y} \right)^2 \tag{12}$$

A chi-square value approximately equal to the number of degrees of freedom indicates a good quality of curve fitting. The number of degrees of freedom is equal to the difference between the number of data points, N, and the number of parameters of the fitted curve to be estimated, m.

The reduced chi-square is expressed by:

$$\chi_v^2 = \frac{\chi^2}{N - m} \tag{13}$$

where N-m is the number of degrees of freedom of the curve fitting. A χ_v^2 value approximately equal to 1 indicates a good fit. Departure from this condition revels that either the chosen curve or the uncertainty assigned to data points is erroneous.

The standard deviation of the fit is:

$$S = \sqrt{\frac{\sum_{i=1}^{N}\left(y_{measured} - y_{fit}\right)^2}{N - m}} \tag{14}$$

3. Results and discussion

The airflow uniformity at the transverse area of the central region of the TA-2 test section is presented in this paper. Velocity distributions obtained with the rake located at P11 to P15 are seen in figures 2a to 2e. The velocity regime is related to the dynamic pressure q = 300 mmH_2O, *i. e.*, the nominal velocity is 75 m/s.

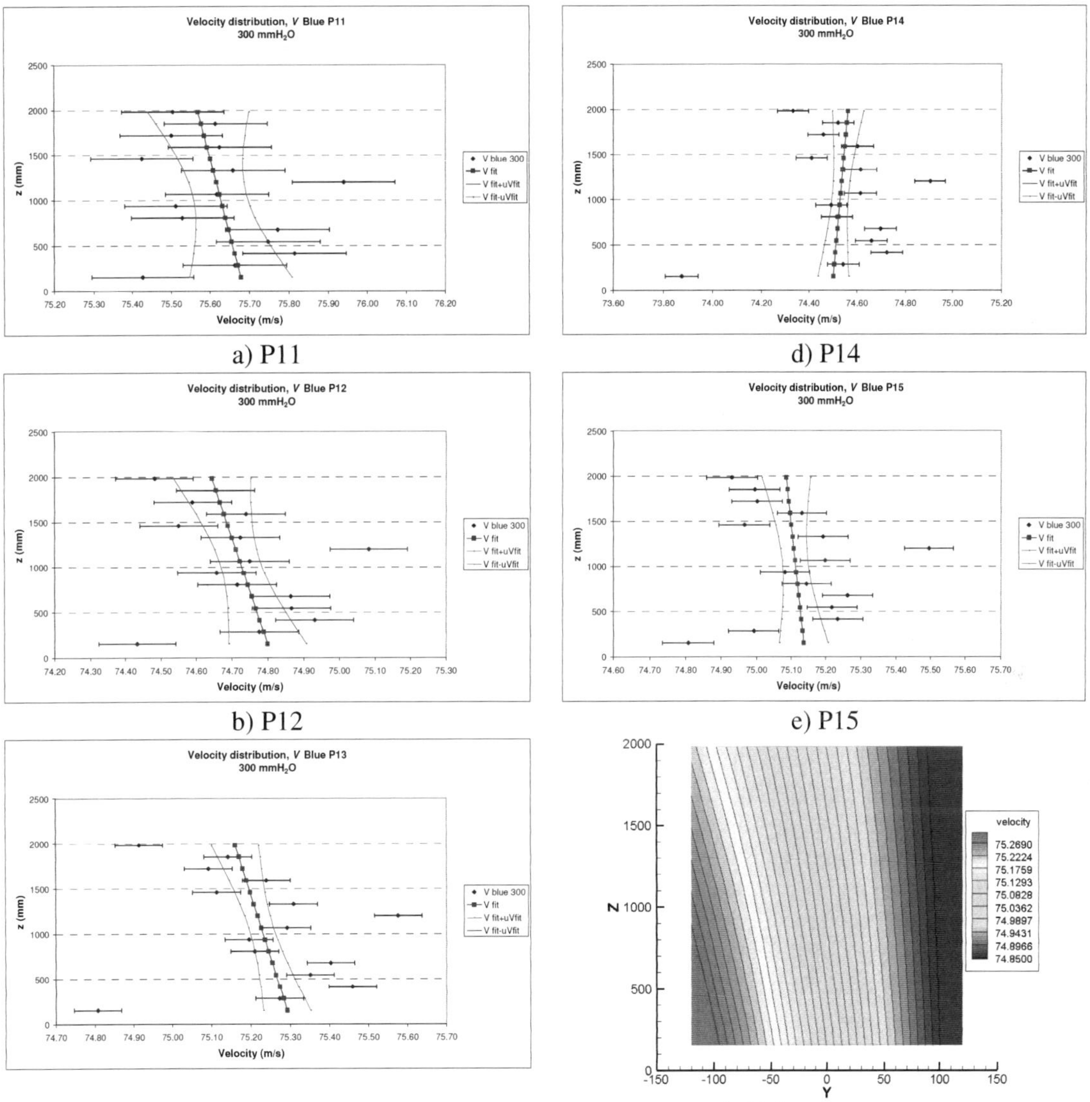

a) P11 d) P14

b) P12 e) P15

c) P13 f) Velocity contour.

Figure 2. Velocity distribution at the TA-2 transverse central section. Nominal velocity is 75 m/s.

The abscissa axis in figures 2a to 2e represents the velocity, V (m/s), and the ordinate axis represents the height of each Pitot tube from the floor, z (mm). Uncertainty limits of the fitted curve are also shown in figures 2a to 2e and correspond to a 68% level of confidence (equation (11)).

The results are grouped together and a graphic where the velocity is shown as color intensity is included to facilitate the visualization of the airflow uniformity at the transverse area of the central region of the TA-2 test section (figure 2f). One observes that the velocity is greater in the negative part of the horizontal axis nearer the floor. The difference between the maximum and the minimum velocity values, considering the uncertainty limits, is approximately 1 m/s. This velocity variation was considered acceptable regarding airflow uniformity requirements.

The calibration curve equations for velocity, Vxz, and pressure coefficient, C_pxz, at position P13 are shown in Table 1, for all airflow regimes. The fitted function was the first degree polynomial because the flow is expected to be uniform (equation (10)). Ideally, the airflow would be free from gradients and the a_1 coefficient value would be equal to zero in the calibration curve.

Table 1. Curve fitting results for velocity, V, and pressure coefficient, C_p. Position P13.

Airflow regime q (mmH$_2$O)	Velocity V (m/s)	Pressure coefficient C_p (dimension one)
50	$V = 30.26(0.021) – 0.3(1.7)\text{x}10^{-5}z$	$C_p = - 0.0073(0.0015) + 2.7(1.2)\text{x}10^{-6}\ z$
100	$V = 43.07(0.07) – 0.0(0.6)\text{x}10^{-4}z$	$C_p = - 0.0116(0.0020) + 21(17)\text{x}10^{-7}\ z$
200	$V = 60.94(0.04) - 4(3)\text{x}10^{-5}\ z$	$C_p = - 0.0119(0.0024) + 12(20)\text{x}10^{-7}\ z$
300	$V = 75.30(0.03) - 72(28)\text{x}10^{-6}\ z$	$C_p = - 0.012(0.003) + 11(25)\text{x}10^{-7}\ z$
400	$V = 87.08(0.04) - 10(4)\text{x}10^{-5}\ z$	$C_p = - 0.0120(0.0029) + 11(24)\text{x}10^{-7}\ z$
500	$V =98.10(0.03) - 99(25)\text{x}10^{-6}\ z$	$C_p = - 0.0131(0.0029) + 11(24)\text{x}10^{-7}\ z$
600	$V =108.042(0.023) - 105(19)\text{x}10^{-6}\ z$	$C_p = - 0.014(0.003) + 9(25)\text{x}10^{-7}\ z$

It was observed that at the position P13 considered as $(x, y) = (0, 0)$ in figure 1a, the velocity decreases as we approach the test section ceiling. At P13, this behavior occurs for all airflow regimes.

The pressure coefficient, C_p, is negative, indicating that the local static pressure, p, is lower than the free stream static pressure, p_∞, downstream of the test section entrance (equation (4)). The positive value of the a_1 parameter in the C_pxz calibration curves means that the local static pressure approximates the value of the free stream static pressure as the height z increases.

Data reduction showed that the quality of the fitting quantified by equation (12) is not satisfied for all test configurations, when using standard deviations of the temporal signals obtained during the runs. Table 2 shows an example. The reduced chi-square values, χ_v^2, were obtained by fitting the experimental velocity data measured at positions P11, P12, P13, P14 and P15. The velocity regime is related to q = 300 mmH$_2$O (nominal V = 75 m/s). One observes that at P13 and P14, the uncertainty assigned to data points are underestimated, resulting in a reduced chi-square around 10, indicating that further investigation of the error sources must be carried out to describe the experimental errors.

Table 2. Reduced chi-square for least squares fitting applied to velocity data. q = 300 mmH$_2$O.

Test section region	P11	P12	P13	P14	P15
χ_v^2	1.28	2.43	10.83	12.88	5.93

The standard deviation of the fitting expressed by equation (14) gives an idea of the value that should be assigned to experimental data, provided that the fitted model is corrected [7]. Table 3 presents the uncertainties associated with the measured velocities, u_V, and the standard deviation of the fitting S, for q = 300 mmH$_2$O. At position P11, the uncertainties u_V are equal to 0.13 for all heights z and the standard deviation S is equal to 0.15; they are of the same order, and the quality of the fit is good (χ_v^2 is around 1). The same applies to P12, where $u_V = 0.11$ and $S = 0.17$. On the other hand, $u_V = 0.06$ and $S = 0.20$ for P13, and $u_V = 0.07$ and $S = 0.24$ for P14, resulting in chi-square values 10 times greater than unity. The results for P15 are $u_V = 0.07$ and $S = 0.17$, and the value of χ_v^2 was considered good.

Table 3. Comparison between uncertainty associated with data points, u_V, and the standard deviation of the fit, S. Airflow regime q = 300 mmH_2O.

Region	P11	P12	P13	P14	P15
	S (m/s)				
	0.14909	0.17112	0.20083	0.23639	0.174434
height, z (mm)	u_V (m/s)				
1982	0.13201	0.11001	0.06118	0.06580	0.07183
1851	0.13182	0.10976	0.06099	0.06564	0.07177
1720	0.13202	0.10986	0.06103	0.06569	0.07176
1590	0.13180	0.10964	0.06091	0.06557	0.07164
1459	0.13215	0.10991	0.06101	0.06573	0.07179
1329	0.13174	0.10966	0.06086	0.06555	0.07158
1198	0.13125	0.10913	0.06064	0.06530	0.07129
1067	0.13181	0.10962	0.06087	0.06555	0.07157
937	0.13200	0.10976	0.06095	0.06566	0.07168
806	0.13196	0.10967	0.06094	0.06564	0.07162
676	0.13154	0.10945	0.06078	0.06548	0.07151
545	0.13158	0.10945	0.06082	0.06552	0.07155
415	0.13147	0.10936	0.06073	0.06546	0.07154
284	0.13173	0.10958	0.06088	0.06562	0.07177
153	0.13215	0.11009	0.06126	0.06621	0.07195

4. Conclusions

The uniformity of the airflow at the central region of the test section of the subsonic wind tunnel TA-2 was analysed. A series of 15 Pitot tubes arranged in a rake were used to measure the local static pressure and the local total pressure.

Calibration curves which represent the pressure and velocity characteristics of the airflow were supplied. The curves were obtained by fitting a first degree polynomial to the experimental data. Only the results for the central area are shown in this paper and the curves are related to the vertical axis.

The chi-square quantity obtained in the curve fitting was good for most of the tests, but further investigation of the error sources must be carried out to describe the experimental errors of some test configurations, as presented in tables 2 and 3.

Although gradients in velocity and pressure distributions were observed for all airflow regimes covered by the tests, the variations in velocity and pressure values were considered acceptable.

5. Acknowledgments

The authors would like to express their gratitude to the São Paulo Research Foundation for the support of this study through grant FAPESP/CEPID 2013/07375-0.

References

[1] Barlow J B, Rae Jr W H and Pope A 1999 *Low-Speed Wind Tunnel Testing* (John Wiley & Sons Inc.) p 713
[2] AIAA R-093-2003 Recommended Practice: *Calibration of Subsonic and Transonic Wind*

Tunnels (American Institute of Aeronautics and Astronautics)

[3] Hudgins M S and Hergert D W AIAA 2005-4280 *Methodology and Results from a Recent Calibration of the Boeing Transonic Wind Tunnel* 41st AIAA/ASME/SAE/ASEE *Joint Propulsion Conf. & Exhibit* (*Tucson, Arizona, USA, 10-13 July 2005*)

[4] Anderson Jr J D 2001 *Fundamentals of Aerodynamics* (New York Mc Graw Hill) p 892

[5] BIPM/JCGM 101:2008 *Evaluation of measurement data – Guide to the expression of uncertainty in measurement – GUM 1995 with minor corrections* (Joint Committee for Guides in Metrology, BIPM, Metrologia, IEC, IFCC, ILAC, ISO, IUPAC, IUPAP and OIML) p 120

[6] BIPM/JCGM 200:2012 *International vocabulary of metrology – Basic and general concepts and associated terms* (*VIM*) (Joint Committee for Guides in Metrology, BIPM, Metrologia, IEC, IFCC, ILAC, ISO, IUPAC, IUPAP and OIML) p 91

[7] Press W H, Teukolsky B P T, Vetterling W T 1990 *Numerical Recipes* (Cambridge University Press) chapter 14 pp 547–565

[8] Bevington P R, Robinson D K 2003 *Data Reduction and Error Analysis for Physical Sciences* (New York Mc Graw Hill Higher Education) p 320

Applying Design Thinking to systemic problems in educational assessment information management

W P Fisher Jr[1], Emily Pey-Tee Oon[2] and Spencer Benson[3]
[1] Research Associate, BEAR Center, Graduate School of Education, University of California, Berkeley, Berkeley California 94702 USA
[2] Faculty of Education, University of Macau, Taipa, Macau SAR, China
[3] Centre for Teaching and Learning Enhancement, University of Macau, Taipa, Macau SAR, China

[1]E-mail: wfisher@berkeley.edu

Abstract. The integrative process of design thinking provides a compelling framework for addressing problems in education involving coherent assessment and instruction. New possibilities for more meaningful developmental coherence within, horizontal coherence across, and vertical coherence between classroom formative and accountability assessments emerge by thinking through and enacting conceptual solutions that begin with the end in mind and that involve end users every step of the way. The nonlinear process involves empathy for students and teachers, the definition of a solvable problem (one with solutions that are feasible, viable, and desirable), the ideation of a brainstormed array of ideal "possibly impossible" resources for creativity (such as a very demanding measurement model and traceability to unit standards), the rapid development of an initial prototype expected to have limitations, and the first of an iterative series of tests that deliberately test the prototype en route to a recycling as needed through any other parts of the process. Applying Design Thinking may lead to new solutions to problems encountered in reconceiving and reconfiguring classrooms as meta-design ecosystem niches of creativity and innovation. An assessment outcome prototype drawn from recent work with teacher/educators that integrates developmental, horizontal, and vertical coherence is presented.

1. Introduction

A broad-based platform for creativity termed Design Thinking [DT] combines aspects of art, engineering, and empathy with persons experiencing systemic problems possibly amenable to new solutions. DT is associated with IDEO and the Hasso-Plattner Institute of Design at Stanford University. Some advocates suggest DT may become so widely adopted it becomes a "New Liberal Arts" [1]. Others insist its fusion of liberal arts and professional education mean DT will come into full flower only at the systems level, across the whole university, breaking down artificial divisions (silos) [2]. This focus on enhanced information flows and communications across horizontal levels has much in common with efforts aimed at increasing the coherence of educational outcome assessments [3-4] and with the ecosystem conception of organizational and interdisciplinary interdependencies and interrelations [5-8].

Toward this end, the five aspects of value creation integrated in DT; "empathy-define-ideate-prototype-test," were employed to devise new approaches to problems associated with coherence across

[1] To whom any correspondence should be addressed.

three primary educational assessment use cases. Developmental coherence involves individual students' growth along learning progressions; horizontal coherence concerns the comparability of learning progressions and outcomes across students, classrooms, and schools; and vertical coherence aligns formative classroom and accountability assessments at regional, national, and international levels [3,4]. Accountability concerns test uses at the individual, class, school and state levels to monitor and assess the degree to which students are meeting learning outcomes and being taught by effective teachers. Though these three domains of assessment cannot be combined by means of a universal test, their common focus on educational processes and outcomes suggests the viability of greater degrees of coherent alignment of available information with the management and policy needs of individual students, classrooms, schools, districts, etc. After all, mathematics ability is referred to as a single thing across curricula, students, tests, schools, even though it is represented incoherently in different terms across and within all of these contexts.

The multilevel nature of meaning across local, midrange, and broader contexts has long been of philosophical interest [5,6]. Communication systems are ecologies in which micro-level processes within individuals are distinct from, but interact with, meso-level processes between individuals, which in turn are distinct from but interact with macro-level group processes. Developmental, horizontal, and vertical coherence issues in educational assessment correspond to these varying levels of complexity in communication systems. Practical applications of these distinctions in epidemiology [7] are leading to new, productive relationships between clinical medicine and public health efforts [8]. The primary justification for taking an ecological approach to problems of coherence in educational assessment is to understand learning in the contexts of its various environments. Contexts affect learning in a variety of ways that cannot be grasped from measures of individual attributes. Broad-scale efforts at sorting out the sources of distinct classes of effects on learning can only begin when the appropriate communications systems are in place.

In the DT context, the viability of more coherent educational assessment information must be complemented by the technical feasibility of the solutions and the desirability end users have for them. The technical challenges are complex, as they involve not only the comparability of different assessments administered to different students, but also comparability across different formative and accountability assessment purposes. Feasibility is supported, however, by a range of available technical solutions, including adaptive administration of item banks [9], theory-driven automatic item generation and inference in the context of metrological traceability [10,11], and test equating methods encompassing a variety of item formats [12].

The desirability of viable and feasible design solutions depends to a large extent on empathy with end users. Empathic identification does not always involve taking end user opinions literally, however. As Henry Ford is reputed to have observed, if he had listened to what people said they wanted in the way of improved transportation, he would have focused on faster horses. It accordingly happens that virtually every major technical innovation, from automobiles to electricity to the telegraph, telephone, television, and the Internet, has emerged in a context in which there was no public demand whatsoever for a line of products that later came to seem indispensable. Can this kind of result can be obtained via the application of DT to the development of new technologies for educational information applications?

2. The DT integrative process

DT is not a linear sequence of steps. One might take up empathy, definition, ideation, prototyping, and testing in almost any order, with variation in emphasis across them, both within and across iterations. The educational example described here moves through an artificial sequence solely for the purpose of orientation to the different aspects of DT integrated into a process.

2.1 Empathy

A number of conceptual, organizational, and technical barriers result in teachers being overburdened with repetitive tasks that follow largely from the need to micromanage the details of every student's

IOP Conf. Series: Journal of Physics: Conf. Series **1044** (2018) 012012 doi:10.1088/1742-6596/1044/1/012012

learning, and of every assessment of that learning. Compared with the information available from well-made formative assessments integrated with learning materials [13], students and teachers alike find it difficult to know what lessons are most appropriate for which students.

The quality of information on educational processes and outcomes that is available in principle is vastly superior to the quality of information actually used by teachers. Low quality information makes it difficult to match readers to texts, but it makes plotting development, comparing results with other teachers, and setting achievable accountability goals virtually impossible. Teachers are being held accountable for managing outcomes they cannot effectively visualize and compare day to day, either within their own classes, or across classrooms. This situation is an example of what Bateson [5,6] called the transcontextual syndrome, a double bind imposed by the figurative schizophrenia of demanding, for instance, a behaviour coherent within a horizontal frame of reference that is simultaneously incoherent in the developmental frame of reference.

2.2 Define

The problem found in empathizing with teachers then concerns the incoherence of the available learning outcome management information systems. What would a coherent education management information system look like? Developmental, horizontal, and vertical coherence are needed to address the learning processes of individual students, to enable comparability over time and across classrooms, and to manage for long term accountability goals. Coherence of this kind demands meaningful and comparable quantitative measures substantively interpretable in terms of learning progressions. Temporarily setting aside concerns for viability, feasibility, and desirability, what kinds of resources might be useful in formulating solutions to these problems of coherence?

2.3 Ideate

Starting from the beginning, how do these coherence problems fit into the larger context and goals of education? Education is supposed to provide students with experiences that enable them to successfully address problems in real life, to create value for themselves and others. No one thinks students need to encounter every possible life problem in the classroom. A curriculum samples from the infinite universe of all possible problems and lessons in ways intended to represent real life well enough to prepare students for adult responsibilities.

Conversely, schools are supposed to produce consistent learning outcomes independent of the particular students in attendance. The usually unstated but widely adopted assumptions are, then, that (a) education takes place over time as a process requiring the conjoint independence of student abilities and the difficulties of the lessons to be learned, such that the probability of student success is conditional on the difference between ability and difficulty, and (b) that this independence is never perfectly realized, however useful it has proven historically as a heuristic fiction.

In this context, how might the problem of educational information coherence be formulated? What not widely considered, “possibly impossible” solutions might be brainstormed as resources for later trial and error applications? Where might audacious models for proceeding be found? Are there approaches in science, engineering and DT that have previously been applied in situations like this, where problems and solutions, questions and answers, exist in similar kinds of mutual, interdependent relations?

Qualitatively speaking, the situation is analogous to the general form of a wide range of measurement models, including, for instance, Newton’s laws [14]. Psychometric models in this form have lately been shown to share important features with models and principles widely used in engineering and science [15,16]. Could the design problem of coherently related educational assessments be successfully addressed in terms of media formatted in a common language and distributed throughout a communications network? Historical and social studies of science have led to an emphasis on the capacity of researchers to coordinate and align their behaviours and decisions virtually, by means of common languages and shared values that do not homogenize but incorporate local variations [17,18]. Extending that work, this kind of ecological networking has been proposed as an alternative approach to addressing contemporary issues in psychology and the social sciences [19], suggesting efforts aimed at devising ecosystems of educational forms of life inhabiting diverse niches may be worth pursuing.

Each form of coherence, and different approaches to configuring their expression, will vary in its parameterization. A model that explicitly states the conjoint relationship of student abilities and item difficulties, and makes the potential portability of that relationship testable, takes the form

$$\ln[P_{nij} / (P_{nij-1})] = B_n - D_{ij}$$

which says that the log-odds of observed success for student *n* on item *i* at partial credit score *j* is equal to the difference between the estimate *B* of person *n*'s ability and the difficulty estimate *D* of item *i* at rating *j* [20,21]. In this kind of model, the group-level construct emerges as a self-organized pattern from the within-individual responses. More complex hierarchical models of multilevel systems are available [22]. A key point is that, in the same way that no drawn triangle conforms perfectly to the Pythagorean theorem, no data ever fit a model of this kind. The question is whether the model is useful.

The solution to the problem of coherence in educational assessment hinges on identifying what is already known about student learning, encapsulating it in a portable technology, and making the information embodied in this device useful in different forms across the developmental, horizontal, and vertical contexts for instructional, quality improvement, and accountability applications. Prototyping is the process in which technical solutions of this kind are proposed.

2.4 Prototype

For students, the problem of knowing what to study next is one of locating themselves on a learning progression and checking for special strengths and weaknesses in need of attention. For teachers, embodying learning progressions on portable tools gauged in common metrics makes it possible to more easily relate student abilities to appropriate learning materials. Figure 1 shows a self-scoring form [23] produced from widely used educational measurement data analysis software. Though these kinds of forms have a long history of use in education [24], they are not widely known. An automated application could work from computerized test data to indicate on this form which questions a student answered correctly and incorrectly, as well as showing the associated correct answer count, measure, uncertainty,

```
360 400   440   480   520   560   600   640   680   720   760        DEVELOPMENTAL SEQUENCE
|-----+-----+-----+-----+-----+-----+-----+-----+-----+-----|  NUM   ITEM CONTENT SUMMARY
0                                    0         :         1  1 100*  MOST DIFFICULT ITEM HERE
0                                  0         :          2   2 109*             ^
|                                                           |                  ^
|                                                           |                  ^
0                           0         :          2          2 105*             ^
0                0        :           1            :        2 108*             ^
0                           0         :         1           1 102*             ^
0                           0         :         2           2 106*             ^
0                        0          :         1             1 103*             ^
|                                                           |                  ^
|                                                           |      INCREASING DIFFICULTY
0               0         :          1                      1 111*             ^
0              0         :         2                        2 110*             ^
|                                                           |                  ^
|                                                           |                  ^
0       :        1        :      2                          2 107*             ^
0     0         :         1                                 1 104*             ^
|                                                           |                  ^
00         :          1                                     1 101*             ^
0          :         1                                      1  99*  LEAST DIFFICULT ITEM HERE
|-----+-----+-----+-----+-----+-----+-----+-----+-----+-----|
360 400   440   480   520   560   600   640   680   720   760 MEASUREMENT SCALE
     0       1   2  3 4    5  6 7 8    9   10   11     12   13  COUNT CORRECT
110    70      50      35     30    35      50     70     110  UNCERTAINTY
                                     1                        \  STUDENT
1        1 21  4    35 3  38 5  27 3 2  74   821   1        1  > MEASURE
6756 7 3 7210 3811109 63478434410317451623 2 5321  6       46 /  DISTRIBUTION
     T            S            M           S            T     MEAN (M), 1 SD (S), 2 SD (T)
0              10   20    30 40  50  70 80   90            99 PERCENTILE
```

Figure 1. Prototype individual student self-scoring form for developmental coherence.

```
360 400   440   480   520   560   600   640   680   720   760  CLASSROOM A-Z MEASURE
|-----+-----+-----+-----+-----+-----+-----+-----+-----+-----|  DISTRIBUTIONS
              P    Y  D
  F   A   M   G LZQCERUC NW H  J    K                          WEEK ONE

                     L    Y  D
       F   G   A   M ZPUCQERU NW H  J    K                     WEEK THREE

               G       P    Y
      F        A     M EZQCLRUS NW D     K        J            WEEK FIVE

                                Q    Y   D  W
                   F  A  P   M LZSRUCE J H N KJ                WEEK SEVEN

                              P   Q  D
                      GA      F  LMSYCERU NJ WK    H           WEEK NINE
|-----+-----+-----+-----+-----+-----+-----+-----+-----+-----|
360 400   440   480   520   560   600   640   680   720   760  MEASUREMENT SCALE
10    7     5      4     3     2      3     5       7     10  UNCERTAINTY (SE CLASS/WEEK MEAN)

     T            S            M          S            T       MEAN (M), 1 SD (S), 2 SD (T)
0     10     20    30   40   50   70      80     90        99  OVERALL STUDENT PERCENTILE
```

Figure 2. Prototype classroom comparison report for horizontal coherence.

```
360 400   440   480   520   560   600   640   680   720   760  END OF SEMESTER
|-----+-----+-----+-----+-----+-----+-----+-----+-----+-----|  ELEVENTH GRADE
                              13 2                            \   OVERALL DISTRICT-WIDE
                    1   2142572791784251 1   1                 \  STUDENT
1 1  1   3 22 4 62735893922385590827032906974 5821  1    1      > MEASURE
6756 7 3 7210 38111098634784344103174516239253219 2  6 89 4 46  /  DISTRIBUTION
     T            S            M          S            T       MEAN (M), 1 SD (S), 2 SD (T)
0             10   20    30 40  50  70 80   90             99  PERCENTILE
110    70     50       35      30      35       50      70    110  UNCERTAINTY

                    1   T )+  6 78   Q    Y   D  W  4
              3 %$ F^& AV*(P2M LZSRUCE J H N KJ!@# 0 5  9      DISTRICT-WIDE CLASSROOMS

                              P   Q  D
                      GA      F  LMSYCERU NJ WK    H           DISTRICT-WIDE SCHOOLS

         D       |        P         |      S                   PROFICIENCY STANDARDS
         15%              53%              32%                 PROFICIENCY %S
                            *                                  LAST YEAR'S MEAN MEASURE
                               *                               MOST RECENT PISA/TIMMS/ICILS
    400             500           600           700        800 SAT EQUIVALENTS
|-----+-----+-----+-----+-----+-----+-----+-----+-----+-----|
360 400   440   480   520   560   600   640   680   720   760  MEASUREMENT SCALE
4     2     1      1     .5     .3     .5     1     2      4   UNCERTAINTY (SE CLASSROOM MEANS)

     T            S            M          S            T       MEAN (M), 1 SD (S), 2 SD (T)
0     10     20    30   40   50   70      80     90        99  OVERALL STUDENT PERCENTILE
```

Figure 3. Prototype multilevel comparison report for vertical coherence.

and percentile rank relative to an overall distribution. Advanced users may want to include data consistency indices (model fit statistics), and other diagnostic, sample, and population statistics. It is important to note here that, in accord with results observed in studies of universals in medicine [18] and other fields, individual student response patterns will rarely conform strictly to the expected pattern.

These applications in the domain of developmental coherence at the micro level within individual students require levels of detail and precision in the information reported not shared by the demands for horizontal and vertical coherence at the meso and macro levels. For horizontally coherent classroom management, teachers need actionable information identifying opportunities for improvement at the student level, as well as from comparisons with other teachers' results (Figure 2). Plots of individual student performance over time, and cross sectional plots of multiple students at a given point in time (as shown in the student measure distribution in Figure 1), will support meaningful comparisons.

Finally, as shown in Figure 3, for vertically coherent management up and down a hierarchy of accountability demands, everyone involved should be able, if needed, to drill down from any given level to the patterns of expected performance on the learning hierarchy for any subgroup of students, or any individual student. The vertical reach of the report could be extended to statewide, national, and international level. Interpretive guidelines will include construct maps and instructions for use.

2.5 Test

Prototype tests are proceeding with focus sessions on the to-date prototype developments. Trial runs with real data will be shown to teachers, who will evaluate the strengths, weaknesses, opportunities, and threats that may conceivably be expected to emerge relative to the proposed tools. A second pass will then be performed using data accumulated by a single teacher over time as a demonstration of the value obtained relative to the pre-existing methods and tools. The results from this second test will be fed into a new focus session in a third pass, and revisions from that test will be used in a pilot deployment.

3. Education ecosystem implications

The coherent coordination of information across levels and silos in social ecosystems implies a view of measured constructs as complex adaptive self-organizing forms of life. Identifying, documenting, and embodying these species' and niches' interdependencies could lead to important new opportunities for the improvement of educational outcomes.

References

[1] Miller P N 2015 *Chronicle Higher Educ* **61** 29
[2] Spinelli S Jr and Nixon N 2015 *Chronicle Higher Educ* **61** 31
[3] Gorin J and Mislevy R 2013 *Inherent measurement challenges in the next generation science standards for both formative and summative assessment* (Princeton, NJ: ETS)
[4] Wilson M 2004 *Towards coherence between classroom assessment and accountability* (Chicago: University of Chicago Press)
[5] Star S L and Ruhleder K 1996 *Information Systems Research* **7** 111-134
[6] Bateson G 1972 *Steps to an ecology of mind* (Chicago: University of Chicago Press)
[7] Susser M and Susser E 1996 *Am J Public Health* **86** 674-677
[8] Bizouarn P 2016 *Médicine/Sciences* **32** 500-505
[9] Barney M and Fisher W P Jr 2016 *Ann Rev Org Psychol Org Behav* **3** 469-490
[10] Stenner A J, Fisher W P Jr, Stone M and Burdick D 2013 *Frontiers in Psychol: Quant Psychol Meas* **4** 1-14
[11] Dawson T 2004 *J Adult Dev* **11** 71-85
[12] Öztürk-Gübes N and Kelecioglu H 2016 *Educational Sciences: Theory & Practice* **16** 715-734
[13] Griffin P 2007 *Studies Educ Eval* **33** 87-99
[14] Fisher W P Jr 2010 *J Physics Conf Series* **238**
[15] Mari L and Wilson M 2014 *Measurement* **51** 315-327
[16] Pendrill L and Fisher W P Jr 2015 *Measurement* **71** 46-55
[17] Latour B 1998 *Remaking reality: Nature at the millennium* ed N Castree and B Willems-Braun (New York: Routledge) pp 221-242
[18] Berg M and Timmermans S 2000 *Configurations* **8** 31-61
[19] Fisher W P Jr and Wilson M 2015 *Pensamiento Educativo* **52** 55-78
[20] Rasch G 1980 *Probabilistic models for some intelligence and attainment tests* (Chicago: University of Chicago Press)
[21] Andrich D 1978 *Psychometrika* **43** 561-573
[22] Beretvas N and Kamata A 2007 *Rasch measurement: Advanced and specialized applications* ed E V Smith Jr and R M Smith (Maple Grove, MN: JAM Press) pp 290-470
[23] Linacre J M 1997 *Physical Medicine and Rehabilitation State of the Art Reviews* **11** 315-324
[24] Connolly A J, Nachtman W and Pritchett E M 1971 *Keymath: Diagnostic Arithmetic Test* (Circle Pines, Minnesota: American Guidance Service)

Biomedical comparison of magnetometers for non-ferromagnetic metallic foreign body detection

L G S Fortaleza[1], E Costa Monteiro[1], C R H Barbosa[1], E C Silva[2] and L A P Gusmão[2]

[1]Postgraduate Programme in Metrology, Pontifícia Universidade Católica do Rio de Janeiro (PUC-Rio), Rio de Janeiro, C.P. 38071, 22451-900, Brazil

[2]Department of Electrical Engineering, Pontifícia Universidade Católica do Rio de Janeiro (PUC-Rio), Rio de Janeiro, C.P. 38071, 22451-900, Brazil

leofortaleza@outlook.com

Abstract. The location and surgical removal of foreign bodies in patients is still challenging, especially for firearm projectiles, which are small and non-ferromagnetic. Conventional location techniques use ionizing radiation, posing health risks while the procedures often last several hours and end unsuccessfully. The use of high sensitivity magnetometers provides a non-invasive and innocuous alternative for metallic foreign body location. The developed technique consists of a primary AC magnetic field generator (a solenoid) inducing eddy currents in non-ferromagnetic metallic foreign bodies, which results in an ultra-low secondary magnetic field that can be measured. This work compares the initially developed theoretical technique using Superconducting Quantum Interference Device (SQUID) magnetometers with the developed prototypes using lower cost alternatives, namely Giant Magnetoresistance (GMR) and Giant Magnetoimpedance (GMI). The comparison is based on biomedical device requirements for widespread clinical application. The proposed GMI location system is deemed the most qualified for clinical use.

1. Introduction

The need for surgical removal of firearm projectiles from patients is commonplace and continues to present challenges for physicians [1]. The only widely available location techniques use ionizing radiation, posing health risks to both patients and staff. Furthermore, these methods are often ineffective, with surgical procedures lasting several hours and typically ending unsuccessfully, especially when dealing with small objects.

The Laboratory of Biometrology (LaBioMet) at Pontifical Catholic University of Rio de Janeiro (PUC-Rio) performs research in the field of non-invasive clinical diagnosis [1-6]. The success in detecting ferromagnetic needles using a Superconducting Quantum Interference Device (SQUID) [2-3] prompted the development of a technique to locate non-ferromagnetic foreign bodies, in particular lead projectiles, which are diamagnetic [1]. This technique was adapted in the development of biomedical devices to locate such bodies while respecting project goals such as low cost for clinical application, innocuousness, non-invasiveness, safety, portability, ease of use and capacity to operate at room temperature [7-11]. In order to comply with such requirements, magnetometers based on the Giant Magnetoresistance (GMR) and Giant Magnetoimpedance (GMI) effects were used as lower cost alternatives to SQUID sensors, which require liquid Helium cooling [12-13].

This work introduces a fundamental comparison between the proposed location systems based on SQUID [1], GMR and GMI sensor elements regarding compliance with project requirements for the clinical application of biomedical devices.

2. Methods

The essential detection technique consists of using a time-varying primary magnetic flux density to induce eddy current loops in the foreign body, which consequently produce a secondary magnetic flux density to be measured by a gradiometric (differential) reading system, composed by two sensor elements [1,8,10]. The gradiometric configuration increases the signal-to-noise ratio (SNR) by minimizing the effect of the environmental and primary magnetic flux densities [1]. Figure 1 is a simplified illustration of this detection concept [10] when using a solenoid as a primary magnetic flux density generator and placing its axis aligned with the centre of the spherical foreign body. This axis is defined as the z-axis, over which the distance or depth values are measured.

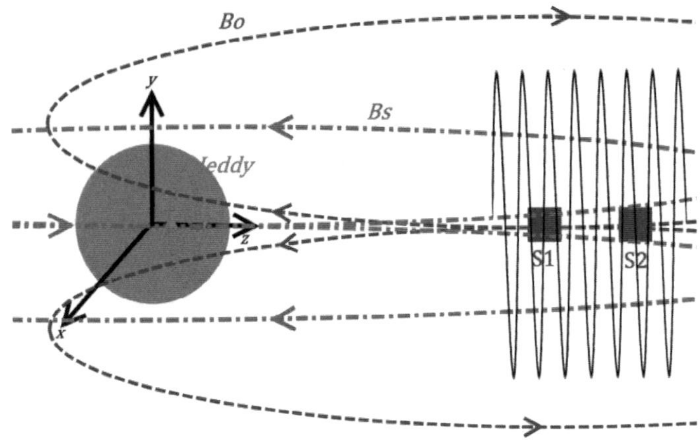

Figure 1. Simplified diagram of the detection procedure. The solenoid generates a time-varying primary magnetic flux density B_o (dashed line) towards the spherical foreign body that induces the also time-varying eddy current loop I_{eddy} (dotted line), in return producing the secondary magnetic flux density B_s (dash-dot line). The result of a gradiometric (differential) reading by sensor elements S1 and S2 is mostly composed by the secondary magnetic flux density [10].

Mathematical work on theoretical models of eddy currents in spherical conductors when applied to this configuration result in equation (1) [1,10,14].

$$\boldsymbol{B}_{\mathbf{smax}}(h, a, f_o) = \boldsymbol{B}_{\mathbf{0}} V(a, f_o)\left(\frac{1}{h^3} - \frac{1}{(h + l_s)^3}\right) \tag{1}$$

Where $\boldsymbol{B}_{\mathbf{smax}}$ is the peak secondary magnetic flux density, $\boldsymbol{B}_{\mathbf{0}}$ is the primary magnetic flux density, h is the depth distance between the foreign body and the sensor element S1, l_s is the baseline (distance between the centres of S1 and S2) and $V(a, f_o)$ is a phasor variable given by equation (2).

$$V(a, f_\mathrm{o}) = a^3\left[\left((2\mu_\mathrm{r} + 1) - (2\mu_\mathrm{r} + \nu^2 + 1)\frac{\tanh \nu}{\nu}\right) \cdot \left((\mu_\mathrm{r} - 1) - (\mu_\mathrm{r} - \nu^2 + 1)\frac{\tanh \nu}{\nu}\right)^{-1}\right] \tag{2}$$

Where μ_r is the relative permeability of the conducting sphere, a is its radius and ν is a factor related to the skin depth in the material, calculated by equation (3).

$$\nu = a(1 + \mathrm{i})(\pi\mu_\mathrm{r}\mu_\mathrm{o}\sigma f_\mathrm{o})^{-1/2} \tag{3}$$

Where μ_o is the permeability of free space, σ is the conductivity of the conducting sphere and f_o is the frequency of the primary magnetic flux density.

These theoretical models were used to estimate secondary magnetic flux density levels for varying distances and foreign body radii [1,10]. These results were associated with sensor specifications in order to design signal processing circuits and conduct complete system simulations and experiments [10]. The GMR sensors used were NVE AA-005-002 and the GMI sensors were Aichi MI-CB-1DJ-M-B commercially available prototypes.

The technical specifications of each sensor element and certain parameters utilized in each system were used for comparison regarding project requirements [7]. Innocuousness was evaluated according to the exposure limits established by the International Commission on Non-Ionizing Radiation Protection (ICNIRP) to electromagnetic fields [15-17] and primary magnetic flux density levels of each detection system. Assuming measurement at the edge of the solenoid, the SQUID system in [1] uses 5.15 mT (rms) at 1 kHz, the GMR system uses a DC component of 4 mT and an AC component of 1.77 mT (rms) at 123 kHz while the GMI system uses 468 nT (rms) at 8 kHz [10]. The magnetic field density decreases rapidly with the distance from the solenoid; therefore a minimum distance to reach the reference levels can be calculated. This analysis, exhibited in Table 1, considers the different reference levels established by ICNIRP for occupational and for public exposure.

Table 1. Innocuousness analysis of the location systems using ICNIRP reference levels [10].

System	Frequency	Primary Magnetic Flux Density (rms)	Occupational Reference Level	General Public Reference Level	Safe Distance Occupational	Safe Distance Public
GMR	DC 123 kHz	4 mT 1.77 mT	2 T [15] 16.3 μT [17]	400 mT [15] 6.25 μT [17]	Compliant 9.6 cm	Compliant 14.0 cm
GMI	8 kHz	468 nT	100 μT [16]	27 μT [16]	Compliant	Compliant
SQUID	1 kHz	5.15 mT	0.3 mT [16]	80 μT [16]	7.86 cm	14.1 cm

Non-invasiveness is determined by the principle of detection used by the location systems while criteria such as portability, ease of use and room temperature operation depend mostly on the sensor element characteristics and the complexity of the electronics involved, in particular the primary magnetic field generator and the signal processing requirements. The main component impacting cost, in this case, is the sensor element. Safety evaluation can be manifold; however, a few critical concerns can be established for each system, including the basic requirements for medical electrical equipment safety.

3. Results

Regarding the innocuousness criteria, the GMI system is the only candidate that uses primary magnetic flux densities below the safety reference levels recommended by ICNIRP at the edge of the solenoid. The other systems require small distances in order to be compliant. Non-invasiveness is satisfied by any magnetic measurement technique, being a common advantage of the three detection systems. The SQUID necessity for cryogenics affects the other requirements, imposing high operational costs and safety concerns, as well as reducing portability, adding complexity of use and preventing room temperature operation. Meanwhile, the GMR and GMI sensors comply with these requirements. The GMR system possesses the lowest cost, however there are some safety concerns due to the high solenoid excitation currents. The low cost of the GMI system should become even more competitive when nanotesla GMI sensors enter mass production. Comparison results are summarized in Table 2.

Table 2. Analysis of the major biomedical device requirements for the considered magnetometers and their respective location systems [10].

Biomedical Device Requirements	SQUID [1]	GMI (Aichi MI-CB-1DJ-M-B)	GMR (NVE AA-005-002)	Most Suitable Magnetometer
Innocuousness (distance for reference levels)	7.86 cm (occ.) 14.1 cm (public)	No limitation	9.6 cm (occ.) 14.0 cm (public)	**GMI**
Non-Invasiveness	Yes	Yes	Yes	**Any**
Low Cost	No	Yes	Very	**GMR**
Safety (main concern)	Cryogenics MEES[a]	MEES[a]	High excitation current MEES[a]	**GMI**
Portability	No (Dewar and Cryogenics)	Yes	Yes	**GMI and GMR**
Ease of Use	No (Requires Cryogenics)	Yes	Yes	**GMI and GMR**
Room Temperature Operation	No (lower than 77 K required)	Yes	Yes	**GMI and GMR**

[a] MEES - Medical Electrical Equipment Safety

4. Conclusions

Although SQUID sensors are currently a benchmark for magnetic measurements by possessing the lowest noise levels, the need for cryogenic temperatures does not comply with several major requirements for clinical use. The analytical comparison between the three proposed non-ferromagnetic metallic foreign body location systems indicates the GMI system as the most suitable for this biomedical application. It is capable of complying with the major biomedical device requirements that enable widespread clinical use, in particular the detection system based on the GMI sensor is the only inherently innocuous according to exposure limits established by ICNIRP guidelines. The GMR system comes at close second with an advantage of even lower cost. These results evidence the potential of introducing new techniques based on magnetic measurements for clinical use, providing innocuous and non-invasive alternatives to traditional techniques and maintaining low costs.

Acknowledgments

This work was supported by the Brazilian funding agencies CNPq, FAPERJ, CAPES and FINEP.

References

[1] Barbosa C H 2004 Localization of firearm projectiles in the human body using a superconducting quantum interference device magnetometer: A theoretical study *Rev. Sci. Instrum.* **75** 6 2098–106

[2] Monteiro E C, Barbosa C H, Lima E A, Ribeiro P C and Boechat P 2000 Locating steel needles in the human body using a SQUID magnetometer *Phys Med Biol* **45** 8 2389–402

[3] Barbosa C H, Monteiro E C, Lima E A, Santos S F, Cavalcanti E G and Ribeiro P C 2001 Improvement of a Technique for Localization of Steel Needles in Humans Using a SQUID Magnetometer *IEEE Trans. Appl. Supercond.* **11** 1 677–80

[4] Pompéia F, Gusmão L A P, Barbosa C R H, Monteiro E C, Gonçalves L and Machado F L A

2008 Ring shaped magnetic field transducer based on the GMI effect *Meas Sci.Tech.* **19** 2 025801

[5] Silva E C, Gusmão L A P, Barbosa, C H, Monteiro E C and Machado F L A 2011 High sensitivity giant magnetoimpedance (GMI) magnetic transducer: magnitude versus phase sensing *Measurement Sci. and Tech.* **22** 3 035204

[6] Silva E C, Barbosa C H, Gusmao L A P, Leipner Y, Fortaleza L G S and Monteiro E C 2014 Point matching: A new electronic method for homogenizing the phase characteristics of giant magnetoimpedance *Rev. Sci. Instrum.* **85** 8 084708

[7] Monteiro E C and Leon L F 2015 Metrological Reliability of Medical Devices *J. Phys.: Conf. Ser.* **588** (2015) 012032

[8] Fortaleza L G S, Barbosa C R H, Monteiro E C, Silva E C and Gusmao L A P 2015 Detecting non-magnetic metallic foreign bodies by GMR sensors through the use of eddy currents. *In 8th Brazilian Congress on Metrology, Bento Gonçalves* 1–4

[9] Fortaleza L G S, Silva E C, Barbosa C R H, Gusmao L A P and Monteiro E C 2015 Electronic circuit for excitation of inductive loads with high currents *Electron. Lett.* **55** 21 1808–9

[10] Fortaleza L G S 2016 *Nonferromagnetic metallic foreign body detection by eddy currents*, MSc Dissertation (Rio de Janeiro: Pontifical Catholic University of Rio de Janeiro)

[11] Louzada D, Monteiro E C, Rivero A M, Fortaleza L G S, Barbosa C R H, Velázquez R T, Silva E C and Gusmao L A P 2016 Quality by Design approach in the development of a magnetic transducer for biomedical measurements: preliminary results on Design Space configuration *J. Phys.: Conf. Ser.* **772** (2016) 012016

[12] Robbes D 2006 Highly sensitive magnetometers - a review *Sens. Actuator* A-Phys. **129** 1–2 Special Issue 86–93

[13] Ripka P 2001 *Magnetic Sensors and Magnetometers* 1. ed. (Norwood, MA: Artech House Publishers)

[14] Smythe W R 1989 *Static and dynamic electricity* 3rd Ed. (USA: Hemisphere Publ. Corp.)

[15] International Commission On Non-Ionizing Radiation Protection 2009 Guidelines on Limits of Exposure to Static Magnetic Fields *Health Phys.* **96** 4 504–14

[16] International Commission On Non-Ionizing Radiation Protection 2010 Guidelines for Limiting Exposure to Time-Varying Electric and Magnetic Fields (1 Hz to 100 kHz) *Health Phys.* **99** 6 818–36

[17] International Commission On Non-Ionizing Radiation Protection 1998 Guidelines for Limiting Exposure to Time-Varying Electric, Magnetic and Electromagnetic Fields (Up to 300 GHz) *Health Phys.* **74** 4 494–522

IOP Conf. Series: Journal of Physics: Conf. Series **1044** (2018) 012014 doi:10.1088/1742-6596/1044/1/012014

Calibration of an industrial data acquisition system used in a solar heating plant

F B Coelho and A F Orlando
Programa de Pós-Graduação em Metrologia,
Pontifícia Universidade Católica do Rio de Janeiro, R. Marquês de São Vicente, 225, Gávea, Rio de Janeiro, Brazil

E-mail: fborges.fc@gmail.com

Abstract. The use of data acquisition systems is essential when it is necessary to measure certain physical quantities involved in a particular process. The equipment must have a certain degree of reliability since all values taken by the acquisition system must be as close as possible to the values obtained by the measuring instruments. In this context, this article assesses the challenge concerning the calibration of data acquisition system used in solar heating plant, in order to determine the metrological reliability of the equipment. The calibration procedure consists of comparing the input signal from a calibrated signal generator, to the indicated value by the acquisition system. The difference between the two values is then treated statistically, resulting in an uncertainty analysis to be used for assessing the equipment reliability.

1. Introduction

The electric power market in Brazil is very competitive. For this reason, many thermoelectric power plants are constantly looking for new technologies aiming to reduce electricity generation costs. The solar heating plant consists of five independent 200 m^2 solar heating plants, each one for preheating one (8,7 MW) of the five selected engines and the fuel oil, out of the 38 engine thermoelectric power plant. In order to keep track of its performance, a data acquisition equipment is used for measuring the output electric current (4 to 20 mA), from temperature and pressure sensors, and frequency (Hz), from flow sensors, installed in selected positions of the hydraulic circuit.

In the thermoelectric power plant of this study, the thermal radiation absorbed by each solar heating system is used to complement the thermal energy necessary to heat the coupled engine and keep it ready for a quick start operation, by preheating the water that circulates between a steam boiler and the engine, thus saving fuel oil. Sometimes, when the water becomes hotter than what is needed because of the excess solar energy, the fuel oil is also preheated, thus improving the energy usage.

2. Operation

The data acquisition system consists of one independent module for each independent solar heating plant, totaling 5 independent modules. Each module has 10 analog input channels, 8 for temperature sensors and 2 for pressure sensors, and 4 digital signal frequency channels for flow sensors. Every 1/60 s, the module scans all channels. The measured values (mA or Hz) are then averaged over a 1 minute period and sent directly to a server, where they are stored in a file. Every hour, a coupled monitoring software reads all values (every minute) and converts them into temperature, pressure or flow rate units. The results are then stored in another file for customer consultation.

3. Test procedures

The test procedure consists of comparing the output values of the data acquisition system and the reference instrument. Reference currents in the range of 4 to 20 mA, simulating the output from temperature and pressure sensors, were injected into the data acquisition system channels. Likewise, reference frequencies in the range of 0 to 100 Hz, simulating the output from flow sensors, were injected into the data acquisition system channels. All reference instruments had been properly calibrated by a Brazilian Calibration Network accredited laboratory.

During the tests, after the equipment having reached the steady state condition, a very stable reference current was injected into the data acquisition system, with its output integrated over a one minute period, simulating the system monitoring frequency. The average value between three consecutive outputs, when the maximum difference between each other was 1 μA, was considered to be the most probable output value from the data acquisition system. The measurement systematic error was calculated as the difference between the most probable value and the reference current. The maximum difference between the output from the data acquisition system during the tests and the most probable value was considered to be the type B uncertainty of current measurement.

The expanded uncertainty of temperature measurement, U, k = 2, by the data acquisition system, in the 4 to 20 mA output current range, was estimated by calculating the square root of the sum of the squares of three components, all of them converted to temperature units. (a) Type B expanded uncertainty of current measurement, $U_{(a)}$, $k = \sqrt{3}$, (b) Expanded uncertainty of converting probe resistance into temperature, according to [2], $U_{(b)}$ = 0.20 °C at 20 °C, $k = \sqrt{3}$, and, (c) Expanded uncertainty of converting probe resistance into current, $U_{(c)}$ = 0.002 mA (0.0125 °C), according to the manufacturer, k = 2.

$$u = \sqrt{{u_{(a)}}^2 + {u_{(b)}}^2 + {u_{(c)}}^2} \tag{1}$$

$$U = k \cdot u \tag{2}$$

A similar procedure was used for testing pressure probes, in the 4 to 20 mA output current range. When estimating the uncertainty of measurement, (b) Expanded uncertainty of converting probe output signal into pressure, $U_{(b)}$ = 10 cm H_2O, obtained from probe calibration curve by the manufacturer, k = 2, and (c) Expanded uncertainty of converting probe signal into current, $U_{(c)}$ = 0.002 mA (1.25 cm H_2O), according to the manufacturer, k = 2.

Likewise, a similar procedure was used to test flowmeters. When estimating the uncertainty of measurement, (b) Expanded uncertainty of converting probe frequency into flow rate, 1% of flow rate, obtained from the probe calibration curve supplied by the manufacturer, $U_{(b)}$, k = 2, (c) Frequency measurement resolution, 1 Hz (±0.17 m^3/h), $U_{(c)}$, $k = \sqrt{3}$.

4. Results

During the tests, all electrical current and frequency data were converted into temperature, pressure or flow rate by the respective conversion curve.

Table 1. Correlation points for conversion from electrical current or frequency to temperature, pressure and flow parameters.

	Temperature		Pressure		Flow Rate	
Point	mA	°C	mA	cm H_2O	Hz	m^3/h
1	4.0	0.0	4.0	0.0	19	3.5
2	20.0	100.0	20.0	1000.0	92	15.6

Figure 1 shows the temperature error measurement by 8 temperature channels in each of the 5 modules. It can be seen that for temperature values smaller than 20 °C the data acquisition system measures it with a large error. For larger temperature values, almost all errors are in the ± 0.3 °C range, which should be considered as a Type B uncertainty of measurement if no distinction is made between channels.

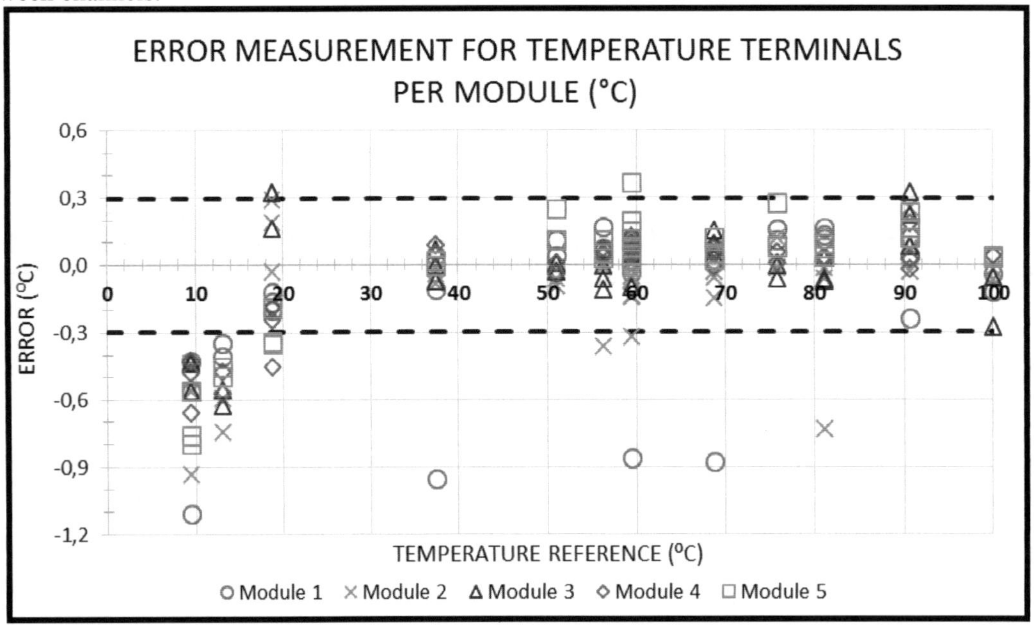

Figure 1. Temperature measurement error for channels in each module.

And figure 2 shows the current error measurement by 2 pressure channels in each of the 5 modules. It can be seen that for pressure values smaller than 200 cm H_2O (0.2 bar) the data acquisition system measures it with a large error. For larger pressure values, almost all errors stays in the ± 2 cm H_2O range, also considered as a Type B uncertainty of measurement.

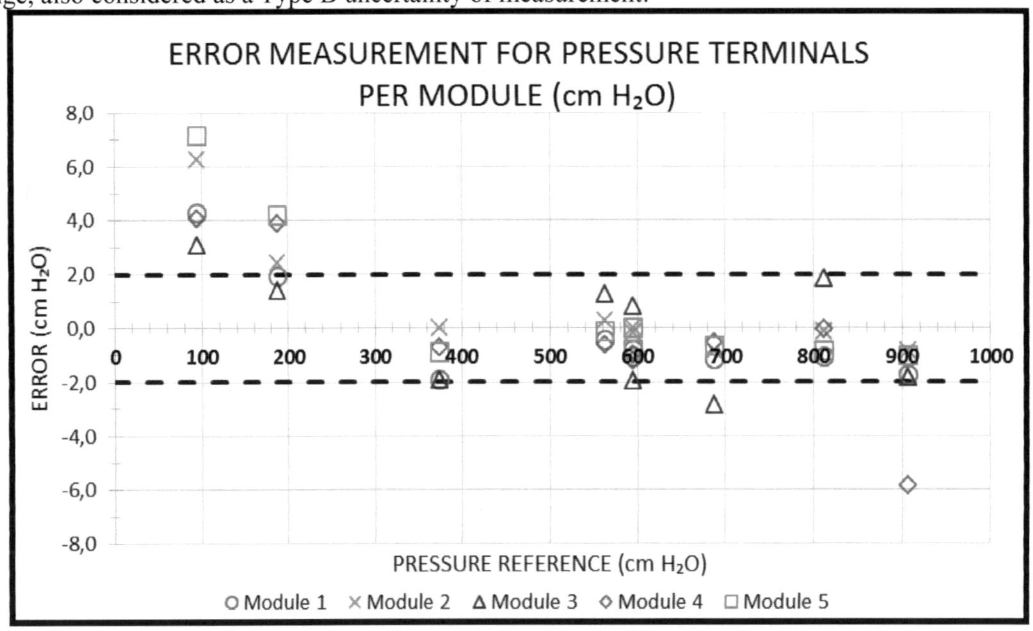

Figure 2. Pressure measurement error for channels in each module.

The frequency error measurement captured by each flow channel of the 5 modules is shown in figure 3. As others parameters, almost all errors flow values are between ± 1 m³/h, considered as a Type B uncertainty of measurement.

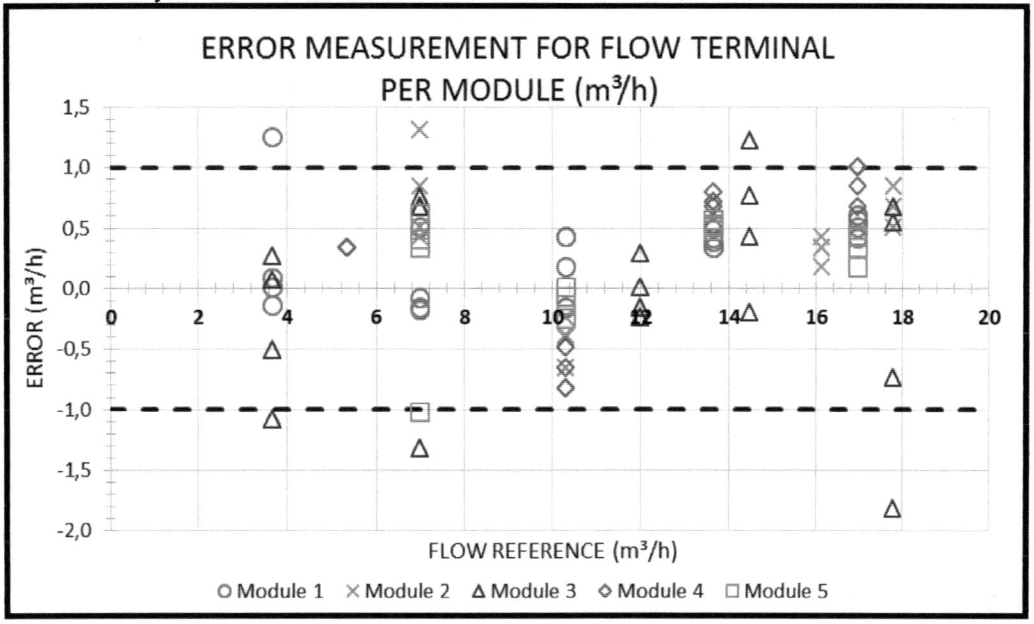

Figure 3. Flow rate measurement error for channels in each module.

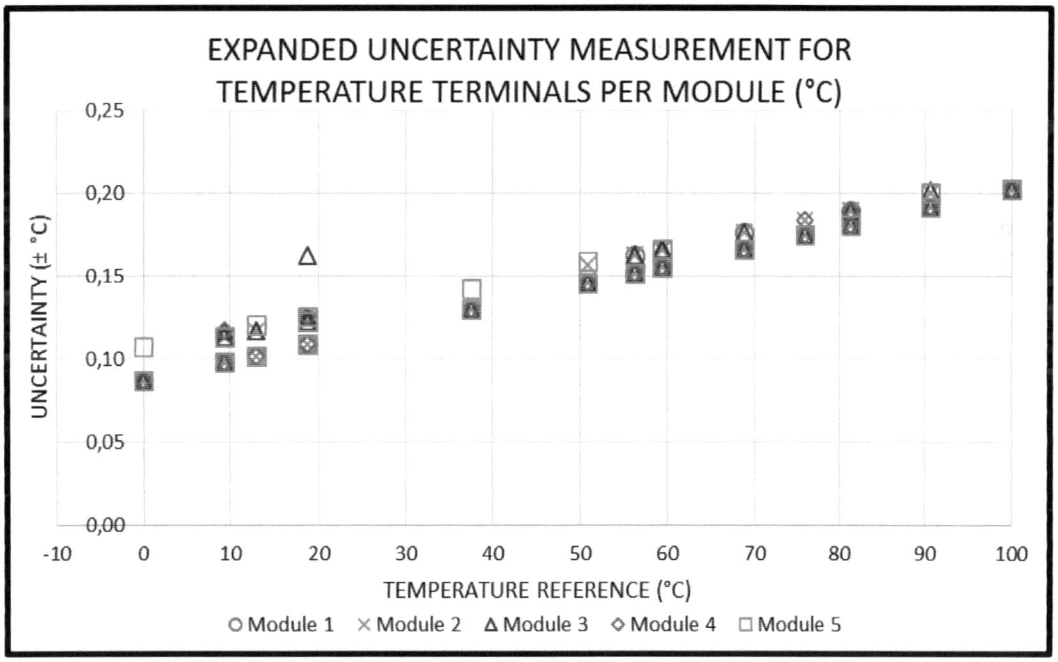

Figure 4. Expanded uncertainty measurement by temperature channels in each module.

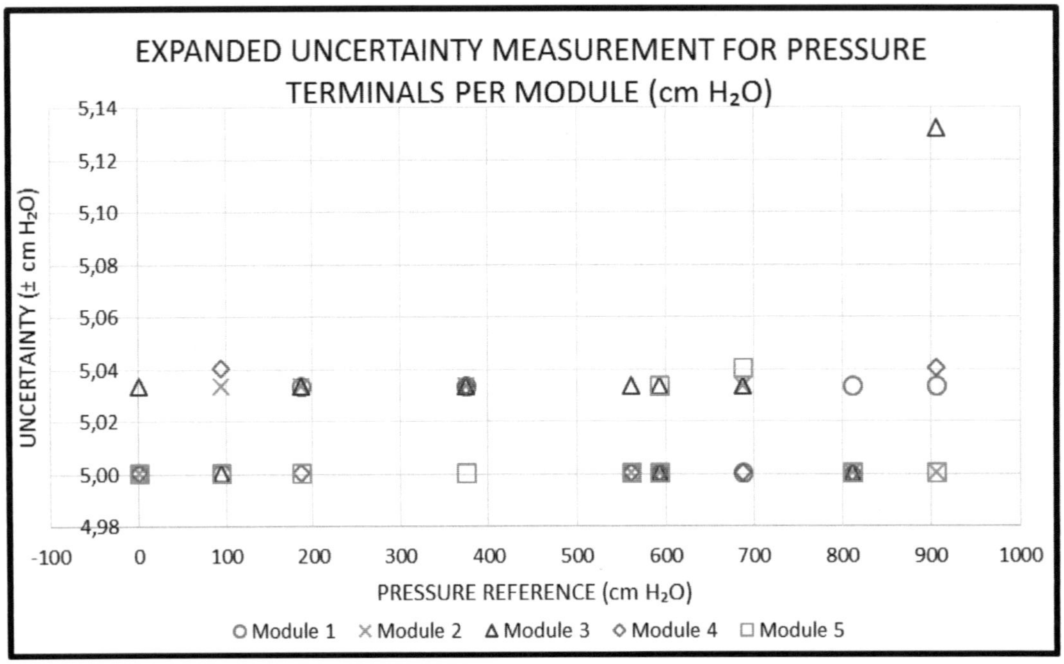

Figure 5. Expanded uncertainty measurement by pressure channels in each module.

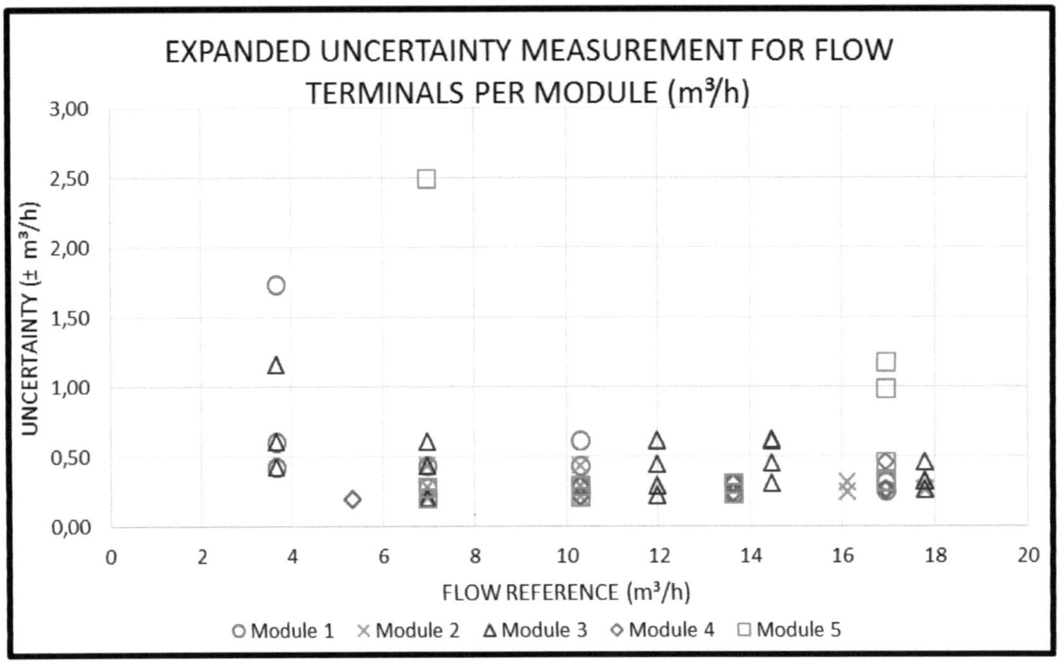

Figure 6. Expanded uncertainty measurement by flow rate channels in each module.

It can be seen that systematic error is larger than expanded uncertainty of measurement. If no distinction is made between channels, no correction can be made, and the maximum error in figures 1, 2 and 3, can be considered as a Type B uncertainty of measurement, which should be added to the calculated expanded uncertainty of measurement in figures 4, 5 and 6 by the root sum square method. If a smaller uncertainty value is required, the output from the data acquisition system must be corrected by the systematic error measured in each channel, and the calculated expanded uncertainty of measurement are given by figures 4, 5 and 6. When in operation, if the measured values are not the same as those in the calibration, a linear interpolation between two neighbor points can be made. More precisely, a least square fit can be used as an interpolation scheme for estimating the most probable value and its uncertainty of measurement.

5. Conclusion

This paper presents a methodology for calibrating a data acquisition system with several channels. It includes estimating the most probable value as a function of a required uncertainty of measurement. (a) If there is no distinction between channels, no correction is made, and the uncertainty of measurement becomes large, (b) The reduction of the uncertainty of measurement requires the calibration of each channel, and, (c) interpolation scheme to be used when the measured point does not coincide with the calibrated points.

After correcting the output from the data acquisition system with calibration data, measurement value coincides with the most probable value to within the uncertainty of measurement, which shows the importance of differentiating sensor output value from the system readout. If no correction is made, results give a misleading information about the uncertainty of measurement with the sensors themselves.

The contribution of this paper is to include the uncertainty of measurement of data acquisition system, usually not considered in the uncertainty budget of the quantity being measured, thus better interpreting the results.

References

[1] ISOGUM 1998 Guia para a expressão da incerteza de medição
[2] IEC 60751 Standard, Edition **2.0** 2008-07 Industrial platinum resistance thermometers and platinum temperature sensors
[3] Environmental Protection Agency M 2008 Data acquisition and information management *QA Handbook.* **V II** Section 14
[4] Jiannong W and Wei W The common data acquisition system based on arm9 2011 *The Tenth International Conference on Electronic Measurement & Instruments (ICEMI 2011)*
[5] Koutroulis E and Kalaitzakis K 2003 Development of an integrated data-acquisition system for renewable energy sources systems monitoring *Renewable Energy 28* (Technical University of Crete - Chania, Greece) p 139-52

Acknowledgments

The authors are indebted to GERA MARANHÃO and GERA AMAZONAS for having sponsored a research and development project for preheating the Diesel engines of a thermoelectric power plant with solar energy.

IOP Conf. Series: Journal of Physics: Conf. Series **1044** (2018) 012015 doi:10.1088/1742-6596/1044/1/012015

Comparison in determination of mass-energy transfer coefficients

G J Teixeira[1,2,4], C H Souza[1], R S Cardoso[1] and J G P Peixoto[1,3]

[1] Instituto de Radioproteção e Dosimetria, Comissão Nacional de Energia Nuclear, Av. Salvador Allende, Jacarepaguá, Rio de Janeiro, BR
[2]Curso Superior de Tecnologia em Radiologia, Universidade Estácio de Sá, Rua André Rocha, 838, Taquara, Rio de Janeiro, BR
[3]Laboratório de Ciências Radiológicas, Universidade do Estado do Rio de Janeiro, Rua São Francisco Xavier, 524, Maracanã, Rio de Janeiro, BR
[4]Faculdade de Ciências Médicas, Universidade do Estado do Rio de Janeiro, Boulevard 28 de setembro, Vila Isabel, Rio de Janeiro, BR

guilherm@ird.gov.br

Abstract. The values obtained through definitions, specific software and measurements of the mass energy transfer coefficients referring to X-ray beams in air using the range of 30 to 150 kV. They allowed the spectra and calculated evaluation the respective mass coefficients. The results shown a decreasing behavior with energy, ranging from 0.3733 to 0.0439 cm^2/g, differing from the behavior of monoergetic beams above 100 keV.

1. Introduction

The entrance surface air kerma and incident air kerma are quantities derived from the air kerma quantity [1]. These quantities are used to verify the given doses in radiodiagnostic practices and compare with reference levels in quality assurance programs implementations. These doses and the equipment performance have great relation too. Following this way, the determination of mass energy transfer coefficients for X-ray spectra to voltage range from 30 to 150 kVp is fundamental.

Researchers and collaborators in quality assurance programs use the mass energy transfer coefficients of the tables present in Attix, being the only reference, but the Attix values come from monoenergetic sources. X-rays are formed by spectra with energies of zero to the maximum beam energy for each selected voltage. It can then be concluded that there must be some discrepancy between the coefficients for monoenergetic and spectral sources.

X-ray diffraction coefficients for multi-energy sources were not considered by Attix [2], that used only mono-energy sources as reference to obtain the coefficients and water-air coefficient ratios for X-rays [3, 4].

Therefore the determination of mass transfer coefficients using X ray spectra has great relevance because nowadays some systems do not use the ionization chamber as a detector. Solid-state detectors like diodes and specific crystals were introduced as dosimetric sets. These solid-state detectors can be calibrated through the energy fluence once the kerma can be considered a function of the energy fluence and energy mass transfer coefficient, for each energy or voltage [2].

2. Materials and Methods

2.1 Spectrum Processor

The Spectrum Processor [4,5] is a software, template are shown in figure 1, capable of generate X-ray spectra, providing photon fluence (photons / mAs mm2) with 0.5 keV intervals, from physical information pre-archived in the matrix, ranging from zero to the maximum beam energy. In addition to the spectrum for each voltage, the program also provides the air kerma at 750 mm from the focal point per kVp selected.

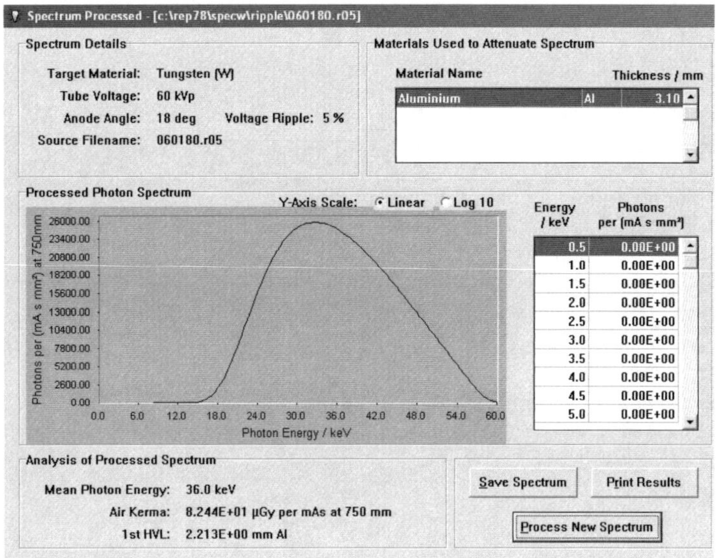

Figure 1. A) Software communication Interface; B) Example of generated spectrum with 60 kVp, anodic angulation of 18 °, total filtration of 3.1 mmAl and 5% of Ripple voltage.

The program was developed by the Institute of Physics and Engineering in Medicine – IPEM [4,5]. The Software interface allows you to enter data such as kVp, total filtering, anode angle, ripple Voltage and target material. Allows generated spectrum to be save in other formats, in order to export the file to spreadsheets. The use of the program requires a 16 or 32 bit operating system and set to British configurations.

2.2 Mass Energy Transfer Coefficients

Almost thirteen spectra from 30 up to 150 kVp were generate in the Spectrum Processor software, with 10 kVp intervals. The air kerma correspondents were obtained for each nominal voltage.

The focal point distance to the measurement point was established as 1000 mm, like clinic practices, by the distance square inverse law, once the standard distance in the Spectrum Processor software is 750 mm. Was established in 2.5 mmAl additional filtration and 20° of anodic angulation at the simulation.

The generated photons fluence for each energy spectra were exported to a spreadsheet and processed. The treatment consisted in calculate the energy fluency for each spectra related and voltage. Once all the energy fluence were calculated to each tube voltage, with equation 1.

$$\psi_E = \int E\phi_E \tag{1}$$

The coefficients for each tube voltage, from 30 to 150 kVp, were calculated through equation 2 using the air kerma obtained with *Spectrum Processor software* and the total energy fluency by each voltage, equation 1.

$$K = \int_{E=0}^{E_{max}} \psi \left(\frac{\mu_{tr}}{\rho} \right)_{E,Z} \quad (2)$$

The maximum energy peaks were observed for each nominal voltage because the effective X ray energy beam does not mean the same voltage applied.

The coefficients were compared with the Attix [2] monoenergetic data and Oliveira interpolated data [2,4,6].

3. Results

The generated mass energy transfer coefficients are shown in figure 2 as function of the beam's main energy and the nominal voltage. The comparison between the Attix data [2], Oliveira [6] and Teixeira [4] are shown in table 1. The difference between the values demonstrates that ionizing sources used at each methodology are fundamentals to obtain these mass energy transfer coefficients to air used by diagnostic X ray.

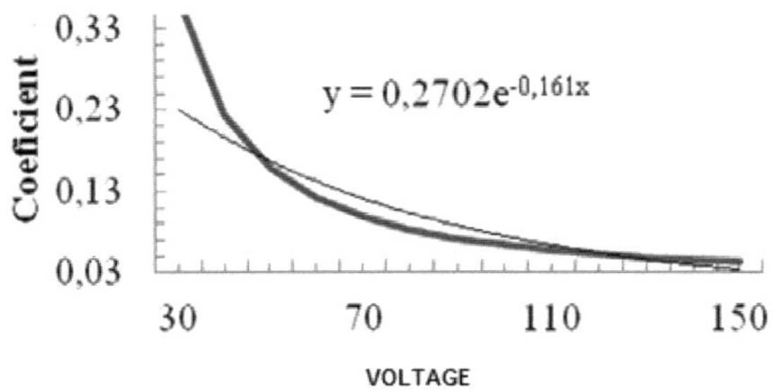

Figure 2. Theoretical air mass-energy transfer coefficients.

4. Discussion

The coefficients obtained behave in a decreasing way as higher is the tube voltage. This phenomenon is due to the fact that when the average energy of the beam is higher the dose deposition will be lower at the same point in the air, since the kerma is directly proportional to the coefficient.

This would also be repeated for materials with other densities, so when the energy is increased the dose deposition decreases at the measurement point but increases at locations beyond the initial measurement point.

Data Attix [2] doesn´t show the coefficients related to 70, 90, 110, 120, 130 and 140 keV to monoenergetic beams. Therefore the linear interpolation [6] had to perform to obtain the coefficients related to these energies.

It should be noted that the main Oliveira's objective [6] was to verify the applicability of NaI (Tl) Scintillation detectors of work was to verify the applicability of NaI (Tl) in air kerma measurements for diagnostic X ray beams and these results were acceptable considering the coefficients lack of multi-energy spectra at the literature, showing the relevance of continue study of these coefficients.

Table 1 shows the coefficients obtained by simulation, data Attix and linear interpolation [2,4,6]. The firsts' data shows the nominal voltage and the second the effective energy related. The effective energy is considered as the most intensity energy in the spectra to each voltage. The last column presents each one monoenergetic nominal beams from data Attix [2] and linear interpolation [6].

Data Attix also shows a behaviour decreasing at higher beam energy, but for 150 keV beam the coefficient shows a 0.0016 cm^2/g increase in relation to 100 keV beam, table 1, that this phenomenon is contrary to the X ray behaviour.

This could be understood due to the possible contribution of the radiative kerma component (K_r). In conventional X ray beams the maximum energy occurs with very low intensity, whereas for a monoenergetic beam, one hundred percent of the photons are obviously as the same energy.

However, the consequence will be that the Compton scattering probability increases to energies greater than 100 keV as a function of the material atomic number, the air in this case.

Table 1. Comparison between the values found in Attix [2]/Oliveira [5] and Teixeira [4].

Voltage (kV)	Effective energy (keV)	μ_{tr}/ρ (cm^2/g)		Monoenergetic Beam (keV)
		Teixeira	*Attix*	
30	23.1	0.3733	0.1480	30
40	27.8	0.2233	0.0668	40
50	31.8	0.1575	0.0406	50
60	35.4	0.1219	0.0305	60
70	38.7	0.0997	*0.0274	70
80	42.2	0.0833	0.0243	80
90	45.4	0.0717	*0.0238	90
100	48.3	0.0636	0.0234	100
110	51.0	0.0575	*0.0237	110
120	53.5	0.0529	*0.0240	120
130	55.8	0.0493	*0.0244	130
140	58.1	0.0463	*0.0247	140
150	60.3	0.0439	0.0250	150

* *mass energy transfer coefficients obtained by Attix*[2] *values linear interpolation.*

The linear interpolations [6] , table 1, shows the same decreasing behavior for 70 and 90 keV, but between 110 and 140 keV, the above mentioned increasing behavior is repeated again, thus bringing about a greater results reliability in the of linear interpolations performed.

Peixoto and Andreo [3] compared the air-water ratios of the mass energy absorption coefficients of some works in the range of 10 to 160 kV. The decreasing behavior and subsequent increase can also be observed, but the increase starts to occur from Half Value Layer - HVL at 2 mmAl, that is, 70 kV approximately, according to the protocols analyzed.

When observe data Attix [2] and linear interpolation [6], the increment probability happens from approximately 100 keV, but in Peixoto and Andreo [3] this behavior is anticipated. This phenomenon was attributed to insertion of the water in the relation, because in this mid the Kerma component of collision (Kc) increases, due to the increased density of one of the materials involved in the measurement.

5. Conclusions

The coefficients generated behave were expected for a radiodiagnostic X ray beam, but presented a decreasing behavior in its values with the voltage increase. For the monoenergetic beams used as reference [2,6], must have criterion in the insertion of these coefficients, because in this way, other effects for a same measurement point may be probable, take into account the uncertainty increasing of the result and obviously the exact measurement point, due to a non-focal dose deposition for energy ranges greater than 100 keV.

Another discovery found is data Attix were the present's values of the mass energy-transfer coefficients and to the mass energy-absorption coefficients to 100 and 150 keV beams were the same

Mass energy transfer coefficients by monoenergetic beams cannot be considered as equal to mass energy absorption coefficients due to secondary electrons and "nonlocal" energy deposition for "higher" energy levels ($\geq$ 100 keV). So these coefficients from monoenergetic beams also cannot be used at kerma in X ray diagnostic simulations measurements.

Appropriate calibration is recommended for chambers or other dosimetric systems of clinical use in diagnostic X ray beams with energies grater than 90 keV and/or voltage greater than 90 kV. It is necessary the correct layer buildup modeling, in order to obtain the desired electronic equilibrium.

References

[1] IAEA 2007 Technical Reports Series 457 - Dosimetry in diagnostic radiology: An international code of practice. Vienna. p 372.

[2] Attix F H 1986 Introduction to Radiological Physics and Radiation Dosimetry. John Wiley & Sons (New York,Wiley-Interscience). p 607.

[3] Peixoto J G and Andreo P 2000 Determination of absorbed dose to water in reference conditions for radiotherapy kilovoltage x-rays between 10 and 300 kV: a comparison of the data in the IAEA, IPEMB, DIN and NCS dosimetry protocols." *Phys. Med. Biol.*, vol. 45, no. 3, pp. 563–575.

[4] Teixeira G J 2013 Coeficientes de Transferência de Energia em Massa para Raios X Diagnóstico: Determinação e Análise. Instituto de Radioproteção e Dosimetria. Rio de Janeiro. p 79.

[5] IPEM 1997 Spectrum Processor Software. UK.

[6] Oliveira L S R 2011 "Avaliação da resposta de detetores cintiladores de NaI(Tl) em medição de kerma no ar em feixes de raios x diagnóstico. Instituto de Radioproteção e Dosimetria. Rio de Janeiro.

Confidence interval for the 100p-th percentile for measurement error distributions

Clarena Arrieta [1], Reinaldo Castro Souza[2], Rodrigo Flora Calili[1]

[1] Postgraduate Program in Metrology, Pontifical Catholic University of Rio de Janeiro, Mârques de São Vicente Street, 225, Gávea, Rio de Janeiro, Brazil

[2] Department of Metrology, Pontifical Catholic University of Rio de Janeiro, Mârques de São Vicente Street, 225, Gávea, Rio de Janeiro, Brazil

E-mail: clare.a18@gmail.com, reinaldo@ele.puc-rio.br, calili@puc-rio.br

Abstract. In metrology, one is usually comparing what indicates an instrument versus what it should be indicated by a reference pattern, resulting in the so called measurement errors. Most errors follow a normal distribution. If one wants to have some confidence about how small is such an error, the traditional confidence intervals for the unknown parameters are computed. This article illustrates, via simulation, the effectiveness of three intervals for estimating the 100p-th *percentile* in a normal distribution, namely, the Traditional (T), the Normal Asymptotic (NA), and the Lawless (LA) intervals. Various samples of different sizes are drawn from a zero mean normal distribution, which resembles the measurement error distributions. For each one of them the 95% confidence interval is estimated by using the three approaches. From the simulation, it is possible to conclude that T and LA intervals are better than NA as they produce shorter length and wider coverage.

Keywords: Metrology, Percentile, Confidence interval, Normal distribution.

1. Introduction

We are often dealing with data survey in measuring procedures and in most of the cases the data come from a normal distribution. Also, we are usually interested in obtaining point or intervals estimates containing some parameter of interest with a given probability which is named confidence level. In practice, we usually choose this level as 95% [1].

A *confidence interval* is a random interval (a, b) obtained from some observed data $X_1, X_2, \ldots, X_n$, in principle different from sample to sample, that frequently includes the value of an unobservable parameter of interest if the experiment is repeated. The *confidence level* $1 - \alpha$ of the interval measures the probability of containing the true parameter, that is, if we repeatedly take random samples using the same method with a 95% confidence level. For one given sample, we do not know whether the confidence interval covers or not the true parameter. The 95% probability only refers to the method that it is used, but not to the individual sample. The value α represents the error obtained when one claims that the probability that the true parameter lies in the interval (a, b) is $1 - \alpha$. In other words, as the estimates computed are random variables, one can expect that for some exceptional cases the interval do not contain the true parameter. It is proposed in this article confidence intervals for the 100p-th percentile $\kappa_p = \mu + Z_p\sigma$, where Z_p denotes the 100p-th percentile from the Standard Normal distribution i.e, $N(0,1)$, based on the work of Chakraborti and Li [2]. The idea is to contribute to the development of confidence intervals that are very much used in metrology studies. It is used 3 intervals coming from 3 different methods:

(i) Maximum-likelihood leading to the minimum variance unbiased estimator, which in turn provides a natural pivotal quantity. Then by approximating critical values (using the t_{n-1} approximation), it is obtained the so called T *interval.*

(ii) Use of a pivotal quantity based on a studentized version of the sample $100p$-th percentile (an order statistic), and the asymptotic normality of order statistics, called the *asymptotic normality interval* (NA).

(iii) A method proposed by Lawless, based directly on the maximum likelihood estimator (MLE); a biased estimator of κ_p, called the *Lawless interval* (LA).

2. Proposed confidence intervals

Let $X_1, X_2, \ldots, X_n$ be iid $N(\mu, \sigma^2)$, and $\kappa_p = \mu + Z_p\sigma$ as before. Chakraborti S. and LI, J. propose the following intervals:

- *The Traditional Interval* (T).

  Let $\bar{X}$ denote the sample mean and $S = \left(\frac{1}{n-1}\sum_{i=1}^{n}(X_i - \bar{X})^2\right)^{\frac{1}{2}}$ the sample standard deviation. Then the MLE of κ_p is $K_p = \bar{X} + Z_pS$ which is biased since $E(\bar{X}) = \mu$ but $E(S) = \frac{\sigma}{C}$, with

$$C = C(n) = \left[\frac{n-1}{2}\right]^{1/2}\left[\Gamma\left(\frac{n}{2}\right)\right]^{-1}\Gamma\left(\frac{n-1}{2}\right)$$

  Therefore $\hat{K}_p = \bar{X} + CZ_pS$ is unbiased for κ_p and its variance is given by

$$Var(\hat{K}_p) = \frac{\sigma^2}{n}\left[1 + nZ_p^2(C^2 - 1)\right]$$

  Finally using the t_{n-1} approximation we obtain the interval

$$\overline{X} + CZ_pS \mp t_{\frac{\alpha}{2},n-1}\frac{S}{\sqrt{n}}\sqrt{1 + nZ_p^2(C^2 - 1)} \tag{1}$$

- *The Asymptotic Normality Interval* (NA).

  This interval comes from the asymptotic normality of the order statistic $X_{(r)}$ with $r = np$, which is a natural point estimator of κ_p [3]. This interval is symmetrically placed like the T interval, but is centered at $X_{(r)}$. The proposed confidence interval is given by

$$X_{(r)} \mp t_{\frac{\alpha}{2},n-1}Se^{Z_p^2/2}\left[\frac{2\pi p(1-p)}{n}\right]^{1/2} \tag{2}$$

- The *Lawless Interval* (LA).

  Lawless [4] considered a confidence interval for κ_p based on the biased MLE $\hat{K}_p = \bar{X} + Z_pS$, by considering a pivotal variable following a noncentral t-distribution with $(n-1)$ degrees of freedom and noncentrality parameter $-\sqrt{n}Z_p$ [5]. The obtained interval is

$$\bar{X} \mp t'_{1-\frac{\alpha}{2},[n-1,(-\sqrt{n}Z_p)]}\frac{S}{\sqrt{n}} \tag{3}$$

3. Evaluating the confidence intervals

In the literature, there are two main tools used to evaluate confidence intervals (a, b), they are the length $b - a$ and the coverage probability $Prob(a \leq \kappa_p \leq b)$. Heuristically, if one wants to increase the probability that the interval contains the true parameter one must increase the length of the interval, and when the confidence level changes, say from 0.9 to 0.95, the confidence interval also changes. In general, the length of a confidence interval increases as the confidence level increases or as the standard error increases, but decreases as the sample size n increases. The same can not be said for the coverage, that is, there exists a trade-off between length and coverage probability of the interval.

The strategy, as usual, is choosing the interval whose length is as small as possible and whose coverage probability is at least equal to the confidence level.

4. Simulation study

It is shown via simulation of different scenarios the comparison of the confidence intervals proposed by Chakraborti and Li [2]. It is used the statistical software R-studio.

4.1. Procedure

Six random samples of sizes $n_1 = 25,\ n_2 = 50,\ n_3 = 100,\ n_4 = 200,\ n_5 = 500,\ n_6 = 1000$ were considered. For each sample size $m = 1000,\ m = 2000$ and $m = 5000$ random sample replications were generated according to the standard normal distribution $N(0, 1)$ and the confidence level used was 0.95. We remark that larger sample sizes do not reduce the inherent variability, but lead to narrower confidence intervals that are a basis for estimating the random component of uncertainty. The considered percentiles are $p_1 = 0.25,\ p_2 = 0.5,\ p_3 = 0.9$.
For the intervals T, NA and LA, denote LT, LNA and LLA as the average lengths, and CT, CNA and CLA as the coverages, respectively.

4.2. Simulation results

The results of the study are summarized below in tables 1, 2 and 3 for $p = 0.25,\ 0.5$ and 0.9 respectively. It is registered the average length and the coverage probabilities when the confidence level is 0.95. The highlighted lines in bold represent the best choice for length and coverage, and, of course, this is due to the fact that the biggest sample size was chosen, in fact when $n = 500$ and $n = 1000$ the coverage probabilities were very close to 95% and shorter lengths, the key point is that is does not depend on the number of replications. Now, for small sample sizes there is no evidence to say that the coverage is close to 95%.

Table 1. Lengths and coverages when the confidence level is 0.95% and $p = 0.25$

				$p = 0.25$			
		Length			Coverage		
n	*Replicas*	LT	LNA	LLA	CT	CNA	CLA
25	1000	0.9081	1.1115	0.9103	95.30	94.60	95.40
50		0.6288	0.7715	0.6295	95.70	95.50	95.90
100		0.4414	0.5423	0.4417	94.70	94.60	94.90
200		0.3087	0.3794	0.3087	94.60	95.70	94.60
500		0.1945	0.2392	0.1945	95.50	94.40	95.30
1000		**0.1375**	**0.1691**	**0.1375**	**95.30**	**94.5**	**95.10**
25	2000	0.9138	1.1185	0.9160	94.30	94.40	94.35
50		0.6298	0.7728	0.6305	95.85	94.75	95.35
100		0.4395	0.5399	0.4397	94.80	94.45	94.60
200		0.3092	0.3800	0.3092	94.30	94.85	94.25
500		0.1946	0.2393	0.1946	94.95	94.55	94.90
1000		**0.1375**	**0.1692**	**0.1376**	**94.90**	**94.75**	**94.78**
25	5000	0.9110	1.1150	0.9132	94.58	94.60	94.60
50		0.6270	0.7693	0.6277	94.46	94.70	94.74
100		0.4384	0.5386	0.4387	94.60	95.14	94.54
200		0.3089	0.3797	0.3090	95.58	95.6	95.52
500		0.1946	0.2393	0.1947	94.80	94.82	94.84
1000		**0.1375**	**0.1691**	**0.1375**	**94.82**	**94.88**	**94.72**

Table 2. Lengths and coverages when the confidence level is 0.95% and $p = 0.5$

				$p = 0.5$			
		Length			Coverage		
n	*Replicas*	LT	LNA	LLA	CT	CNA	CLA
25	1000	0.8101	1.0153	0.8101	94.70	93.20	94.70
50		0.5685	0.7125	0.5685	95.90	95.40	95.90
100		0.3960	0.4963	0.3960	95.80	94.90	95.80
200		0.2781	0.3486	0.2781	95.20	95.60	95.20
500		0.1757	0.2202	0.1757	95.10	95.30	95.10
1000		**0.1240**	**0.1554**	**0.1240**	**95.10**	**95.00**	**95.10**
25	2000	0.8137	1.0198	0.8137	95.45	93.60	95.45
50		0.5653	0.7085	0.5653	95.05	95.00	95.05
100		0.3947	0.4947	0.3947	95.20	95.30	95.20
200		0.2783	0.3488	0.2783	95.45	95.40	95.45
500		0.1756	0.2201	0.1756	95.60	94.55	95.60
1000		**0.1240**	**0.1555**	**0.1240**	**95.45**	**95.55**	**95.45**
25	5000	0.8166	1.0234	0.8166	95.24	94.04	95.24
50		0.5646	0.7077	0.5646	94.88	95.22	94.88
100		0.3957	0.4960	0.3957	95.54	94.88	95.54
200		0.2784	0.3489	0.2784	95.12	95.44	95.12
500		0.1756	0.2201	0.1756	94.72	95.02	94.72
1000		**0.1241**	**0.1555**	**0.1241**	**95.28**	**94.80**	**95.28**

Table 3. Lengths and coverages when the confidence level is 0.95% and $p = 0.9$

				$p = 0.9$			
		Length			Coverage		
n	*Replicas*	LT	LNA	LLA	CT	CNA	CLA
25	1000	1.1093	1.3889	1.1170	95.20	92.20	95.20
50		0.7644	0.9627	0.7669	92.60	93.60	93.60
100		0.5348	0.6755	0.5357	94.10	93.90	94.10
200		0.3770	0.4768	0.3773	94.70	95.20	95.00
500		0.2373	0.3004	0.2373	94.50	94.40	94.80
1000		**0.1675**	**0.2120**	**0.1675**	**95.70**	**94.60**	**96.10**
25	2000	1.1174	1.3990	1.1252	95.30	91.55	95.35
50		0.7674	0.9664	0.7699	94.85	94.10	94.85
100		0.5369	0.6782	0.5378	95.30	94.45	95.45
200		0.3765	0.4763	0.3768	95.35	94.80	95.40
500		0.2370	0.3000	0.2371	95.60	94.15	95.65
1000		**0.1674**	**0.2119**	**0.1674**	**94.50**	**94.85**	**94.75**
25	5000	1.1149	1.3958	1.1227	94.70	91.40	94.46
50		0.7681	0.9673	0.7706	95.58	94.88	95.56
100		0.5357	0.6766	0.5365	95.08	94.32	94.94
200		0.3762	0.4759	0.3765	95.42	95.08	95.24
500		0.2372	0.3003	0.2373	95.36	95.08	95.24
1000		**0.1675**	**0.2121**	**0.1676**	**95.12**	**94.92**	**95.08**

In order to visualize better, the results are displayed in figure 1 below. From this figure 1, it is possible to see that in terms of length, the worst interval is NA as it has a wider length, but the other two have quite similar behavior. Moreover, figure 2 shows that the T and LA intervals have better coverage than the NA since they have less variability in coverage and are concentrated around the level of significance.

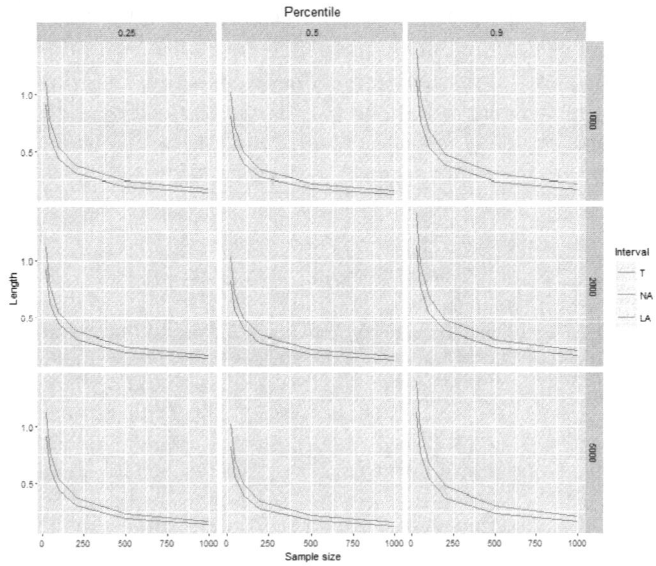

Figure 1. Length vs Sample size.

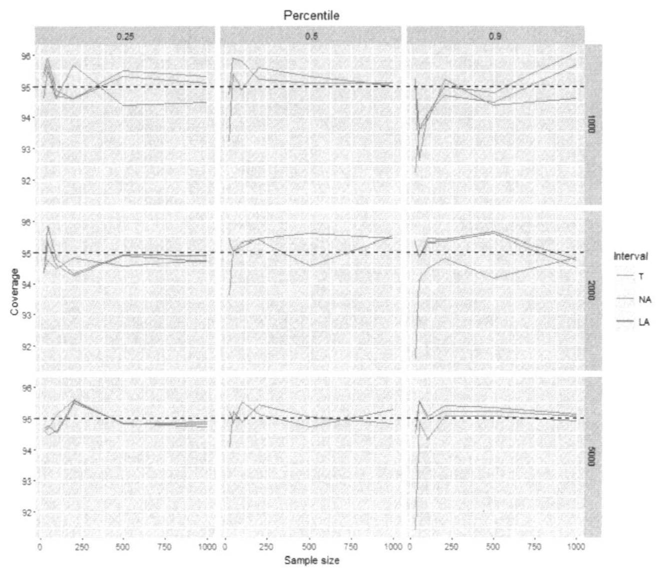

Figure 2. Coverage vs Sample size.

IOP Conf. Series: Journal of Physics: Conf. Series **1044** (2018) 012016 doi:10.1088/1742-6596/1044/1/012016

5. Conclusion

Based on the results obtained by simulation, in this study, it is possible to conclude that the least efficient interval among all three considered in this paper is the NA interval. Therefore, we recommend to use the T an LA intervals which have a better performance in the sense that they have shorter length and wider coverage.

However, for theoretical and practical aspects related to measurements it should be considered all possibilities contributing to the development of confident results in metrology applications. For example, the quality of measurement is quantified by means of evaluation of uncertainty of such a measurement [6], in particular when one is dealing with uncertainty problems where data follows a normal distribution. In this case, one can easily work with the intervals T and LA, as they seem to produce better and satisfactory results. For future works it is recommended the use the proposed methodology applied in real data rather than simulated data.

References

[1] ABNT 2010 Estatística-Vocabulário e símbolos Parte 1: Termos estatísticos gerais e termos usados em probabilidade *NBR ISO, RJ* **3534-1** 69

[2] Chakraborti S and Li J 2007 Confidence Interval Estimation of a Normal Percentile *Amer. Statist.* **61** 331-336

[3] Gibbons J D and Chakraborti S 2003 Nonparametric Statistical Inference *New York: Marcel Dekker* **168** xxiv+645

[4] Lawless, Jerald F 2003 Statistical Models and Methods for Lifetime Data *Wiley-Interscience [John Wiley & Sons], Hoboken, NJ* **2** xxii+630

[5] Owen D B 1968 A Survey of Properties and Applications of the Noncentral t-Distribution, *Technometrics* **10** 445-478

[6] Pérez H M 2012 Estimación de incertidumbres Guía GUM *Revista Española de Metrología* **13** 114-130.

IOP Conf. Series: Journal of Physics: Conf. Series **1044** (2018) 012017 doi:10.1088/1742-6596/1044/1/012017

Control chart analysis for interferometric length measurement

L Alvarez[1], E Beer, G Giarmana and K Bastida
Instituto Nacional de Tecnología Industrial (INTI), Av. Gral. Paz 5445, San Martín, Buenos Aires, Argentina

E-mail: alvarez@inti.gob.ar

Abstract. INTI is the national metrology institute of Argentina and the responsible for maintenance and the dissemination of the national length standard. The practical realization of this reference is performed according to the optical wavelength standards established and published by the International Committee of Weights and Measures (CIPM). The length unit is materialized by artefacts known as gauge blocks which length is measured by optical interferometric methods. For this purpose INTI has a commercial interferometer based on a Twyman-Green configuration. By this system INTI provides length traceability to national calibration laboratories and others national metrology institutes of the region. Even though the equipment that forms the interferometric system is calibrated periodically, it is convenient to establish quality control procedures for monitoring not only the system as a whole but also the calibration process. We report on a 5-year study of two control steel gauge block sets, the statistical techniques applied on control chart and its selection criteria. The pool of data allows long term stability analysis of these standards as well as the operating examination of the interferometer.

1. Introduction

According to the International System of Units (SI), the metre is defined as "the length of the path travelled by light in vacuum during a time interval of 1/299 792 458 second"[1]. In Argentina, INTI maintains and disseminates the national length standard. The practical realization of this unit is performed according to the optical wavelength standards established and published by the International Committee of Weights and Measures (CIPM) [2]. The materialization of the metre is carried out by measuring gauge blocks of different nominal lengths applying interferometric techniques which use optical radiations of known wavelength traceable to the realization of the metre. This is really the first link in the length traceability transfer for the national and regional industry [3], impacting the quality of length measurements associated with industrial processes where demanding mechanical tolerance values are required, such as the automotive, air and nuclear industries, among others.

INTI offers an interferometric gauge block calibration service by operating a commercial interferometer based on a Twyman-Green configuration. This NPL-TESA automatic gauge block

[1] To whom any correspondence should be addressed.

interferometer (AGI 1/300) uses two stabilized He-Ne lasers of wavelengths of 632 nm and 543 nm. To measure the central length of gauge blocks by this technique one of the gauge face has to be wrung on an auxiliary platen [4]. The temperature has a measurable effect on the gauge block length, so in order to guarantee thermal stabilization at 20 ºC of this assembly inside the interferometer a temperature control system is used in laboratory. The expanded uncertainties considered for a single measurement of a gauge block correspond to that published in Appendix C database at BIPM website (http://kcdb.bipm.org/appendixC/). The measurement uncertainty is evaluated taking into account components attributed to factors which influence the measurement [5].

Interferometric techniques are highly sensitive to environmental conditions since light beams propagate in air. For this reason it is necessary to use high precision instruments that allow monitoring relevant environmental conditions inside the interferometer. Even though these equipments are periodically calibrated, it is convenient to have certain quality control procedures for monitoring and detecting anomalies in the measurement system immediately [6]. We report on a 5-year study of two control steel gauge block sets. The control chart analysis is based on the warning and control limits determined by statistics. The pool of data allows long term stability analysis of the control gauge block sets as well as the operating examination of the interferometer, i.e. whether the measurement system is in-control or out-of-control state.

2. Control gauge block sets

Two control steel gauge block sets were implemented in order to carry out the control chart analysis. High accuracy gauge blocks in grade K were used for the selection of the sets and a series of tests for evaluating the wringability were performed. These sets of gauge blocks are measured periodically and the results are analyzed for establishing the warning and control limits corresponding to each block in the sets.

2.1. Fixed wringing control gauge blocks (FWCGB)

It is a set of ten gauge blocks in the range of nominal length from 0.5 mm to 100 mm that are wrung to a specific auxiliary reference platten which is permanently left inside the calibration cabin of the interferometer. In this way the calibration face is always the same assuming also a constant wringing film.

2.2. Variable wringing control gauge blocks (VWCGB)

Initially it was a set of two gauge blocks with 100 mm-nominal length and three with 300 mm-nominal length. Nowadays we have included five gauge blocks more in the range from 100 mm to 300 mm. Unlike the previous set, these gauge blocks are not permanently wrung to a reference auxiliary platten. Due to their aspect ratio most of these gauge blocks are more susceptible to possible turning over than short gauge blocks, therefore they are wrung to the reference surface with the platen localized into the interferometer measurement cabin every time they are measured. The calibration of each of the faces is also performed periodically and alternating the measuring faces between calibrations.

3. Control charts, criteria and results

Statistical treatment of the measurements is carried out in an analogous way for both control gauge block sets. Most of the measurements were made by the same operators and each of them was measured only once in each series. In addition, dimensional stability limits established by ISO 3650 are taken into account. The specification for relative stability, for K-grade gauge block, is 0.25×10^{-6} per year, so it takes a long time to measure dimensional stability in this kind of gauge blocks.

The following criteria for the control charts analysis are considered:

1. If the measured value falls outside the established control limits, it would be an outlier so the calibration is repeated. If the outlier is obtained again, others gauge blocks of the same nominal

IOP Conf. Series: Journal of Physics: Conf. Series **1044** (2018) 012017 doi:10.1088/1742-6596/1044/1/012017

length are calibrated. If the outlier does not repeat, a problem intrinsic to the first block measured is assumed and the first measurement then is discarded. If the same behaviour is observed then proceed according to 2.

2. If the results of the whole set of control gauge blocks fall outside the established control limits, i.e. all the calibrations show outliers, the service is suspended and the operation of the interferometric system as a whole is evaluated.

3.1. General control charts

Assuming a normal distribution, the warning (WL) and control limits (CL) are determined by σ and 2σ, respectively, i.e. the probability of the measured value falling within the upper and lower limits is 68 % and 95 % [6-7]. The centreline (CL) is determined by the mean value. Figure 1 shows the measurement history in the deviation of nominal length for the 0,5 mm steel gauge block between 2012 and 2017.

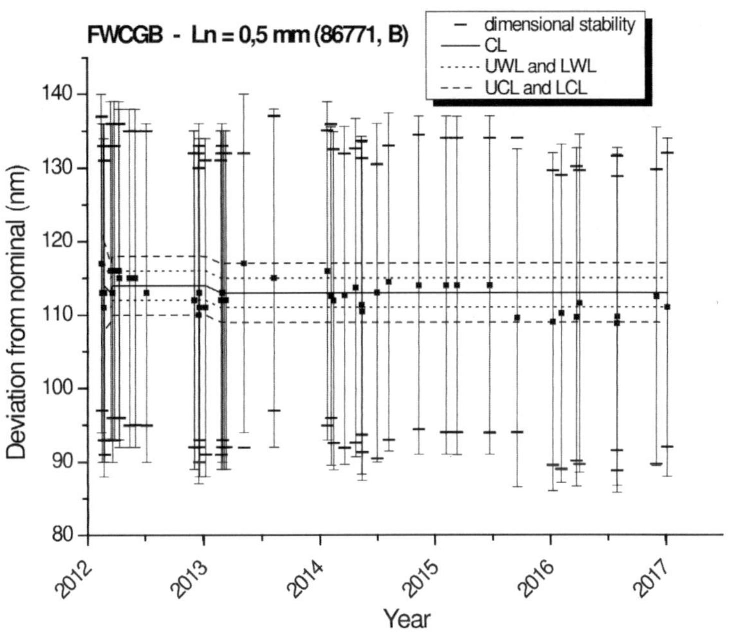

Figure 1. Measurement history for the 0,5 mm steel fixed wringing control gauge block. Error bars show expanded uncertainty (k=2).

This fixed wringing control gauge block was manufactured several years before this control measurement was started hence the data show the 0,5 mm control gauge block is very stable, as it was expected.

3.2. Control charts with drift

In some cases the analysed control charts show temporal drift. In this case the criteria of 3.1 will not be effective to detect anomalies in the measurement system hence the control chart is determined by standard regression analysis. All the data with drifts are analysed assuming that the distribution of the *y*-variable at each value of *x* is normal with the same variance, but no assumptions are made about the distribution of the *x*-variable. The drift is not included in the measurement uncertainty of the deviation from the nominal length. The estimated standard errors of the slope, $\hat{\beta}_1$, and the fitted intercept coefficient, $\hat{\beta}_0$, are calculated by:

IOP Conf. Series: Journal of Physics: Conf. Series **1044** (2018) 012017 doi:10.1088/1742-6596/1044/1/012017

$$S\hat{E}(\hat{\beta}_1) = \sqrt{s^2/\sum_{i=1}^{n}(X_i - \bar{X})^2} \quad \text{and} \quad S\hat{E}(\hat{\beta}_0) = \sqrt{s^2\left[\frac{1}{n} + \frac{\bar{X}^2}{\sum_{i=1}^{n}(X_i - \bar{X})^2}\right]} \tag{1}$$

Two types of intervals are determined:

- Confidence bands. They combine the confidence intervals of the slope and intercept of the regression function. It determines how precisely the data define the best-fit line. Thus the confidence interval bounds at $X = X_0$ are given by:

$$\hat{Y}_0 \pm t_c\sqrt{s^2\left[\frac{1}{n} + \frac{(X_0-\bar{X})^2}{\sum_{i=1}^{n}(X_i-\bar{X})^2}\right]} \tag{2}$$

  where the square root term is the standard error that accounts for the uncertainty in estimating the true mean of Y at $X = X_0$, and t_c is the appropriate value from the t distribution with $n - 2$ degrees of freedom.

- Prediction bands. This interval is wider than the associated confidence interval because the scatter of data about the regression line is more important. These intervals represent the uncertainty of predicting the value of a single future observation, the new observation, and the uncertainty associated with the regression parameters as well. The prediction interval bounds at $X = X_0$ are given by:

$$\hat{Y}_0 \pm t_c\sqrt{s^2\left[1 + \frac{1}{n} + \frac{(X_0-\bar{X})^2}{\sum_{i=1}^{n}(X_i-\bar{X})^2}\right]} \tag{3}$$

The most cases which present drifts correspond to gauge blocks of nominal lengths below of 10 mm. Figure 2 shows the measurement history in the deviation of nominal length for the 5 mm steel gauge block between 2011 and 2017.

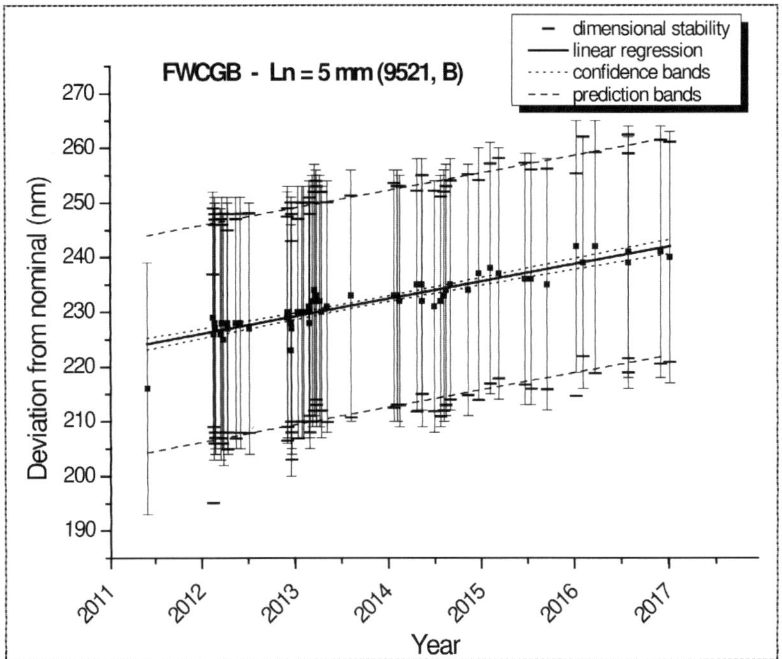

Figure 2. Measurement history for the 5 mm steel fixed wringing control gauge block. Error bars show expanded uncertainty (k=2).

There appears to be a long-term increase in length since this gauge block exhibits lengthening of approximately +3 nm year^{-1}.

Length instabilities in gauge block measurement could be attributed to some details on the manufacturing and hardening process of the gauge block, in which case the gauge blocks would exhibit lengthening [8]. Contrary to the observed on previous figure, Figure 3 shows a contraction of around -3 nm year^{-1}. This long-term decrease in length corresponds to the 9 mm steel gauge block between 2011 and 2017.

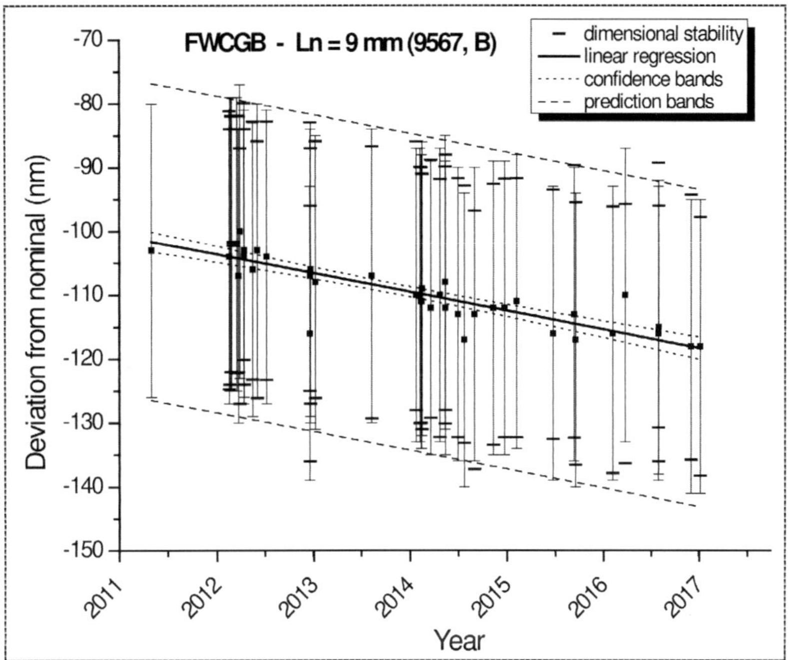

Figure 3. Measurement history for the 9 mm steel fixed wringing control gauge block. Error bars show expanded uncertainty (k=2).

Drifts associated with contractions of length are more uncertain. The drifts observed in the fixed wringing control gauge blocks vary between -1 nm year^{-1} and +3 nm year^{-1}. Since all the FWCGB were manufactured several years before this control measurement was started, drifts are suspected to be caused by variation of the wringing film over the time. In order to verify this hypothesis some FWCGB with temporal drift will be removed and re-wrung on the platen to be measured again.

3.3. Other control charts

Some VWCGB were very new when they were included as control gauge blocks. In these cases the results show how stability length improves with age. Figure 4 shows the measurement history in the deviation of nominal length for one of the 300 mm steel gauge block between 2011 and 2017. The first data point corresponds to manufacturing certificate and it is included only as a reference value. Between the end of 2012 and the end of 2014 the results evidence a negative variation in length of approximately -50 nm year^{-1}. In the last years the length seems to be reaching temporal stability. Both criteria 3.1 and 3.2 are applied in this case as the control limits show in Figure 4. In this case, the confidence and prediction bands coincide, i.e. due to the few data points considered for the regression, scatter is the main source of uncertainty hence the width of the confidence limits enlarges.

On the other hand the analysis of this particular control chart allowed anomaly detection on the temperature sensors in 2012. The event was verified by outliers measured for all the VWCGB and proportional to nominal lengths.

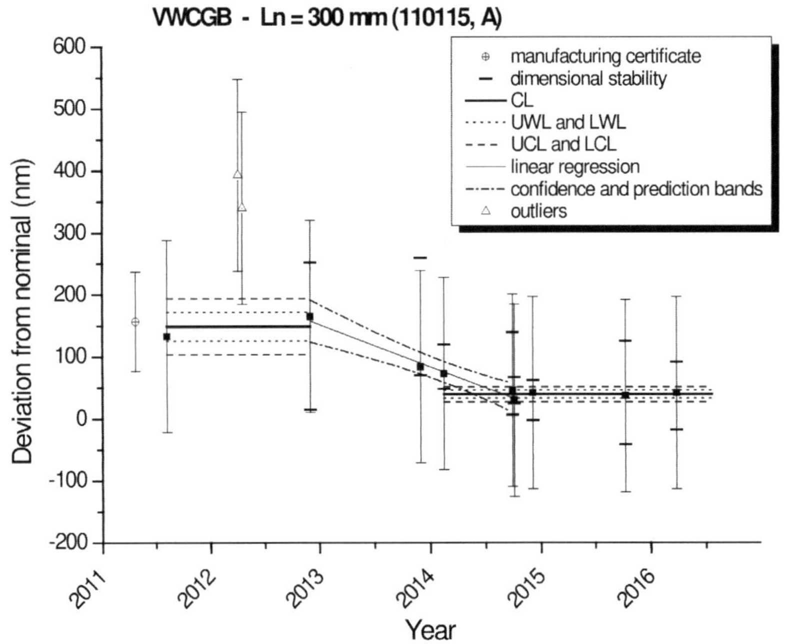

Figure 4. Measurement history for the 300 mm steel variable wringing control gauge block. Error bars show expanded uncertainty (k=2).

4. Conclusion

A quality control procedure was effectively established for monitoring interferometric length measurement. The performance of two control gauge block sets over 5 years has been analysed and selection criteria to accept or reject a measured value based on historical measurements have been established. The pool of data also allows for quantitative examination of temporal stability of the gauge blocks in the set analysed as well as the long-term performance of the interferometer as a global system.

5. References

[1] Documents concerning the new definition of the metre 1984 *Metrologia* vol 19 pp 163-177

[2] Quinn T J 1992 *Mise en pratique* of the definition of the metre *Metrologia* vol 30 pp 523-541

[3] BIPM, IEC, IFCCC, ILAC, ISO,, IUPAC, IUPAP and OIML 2007 International Vocabulary of Metrology-Basic and General Concepts and Associated Terms (VIM) ISO/IEC Guide 99:2007 (Geneva, Switzerland: International Organization for Standardization)

[4] ISO3650 1998 Geometrical Product Specifications (GPS)-Length Standards-Gauge Blocks (Geneva, Switzerland: International Organization for Standardization)

[5] Decker J E and Pekelsky J R 1996 *Metrologia* vol 34 pp 479-493

[6] Doiron T and Beers J S 1995 The Gage Handbook US Department of Commerce, Technology Administration, National Institute of Standards and Technology

[7] Chen-Yun H, Gwo-Sheng P and Kam-Wa L 2014 How to Select the Appropriate Type of Control Chart in Metrology NCSL International Workshop & Symposium

[8] Lewis A J, Hughes B, and Aldred P J 2010 Long-term study of gauge block interferometer performance and gauge block stability *Metrologia* vol 47 pp 473-486

IOP Conf. Series: Journal of Physics: Conf. Series **1044** (2018) 012018 doi:10.1088/1742-6596/1044/1/012018

Convergence analysis of the extended Kalman filter used in the ultrasonic time-of-flight estimation

R Gouveia, J Villanueva, F Santos and J. Silva
Department of Electrical Engineering, Federal University of Paraiba, JP 58051-900, Brazil
E-mail: rafael.gouveia@cear.ufpb.br, jmauricio@cear.ufpb.br, felipeaugusto.santos@cear.ufpb.br and joao.junior@cear.ufpb.br

Abstract. The ultrasonic Time-of-Flight (ToF) estimation may be achieved using different algorithms, such as signal processing techniques, artificial intelligence and, statistical estimators, among others. These algorithms have a highlight point, which is the guarantee of the estimation reliability through the convergence of the results and low estimation uncertainty. Hence, this work aims to perform the convergence analysis of the Extended Kalman Filter (EKF) algorithm to ToF estimation, with application in wind speed measurement. Therefore, a state space model of a delayed sine wave was constructed. The modeling used shows the influence of time-varying parameters, which determine the convergence of the states. By analyzing the convergence of the algorithm, it was possible to determine the range of variation of the model parameters to guarantee the final estimation results. Through the construction of a computational model using Matlab@Simulink are presented the simulation results to wind speed measurement.

1. Introduction

In the energy sector, there are numerous sources of energy available that can be used, standing out in the renewable ones, such as wind. Wind energy uses the strength of the wind to generate electricity through wind turbines installed in strategic locations. To determine the best locations for the installation of wind turbines, it is essential to monitor the wind speed, thus determining statistical parameters and constructing a model of it. However, it is required that the sensors used have characteristics of low uncertainty, less time constant for the detection of gusts of wind and with possibilities of operation in severe environments, such as rainfall, snow, and solar radiation. Thus, among the wind speed measurement systems that stand out are the ultrasonic anemometers. These instruments indirectly determine the wind speed by means of the ToF ultrasonic transit time, which is defined as the travel time of the ultrasonic wave from its emission on a transmitting transducer to its detection on a receiving transducer. In the bibliography, several methods are found for ToF estimation, such as threshold detection, phase difference, cross correlation, and Kalman filter.

In [1] and [2], the ToF estimation was approached using the Extended Kalman Filter (EKF) algorithm, by means of the estimation of states associated with a delayed sine wave. One of the limitations of this approach was the model's dependence on time-varying parameters. In this scenario, the objective of this work is to analyze the convergence of the EKF regarding the time-varying parameters of the model, with the purpose of adjusting the parameters of the model to a ToF estimation with low uncertainty.

IOP Conf. Series: Journal of Physics: Conf. Series **1044** (2018) 012018 doi:10.1088/1742-6596/1044/1/012018

2. Proposed procedure: State space model

Figure 1 shows a measurement configuration using ultrasonic transducers, positioned in a test section of a wind tunnel with laminar flow and Reynolds number less than 2000 [3]. In this configuration, the wind speed can be found by:

$$\vartheta = \frac{1}{\cos\theta}\left(\frac{L}{t_{AB}} - C\right) \tag{1}$$

where: L is the distance between the transducers, θ is the angle between the wind direction and the alignment of the transducers, t_{AB} is the time of flight, C is the speed of sound in the air that depends on the temperature (T_K in Kelvin) given by $C = 20.074\sqrt{T_K}$.

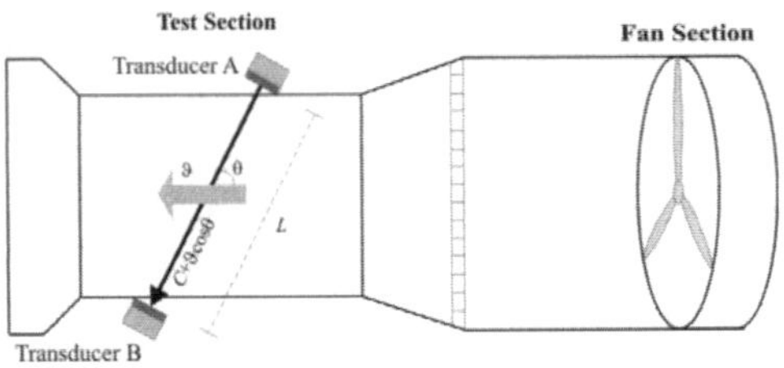

Figure 1. Wind tunnel – Test section for wind speed measurement.

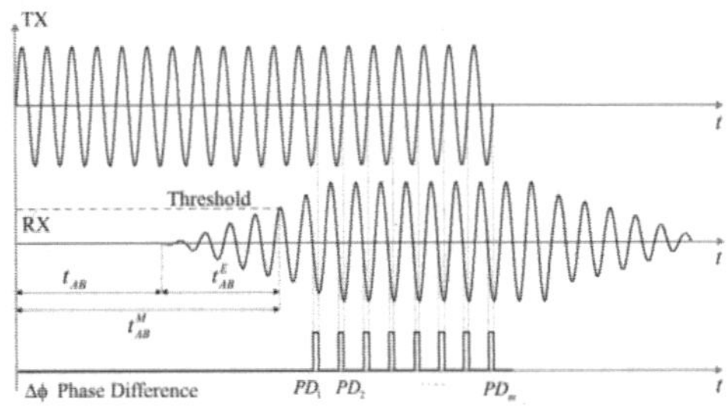

Figure 2. Measurement ToF using threshold detection and phase difference.

Figure 2 shows the transmitted (TX) and received (RX) signals. In this figure, it is possible to identify the ToF and the techniques for its estimation (threshold detection and phase difference [2] [5]). Considering that the signal received on the transducer RX, $u(t)$, is a sum of sinusoidal signals composed by: a main ultrasonic signal, $y(t)$, reflected ultrasonic signals with attenuation and delay, and additive noise, given by:

$$\begin{aligned} u(t) &= y(t) + \sum_{i=1}^{\infty} A_i \sin\left(\omega_i (t - t_{ToFi}) + \phi_i(t)\right) + \eta(t) \\ y(t) &= A(t)\sin\left(\omega(t)(t - t_{ToF}(t)) + \phi(t)\right) \end{aligned} \tag{2}$$

Where: A, ω, ϕ and t_{TOF} are the amplitude, frequency, phase and time-of-flight of the main ultrasonic signal; A_i, ω_i, ϕ_i and t_{TOFi} are the amplitude, frequency, phase and time-of-flight of the reflected ultrasonic signal; t is the time and η is the additive noise.

From equation (2), we have built a nonlinear state space model [1], [4], given by:

$$\begin{aligned} \dot{\mathbf{x}}(t) &= \mathbf{f}(\mathbf{x}, t) + \mathbf{w}(t) \\ \mathbf{y}(t) &= \mathbf{g}(\mathbf{x}, t) + \mathbf{v}(t) \end{aligned} \tag{3}$$

Where: the states of the model are amplitude $x_1 = A$, frequency $x_2 = \omega$, phase $x_3 = \phi$ and the time-of-flight $x_4 = t_{ToF}$; $\mathbf{w}$ is the noise of the states with mean zero and standard deviation σ_w; $\mathbf{v}$ is the noise of the measurements with mean zero and standard deviation σ_v, and the nonlinear functions $\mathbf{f(x}, t)$ and $\mathbf{g(x}, t)$ are given by:

$$\mathbf{f}(\mathbf{x}, t) = \begin{bmatrix} \mu_1 (u - x_1 \sin\beta)\sin\beta \\ \mu_2 x_1 (u - x_1 \sin\beta)\cos\beta \\ \mu_3 x_1 (u - x_1 \sin\beta)\cos\beta \\ \dfrac{\mu_3 - \mu_2 x_4}{x_2 . x_1^{-1}}(u - x_1 \sin\beta)\cos\beta \end{bmatrix} \quad \text{and} \quad \mathbf{g}(\mathbf{x}, t) = \begin{bmatrix} x_1 \sin\beta \\ \hat{t}_{ToF} \end{bmatrix} \quad \text{and } \beta = x_2(t - x_4) + x_3 \tag{4}$$

In Equation (4) was defined the variables μ_1, μ_2, and μ_3, which are associated with the convergence of the Extended Kalman Filter (EKF) used for the estimation of the states from the measurements of the received signal amplitude and the ToF obtained by the phase difference technique [1]. In this work, it will be analyzed and defined the criteria to obtain the ranges of values of the variables which lead the estimator to a quick convergence and with the lowest standard deviation of the estimation.

3. Simulation results: ToF estimation using EKF

In the interest of implementing the Kalman Filter algorithm for ToF estimation, the Equation (4) was discretized considering a sampling period, T_s. Hence, a discrete state space system was obtained, as illustrated below:

$$\begin{aligned} \mathbf{x}_{k+1} &= \mathbf{f}(\mathbf{x}_k, k) + T_s \mathbf{w}_k \\ \mathbf{y}_k &= \mathbf{g}(\mathbf{x}_k, k) + T_s \mathbf{v}_k \end{aligned} \quad (5)$$

Where: $\mathbf{x}_k$ is the n-dimensional vector of process states; $\mathbf{y}_k$ is the m-dimensional vector of current measurements; $\mathbf{w}_k \sim N(0, \mathbf{Q}_k)$ and $\mathbf{v}_k \sim N(0, \mathbf{R}_k)$ are the sequences of uncorrelated Gaussian noise; $\mathbf{f}(\mathbf{x}_k, k)$ and $\mathbf{g}(\mathbf{x}_k, k)$ are nonlinear functions in discrete-time kT_s, given by:

$$\mathbf{f}(\mathbf{x}_k, k) = \begin{bmatrix} x_1 + T_s u \sin\beta_k - T_s x_1 \sin^2\beta_k \\ x_2 + T_s x_1 u \cos\beta_k - T_s x_1^2 \sin\beta_k \cos\beta_k \\ x_3 + T_s x_1 u \cos\beta_k - T_s x_1^2 \sin\beta_k \cos\beta_k \\ x_4 + T_s x_1 \left(\dfrac{1 - x_4}{x_2} \right) (u - x_1 \sin\beta_k) \cos\beta_k \end{bmatrix} \quad \text{and} \quad \mathbf{g}(\mathbf{x}_k, k) = \begin{bmatrix} x_1 \sin\beta_k \\ x_4 \end{bmatrix} \quad (6)$$

Through the equations (5) and (6) it concludes that the state space system is nonlinear; therefore, to perform the estimation of the states, the extended Kalman filter (EKF – Extended Kalman Filter) should be used. Table 1 shows the EKF algorithm to Time-of-Flight (ToF) estimation, where the states are defined by the amplitude (A), frequency (ω), phase (ϕ) and time-of-flight (ToF).

Table 1. EKF algorithm for the ToF estimation.

Initialization of states: $\mathbf{x}_0 = \begin{bmatrix} A_0 \\ \omega_0 \\ \phi_0 \\ ToF_0 \end{bmatrix}$ $\mathbf{P}_0 = \begin{bmatrix} \sigma^2_{\varepsilon A} & 0 & 0 & 0 \\ 0 & \sigma^2_{\varepsilon\omega} & 0 & 0 \\ 0 & 0 & \sigma^2_{\varepsilon\phi} & 0 \\ 0 & 0 & 0 & \sigma^2_{\varepsilon ToF} \end{bmatrix}$
State estimate propagation: $\hat{\mathbf{x}}_{k\|k-1} = \mathbf{f}(\hat{\mathbf{x}}_{k-1})$ Error covariance propagation $\mathbf{P}_{k,k-1} = \left[\dfrac{\partial \mathbf{f}(\hat{\mathbf{x}}_{k-1})}{\partial \mathbf{x}_{k-1}} \right] \mathbf{P}_{k,k-1} \left[\dfrac{\partial \mathbf{f}(\hat{\mathbf{x}}_{k-1})}{\partial \mathbf{x}_{k-1}} \right]^T + \mathbf{H}(\hat{\mathbf{x}}_{k-1}) \mathbf{Q}_{k-1} \mathbf{H}^T(\hat{\mathbf{x}}_{k-1})$ Kalman gain matrix $\mathbf{G}_k = \mathbf{P}_{k,k-1} \left[\dfrac{\partial \mathbf{g}(\hat{\mathbf{x}}_{k\|k-1})}{\partial \mathbf{x}_k} \right]^T \cdot \left[\left[\dfrac{\partial \mathbf{g}(\hat{\mathbf{x}}_{k\|k-1})}{\partial \mathbf{x}_k} \right] \mathbf{P}_{k,k-1} \left[\dfrac{\partial \mathbf{g}(\hat{\mathbf{x}}_{k\|k-1})}{\partial \mathbf{x}_k} \right]^T + \mathbf{R}_k \right]^{-1}$ Error covariance update $\mathbf{P}_{k,k} = \left[\mathbf{I} - \mathbf{G}_k \left[\dfrac{\partial \mathbf{g}(\hat{\mathbf{x}}_{k\|k-1})}{\partial \mathbf{x}_k} \right] \right] \mathbf{P}_{k,k-1}$ State estimate update $\hat{\mathbf{x}}_{k\|k} = \hat{\mathbf{x}}_{k\|k-1} + \mathbf{G}_k \left[\mathbf{y}_k - \mathbf{g}(\hat{\mathbf{x}}_{k\|k-1}) \right]$

Figure 3 shows the result of the application of the EKF algorithm on a received signal $u(t)$ and the estimation of the parameters of the ultrasonic signal $y(t)$. The estimated parameters by the EKF algorithm are associated with the estimated signal given by equation (2), such as: $x_1 = A$, $x_2 = \omega$, $x_3 = \phi$ and $x_4 = ToF$. Even with satisfactory simulation results, the convergence of the EKF algorithm may depend on the values assigned to the variables μ_1, μ_2, e μ_3, which transform the system into a time variant system. It can be noticed that even if the received signal has the influence of noise, the Kalman filter estimates the states in a robust way.

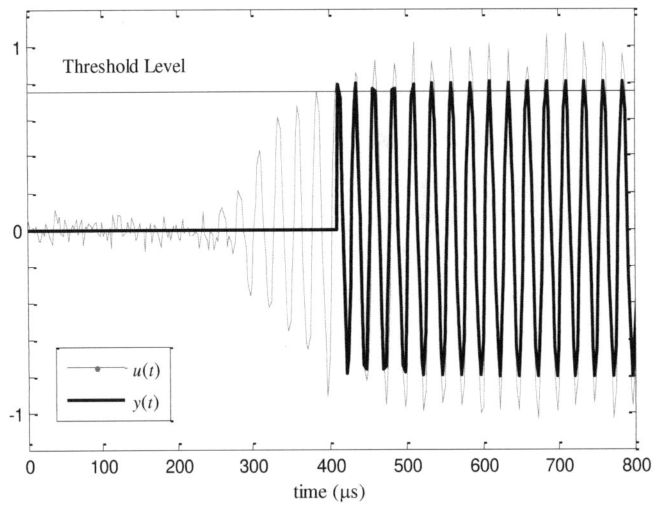

Figure 3. Parameters estimated of the ultrasonic signal using the EKF algorithm.

4. Convergence analysis of the EKF algorithm for ToF estimation

To perform the convergence analysis of the EKF with respect to the model parameters in state spaces, μ_1, μ_2, and μ_3, a computational model was constructed to measure wind speed using Matlab@Simulink. The ultrasonic transducers were modeled by means of bandpass Butterwoth filters. An additive noise of the Gaussian type with a standard deviation of 0.01 mv was considered. ToF represents the delay of the ultrasonic wave from the emission until its detection. In the simulations, a ToF equal to 228.931 μs was used (Equation (1)), for a wind speed of 10 m/s, temperature of 40°C, distance between the transducers of 83 mm, and angle of inclination of $\pi/4$.

Figure 4 the results of the ToF estimations are illustrated by the EKF algorithm. The variations of μ_1, μ_2, and μ_3 were considered in the range of 1 to 10000. It is observed that the ToF estimates do not change even with the variations of μ_1 and μ_2. In these simulations, an uncertainty was obtained in the ToF estimation of 0.5 μs, defining a confidence interval in which the result of the theoretical ToF is contained [6].

On the other hand, in the same figure, it was observed that the ToF estimates undergo changes in their estimation in relation to the increase of μ_3, increasing the uncertainty in the estimation as well. In this simulation, an uncertainty of 2 μs was obtained in the ToF estimation. From this result, it can be verified that the convergence of the EKF algorithm depends on the values of the model parameters (Equations (5) and (6)). Consequently, the uncertainty associated with the state estimate $x_4 = ToF$ can be compromised for certain values of these parameters.

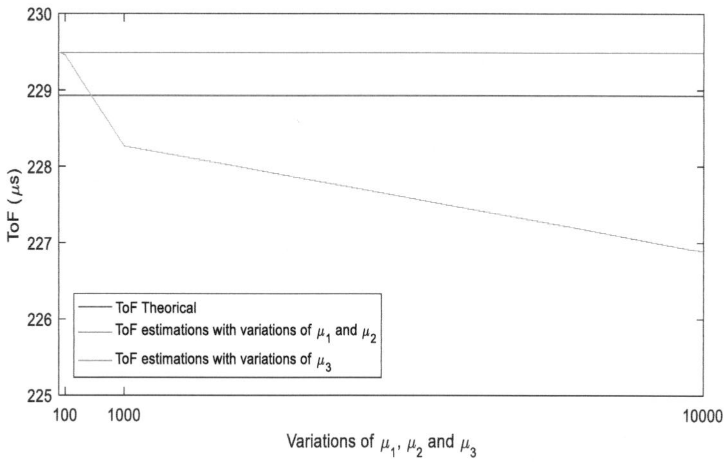

Figure 4. Estimation of ToF by EKF for different variations of μ_1, μ_2, and μ_3.

In figure 5, the convergence of the EKF for the estimation of the state associated with the ToF is shown. Considering $\mu_1 = 1$, $\mu_2 = 1$ and variations of $\mu_3 = \{1, 10^3, 10^4, 10^6, 10^7\}$. It is possible to observe that the standard deviation decreases according to the number of EKF iterations; however, for greater values of μ_3 it is realized that the EKF algorithm does not converge to a minimum value of the state variance x_4 (ToF). It may also be noted that the minimum standard deviation value was obtained for $\mu_3=1$.

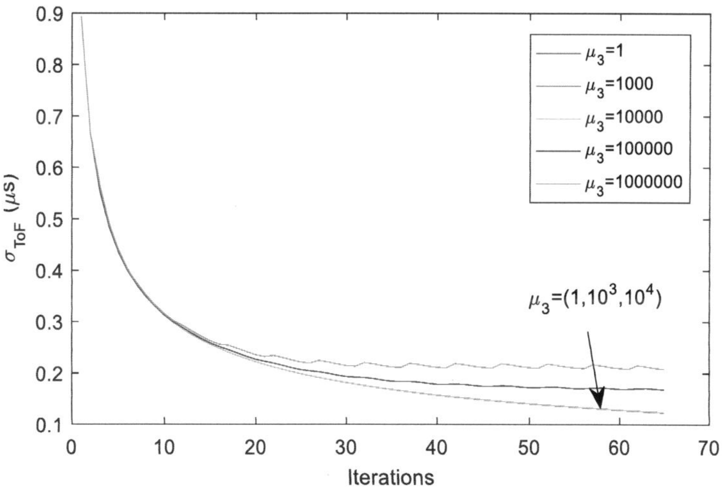

Figure 5. Standard deviation of ToF estimation through EKF.

5. Conclusions

In this work, the algorithm for the estimation of the ultrasonic transit time based on the extended Kalman filter (EKF) was developed. It was observed that the state model is nonlinear and time variant. With respect to the nonlinear characteristic, the extended Kalman filter allows an approximation by means of a linearization process. Regarding the time-varying characteristic, the model depends on the variables μ_1, μ_2, and μ_3, inducing the result obtained in the ToF estimation process, as well as showing the increase in the uncertainty of the estimated value. This result can be explained by the fact that the EKF algorithm does not converge for certain values of these variables, making it unstable. As result of this analysis is the determination of the possible variations of variables μ_1, μ_2, and μ_3, in which the EKF algorithm presents convergence to a ToF value and with low uncertainty.

Acknowledgments
The authors are thankful to the Pró-Reitoria de Pesquisa da Universidade Federal da Paraíba (PROPESQ) for the financial support to develop this research.

References

[1] Villanueva J, Catunda S, Freire R, Costa M and Ingaroca 2013 Wind speed measurement based on time-of-flight estimation using Extended Kalman Filter *IEEE Intern. Inst. and Meas. Technology Conf.* pp 1182-7

[2] Villanueva J, Catunda S and Tanscheit R 2009 Maximum-Likelihood Data Fusion of Phase-Difference and Threshold-Detection Techniques for Wind Speed Measurement. *IEEE Transactions on Instrumentation and Measurement.* **58** 2189-7

[3] Pereira M 2009 Flow Meters: Part 1 *IEEE Instrumentation and Measurement Magazine.* **12** 18-9

[4] Ziarani A and Konrad A 2004 A method of extraction of nonstationary sinusoids *Signal Processing* **84** 1323-14

[5] Duarte R, Villanueva J, Costa M, Freire R, Catunda S and Costa W 2013 Ultrasonic time-of-flight estimation for wind speed measurement based on time-frequency domain using STFT *19th IMEKO TC 4 Symposium* (*Barcelona, Spain, 18-19 July 2013*) pp 72-6

[6] JCGM 200:2012 *International vocabulary of metrology – Basic and general concepts and associated terms* (*VIM*).

IOP Conf. Series: Journal of Physics: Conf. Series **1044** (2018) 012019 doi:10.1088/1742-6596/1044/1/012019

Design and development of a hyperspectral data measurement system used in precision agriculture

M C Luculescu, L Cristea, S C Zamfira and A Boer[1]
29 Eroilor Blvd, Transilvania University of Brasov, 500036, Romania

E-mail: boera@unitbv.ro

Abstract. The paper presents the design and development of a hyperspectral data acquisition system which provides high precision measurements for vegetation reflectance. The system was developed using four spectrometers, two of them for the visible domain and two for near-infrared. They measure the incident radiation and the reflected one simultaneously. In order to establish the optimal value of the integration time, an iterative algorithm was used for each spectrometer. A software platform was developed on the Raspberry Pi microcontroller system for data acquisition, data processing and remote control. The system provides real-time data processing and graphical representation of the acquired data.

1. Introduction

The physical characteristics of the vegetation cause the electromagnetic radiation to be reflected or absorbed in a way that is unique to each type of vegetation and may indicate its development and health status. Within the electromagnetic spectrum were considered mainly the visible (VIS) and near-infrared (NIR) domains.

In the last decades several vegetation indexes were introduced which are defined as combinations of the measured reflectance properties at two or more wavelengths. The Normalized Difference Vegetation Index (NDVI) is the most well-known index, which normalizes green leaf scattering in the NIR domain and chlorophyll absorption in the red wavelength [1], [2], [3].

The majority of optical sensors which are used for reflectance and NDVI measurements are of multispectral nature, i.e. they measure the radiation at discrete wavelengths along the VIS and NIR spectrum. The number of wavelengths is usually between 12 and 16. On the other hand hyperspectral sensors provide an almost continuous reading, the number of wavelengths being around 1000 or more [4], [5].

2. Hardware setup

A remotely controlled data acquisition system for hyperspectral measurements of vegetation parameters was designed and developed. The hardware setup consists of four Ocean Optics STS spectrometers (two for VIS and two for NIR) connected to a Raspberry Pi single-board computer via USB interface. The STS spectrometers provide hyperspectral measurements for 1024 wavelengths in the VIS and NIR domains. This allows us to make very precise measurements for the domain of wavelengths between 400 and 1000 nm, practically eliminating the need of interpolation parameters and minimizing the data processing after acquisition.

[1] Corresponding author.

Two STS spectrometers measure the incident radiation and the other two the reflected one. The spectrometer that measures the incident radiation uses a corrector cosine, which collects signal form 180 degree field of view. The spectrometer used for reflected radiation has a smaller field of view [4], [5].

3. Software setup

The software setup consists of the following components:

- Applications which access the STS device through the SeaBreeze driver interface [6];
- Ruby scripts for launching the above mentioned applications;
- Web server based on the Ruby on Rails framework [7] which allows the remote access through a web interface;
- Real-time graphical representation of the acquired data using the Flot Javascript library.

Our applications, which access the STS spectrometers through the SeaBreeze driver interface [6], were written in the C programming language. These applications set the acquisition parameters, such as integration time and number of iterations. An iterative algorithm which finds the optimal value of the integration time for each spectrometer was used. The pseudocode for the respective algorithm is presented in Listing 1.

Listing 1. Pseudocode for the iterative procedure which finds the optimal value of the integration time

```
maxval ← 16383;
tint ← 1200;
while maxval == 16383 do
|  if tint > 500 then
|  |  tint ← tint - 200;
|  end
|  else if tint > 200 then
|  |  tint ← tint - 100;
|  end
|  else if tint > 10 then
|  |  tint ← tint - 10;
|  end
|  else
|  |  tint ← tint - 1;
|  end
|  for i ← 1 to niter do
|  |  get formatted spectrum;
|  end
|  compute averages for each wavelength;
|  maxval ← Max (averages);
end
tintval ← tint;
write tintval to file;
```

For easier debugging we created three standalone applications for the dark measurement, reference spectra and the actual measurement. These applications can be launched from the command line (via SSH access to the Raspberry Pi) or through the web interface.

The acquired data is written to files with the name *VIS-index.txt*, respectively *NIR-index.txt*, where the index is automatically incremented. Each file contains the values for the wavelengths and the corresponding reflectance/ transmittance. The raw data can be transferred to computer via SFTP for further analysis and graphical representation.

Figure 1 shows the components of our software setup and their interactions.

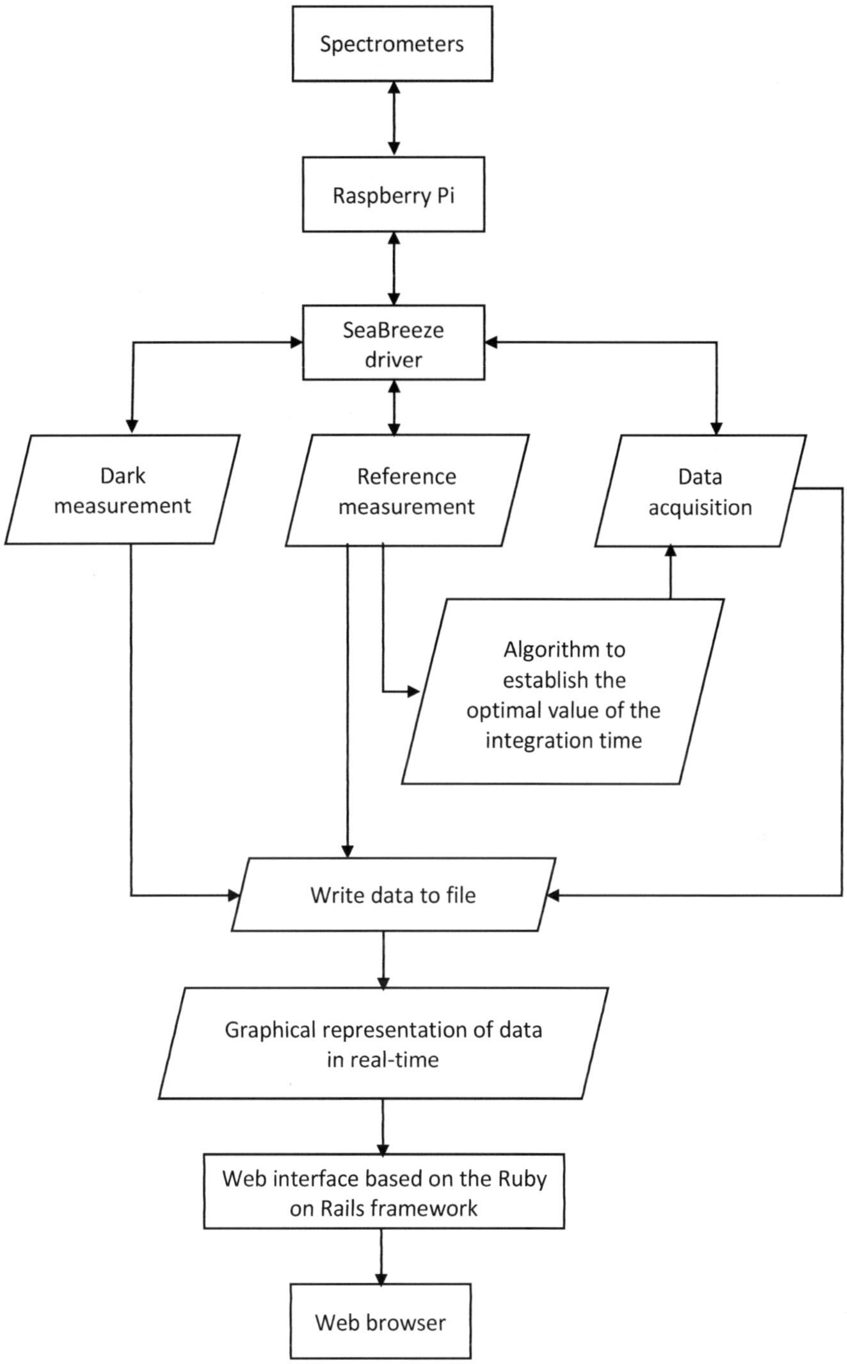

Figure 1. Components of the software setup and their interactions.

IOP Conf. Series: Journal of Physics: Conf. Series **1044** (2018) 012019 doi:10.1088/1742-6596/1044/1/012019

The reflectance values for each wavelength can be computed by means of the following relation:

$$R(\;)=\frac{R_{\text{white}}(\;)}{F(\;)}\frac{I_{\text{down}}(\;)}{I_{\text{up}}(\;)} \tag{1}$$

where

$$F(\;)=\frac{I_{\text{ref, down}}(\;)}{I_{\text{ref, up}}(\;)} \tag{2}$$

In the above equations $R_{\text{white}}(\;)$ represents the reflectance value for the white reference panel, $I_{\text{down}}(\;)$ and $I_{\text{up}}(\;)$ are the dark subtracted intensities measured as counts by the spectrometers pointing downward and upward respectively. $I_{\text{ref, down}}(\;)$ and $I_{\text{ref, up}}(\;)$ denote the intensities measured in the case of the white reference panel.

The reflectance values for the white reference panel were obtained in laboratory conditions using a reflection probe with holder and a white reflectance standard (barium sulfate). We made several measurements and performed averages for each wavelength. Finally we applied a smoothing algorithm using Bezier curves [8]. The reflectance values were stored in a file and uploaded to the Raspberry Pi microcontroller board. Figure 2 shows the plot of the reflectance as function of wavelength for the white panel.

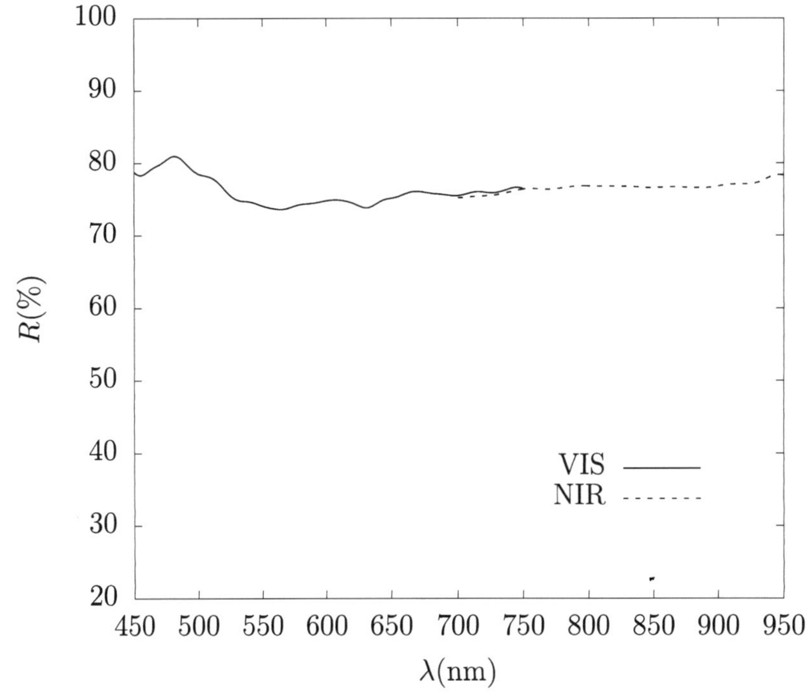

Figure 2. Reflectance vs. wavelength for the white reference panel.

4. Testing the data acquisition system

In order to test the accuracy of the developed data acquisition system for hyperspectral measurements, transmittance measurements in laboratory conditions were made. We measured the maximum value of

IOP Conf. Series: Journal of Physics: Conf. Series **1044** (2018) 012019 doi:10.1088/1742-6596/1044/1/012019

the transmittance for some optical filters, as well as the wavelength corresponding to the transmittance maximum. We computed also the error in the estimation of the above mentioned parameters. The obtained data is presented in the table 1.

Table 1. Transmittance maximum and the corresponding wavelength for some optical filters.

λ (nm)	Δλ (nm)	T_{max} (%)	ΔT_{max} (%)
419.050	0.460	51.153	0.356
470.225	0.470	42.757	0.006
575.775	0.470	29.377	0.137
577.440	-	35.843	0.079
625.200	-	18.158	0.815
715.015	0.470	36.221	1.390

These preliminary results show that in laboratory conditions the error in the evaluation of the transmittance maximum is smaller than 3.84%. For the corresponding wavelength estimation the error is smaller than 0.5 nm, comparable with the resolution of the optical sensor in the STS spectrometers. The results presented in the above table were obtained by averaging the experimental results from 5 independent measurements.

For vegetation reflectance measurements the reference spectra must be acquired using a white reflectance standard. The plots from figure 3 and figure 4 show the reflectance values as function of the wavelength for green and dry vegetation.

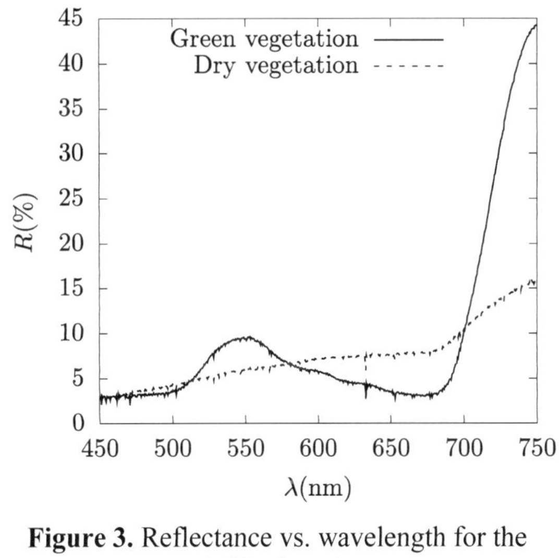

Figure 3. Reflectance vs. wavelength for the visible domain.

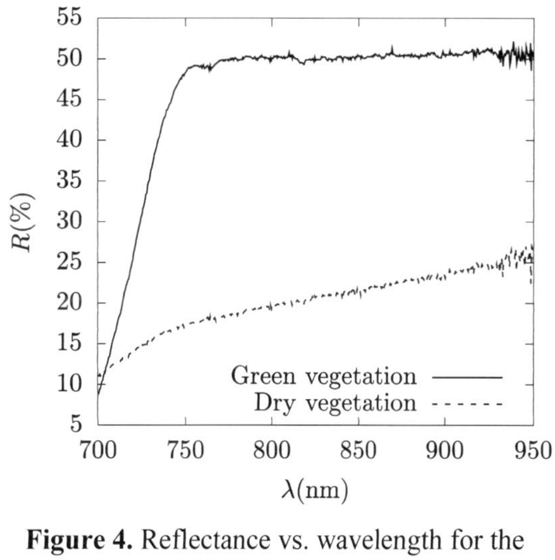

Figure 4. Reflectance vs. wavelength for the near-infrared domain

In order to ensure repeatability we made successive independent measurements and we compared the obtained values for the reflectance. The uncertainty for the reflectance was smaller than 3.98% for the visible spectrum (at 681.6 nm), respectively 2.13% for the NIR spectrum (at 938.33 nm). The results presented here were obtained as averages over 5 measurements.

The main sources of errors are:

- variation of solar intensity;
- instability of the incident angle;

- air attenuation.

In order to reduce the influence of the solar intensity variation we must repeat the reference measurement if it is necessary (cloudy sky). Air attenuation influence can be minimized if we ensure that all measurements are made at the same height (for example 5 m over ground). For optimal results the integration time should be located in the interval [10, 500] ms.

5. Conclusions

The proposed measurement system provides high precision data due to the acquisition of a larger amount of experimental values, which allows the computation of the reflectance for the visible and near-infrared domain (1024 wavelengths in each domain). The method uses four spectrometers, two for the visible and two for the near-infrared domain, which measure the incident, respectively the reflected radiation simultaneously. The system provides real-time data processing and graphical representation of the acquired data

References

[1] Carlson T N and Ripley D A 1997 On the relation between NDVI, fractional vegetation cover, and leaf area index, *Remote Sens. Environ.* 62.3 241-252.

[2] Burkart A 2015 Multitemporal assessment of crop parameters using multisensorial flying platforms Ph. D. Thesis Fakultät der Rheinischen Friedrich-Wilhelms-Universität Bonn

[3] Montandon L M and Small E E 2008 The impact of soil reflectance on the quantification of the green vegetation fraction from NDVI *Remote Sens. Environ.* 112.4 1835-45.

[4] Burkart A et. al. 2014 A novel UAV-based ultra-light weight spectrometer for field spectroscopy *IEEE Sens. J.* 14(1) 62-67.

[5] Larsolle 2003 A Instantaneous Measurement of Reflectance Spectra in the Open Field using Diode Array Spectrometers Biosyst Eng 86(1) 1-8.

[6] SeaBreeze API, https://oceanoptics.com/api/seabreeze/

[7] Thomas D and Heinemeier H D 2005 Agile web development with Rails, The Pragmatic Programmers LLC.

[8] Sambhunath B and Lovell B C 2007 Bézier and splines in image processing and machine vision, Springer Science & Business Media.

Acknowledgments

This paper was realized within the Partnerships Programme in priority domains-PN-II, which runs with the financial support of MEN-UEFISCDI, Project no. PN-II-PT-PCCA-2013-4-1629 and it was also supported by a grant of the Romanian National Authority for Scientific Research and Innovation, CNCS/CCCDI – UEFISCDI, project number PN-III-P2-2.1-BG-2016-0132, within PNCDI III.

IOP Conf. Series: Journal of Physics: Conf. Series **1044** (2018) 012020 doi:10.1088/1742-6596/1044/1/012020

Design, confection and calibration of a flow phantom with ultrasound applications

DP Oliveira[1], **WCA Pereira**[2], **RPB Costa-Félix**[1, a]
[1] Laboratory of Ultrasound, Inmetro, Duque de Caxias, RJ, Brazil
[2] Ultrasound Laboratory, COPPE-UFRJ, Rio de Janeiro, RJ, Brazil

E-mail: debora_poliveira@hotmail.com

Abstract. Ultrasound phantoms are standardized biological tissue models with well-defined acoustic properties, dimensions, and internal characteristics. Despite blood flow is a relevant factor in heat transfer in biological tissues, flow phantoms applied to therapeutic ultrasound are scarce in the literature. Therefore, this work aims to develop an ultrasound flow phantom capable of generating continuous flow with values similar to brachial artery. The phantom was made of agar based material with a tube of 4.6 mm of inner diameter inserted within, including a region of interest. The flow was produced with a linear flow propulsion system, generated by a peristaltic pump and an equalizing pressure system of milimetric graduated columns.
The phantom was capable of generating adjustable and continuous flows rate. The flow rate generated by flow phantom propulsion system range between 20.85 ± 0.95 $m^3.s^{-1}$ and 118.36 ± 1.18 $m^3.s^{-1}$. Those values are similar to flow rate values in the brachial artery. We expect to apply the ultrasonic flow phantom to therapeutic ultrasound safety evaluation in testing houses.

1. Introduction

Phantoms are standardized biological tissue models with well-defined properties, developed to behaviour as a simplified biological organism. In order to better mimetize biological system for ultrasound phantom some acoustic properties as speed of sound and acoustic attenuation equivalent to human tissues are required.
Ultrasound phantoms are intended to many applications, such as calibrating ultrasound diagnostic and therapeutic equipment, training medical personnel in ultrasound imaging, evaluating heating generated by ultrasonic transducers, and others.

The bioheat transfer equation is a model of heat exchange in biological tissues [1]. The bioheat equation describes that the temperature elevation in a biological tissue is due to the energy offered by the source of heat and metabolism, and heat dissipation due to blood flow. Ultrasonic phantoms that mimic blood flow applied to diathermy or hyperthermia assays were not found in the literature. In spite of being an important variable on tissues heat exchange with direct influence on heating patern, blood flow has been actually neglected on experimental models to assess therapeutic ultrasound safety. This work reports the confection and calibration of an ultrasound flow phantom with continuous flow range values and dimensions compatible with brachial artery [2-3].

[a] rpfelix@inmetro.gov.br.

IOP Conf. Series: Journal of Physics: Conf. Series **1044** (2018) 012020 doi:10.1088/1742-6596/1044/1/012020

2. Metodology

The flow phantom (figure 1) consists of three elements: a pressurization system, the fluid (1000 mL of degassed distilled water), and a region of interest (*ROI*).

The components of the phantom are: (a) A peristaltic pump (102FD / R , Watson Marlow Bredel, USA) activated by a continuous electric motor (6215A Power Supply, HP, USA), the pump is connected with a silicone tube (*a-b)* of 5.1 mm inner diameter and 550 mm length; (b) Two cylindrical reservoirs (*B1* and *B2*) with connections located at their bases, which are plastic beakers of 1000 ml with 64.5 mm inner diameter and 420 mm height; (c) Two silicone interconnecting tubes (*c-d*, *e-f*) with 7 mm internal diameter and 600 mm length; and (d) Fluid.

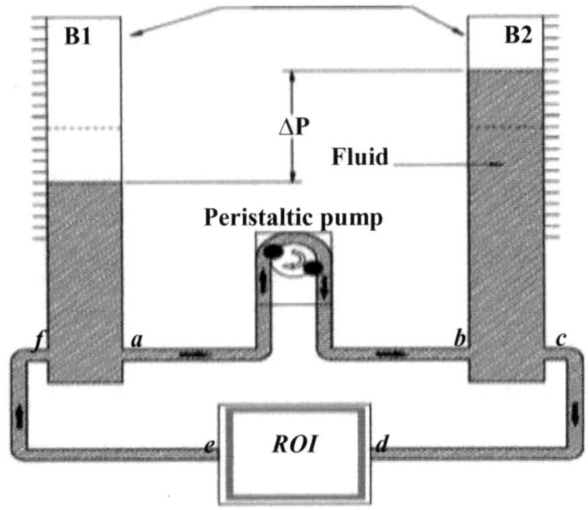

Figure 1. Hydraulic circuit schematic representation for the flow phantom in continuous mode.

The ROI consists of an acrylic frame (160 mm × 160mm × 52 mm) filled with agar based tissue mimic material [4] with a thin walled (0.2 mm) silicone duct of 4.6 mm internal diameter compatible with brachial artery inner diameter. The ROI is connected to rest of pressurization system with two connectors of 3.9 mm inner diameter with internal angle of approximately 6° minimizing turbulence ensuring a smooth flow transition between interne and external ROI ducts. Thermal sensors were positioned in the connectors and on surface of ROI.

The peristaltic pump flow rate ($m^3 \cdot s^{-1}$) was calculated weighing the mass of water in grams ejected by the pumped in one minute with a calibrated balance model Adventure AR 3130 (Ohaus, Parsippany, New Jersey, USA), with expanded uncertainty ($p = 0.95$) of 0.001g. The elapsed time was measured using a digital timer with a resolution of ± 0.05 seconds.

The pressure (Pa) in the flow phantom pressure system was obtained by measuring the level difference between the fluids of columns B1 and B2 using millimeter scales fixed to the columns with 1 mm resolution and a stainless steel ruler of 300 mm with 0.5 mm resolution calibrated with an expanded uncertainty of 0.03 mm ($p = 0.95$).

All experiments were repeated five times in repeatability conditions for each voltage of the pump power supply (3V to 12V) monitored with a digital multimeter model 33410A (Agilent Technologies, Santa Clara-CA, USA) calibrated with U = 0.04 V ($p = 0.95$). The ambient temperature was monitored with an analogic thermometer (Thermo Shneider, Wertheim, Germany) calibrated with U = 0.02°C ($p = 0.95$).

The measurement uncertainties for all experimental setup were assessed according to JCGM 100:2008 [5].

The following sources of uncertainty were considered in measuring the driven voltage to the peristaltic pump electric motor power supply: type A (standard deviation between the five repetitions of the experiment); type B was obtained from the digital multimeter calibration certificate (0.04 V).

Furthermore, the following sources of uncertainty were considered for the flow rate within the phantom: type A (standard deviation of the mean); type B evaluation of the mass uncertainty was obtained from the balance calibration certificate (0.001 g) and type B uncertainty was obtained by evaluation of the time the chronometer resolution (0.05 s).

The following sources of uncertainty were considered in measuring the pressure in the phantom: type A (standard deviation of the mean) for the level difference between the fluids on reservoirs B1 and B2 in the five repetitions of the experiment; type B evaluation of the the level difference between the fluids on reservoirs B1 and B2 uncertainty was obtained from the ruler calibration certificate (0.03 mm).

The expanded uncertainty (U) of pressure, flow rate, resistance and tension were calculated by multiplying the combined standard uncertainty (u_c) and the coverage factor (k) based on a t-distribution for a level of confidence interval of 95%.

The Reynolds number (Re), the theoretical system pressure (ΔP) and theoretical system resistance (Rc and Ra) in ROI were expressed by equations (1) to (4) [6]. The theoretical fluid dynamics properties in the phantom were calculated to better evaluate fluid behavior in ROI.

$$\mathrm{Re} = \frac{4 \cdot \rho \cdot Q}{\mu \cdot \pi \cdot D}, \tag{1}$$

In which ρ is density, Q is the flow rate, μ is the fluid dynamic viscosity and D is the tube internal diameter.

$$\Delta P = Q^2 \cdot (R_c + R_a), \tag{2}$$

In which the difference in pressure (ΔP) between the reservoirs (*B1* and *B2*) is equal to the flow rate squared (Q) multiplied by the circuit intrinsic resistance (R), which can be defined as the resulting resistance from the continuous straight pipe (R_c) and the accessories (R_a) that connect them.

In the laminar flow case, R_c can be calculated by equation (3) which expresses the pressure fall in a horizontal tube.

$$R_c = \frac{128 \cdot \mu \cdot L}{\pi \cdot D^2 \cdot Q}, \tag{3}$$

In which μ is the fluid dynamic viscosity, L is the tube length, Q is the flow rate, and D is the tube internal diameter.

$$R_a = \frac{K \cdot 8 \cdot \rho}{\pi^2 \cdot D^4}, \tag{4}$$

In which K is the dimensionless loss coefficient ρ is density, and D is the tube internal diameter.

The experimental flow resistance value (R_c+ R_a) was calculated using equation (2). The measurement uncertainty was estimated in accordance with JCGM 100:2008 [5] and the following sources of uncertainty were considered in measuring the resistance in the phantom: the standard uncertainty of experimental pressure due to the level difference between the fluids of columns B1 and

IOP Conf. Series: Journal of Physics: Conf. Series **1044** (2018) 012020 doi:10.1088/1742-6596/1044/1/012020

B2; type B evaluation of the mass uncertainty was obtained from the balance calibration certificate (0.001 g) and type B uncertainty was obtained by evaluation of the time the chronometer resolution (0.05 s).
The theoretical fluid dynamics properties calculations were done assuming a K of 0.35[6], a density (ρ) of 1000 kg.m^{-3} and a dynamic viscosity (μ) of 10^{-3} kg.(m.s)$^{-1}$ for the fluid (deionized water).

3. Results

The flow phantom was calibrated by means of the voltage (V) applied to the peristaltic pump motor, flow rate (m³·s^{-1}) and pressure (Pa) with their respective uncertainties as shown in table 1. Furthermore, experimental flow resistance (Pa.s².m^{-6}) and theoretical values of flow resistance (Pa.s².m^{-6}) with respective uncertainty were also calculated as seen in table 2.

Table 1. Experimental values of average flow rate (ml.min^{-1}) with expanded uncertainty, average peristaltic pump tension (V) with expanded uncertainty and average pressure (Pa) with expanded uncertainty.

Nominal tension (V)	Voltage (V)	u_c (V)	k	Flow rate (10^{-7} ·m³·s^{-1})	u_c (10^{-7} ·m³·s^{-11})	k	Pressure (Pa)	u_c (Pa)	k
12	12.017 (±0.207)	0.103	2	19.727 (±3.883)	1.942	2	201.037 (±6.053)	2.402	2
11	11.000 (±0.204)	0.102	2	17.841 (±3.538)	1.769	2	172.597 (±3.010)	1.387	2
10	10.013 (±0.064)	0.031	2	15.851 (±3.156)	1.578	2	158.868 (±5.526)	2.193	2
09	8.980 (±0.077)	0.036	2	14.092 (±2.784)	1.392	2	139.255 (±5.526)	2.193	2
08	7.977 (±0.033)	0.016	2	12.353 (±2.452)	1.226	2	119.641 (±5.526)	2.193	2
07	7.003 (±0.033)	0.016	2	10.549 (±2.098)	1.049	2	104.931 (±3.489)	1.551	2
06	5.977 (±0.031)	0.015	2	8.907 (±1.781)	0.890	2	95.125 (±3.489)	1.551	2
05	4.970 (±0.029)	0.014	2	7.154 (±1.449)	0.725	2	74.531 (±6.876)	2.595	3
04	3.993 (±0.024)	0.012	2	5.404 (±1.087)	0.543	2	57.859 (±3.010)	1.387	2
03	2.963 (±0.027)	0.013	2	3.475 (±0.698)	0.349	2	42.169 (±3.489)	1.551	2

Table 2. Experimental average resistance (Pa·s²·m^{-6}) with expanded uncertainty and theoretical resistance – Ra and Rc (Pa·s²·m^{-6}).

Nominal Voltage (V)	Experimental Resistance - (R_c +R_a) (10^{+14} ·Pa·m^{-6}·s²)	u_c (10^{+14} ·Pa·m^{-6}·s²)	k	Theoretical Resistance - R_c (10^{+13} ·Pa·m^{-6}·s²)	Theoretical Resistance – R_a (Pa·m^{-6}·s²)
12	0.517 (±0.018)	0.006	3	0.672	2.580·10^{-6}
11	0.542 (±0.012)	0.004	3	0.743	2.580·10^{-6}
10	0.632 (±0.025)	0.009	3	0.836	2.580·10^{-6}
09	0.701 (±0.032)	0.011	3	0.941	2.580·10^{-6}
08	0.784 (±0.041)	0.014	3	1.073	2.580·10^{-6}
07	0.943 (±0.040)	0.014	3	1.256	2.580·10^{-6}
06	1.199 (±0.056)	0.019	3	1.488	2.580·10^{-6}
05	1.456 (±0.145)	0.051	3	1.853	2.580·10^{-6}
04	1.981 (±0.136)	0.047	3	2.453	2.580·10^{-6}
03	3.492 (±0.368)	0.128	3	3.814	2.580·10^{-6}

The number following the symbol ± is the numerical value of expanded uncertainty (U), obtained by multiplying the combined standard uncertainty (**u_c**) and the coverage factor (**k**) based on a t-distribution for a coverage probability of 0.95.

IOP Conf. Series: Journal of Physics: Conf. Series **1044** (2018) 012020 doi:10.1088/1742-6596/1044/1/012020

Flow phantoms are hydraulic circuits capable of generating known and adjustable flow pattern pproviding a controllable experimental system against which hypotheses can be tested or measurements validated [7]. A flow phantom will typically consist of a vessel (generally confectioned in a rubber based material) suspended in a tissue mimic material and a fluid, which is pumped through the phantom. An important consideration in the flow phantom design is that the vessel should be similar in size and position to those of the vessel *in vivo* [8]. In the present work a phantom with similar dimensions and capable of generating continuous laminar flow values similar to those observed in brachial artery was designed and built.

The measurement of flow rate and pressure in ROI covers the entire range of voltages (3V to 12 V) to be applied to the peristaltic pump. The flow rates in ROI ranged from $3.475 \cdot 10^{-7} \pm 0.698$ $m^3.s^{-1}$ and $19.727 \cdot 10^{-7} \pm 3.883$ $m^3.s^{-1}$. A flow rate of $12.353 \cdot 10^{-7} \pm 2.452$ $m^3.s^{-1}$ was obtained for 8V, which is similar to the flow rate value for average brachial artery flow rates ($12.286 \cdot 10^{-7}$ $m^3.s^{-1}$) [2-3].

The pressures in ROI ranged from 42.167 ± 3.225 Pa to 201.031 ± 5.595 Pa. Despite the pressure values are not compatible with those from brachial artery, the achievement of low pressure values was desirable to minimize the risk of TMM damage.

The use of the reservoirs (B1 and B2) dampens oscillations from the peristaltic pump flow, and minimizes variation of the water level in the column during the pump cycle, which makes the flow laminar and continuous. The continuous flow system was chosen because it facilitates obtaining a laminar flow that is more common in small arteries [6-7].

The regime of flow in the phantom is laminar and continuous, but the model can be adapted to mimic a different anatomic region with large size vessels and pulsatile flow, eliminating B1 or B2 reservoirs.

The Reynolds number (Re) was obtained by equation (1). Re is directly proportional to the flow rate (Q) and inversely proportional to the tube diameter. Moreover, in laminar flow $Re \leq 2000$ and the fluid flows in lamina or parallel layers. In turbulent flow $Re \geq 3000$ and the fluid flows in a nonlinear behavior. In other cases, for $2000 \leq Re \leq 3000$ one of two types of flow (laminar or turbulent flow) can occur [6]. The higher Re was 547, obtained from the higher $Q = 19.727 \cdot 10^{-7} \pm 1.409$ $m^3.s^{-1}$, which leads to $Re \leq 2000$. Therefore, the flow was considered laminar within the ROI.

The theoretical and experimental resistance values are compatible with literature [7]. As the theoretical resistance values of R_a are smaller when compared to R_c, it can be assumed the influence of the accessories on flow resistance is minimum, so the flow tends to be laminar in the phantom.

The theoretical values of resistance are smaller when compared to experimental resistance values. Thereby, it can be assumed that theoretical resistance values are underestimated for the phantom, especially for the values of R_a as irregularities in accessories (connectors) inner can increase the flow resistance.

Flow phantoms are intended to many applications, such as calibrating ultrasound equipment, training medical personnel skills on blood vessel access, evaluation of vessels and parenchyma perfusion, and ultrasound imaging applications [8-9]. Diathermy or hyperthermia assays were not listed in the applications of ultrasound flow phantoms in literature. Besides that, future works aims to apply the flow phantom built to improve experimental models for studying the heating pattern in biological tissues.

4. Conclusion

A phantom with a laminar flow was confectioned. The dimension and flow rate values on phantom are compatible with brachial artery [2-3]. The pressure values on the phantom were compatible with similar models of literature [6-7]. This work aims to improve the experimental models for studying the heating pattern in biological tissues, especially therapeutic ultrasound safety tests. As future perspective, is to apply the flow phantom to therapeutic ultrasound essential performance testing. Furthermore, improvements are proposed in the phantom as the use pulsatile flow to better mimic larger artery flow.

References

[1] Pennes H H 1948 Analysis of tissue and arterial blood temperature in the resting human forearm. *J. Appl. Physiol.*, **1** 93-102

[2] Gokce N, Vita J A, Bader D S, Sherman D L, Hunter L M, Holbrook M, O´Malley C, Keaney, J F J and Balady G J 2002 Effect of exercise on upper and lower extremity endothelial function in patients with coronary artery disease *Am. J. Cardiol.* **90** 124-127

[3] Gault J H, Ross J J and Mason D T 1966 Patterns of Brachial Arterial Blood Flow in Conscious human Subject with and without Cardiac Dysfunction. *Circulation* **34** 833-848.

[4] IEC 60601-2-5:2009, *Medical electrical equipment - Part 2-5: Particular requirements for the basic safety and essential performance of ultrasonic physiotherapy equipment,* International Electrotechnical Commission (Geneva, Switzerland)

[5] JCGM100:2008, *Evaluation of measurement data - Guide to the expression of uncertainty in measurement*, Bur. Int. Poids et Mesures (Sèvres, France)

[6] Fox R W, McDonald A T 2010 *Introdução à mecânica dos fluidos* (Rio de Janeiro: LTC)

[7] Silva F A M, Krüger M A and Pereira W C A 2014. Continuous flow phantom for the calibration of an ultrasonic transit time flowmeter *Rev. Bras. Eng. Biomed.* **30** 3-10

[8] Kenwright D A, Laverick N, Anderson T, Moran C M and Hoskins P R 2014. Wall-less Flow Phantom for High-Frequency Ultrasound Applications *Ultrasound Med Biol.* **41**890-897

[9] Hudson J M, Williams R, Milot L, Wei Q, Jago J and Burns P N 2016. In vivo validation of volume flow measurements of pulsatile flow using a clinical ultrasound system and matrix array transducer *Ultrasound Med Biol.* **43** 579-585

Acknowledgments

Authors acknowledge CNPq, CAPES and FAPERJ for financial support.

IOP Conf. Series: Journal of Physics: Conf. Series **1044** (2018) 012021 doi:10.1088/1742-6596/1044/1/012021

Development of an automated system based on the concept of evolutionary hardware to determine the optimal operating point of GMI sensors

J D Benavides Mora and E Costa da Silva
Department of Electrical Engineering, Pontifical Catholic University of Rio de Janeiro Rio de Janeiro, RJ, Brazil

E-mail: jdabem@gmail.com, edusilva@ele.puc-rio.br

Abstract. The sensitivity of a magnetometer is directly associated with the sensitivity of its sensor elements. In the case of GMI sensors, this sensitivity is optimized by maximizing the variation of their impedance magnitude or phase as a function of the magnetic field to which the sample is subjected. Previous studies showed that the sensitivity of $Co_{70}Fe_5Si_{15}B_{10}$ samples is most affected by four parameters: sample length, biasing magnetic field, DC level and the frequency of the excitation current. However, the search for the set of parameters that optimizes the sensitivity of the samples is usually empirical and very time consuming. Thus, this paper proposes a new optimization technique, based on the use of genetic algorithms evolving in hardware, in order to define the set of parameters responsible for maximizing the samples sensitivity.

1. Introduction

Giant Magnetoimpedance (GMI) sensors are a new family of magnetic sensors that exhibit a huge variation of their impedance when subjected to external magnetic fields. These sensors have been used in the development of high sensitivity magnetometers, aimed at measuring ultra-weak magnetic fields. In turn, the sensitivity of a magnetometer is directly associated with the sensitivity of their sensor elements. In the case of GMI samples, this sensitivity is optimized by maximizing the variation of the impedance magnitude or phase as a function of the magnetic field applied to the sample [1–2].

The sensitivity of GMI sensors is affected by a set of conditioning parameters. However, currently, there is no sufficiently general physical model capable of modeling the impedance sensitivity in function of the entire set of parameters that affected it. Thus, the search for the optimal combination of these parameters is usually empirical [3].

Recent researches indicate that GMI magnetometers based on impedance phase readings have significantly higher sensitivities than those based on impedance magnitude readings [4–6]. Consequently, this paper focus on the optimization of the phase sensitivity of GMI samples. In previous studies, experimental tests were performed with $Co_{70}Fe_5Si_{15}B_{10}$ ribbon-shaped GMI samples in order to determine the set of parameters that most affects the sensitivity of the analyzed samples. The obtained results have showed that the most critical parameters are: length of the sample (l), DC level (I_{DC}) and frequency (f) of the excitation current, in addition to the biasing magnetic field (H_{pol}) [7].

The search for the set of parameters that optimizes the sensitivity of the samples is usually empirical and very time consuming. Thus, this works proposes a new optimization technique, based on the use of genetic algorithms evolving on hardware, in order to define which set of parameters is responsible for maximizing the sensitivity of the samples.

2. Automatic Optimization System

The automatic optimization system developed in this work allows defining the set of parameters (I_{DC}, f, H_{pol}) that leads to the maximization of the impedance phase sensitivity (S_{phase}) of GMI sensors with different chemical compositions and geometric shapes (length, thickness and width). The developed system is controlled by a computational model based on genetic algorithms, implemented so that the genes of the individuals of the population are represented by real numbers associated to each of the parameters that affect the sensitivity of the sensors. The equipment used in the system have their values adjusted by the genes of the individuals, so that the sensitivity corresponding to each individual can be measured experimentally. In turn, the developed GA uses the experimental measurements of the sensitivity as its fitness function, in order to estimate the aptitude of each individual.

The hardware of the system for automatic optimization of GMI samples have six main modules: LCR meter (4285A, Agilent); Current source (E3648A, Agilent); Helmholtz Coil; Polarity inverter, data acquisition board - DAQ (NI USB-6221, National Instruments) and a personal computer (PC). The LCR meter and the current source are interconnected to the PC via a GPIB-USB interface. In turn, the DAQ is connected to the PC via a USB interface. The LCR meter is responsible for exciting the samples with the desired electrical current and performing measurements of the impedance phase. The Current source is used to excite the Helmholtz Coil so that it can generate the external magnetic field to which the sample is subjected. On the other hand, the polarity inverter is used to invert the polarity of the current that flows through the Helmholtz Coil, inverting the direction of the generated magnetic field. The polarity is controlled by a digital output of the DAQ.

The system software runs on the PC and it is composed by two main modules: the control and measurement module, implemented in LabVIEW, and the optimization module, implemented in MatLab. The first module, is responsible for communicating with the equipment used in the system in order to control them and obtain the measurements of interest. The second module, is responsible for executing the optimization algorithm, based on genetic algorithms. The communication between these two modules is performed via TCP / IP protocol. Figure 1 shows a block diagram of the developed system.

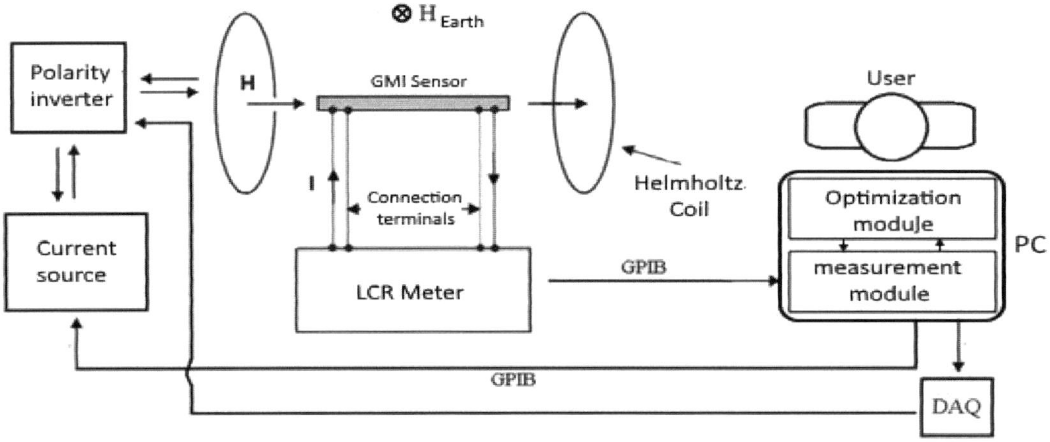

Figure 1. Basic setup of the system for automatic optimization of GMI samples sensitivity.

IOP Conf. Series: Journal of Physics: Conf. Series **1044** (2018) 012021 doi:10.1088/1742-6596/1044/1/012021

The control and measurement module receives the following parameters from the optimization module: I_{DC}, f and H_{pol}. These parameters are the genes of each individual present in the GA population. In this way, the control and measurement module can properly configure the system equipment. In turn, the control and measurement module sends to the optimization module the phase readings of the sensor sample impedance, carried out under the conditions established by the optimization module. Two phase values are obtained for each individual, in order to estimate its phase sensitivity (S_{phase}). The phase measurements are performed at $H_{pol} - \Delta H$ and $H_{pol} + \Delta H$.

The phase sensitivity S_{phase} is given by the derivative of the impedance phase (θ) in relation to the applied external magnetic field (H). In turn, considering the characteristics of the problem addressed, S_{phase} is defined as the fitness function of the genetic algorithm, and the higher the value of S_{phase}, the better the evaluated individual. As the analytical expressions of θ as a function of H are not known, the derivative was approximated by the technique of numerical differentiation based on central differences, as described by equation (1).

$$S_{phase} = \frac{\theta(I_{DC}, f, H_{pol} + \Delta H) - \theta(I_{DC}, f, H_{pol} - \Delta H)}{2\Delta H} \qquad (1)$$

All results presented in this work were calculated considering ΔH equal to 0.01 Oe.

The limits of each variable of the search space were stablished taken into account the working range of the equipment used in the optimization system. The bounds of the variables were settled to: 75 kHz $\leq f \leq$ 30 MHz, 0 mA $\leq I_{DC} \leq$ 100 mA and -2 Oe $\leq H_{pol} \leq$ 2 Oe.

The amplitude (I_{AC}) of the AC current was kept in 15 mA, because it has been previously seen that such parameter did not significantly affect the sensitivity of the GMI samples. The genetic algorithm was implemented in MatLab, using a population size of 100 individuals, maximum number of generations of 100, elitism criterion of 10 and the genetic operators chosen were heuristic crossover, adaptive mutation and tournament selection.

2.1. Dynamics of Operation

As shown in figure 2, the genetic algorithm is configured according to parameters defined by the user. Then, the genetic algorithm generates a random set of individuals that compose the initial population. These individuals are sent sequentially to the measurement and control system, which measures two phase values (θ) to each of them, one at $H_{pol} - \Delta H$ and other at $H_{pol} + \Delta H$. The results of the measurements are sent back to the GA, that evaluates the aptitude of each individual by calculating their respective evaluation function – equation (1). At the end of this process, if a predefined stopping criterion is not reached, the individuals of the population are subjected to genetic operators in order to generate a new population, which will be evaluated by the same process here in described. The process is repeated until the genetic algorithm reaches the established stopping criterion. At the end of the GA evolution, the combination of parameters (I_{DC}, f *and* H_{pol}) responsible for the maximum phase sensitivity is considered the best solution for the optimization problem.

Once the solution has been found, a characterization process is executed in order to evaluate the behaviour of the impedance phase of the analysed GMI sample in function of the magnetic field, for the optimal values of I_{DC}, f and H_{pol} returned by the genetic algorithm.

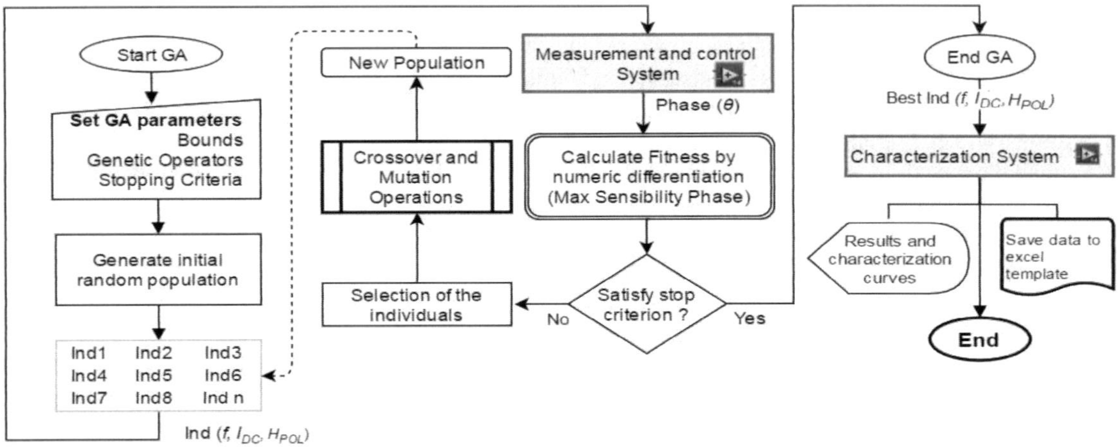

Figure 2. Flow diagram of the developed system.

3. Results and Analysis

The experimental measurements were carried out in a ribbon shaped $Co_{70}Fe_5Si_{15}B_{10}$ sample with 3 cm length, an average width of 1.5 mm and a thickness of 60 μm. This same sample was characterized in previous works [6], by other methodologies. In this way, the results obtained in the present work can be compared to the ones achieved in previous studies, presented in literature.

The developed genetic algorithm found a maximum sensitivity of 23.09°/Oe and took 31 generations to converge, spending an approximated time of 3.8 hours. The set of parameters of the individual that led to the maximization of the sensitivity is shown in table 1.

Table 1. Results returned by the GA.

f	16.71	MHz
I_{DC}	31.07	mA
I_{AC}	15	mA
H_{pol}	-1.31	Oe
S_{phase}	***23.09***	°/Oe

Figure 3 shows the characterization curve of the analysed GMI sample, excited by a current with DC level of 31.07 mA, amplitude of 15 mA and frequency of 16.71 MHz.

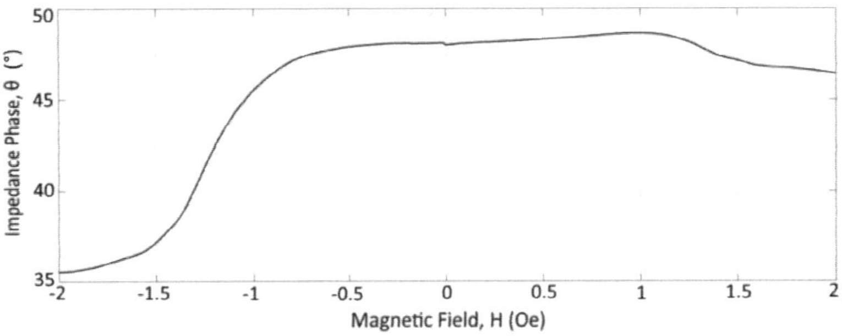

Figure 3. Impedance phase of the GMI sample in function of the external magnetic field, for an excitation current with 31.07 mA DC level, 15 mA amplitude and 16.71 MHz frequency.

In turn, the phase sensitivity S_{phase} can be obtained by the derivative of the impedance phase with respect to the magnetic field. In this way, a numerical differentiation was applied to the phase curve presented in figure 3, aiming at obtaining their respective point-to-point sensitivity values, shown in figure 4. The results presented in figure 4 shows that the highest sensitivity achieved was 23.13 °/Oe at

IOP Conf. Series: Journal of Physics: Conf. Series **1044** (2018) 012021 doi:10.1088/1742-6596/1044/1/012021

H = -1.31 Oe. As expected, these values have good agreement with the values returned by the GA, which predicted a maximum sensitivity of 23.09 °/Oe at H = -1.31 Oe.

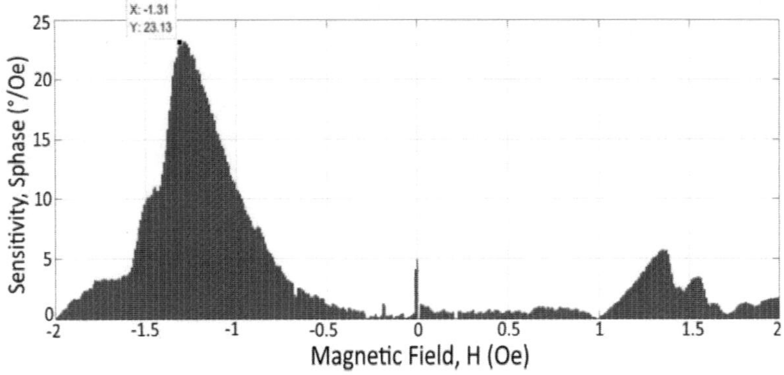

Figure 4. Phase sensitivity in function of the external magnetic field, for the GMI sample subjected to a current with 31.07 mA DC level, 15 mA amplitude and 16.71 MHz frequency.

In order to evaluate the repeatability of the system, the same optimization process that led to the results indicated in table 1 was repeated two more times. The results obtained in each of the three performed tests are presented in table 3. Additionally, to the S_{phase} and H_{pol} values returned by the GA, table 3 also presents the maximum phase sensitivities obtained by the characterizations carried out after each optimization process ($S_{characterization}$), as well as the values of magnetic fields ($H_{characterization}$) corresponding to each $S_{characterization}$ obtained. The number of generations until the convergence of GA and the total computational processing time are presented for all tests performed.

Table 2. Analysis of the repeatability of the results obtained by the developed system.

	Test 1	Test 2	Test 3	Units
S_{phase}	*23.09*	*22.96*	*22.87*	°/Oe
$S_{characterization}$	23.13	23.57	23.83	°/Oe
f	16.71	16.78	16.76	MHz
I_{DC}	31.07	34.58	40.49	mA
I_{AC}	15	15	15	mA
H_{pol}	-1.31	-1.33	-1.34	Oe
$H_{characterization}$	-1.31	-1.32	-1.34	Oe
Number of generations to convergence	31	20	21	generations
Processing time	3.8	2.4	2.7	hours

According to the obtained results, the sensitivity values returned by the GA (S_{phase}) were always about 23°/Oe. Besides, the sensitivity values obtained by the characterization curves ($S_{characterization}$) were satisfactorily close to their respective S_{phase} values. These results indicate a good repeatability of the developed algorithm.

In all of the tests performed, the frequency (f) was around 16.7 MHz; the biasing magnetic field (H_{pol}) was always close to -1.30 Oe and the amplitude of the alternating current (I_{AC}) was exactly the same, 15 mA, because this parameter was kept fixed. One can also notice that, as expected, $H_{characterization}$ was always near enough to H_{pol}. On the other hand, the DC current level have varied between 30 mA and 40 mA, which indicates that multiple values of I_{DC} can lead to maximum phase sensitivities. Furthermore, table 2 indicates that the algorithm takes about 3 hours to converge to a solution.

Figure 5 shows the impedance phase characterization curves as a function of the magnetic field, for each of the three tests performed. Each characterization curve was obtained by adjusting to the excitation current (i_c) of the GMI sample according to the values of I_{DC}, I_{AC} and f specified in table 2.

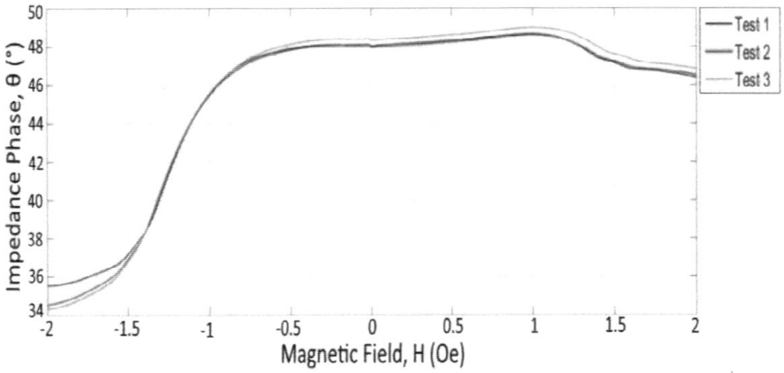

Figure 5. Experimental characterization curves of the impedance phase as a function of external magnetic field, adjusting the excitation current of the GMI sample according to the values of I_{DC}, I_{AC} and f specified in Table 2, for each test.

4. Conclusions

The developed optimization system is based on the concept of evolutionary hardware. It allows to adjust the operational point of GMI sensors, inserted in magnetometer circuits, aiming at improving their sensitivity. The system is controlled by a computational model based on genetic algorithms, implemented so that the genes of the individuals of the population represent the parameters of interest (I_{DC}, f and H_{pol}) that affect the sensitivity of the sensors (S_{phase}). The evaluation of the aptitude of each individual is done by experimental measurements. The system evolves towards the best combination of variables of interest (I_{DC}, f and H_{pol}) responsible for the maximization of the phase sensitivity.

The maximum sensitivity obtained was 23.83 °/Oe, which is about 2.5 times higher than the maximum sensitivity obtained in previous works published in literature (9.48 °/Oe) [6]. The developed system shows a good repeatability of the results and takes about 3 hours to converge to a solution. The achieved results allow to conclude that the developed system have implemented an effective automatized tool to define the set of variables that optimizes the phase sensitivity of GMI samples.

Acknowledgments
The authors thank CAPES and CNPq for financial support to the development of this work.

References

[1] Machado F L A and Rezende S M 1996 A theoretical model for the giant magnetoimpedance in ribbons of amorphous soft-ferromagnetic alloys *J. Appl. Phys.* **79** 6558

[2] Gonçalves L A P, Soares J M, Machado F L A and Rodrigues A R 2006 Hall and giant magnetoimpedance effects in the Co70Fe5Si15B10 metallic glass *J. Non. Cryst. Solids* **352** 3659–62

[3] Silva E C da, Vellasco M M B R, Barbosa C R H, Monteiro E C, Gusmão L A P de, Costa E, Silva D, Vellasco M M B R, Barbosa C R H, Monteiro E C and De Gusmão L A P 2012 Modelagem da sensibilidade de amostras GMI por redes neurais *Sba Control. Automação Soc. Bras. Autom.* **23** 636–48

[4] Costa Silva E, Gusmão L A P, Hall Barbosa C R and Costa Monteiro E 2014 An enhanced electronic topology aimed at improving the phase sensitivity of GMI sensors *Meas. Sci. Technol.* **25** 115010

[5] Costa Silva E, Gusmão L A P, Hall Barbosa C R and Costa Monteiro E 2009 High sensitivity triaxial magnetic field transducer, based on the phase characteristics of the GMI effect *19th IMEKO World Congr. 2009* **4** 2572–6

[6] Silva E C, Gusmão L a P, Barbosa C R H, Monteiro E C and Machado F L a 2011 High sensitivity giant magnetoimpedance (GMI) magnetic transducer: magnitude versus phase sensing *Meas. Sci. Technol.* **22** 35204

[7] Phan M-H and Peng H-X 2008 Giant magnetoimpedance materials: Fundamentals and applications *Prog. Mater. Sci.* **53** 323–420

Development of digital Open Educational Resource for metrology education

L S Alves and J M Granjeiro
Postgraduate Program in Biotechnology – Inmetro
Av. Nossa Senhora das Graças, 50, Xerém, Duque de Caxias/RJ - Brazil

E-mail: lsalves@inmetro.gov.br

Abstract. Open Educational Resources (OER) designates teaching, learning and investigation materials in any supports, digital or others, located either within the public domain or disclosed under an open license that allows access, use, adaptation and redistribution, free of charge, by third parties with few or no restriction. The general paper purpose is to submit a digital OER proposal, an educational resource called "Introduction to Metrology", planned for a five steps life cycle production scheme: to find, to create, to adapt, to share, to use. The work applies a qualitative approach, with the survey of all educational resources identified in the International Bank of Educational Objects (BIOE), from the research of terms "metrology", "measurement" and "measurement unit" as keywords; besides the survey of materials identified on INMETRO's website. Five videos-type resources identified in BIOE and one text in INMETRO's website were strategies to deliver contents. Moreover, the production of four unpublished exercises blocks with software Hot Potatoes® and CMapTools®. OER is available as a course inside Moodle®. STEM Education needs enrichment with metrological concepts to establish a common and standardized language. OER could support metrology education, therefore propose didactic sequences and activities are the way to encourage its use in the classroom.

1. Open Educational Resources

UNESCO's Forum "Impact of Open Courseware for Higher Education in Developing Countries", which occurred in 2002 in Paris, recommended the term "Open Educational Resources" (OER) to nominate "the open provision of educational resources, enabled by information and communication technologies, for consultation, use and adaptation by a community of users for non-commercial purposes." [1]

The general purpose of this paper is to submit a digital OER proposal, called "Introduction to metrology", planned for a five steps life cycle production scheme: to find, to create, to adapt, to share and to use. [2] The description of each one of the steps contains the specific objectives of this paper and offers an explanation about the OER proposal.

The step "to find" concerns the search for resources capable of coming useful for the needs of teaching and learning processes. [3] The access to information in digital culture makes this work easier. The Brazilian digital didactic material, called International Database of Educational Objects (BIOE) was the main search site. Its purpose is to keep and share digital educational resources of free access, as well as more elaborated ones, in different formats – such as audio, video, animation, simulation, educational software, image, map and hypertext. It was a creation of the Ministry of

Education, in partnership with the Ministry of Science and Technology, Latino American Network of Educational Portals (RELPE), Organization of the Ibero-American States (OEI) and others [4]. Digital didactic materials researched at the website of the National Institute of Metrology, Quality and Technology (INMETRO/Brazil) complement the content.

The step "to create" can happen without the utilization of another resource (from scratch) or with the combination of resources found to assemble a new resource, thus expanding the concept of authorship [3]. We employ both strategies in the OER proposal. The only strategy to deliver contents is presenting objects found in two websites (videos and texts type resources). Furthermore, we apply software - Hot Potatoes® and CMapTools® - to make four blocks of unpublished exercises.

The Hot Potatoes® is freeware created and developed by professors and technicians associated with University of Victoria (Vancouver, CA). Hot Potatoes® suite includes six applications, enabling you to create interactive multiple-choice, short-answer, jumbled-sentence, crossword, matching/ordering and gap-fill exercises for the World Wide Web [5]. The Cmap software is a result of researches conducted at the Florida Institute for Human & Machine Cognition (IHMC). It empowers users to construct, navigate, share and criticize knowledge models represented as concept maps. [6]

The step "to adapt" intends to adequate the resource to the specific context by correcting, improving and contextualizing the material by means of the insertion or removal of components; changing the sequence of activities; edition or alteration of images, texts, audio, video. [3] The fulfillment of this step involved the production of new exercises in order to come useful for the specifications of the qualification proposals.

The step "to share" depends on the availability of materials to the global community under an open license and is the OER's defining step [3]. Consequently, OER is available as an open course inside Moodle®, which is a learning platform designed to provide educators, administrators and learners with a single robust, secure and integrated system to create personalized learning environments. Built for global learning, it is easy to use and free, with no licensing fees, as open source software [7]. URL address to access the website for OER is http://ctmetrologia.acrisp.org/moodle/course/view.php?id=2

Finally, the step "to use", which depends on the utilization of OER in teaching and learning processes [3], will be developed among groups of Science teachers and students from technical courses such as metrology and biotechnology.

The work methodology is based on a qualitative approach, with the survey of all educational resources identified in the BIOE from the research of the terms "metrology", "measurement" and "measurement unit" as keywords; besides the selection of the appropriate resources among those on INMETRO's websites.[8-10]

2. Open Educational Resources on metrology

The research of the terms "metrology", "measurement" and "measurement unit" in the BIOE resulted in 70 educational resources related to metrology education: 3 in English, 13 in Spanish, 54 in Portuguese. The INMETRO's website has 21 Videos [8], 183 editions of the bulletin "Inmetro Informação", 148 editions of the internal newspaper "Na Medida", 13 Books, 3 Manuals, 3 Folders and 18 Booklets for online access [9 and 10]

2.1 The new Open Educational Resource (OER) "Introduction to Metrology"

The overall goal of the new OER "Introduction to Metrology" is to introduce the topic "measurements", the application areas and the importance of the metrological science. The target audience is formed by basic education teachers and students starting high school.

To plan the new OER, five videos-type resources identified in BIOE and one text in INMETRO's website were strategies to deliver contents -"International Vocabulary of Metrology" [11]. Moreover, the production of four unpublished exercises blocks with software Hot Potatoes® and CMapTools®. Table 1 organizes the OER planning with the description of each resource – title, type, link and

IOP Conf. Series: Journal of Physics: Conf. Series **1044** (2018) 012022 doi:10.1088/1742-6596/1044/1/012022

contents. Figure 1 shows the concept map proposed as an exercise to build a graphic summary of all contents.

Table 1: Planning of OER "Introduction to metrology"

Title, Type and links	Contents
Measurements – Introduction Type: video (duration 4 min) http://objetoseducacionais2.mec.gov.br/handle/mec/22210	Units and Quantity-value scale and numerical quantity value; Importance of measurement standard; Quantities: time, mass, temperature and volume;
International Vocabulary of Metrology: Basic and general concepts and associated terms – JCGM (VIM 2012). Type: text http://www.inmetro.gov.br/inovacao/publicacoes/vim_2012.pdf	Quantities and Units; Measurement; Devices for measurement; Properties of measuring devices; Measurement standards (Etalons); Metrology
HotPotatoes® interactive matching/ordering and gap-fill exercises	Quantities and Units; Measurement; International System of Quantities (ISQ); Base units.
Measurements – measurement standards ; Type: video (duration 2 min) http://objetoseducacionais2.mec.gov.br/handle/mec/22424	Measurements; Reference measurement standard; Comparison of measurement patterns; Mass etalon.
Measurements - Measurement example analysis Type: video (duration 3 min) http://objetoseducacionais2.mec.gov.br/handle/mec/22425	Measuring instruments; Instrument's quantity-value scale; Measurement error; Reference measurement standard; Importance of Calibration.
International Vocabulary of Metrology: Basic and general concepts and associated terms – JCGM (VIM 2012) Type: text http://www.inmetro.gov.br/inovacao/publicacoes/vim_2012.pdf	Quantities and Units; Measurement; Devices for measurement; Properties of measuring devices; Measurement standards (Etalons); Metrology
HotPotatoes® interactive jumbled sentence and gap-fill exercises	Measurement; Devices for measurement; Reference measurement standard
Measurements - International System of Units (SI) Type: video (duration 3 min) http://objetoseducacionais2.mec.gov.br/handle/mec/22428	Scope of International System of Units (SI); English Measurement System; Base quantities (time, mass and length) and measurement units; Definition of the second; Derived quantities; The international prototype of the kilogram; BIPM; INMETRO
Measurements – prefixes within the International System of Units (SI) Type: video (duration 2 min) http://objetoseducacionais2.mec.gov.br/handle/mec/22429	Multiples and submultiples of units; Conversion factor between units.
HotPotatoes® interactive matching/ordering exercise	International System of Units (SI); Base quantities (time, mass and length) and measurement units; The international prototype of the kilogram; BIPM; INMETRO; Multiples and submultiples of units; Conversion factor between units; prefixes.
Concept map	Graphic summary of all contents

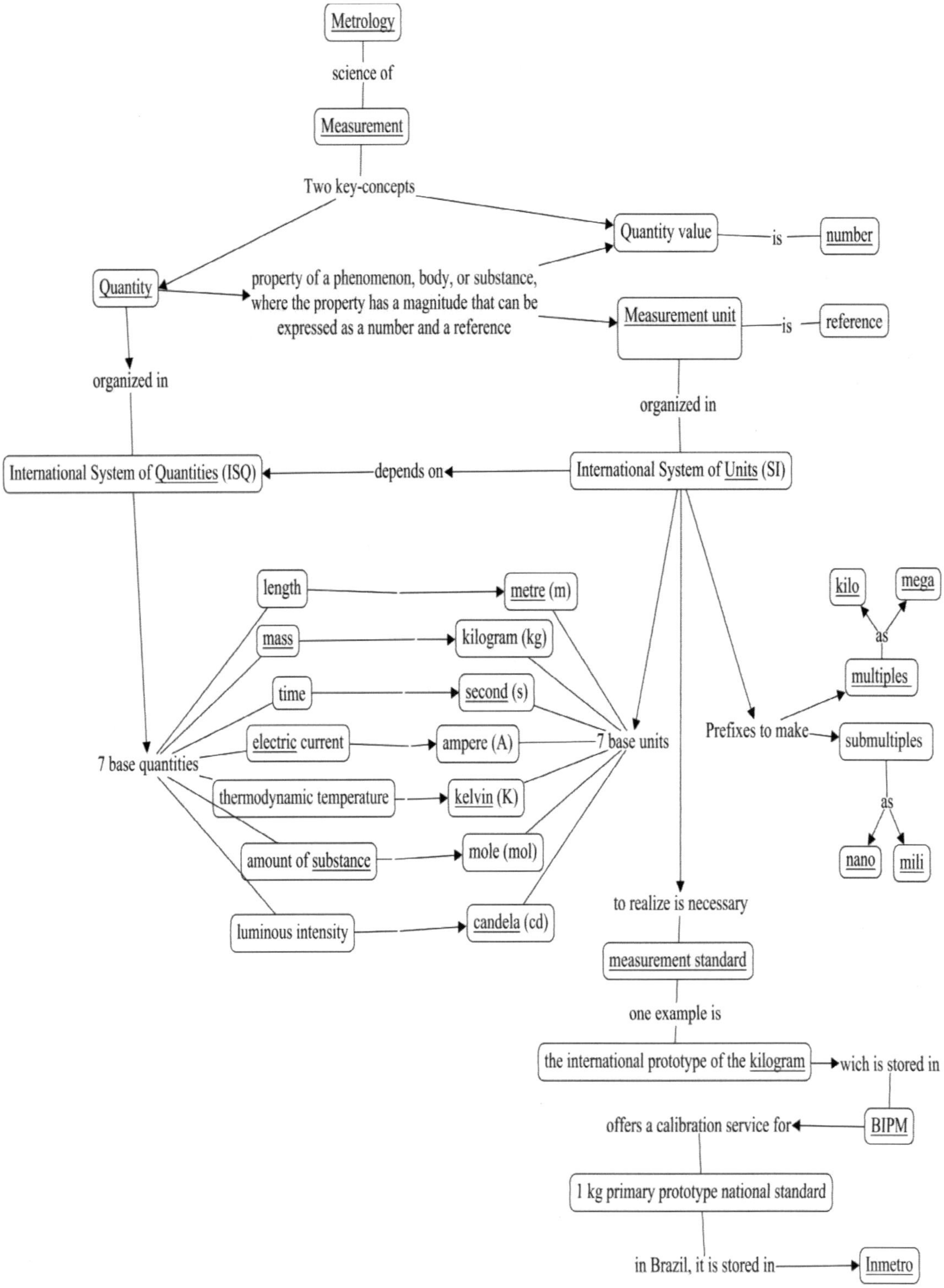

Figure 1: Concept map proposed as an exercise to build a graphic summary of all contents. Underlined words are gaps to fill by students.

The OER proposal is a digital didactic sequence.

Types of five interactive exercises that compose the three blocks are matching/ordering, gap-fill and jumbled sentence. These exercises have the purpose of highlighting the main metrology concepts mentioned in the videos. Furthermore, all exercises stimulate the reading of the "International Vocabulary of Metrology" [11]. We intend to show this material as a potentially didactic material.

Inside de OER proposal, the fourth exercise is a concept map available as a PDF file to be printed, with gaps to be filled by students. Drawing concept map spends a long time. Consequently, this choice decreases the activity time, meanwhile presenting concept maps to the audience. Even with the spreading of digital culture, we believe in the importance of a printable document as a summary of all contents to finish digital didactic sequences.

Concept maps are graphical tools for organizing and representing knowledge. They include concepts, usually enclosed in circles or boxes of some type, and relationships between concepts indicated by a connecting line linking two concepts. Words on the line, referred as linking words or linking phrases, specify the relationship between the two concepts. [12]

3. Considerations on BIOE

All educational resources found in the research have been submitted between the years of 2008 and 2012. The link "recent submissions" on the website shows the last submission on March 2015, which seems to have been the lifetime of the BIOE. [13]

We realized that the main difficulty on BIOE was the impossibility to make a previous selection of the resources based on the available description, thus forcing the opening of all researched resources. The description of them needs categories that describe the specific contents, rather than classification categories such as disciplines and teaching level. Curricular changes and distinct pedagogical projects make these unnecessary, while content description could expand the research possibilities, speed up the selection and potentiate the utilization of resources as the teacher can see the usage opportunities in a better and faster way. Another difficulty was finding disabled files.

Even though, at first glance, we did not find some resources identified in BIOE in other repositories, more updated and specific as portals EduCapes and the Bank of Objects in Metrology (BOM). This was the reason for the choice of BIOE as a repository for research.

EduCapes opened in September 2016, managed by CAPES' Distance Learning Board (Brazilian Ministry of Education). [14] BOM is a digital archive of contents on metrology developed by the Brazilian Metrology Society from the financing of the National Council of Scientific and Technologic Development (CNPq) by means of the South American Program to Support Cooperating Activities in Science and Technology (PROSUL). [15]

One interesting result was the personal identification of the author "Anderson Coser Gaudio", a professor at the Physics Department of the Federal University of Espírito Santo (UFES). He produced a series of seven videos on the topic "Measurement". This information could be important to expand the small Brazilian community involved in metrology education

4. Considerations on available materials on INMETRO's website

Videos from Inmetro are available in its Youtube® page. [8] This is a problem for Brazilian students and teachers because many institutions block access to this web page. All folders and booklets elucidate the conformity assessment.

Only the book "International Vocabulary of Metrology: Basic and general concepts and associated terms" [11] we considered adequate for the target audience and overall goal of the new OER "Introduction to metrology".

5. Conclusion

The STEM Education, which stands for Science, Technology, Engineering and Mathematics, needs to be enriched with metrological concepts that are fundamental in the modern society as it establishes a common and standardized language, as well as procedures that ensure the reliability in measurement result.

The research identified digital educational resources on metrology available for immediate use, produced new resources in the form of activities and proposed a new resource.

It was possible to perceive the importance of BIOE, since all videos-type resources chosen were only available there. Although new digital platforms and banks with educational materials are important to disseminate scientific information, they must include all materials already identified and improve their categories for materials classification.

As an illustration, we think about the inclusion of all metrology's OER identified on the BIOE and videos from Inmetro's Youtube® page to BOM. Others suggestions are:

- making simple materials on metrology, as folders and booklets;
- improving the classification categories of digital objects;
- and proposing didactic sequences for their use.

The movement "Open Educational Resource – OER" can contribute to disclosure metrology by stimulating the reutilization of available materials with open licenses. However, experts must first analyse the quality of such materials in order to provide a technical approval and, mainly, propose didactic sequences and activities to encourage its utilization in the classroom.

6. References

[1] UNESCO 2002. Forum on the Impact of Open Courseware for Higher Education in Developing Countries. Final report. <http://unesdoc.unesco.org/images/0012/001285/128515e.pdf>
[2] OER Handbook for Educators. <http://wikieducator.org/OER_Handbook/educator>
[3] EDUCAÇÃO ABERTA 2011. Recursos Educacionais Abertos (REA): Um caderno para professores. <http://educacaoaberta.org/cadernorea/o_que>
[4] Brazil. Ministry of Education. International Database of Educational Objects. (BIOE) Mission. <http://objetoseducacionais2.mec.gov.br/staticspages?t=0&locale=pt_BR>
[5] HotPotatoes®. What is Hot Potatoes? <https://hotpot.uvic.ca/>
[6] CMapTools® <http://cmap.ihmc.us/>
[7] Moodle®. About Moodle. <https://docs.moodle.org/33/en/About_Moodle>
[8] Inmetro's Youtube® page. <https://www.youtube.com/user/tvinmetro>
[9] Brazil. Ministry of Industry, Foreign Trade and Services. INMETRO. Publicações – Introdução. <http://www.inmetro.gov.br/inovacao/publicacoes.asp>
[10] Brazil. Ministry of Industry, Foreign Trade and Services. INMETRO. Publicações Avulsas. <http://www.inmetro.gov.br/inovacao/publicacoes_avulsas.asp>
[11] International Vocabulary of Metrology: Basic and general concepts and associated terms – JCGM 200:2012. 3rd. ed. 2012. 1a edição luso-brasileira do VIM 2012 (JCGM 200:2012) <http://www.inmetro.gov.br/inovacao/publicacoes/vim_2012.pdf
[12] Cañas A J and Novak J D 2009 What is a Concept Map? <http://cmap.ihmc.us/docs/conceptmap.php>
[13] Brazil. Ministry of Education. International Database of Educational Objects. (BIOE) Recent Submissions. <http://objetoseducacionais2.mec.gov.br/?locale=en>
[14] Brazil. Ministry of Education.eduCAPES. O que é o eduCAPES? <https://educapes.capes.gov.br/redirect?action=about>
[15] Bank of Objects in Metrology (BOM). Sobre o Bom. <http://repositorio.bom.org.br:8080/jspui/#openModal>

Acknowledgments

We would like to acknowledge Gelson Martins da Rocha, metrology research, our colleague in Inmetro, who opens his Moodle platform to our course.

Development of Ultrasound Thermometry Technique Using Tissue-Mimicking Phantom

Aleksander E. Berkovich, Evgueni M. Smirnov, Andrey D. Yukhnev, Yakov A. Gataulin, Daria E. Sinitsyna, Dmitriy A. Tarkhov
Peter the Great St. Petersburg Polytechnic University, 29 Politecnicheskaya St., St. Petersburg, Russia
E-mail: a.yukhnev@mail.ru

Abstract
A phantom with heating of tissue-equivalent material has been designed to elaborate the method of ultrasound thermometry. Algorithms have been developed to evaluate backscatter ultrasound signals shift caused by temperature-sensitivity of the speed of sound. Time-dependent material temperature field needed for estimation of the method accuracy was measured with thermistor probes.

1. Objectives

High Intensity Focused Ultrasound is widely used in various fields of medicine [1]. Ultrasonic devices for blood coagulation in therapy of vascular injury and internal organs, for veins obliteration, as well as for ablation of breast and thyroid tumors are currently under extensive development worldwide. Our experience in this area was partially presented in [2], where special equipment for testing of focused ultrasound devices are described.

For carrying out of hyperthermia process and for its efficiency control a real-time temperature monitoring is needed. Currently, the magnetic resonance method is the only one available for non-invasive thermometry, but it requires rather expensive equipment and has a low temporal resolution. Ultrasound thermometry considered as an alternative, cheap and fast, non-invasive technique is under development for about 20 years [3]. It has already shown good results during tests using various ultrasonic phantoms [4], as well as biological tissues and living organisms, but clinical applications of the technique is yet to come.

The present contribution covers description of equipment used for development and optimization of the ultrasound thermometry, as well as algorithms for backscatter ultrasound signal data-processing. It is a step to get an effective and low-cost high-precision medical tool via combination of therapeutic and diagnostic ultrasound devices.

2. Methods

Ultrasound thermometry technique is based on phenomenon of the speed of sound change due to a temperature increase. This phenomenon causes a time shift of the backscatter ultrasound signals from a heated material relative to the reference signal of the unheated material. The temperature increase, ΔT, at a current point can be calculated as $\Delta T = K\varepsilon$, where ε is the so-called ultrasound signal thermal shift strain defined as $\varepsilon = \frac{d}{dz}(\Delta d)$, where z is a coordinate along the ultrasound beam direction and K is a material dependent coefficient [3].

Phantom with a tissue-equivalent material (agar-agar filled with graphite) is used in the present work. The material simulates density (1050 kg/m^3), speed of sound (1550 m/s), attenuation coefficient (0.7 dB/(cm·MHz)), heat capacity (3700 J/kg·°C) and heat conductivity (0.59 W/m·°C) of a biological tissue. Nearly two-dimensional temperature field is created by heating a nichrome wire, 0.3 mm diameter and 150 mm length, located inside the material at the depth of 20 mm (Figure 1).

During heating the wire the material temperature at chosen distances from the wire is measured with six thermistor probes. The heating time, t, is counted from switching-on the DC power supply. The temperature measurement uncertainty is estimated as 0.1 °C. Simultaneously with measurements of time-dependent material temperature (results are illustrated in Figure 2), ultrasound signals are

IOP Conf. Series: Journal of Physics: Conf. Series **1044** (2018) 012023 doi:10.1088/1742-6596/1044/1/012023

registered near the heated wire in an ultrasound-linear-transducer scanning plane, 20 mm width and 10 mm depth, perpendicular to the wire. An array of data for one record includes 161x512 values of ultrasound signal. In order to eliminate the temperature probes influence on the ultrasound signal and taking into account the temperature field uniformity along the wire, zones for temperature measurement and for ultrasound scanning are separated in space, as shown in Figure 3.

Figure 4 shows the temperature increase profile $\Delta T(z)$ measured at t=20s with six probes positioned in the y=0 plane (Figure 3). The depth coordinate, z, as well as the transversal coordinate, y, are counted from straight lines passing through the heated wire center. The profile covers experimental points obtained from measurement data averaged over 5 experiments. The approximation/interpolation curve shown was obtained with the least-squares smoothing procedure.

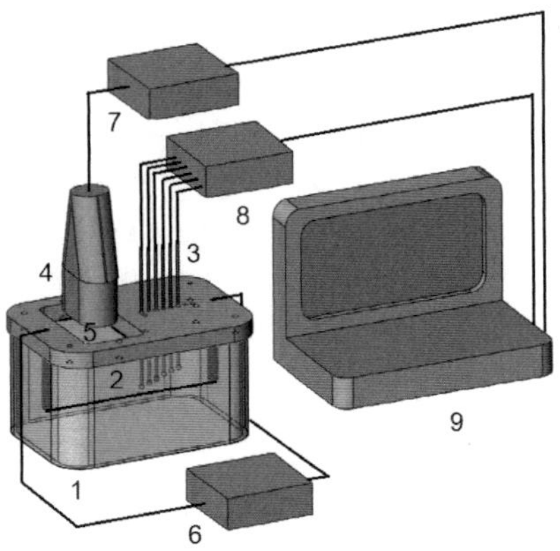

Figure 1. Experimental setup: 1 – tissue-mimicking phantom; 2 - heated nichrome wire, 3 - temperature probes, 4 - ultrasound probe, 5 - acoustic window, 6 - DC power supply, 7 - ultrasound scanner, 8 – ADC, 9 – PC

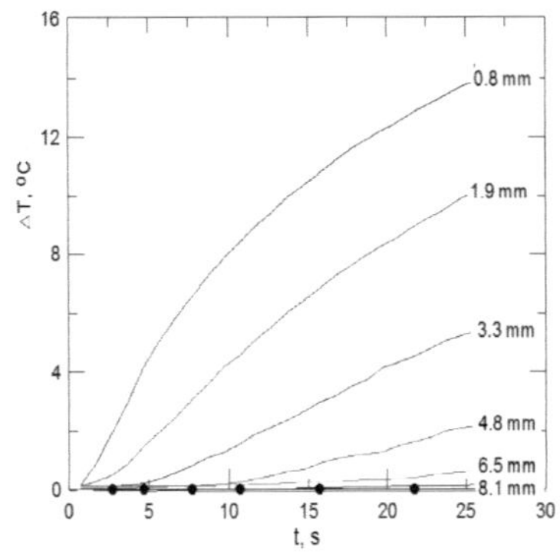

Figure 2. Temperature temporal variations measured at different distances from the heated wire center. Time-points of backscatter ultrasound signal record are put on the t-axis

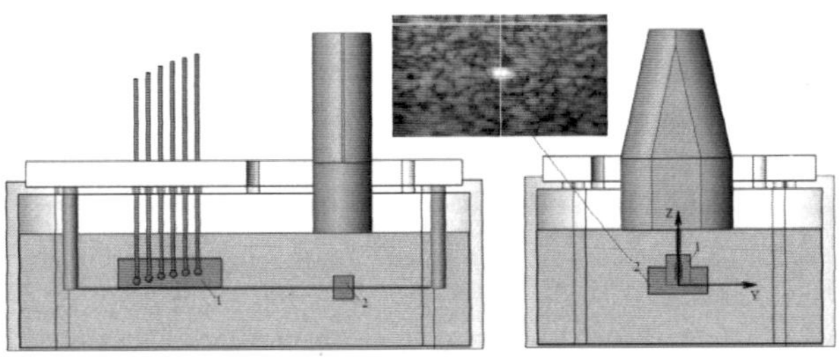

Figure 3. Location of zone 1 for temperature measurements with six probes and zone 2 for backscatter ultrasound signal record (two sections of the phantom are shown)

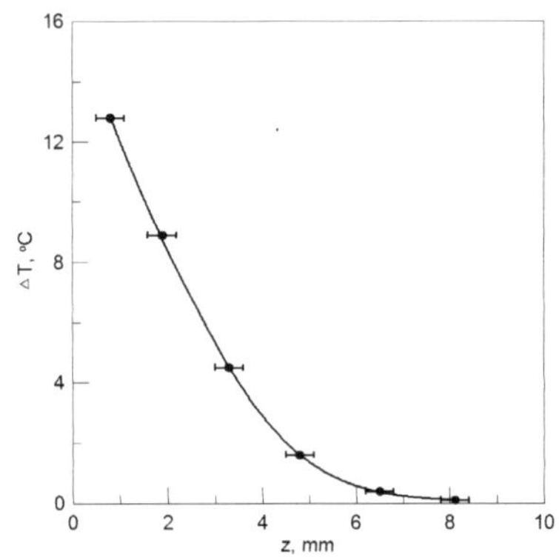

Figure 4. The measured temperature increase profile in the vicinity of the heated wire at instant of 20s data of 6 temperature probes and an approximation curve

Figure 5 illustrates the time shift of the ultrasound signal of the heated material relative to the reference signal. In [3,5] the time-shift is evaluated by a cross-correlation procedure. We suggest an alternative procedure, which demands less calculation and is accurate enough. The main idea is to evaluate the time-shift Δd as a difference between the average values of the direct and reflected signal for each half-cycle of oscillations.

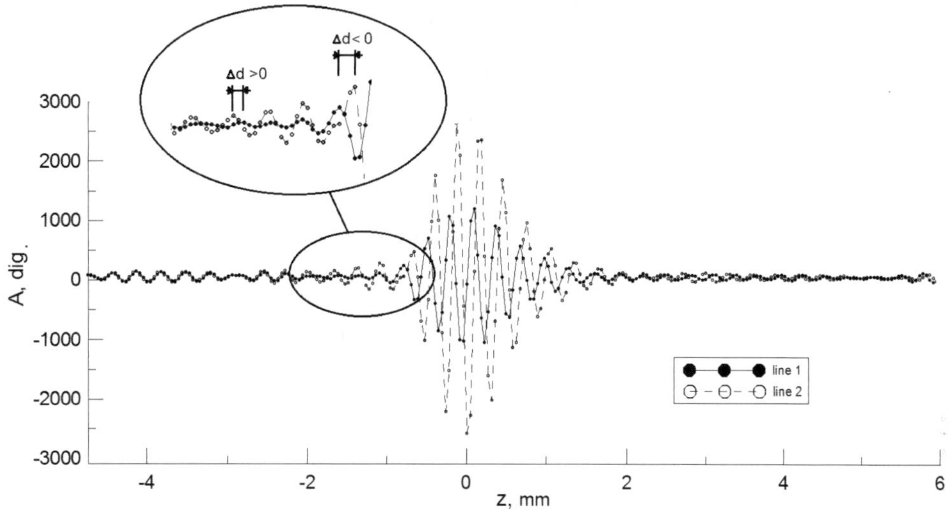

Figure 5. Illustration of the ultrasound signal shift: line 1 - before heating, line 2 – at a heating instant

Generally, difficulties in calculating the signal shift are caused by variations of the signal amplitude and by a rather limited number of recorded points for each half-wave of the curve (Figure 5). For more accurate and stable evaluation of the shift of the i-oscillation, it was determined as a difference, Δd_i , between centers of the corresponding groups of points (taken from the data obtained before and after heating), with the same sign of deviation from the mean value. The group center depth coordinate was calculated as $z_i = \sum_{k=1}^{N_i} A_k z_k \Big/ \sum_{k=1}^{N_i} A_k$, where A_k is the recorded ultrasound signal value, index k runs over all the points (1 to N_i) corresponding to a current half-oscillation.

Unfortunately, experimental signals can contain various kinds of defects: splitting an oscillation into two oscillations, occurrence of an unformed oscillation etc (Figure 5). The presence of these defects can lead to considerably errors in evaluations of the half-oscillation coordinates, and hence in the evaluation of the signal-shifts. As a rule, considerable defects arise near the heated wire. In order to reduce the influence of these defects, a special procedure is introduced for data conditioning. The procedure is based, first, on introducing a condition on the rate of the shift change: $|f(z_{i+1}) - f(z_i)| < c(z_{i+1} - z_i)$, where $f(z)$ is the magnitude of the shift Δd_i at point z_i, and, second, on a limitation of the difference of the shifts for two neighbor points: $|f(z_{i+1}) - f(z_i)| < g$.

For evaluation of the ultrasound signal thermal shift strain ε demanding application of a differencing procedure, the conditioned data for Δd_i should be approximated with an appropriate function. Some researchers [5] use the probability integral assuming that the temperature field satisfies the constant-properties heat conductivity equation. We use the arctangent function $\varphi(z_i) = a(\pi / 2 + arctg(bz_i))$ for this purpose, with two coefficients, a and b, that are defined using the least-squares method involving an iterative procedure to remove points with excessive deviations. The condition for removing such points is as follows: $\left|f_j(z_i) - \varphi_j(z_i)\right| > \gamma\sigma_j$, where $\sigma_j = \sum_{i=1}^{N_j} \sqrt{(f-\varphi)^2 / N}$ - is the root-mean-square deviation, $\varphi_j(z_i) = a_j(\pi / 2 + arctg(b_j z_i))$ is the approximation curve for the j-iteration. The experience of using the conditioning/approximating procedures developed has showed that the best values of the parameters introduced are as follows: c = 0.1, g = 4x(10/512) mm, γ = 1.5, j = 1÷4.

3. Results

Figure 6 presents an example of results from thermal shift data processing by the above-described procedures. The ultrasound data were taken for the beam passing 0.8 mm away from the center of the heated wire. To relate these data with values of temperature increase, the assumption of axial symmetry of the temperature field was adopted, so that $\Delta T(y$=0.8 mm, z=0) is assumed to be equal to $\Delta T(y$=0, z=0.8 mm) measured by a temperature probe positioned in zone 1 (Figure 3). Symbols in Figure 6 show experimental values of the signal shift that have remained after application of the data conditioning procedure and after removing points with excessive deviations from the approximating curve shown by solid line.

The same data processing procedure was applied to the data recorded for all the beams (of 161) emitted by the ultrasound transducer. The ultrasound signal thermal shift strain is evaluated as $\varepsilon = \frac{d}{dz}(\Delta d) = \frac{ab}{1+(bz)^2}$, according to the adopted approximating (arctangent) function. The thermal strain map obtained for the instant of 20 s after the wire heating start is shown in Figure 7 (as previously, the co-ordinate origin is located at the center of the wire). Generally, the map corresponds to axisymmetric temperature field arising during the unsteady heating of the tissue-mimicking material. However, a notable deviation from the axial symmetry is observed in the central part of the map that might be attributed to interaction of ultrasound beams with the wire.

As mentioned in Section 2, the temperature increase ΔT at a current point of ultrasound measurements can be calculated as $\Delta T = K\varepsilon$. Since the factor K is unknown for the used tissue-mimicking material, a part of the obtained experimental data (the smaller one) can be used to estimate this coefficient, and the remaining part (the bigger one) can be used to check an overall consistency of the measurement results obtained with thermometer probes and with the ultrasound technique.

In the present work, the data corresponding to the points positioned along the z=0 line at 0.8 mm< $|y|$ < 5.0mm was used for the purpose of the K-factor estimation. Figure 8 presents correlation of the thermal shift strain evaluated at these points and the temperature increase measured with temperature probes (here again the assumption of the temperature-field axial symmetry was used). Assuming a linear relation between ΔT and ε, we have evaluated the value of the K-factor for the tissue-equivalent material used as 540 °C.

The K-factor obtained was used to convert the thermal shift strain map given in Figure 7 into a temperature increase map shown in Figure 9. The central stripe-like zone in the temperature increase map, with $|z|$ < 0.8 mm has been omitted due to unreliability of the data for this zone where the wire produces strong distortions of the ultrasound field. Comparing the temperature increase map obtained with the ultrasound thermometry with the map reconstructed from the temperature probes measurements (Figure 10), one can conclude that the ultrasound thermometry technique developed produces reasonable results. Additionally, Figure 11 presents difference between the temperature increase values obtained with the two methods. In the main, the difference does not exceed 2 °C .

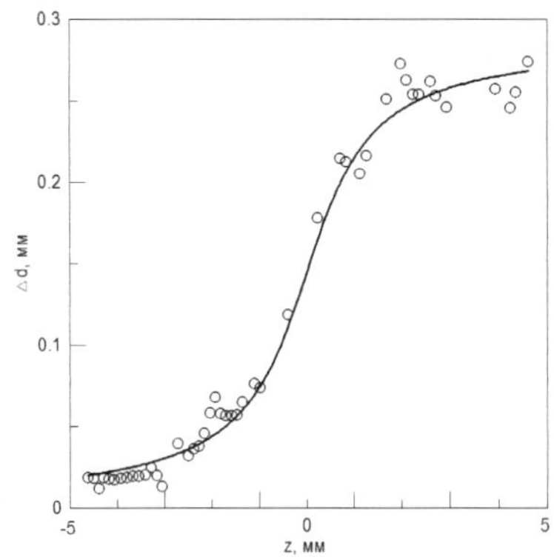

Figure 6. Ultrasound signal shift versus depth for the ultrasound beam positioned at y=0.8mm: (symbols) experimental data for t=20 s and (line) approximation by arctangent function

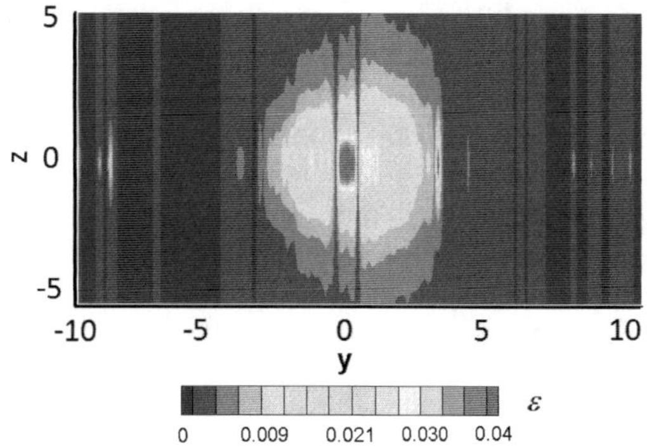

Figure 7. Map of ultrasound signal shift strain for t = 20 s

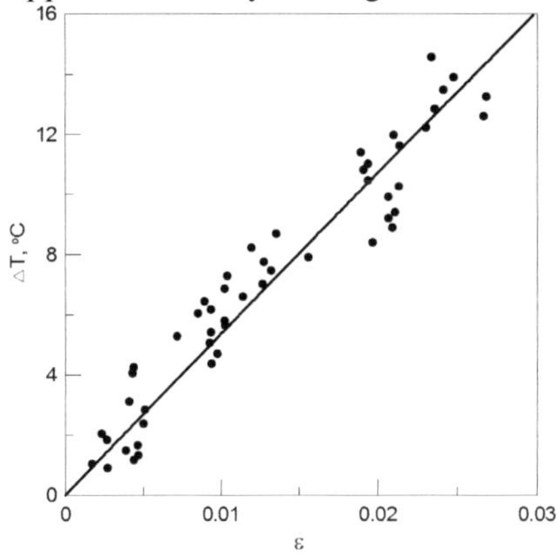

Figure 8. Correlation of the temperature increase and the ultrasound signal thermal shift strain: experimental data (z=0 mm) are approximated with a linear function: K = 540, R^2=0.88

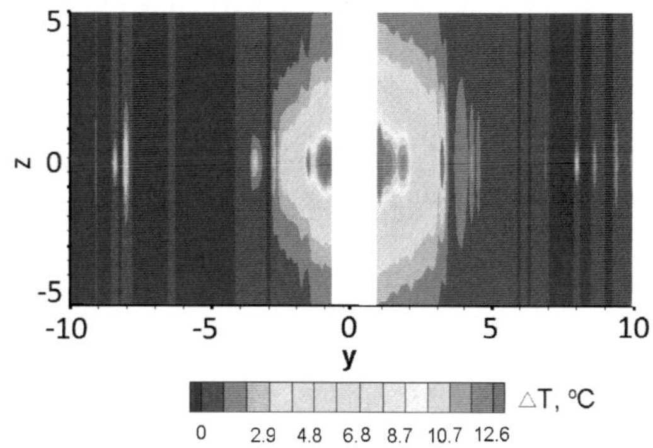

Figure 9. Temperature increase map obtained with the ultrasound thermometry for the instant of 20 s

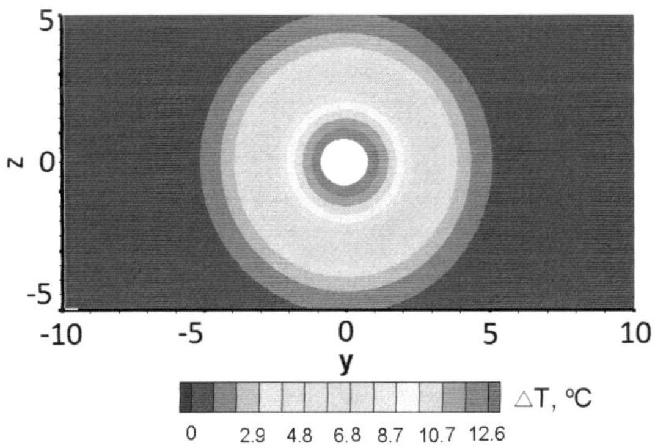

Figure 10. Axisymmetric temperature increase map reconstructed from the temperature probes measurements for the instant of 20 s

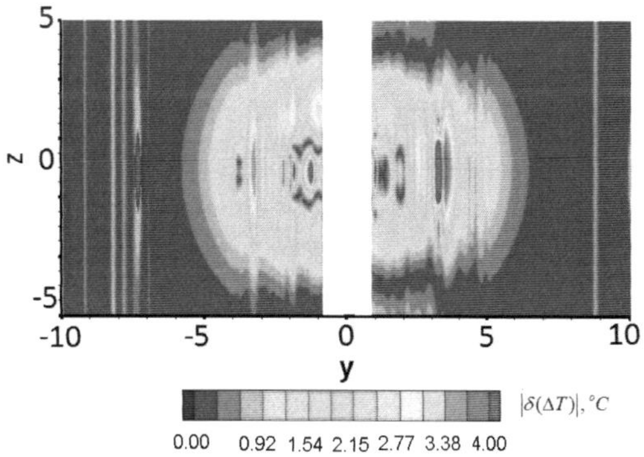

Figure 11. Map of difference between temperature increase obtained with the ultrasound thermometry and measured with the temperature probes

4. Conclusions

A phantom of tissue-mimicking material has been designed to elaborate in detail the method of ultrasound thermometry based on phenomenon of temperature-sensitivity of the speed of sound. A long heated wire, 0.2 mm diameter, was used to get time-dependent axisymmetric temperature increase relative to the reference stationary state of the phantom material.

Analysis of the primary backscatter ultrasound data used for evaluation ultrasound signals shift due to temperature increase has shown that typically these data contain various kinds of defects, mostly pronounced in the vicinity of the wire and in the region of relatively small temperature increase. Original algorithms have been developed for the primary data conditioning and their approximation of a smooth function needed for quantitative evaluation of the thermal shift strain.

Synchronized test data for time-dependent material temperature field was obtained with a set of thermistor probes. These data were used both for the evaluation of the *K*-factor in the adopted linear relation between values of local thermal shift strain and temperature increase and for estimation of an overall accuracy of the developed ultrasound thermometry technique. For the tissue-mimicking material used, a value of the *K*-factor obtained by correlation of the mostly reliable part of the thermal shift strain data and the test temperature data is equal to 540 °C.

Overall comparison of the temperature maps obtained with the ultrasound thermometry with temperature probes measurements has shown a reasonably accuracy of the developed method of ultrasound diagnostics of temperature field. In the main, the difference does not exceed 2 °C.

Further work will be devoted to additional improvement of accuracy and robustness of the technique presented, as well as to its application to the case of anisotropic tissue-mimicking materials that approximate biological objects more properly.

Acknowledgments. The study has been carried out with the financial support of the Ministry of Education of the Russian Federation (grant No.2015-2018-07-044 "Multifunctional complex for the diagnosis and therapy of breast and thyroid tumors").

References

[1] Khokhlova V.A., Crum L.A., ter Haar G., Aubry J.-F. (Eds.) 2017 High Intensity Focused Ultrasound Therapy (Berlin: Springer)
[2] Berkovich A.E., Bursian A.A., Senchik K.U. et al. Biomed Eng (2016) 50: 96
[3] Maass-Moreno R., Damianou C.A., Sanghvi N.T. J. Acoust. Soc. Am. (1996) 100: 2522
[4] Fuhrmann T.A., Georg O., Haller J. et al. J. Therapeutic Ultrasound (2016) 4: 28
[5] Anand A., Kaczkowski P. J. Ultrasound in Med. & Biol., (2008) 34, 1449

Digital transformation of education & training in photonical measurement engineering & quality assurance (PMQ)

D Hofmann[1], R Margull[1] and P-G Dittrich[1,2]

[1]SpectroNet c/o Technologie- und Innovationspark Jena GmbH
Wildenbruchstr. 15, 07745 Jena Germany

[2]Technische Universität Ilmenau,
Faculty of Mechanical Engineering,
Department of Quality Assurance and Industrial Image Processing
Gustav-Kirchoff-Platz 2, 98684 Ilmenau Germany

E-mail: d.hofmann@spectronet.de

Abstract.
Aim of the paper is the promotion of the **worldwide new situation** that

- the contactless acquisition of measured quantities becomes more convenient + reliable + affordable (c+r+a) by **photonical** sensor systems,
- the mobile processing of measured values becomes more c+r+a by **mobile** microcomputers for example smartphones,
- the **substantiation** of quality assurance on objective acquired (measured) data becomes more c+r+a by digital transformation,
- the **unified** measurements of physical, physiological and chemo-analytical measurement quantities become more c+r+a by photonical measurement sensor systems with filters on CMOS chips to determine for example
    - **geo-physical** quantities (points, lengthes, areas and spaces of solids which nowadays can be acquired contactless with laser beams),
    - **medico-physiological** quantities (color sensitivities of test persons in accordance with standardized eye color sensitivity curves based on normally 7 Mio red-green-blue eye cuppings which are identifying up to 600.000 color shades) and
    - **chemo-analytical** quantities that means compositions of solids, fluids and gases which are measurable with spectrometers or nowadays with hyperspectral cameras.

1. Introduction into a new situation by historical road mapping

The **19th Century** was the century of mechanical MATTER processing. The social progress driver was the "artificial" industrial processing of matter. Typical examples are steel factories and heavy machine industry. The clustering of industrial centers has been accomplished by country roads, water ways and rail ways.

The **20th Century** was the century of electrical ENERGY processing. The social progress driver was the "artificial" industrial processing of electrons. Typical examples are electric power stations and electric motors. The clustering of industrial power centers has been accomplished by overhead power lines and urban electricity networks. A parallel development was the dawn of information processing by mechanical, electrical and electronical computers and the information transfer by telephone, radio and television – analogue, wire-bound and wire-less.

The **21st Century** is the century of photonical INFORMATION processing. The social progress driver is the "artificial" industrial and individual processing of photons. Typical examples are solar power stations and mobile information machines for imaging. The clustering of industrial and personal information centers is accomplished by smartphones with cameras and cloud-based image processing – digital, fiber-bound and fiber-less.

2. Digital and cloud-based education and training in photonical measurement engineering and quality assurance

The **SPIE Press catalog** [1] is the world's largest collection of information on optics & photonics applied research. It contains > 450.000 research papers and 10 scholarly journals with around 18.000 new papers added each year, and > 250 eBooks. It is a valuable source for education and training in optics & photonics also.

The **SpectroNet** cluster [2] contains > 3.000 papers and > 2.000 videos dealing with photonic measurement engineering and quality assurance.

The **Industry** maintains academies, summer schools, hubs and training centers with valuable contributions for a digital transformed education & training in photonical measurement engineering and quality assurance. Selected examples are:

- European Imaging Academy [3],
- Photonics21 Summer Schools [4],
- Basler AG Imaging Hub [5],
- Fluke Open Imaging Training Center [6].

3. Systematizations in photonical measurement engineering and quality assurance

For easier understanding of new contents in photonical measurement engineering and quality assurance selected HOW's and HOW-TO's will be illustrated.

When speaking about photonical measurement engineering every part of the measurement circle must be considered. Applicable solutions for quality assurance through methods, hardware and software are essential (Figure 1).

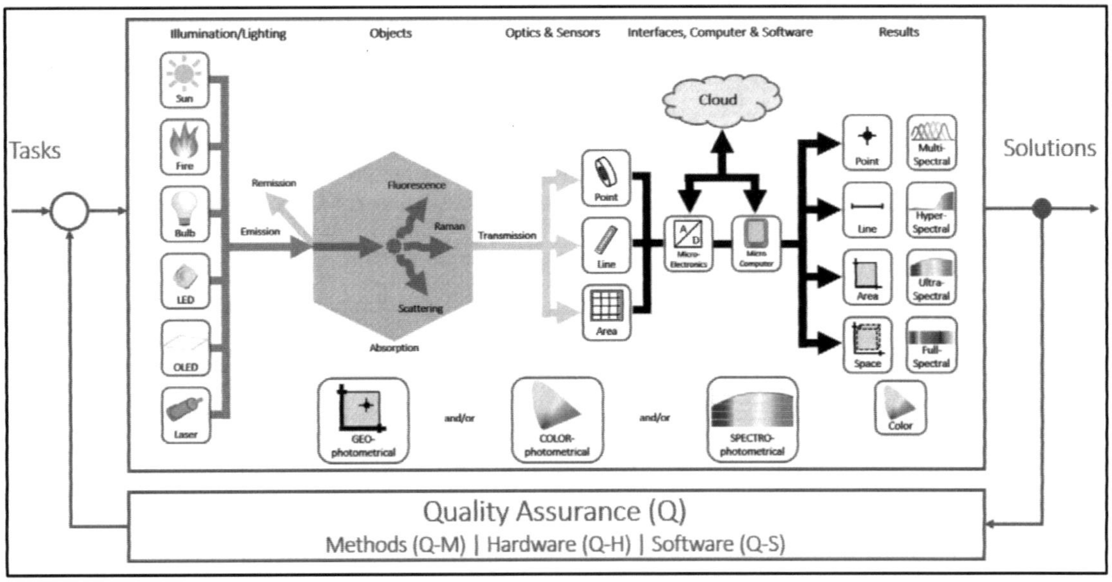

Figure 1. HOW and HOW-TO photonical sensor systems can be systematized.

Nowadays the so called "PhotoSensors 4.0" enable the technical and objective measurement acquisition of geo-, color- and spectro-photometrical information (Figure 2).

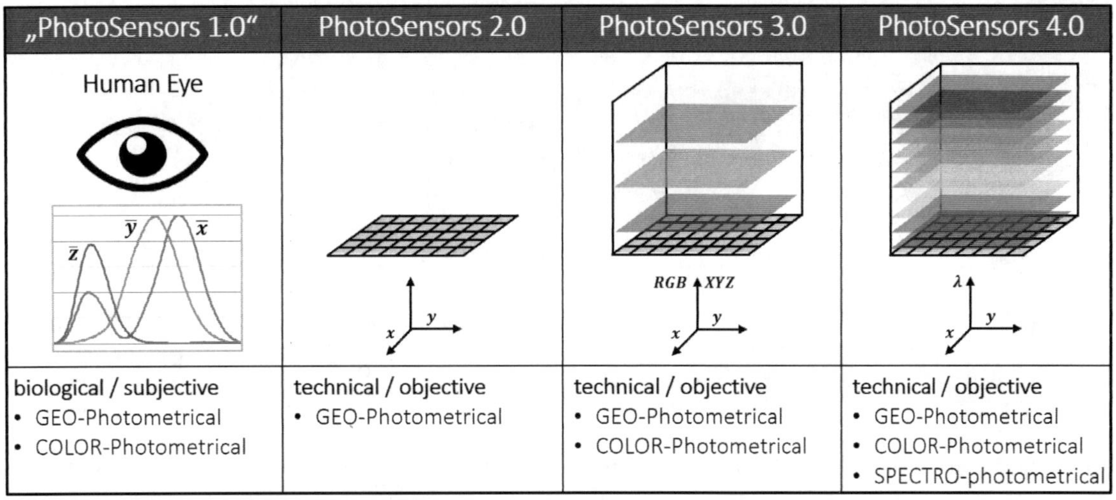

Figure 2. HOW and HOW-TO photo sensors can be arranged.

Concerning the application of PhotoSensors 4.0 there are different sensor principles which enable real-time acquisition (Figure 3).

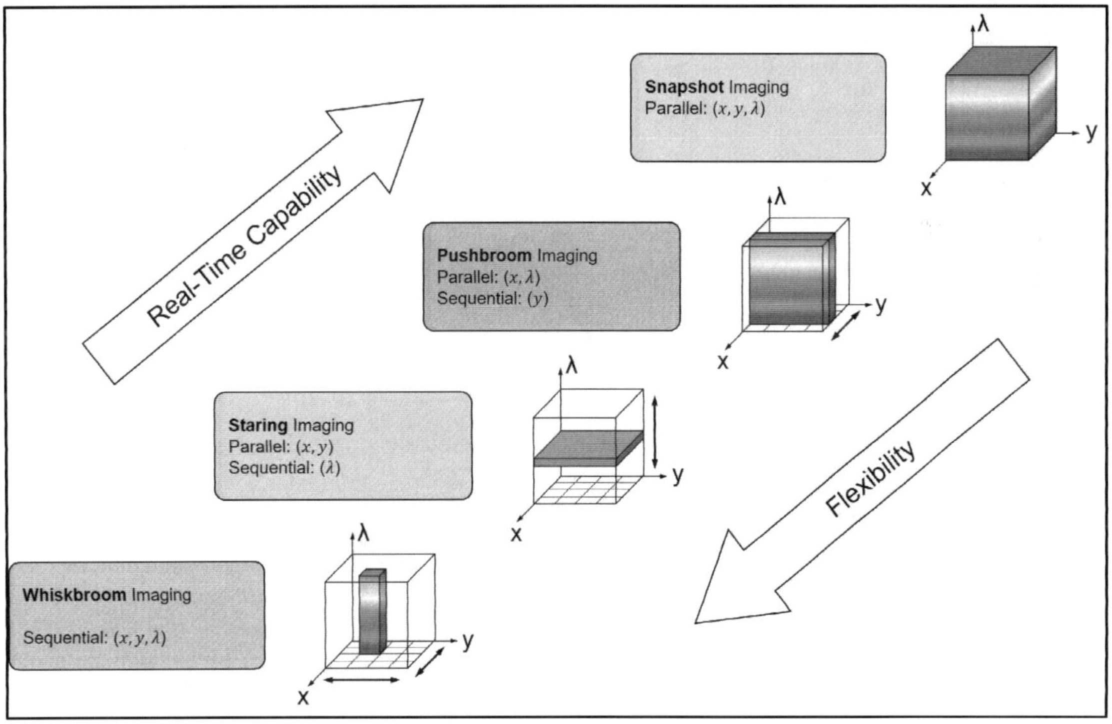

Figure 3. HOW and HOW-TO the real-time capability of photo sensors is construction dependent.

Latest developments of Photosensors 4.0 work with Filter-On-Chip (FOC) CMOS-Sensors. This enables the parallel acquisition of geo-, color- and spectro-photometrical information via video rates (Figure 4).

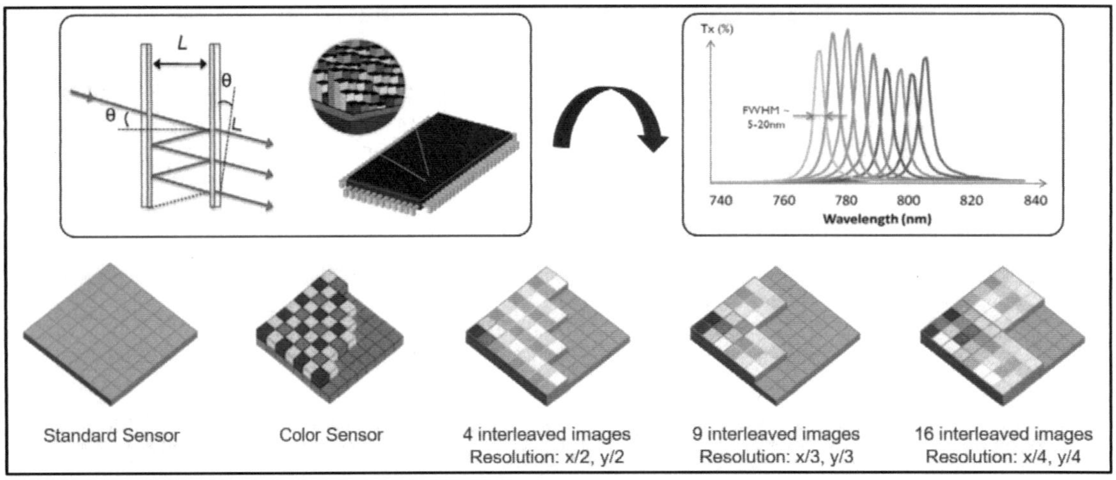

Figure 4. HOW and HOW-TO filter-on-chip solutions can be classified [7], [8].

Every developments in measurement engineering and quality assurance need their time to become usable as mass products. The current breakthrough of PhotoSensors 4.0 as multispectral Filter-On-Chip CMOS-camera systems started 40 years ago, by the "MKF 6" first multispectral camera which was developed and manufactured by Carl Zeiss Jena in 1976. The essential difference between latest developments and the first multispectral camera in 1976 is the reduction of mass (Figure 5).

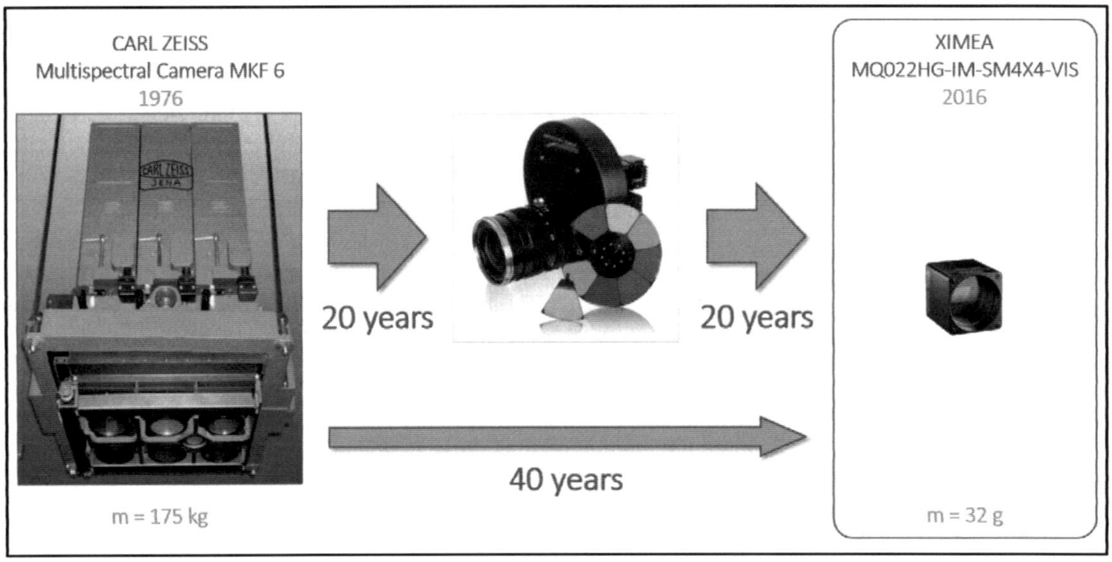

Figure 5. HOW and HOW-TO multispectral cameras lost their mass (weight) as a function of time.

Another main thing to realize is that photonical measurement engineering and quality assurance becomes applicable via hand-held photonical measuring systems and move into the mass markets (Figure 6).

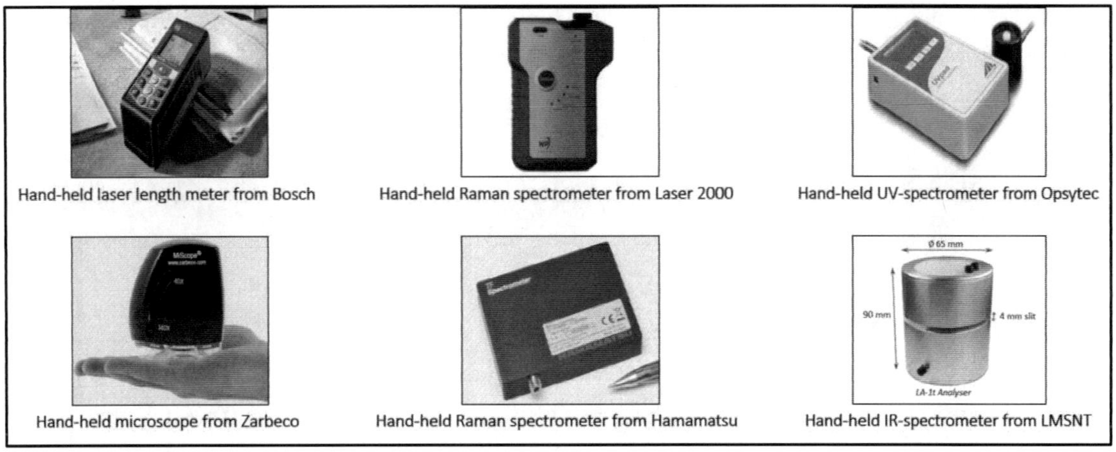

Figure 6. HOW and HOW-TO hand-held photonical measuring systems move into the mass market [9], [10], [11], [12], [13], [14].

Through latest developments in photonical sensors and their usable detector ranges as well as optical materials show the extension of the measurement regions from UV via VIS to NIR (Figure 7).

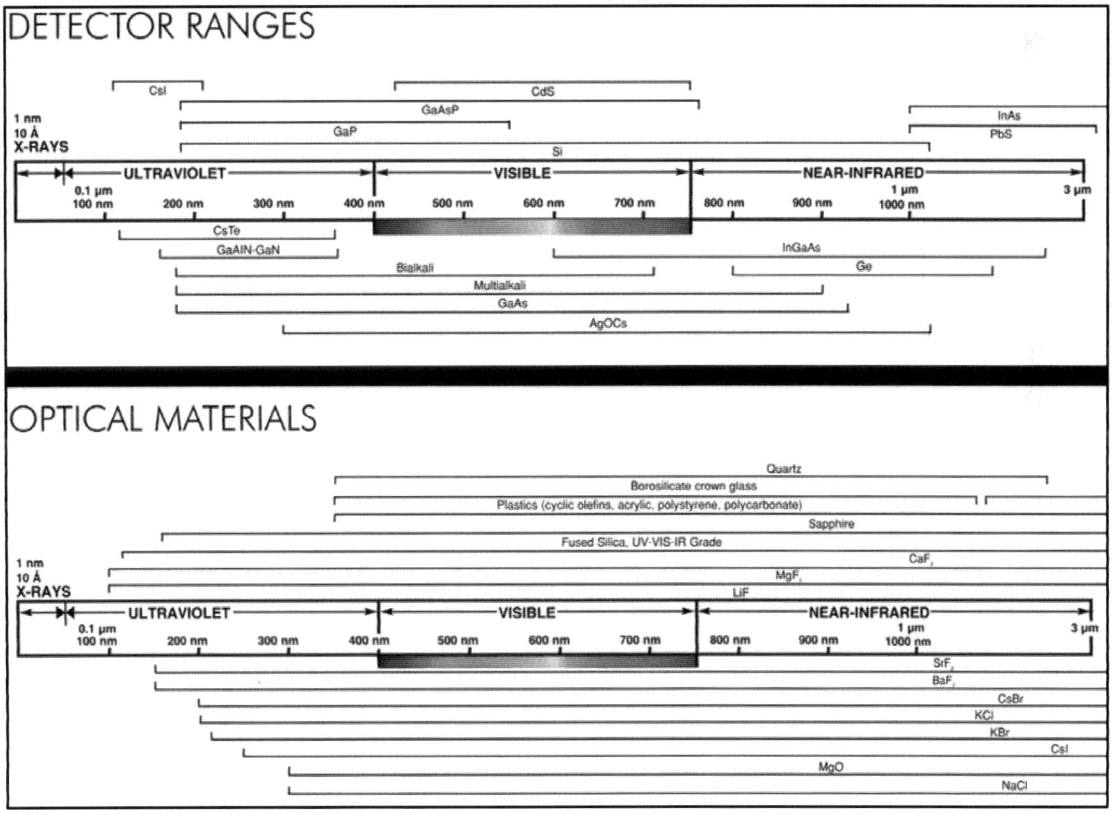

Figure 7. HOW and HOW-TO the measuring ranges are extended from visible to ultraviolet (left) and infrared (right) [15].

To overcome higher sensor complexity and the necessary complex data interpretation a big amount of software packages can be used for image processing (Figure 8).

Figure 8. HOW and HOW-TO image processing can be done.

When tackling complex sensors and systems education and training in photonical measurement engineering for employees, partners and customers is becoming more and more important. To tackle these facts one digital platform for self-service-learning by online accessible presentations and videos can be mentioned (Figure 9).

Figure 9. HOW and HOW-TO an individual digital education & training program can be done.

4. Summary & Conclusions

Due to the well-known and accepted proverb "A picture is worth a thousand words" the 21st Century as Century of Photonics will be full of "enlightenments" in pictures. Selected HOW's and HOW-TO's have been illustrated to facilitate the understanding of the significant influence of photonics on daily life.

The digital transformation proves to be also a valuable assistant for education and training in photonical measurement engineering and quality assurance. The digital transformation makes education and training both self-controllable and c+r+a (convenient + reliable + affordable).

References

[1] „SPIE Digital Library: The world's largest collection of optics & photonics research," [Online]. Available: http://spiedigitallibrary.org/. [Zugriff am 24 March 2017].

[2] „SpectroNet - International Collaboration cluster," [Online]. Available: http://www.spectronet.de/. [Zugriff am 24 March 2017].

[3] „European Imaging Academy (EIA) | STEMMER IMAGING," [Online]. Available: www.stemmer-imaging.de/de/european-imaging-academy/. [Zugriff am 24 March 2017].

[4] „Photonics21," [Online]. Available: http://www.photonics21.org/TrainingEducation/PhotonicsSummerSchools.php. [Zugriff am 24 March 2017].

[5] „Basler AG on Imaginghub," [Online]. Available: https://imaginghub.com/partners/51-basler-ag. [Zugriff am 24 March 2017].

[6] „Thermal Imaging Training Centre," [Online]. Available: http://www.fluke.com/fluke/auen/training/training/titraining/default. [Zugriff am 24 March 2017].

[7] „Imec - Hyperspectral imaging," [Online]. Available: http://www.imec-int.com/en/hyperspectral-imaging. [Zugriff am 24 March 2017].

[8] „SILIOS Technologies diffractive optical element | CS Filters for Multispectral Imaging," [Online]. Available: http://www.silios.com/cs-filters-for-multispectral-imaging. [Zugriff am 24 March 2017].

[9] Bosch, „Laser-Entfernungsmesser Messtechnik | Bosch Professional," Bosch, [Online]. Available: https://www.bosch-professional.com/de/de/laser-entfernungsmesser-101300-ocs-c/. [Zugriff am 04 July 2016].

[10] Zarbeco, „MiScope • Handheld Digital Microscope," Zarbeco, [Online]. Available: http://zarbeco.com/products-misc.shtml. [Zugriff am 04 July 2016].

[11] Laser 2000, „Handheld Raman Spectrometer," Laser 2000, [Online]. Available: https://www.laser2000.de/Photonics/Optische-Messtechnik/Spektroskopie/Ramanspektrometer/Handheld-Raman-Spectrometer.html?force_sid=8is53qbp9notid9f9m7rrncvq5. [Zugriff am 04 July 2016].

[12] Hamamatsu Photonics, „Mini-spectrometer TF series C13054MA," Hamamatsu Photonics, [Online]. Available: http://www.hamamatsu.com/eu/en/product/category/5001/4016/C13054MA/index.html . [Zugriff am 04 July 2016].

[13] Opsytec, „Spektralradiometer UVpad E - mit externem Sensorkopf für Bestrahlungsstärke," Opsytec, [Online]. Available: http://www.opsytec.de/produkte/spektrometer/uvpad-e/. [Zugriff am 04 July 2016].

[14] LED Microsensor NT, „LA-1t LED Analyser," LED Microsensor NT, [Online]. Available: http://lmsnt.com/evalkits/LA1t/. [Zugriff am 04 July 2016].

[15] EDU.Photonics.com, „The Photonics Spectrum Reference Chart | Photonics Handbook® |," EDU.Photonics.com, 13 March 2009. [Online]. Available: https://www.photonics.com/EDU/Handbook.aspx?AID=25111. [Zugriff am 20 June 2017].

Ecologizing vs modernizing in measurement and metrology

W P Fisher Jr[1] and A J Stenner[2]
[1] Research Associate, BEAR Center, Graduate School of Education, University of California, Berkeley, Berkeley California 94702 USA
[2] Senior Scientist, MetaMetrics, Inc., Durham North Carolina 27713 USA

[1]E-mail: wfisher@berkeley.edu
[2]E-mail: jstenner@lexile.com

Abstract. The complex multilevel nature of language is a significant barrier to the conceptualization, design, and implementation of effective large scale measurement communications systems in education. The barrier's existence and challenges are obscured, first, by the modern emphasis on data as the hallmark criterion of an objective basis for decision making. The situation is further obscured by the postmodern observation that attention is focused by linguistic, historical, and cultural concerns. Though it is true that data are salient only to the extent implicit or explicit theory is brought to bear, postmodernism does not follow through with its insights into how to create effective information infrastructures. Data, theory, and instruments function interdependently in the historical successes of science to bring things into language. The ecological economy of language is realized by the efficiencies obtained when new concepts produced by collective learning processes are embodied in tools distributed throughout cognitive ecosystems. These ecosystems are comprised of discontinuous but manageably coherent levels of increasingly complex forms of information, following the pattern of everyday language. An emerging reading measurement ecosystem of this kind offers a developmentally, horizontally, and vertically coherent language for managing literacy education across niches defined by classrooms, schools, districts, homes, offices, libraries, test and assessment agencies, and book publishers.

1. Introduction

The fact that language functions at varying levels of complexity and abstraction has long been recognized and addressed by philosophers and linguists [1-4]. Only recently, however, have information systems designers begun to address the conceptual and practical problems associated with the multilevel nature of language [5-11]. The problems encountered are particularly acute in the context of education's need for developmental, horizontal, and vertical coherence across the multiple kinds of assessments employed [12-14]. Many problems encountered in the design of educational measurement information systems are rooted in the unrecognized and unaddressed complexities of language's multiple levels of abstraction. Until these problems are confronted, little in the way of effective measurement communications in education and other fields can be expected.

Questions that arise in this context ask if it possible to reconcile the tension between the need in education and other areas, such as health care, for local, personal, and customized assessment

[1] To whom any correspondence should be addressed.

technologies, on the one hand, and the need for those technologies to be traceable to common standards. Must standards always require an imposition of irrelevant or misguided criteria on local idiosyncracies? Will the culture wars between modern universalism and postmodern relativism become a permanent feature of intellectual life? Or will embedded relationality prove its capacity to integrate the opposition of modern universalism and postmodern relativism? Is there some way in which we can care for our educational and healthcare measurement technologies as we do our children [15], with the aim of preventing the creation of new well-intentioned but nonetheless monstrous effects? Can the unique particulars of local experience be meaningfully and usefully integrated into our institutions' information infrastructures [6]? Is it possible for the values and experiences of individuals to be incorporated deeply and authentically into planned, built, and legislated forms of social life? Could the skills, knowledge, and responsibilities of those individuals be enhanced by their participation in those forms of social life?

Positive answers to these questions are suggested by recent developments in philosophy and the history and social studies of science [9-11, 15-19], and in measurement theory synthesizing engineering and psychometric perspectives [20-24].

Because the multilevel complexities of language are not taken into account, education, health care, human resource management, social services, and many other areas of life are marked by a kind of schizophrenia [4-5] that emerges in terms of the dissonance between a caring focus on individual needs for learning and healing, on the one hand, and demands for accountability focused on standards and comparability. Support for irrevocable concerns with the individual student's and patient's spontaneous processes of development and healing stands as an immovable thesis that is increasingly in opposition to the larger social antithesis of an imposed demand for evidence proving the achievement of quality standards. How might new institutional forms of social life resolving the schizophrenic break be formed at a higher level of information infrastructure complexity? How might those measured forms of life synthetically integrate the necessary concern for development and healing with a new ecologizing bottom-up approach to accountability and standards that authentically embodies individual uniqueness?

Linguistic communication systems incorporate within-individual processes distinct from, but interacting with, mid-level processes between individuals, and which in turn are distinct from but interacting with high-level group processes. These levels of complexity in communication have informed practical applications in epidemiology [7] leading to new, productive relationships between clinical medicine and public health efforts [8].

Might not a similar kind of productivity be possible in education if we apply similar ecological approaches to conceiving, designing, and implementing developmental, horizontal, and vertical coherence [12-14] in educational assessment? Cognitive and social ecologies contextualize problems of coherence in educational assessment in ways that may be key to understanding learning in each distinct niche of the various environments in which it lives. Learning varies across these levels of complexity in ways that cannot be grasped directly from individual measures. What form might conceivably be taken by educational assessment communications systems capable of supporting broad-scale efforts at sorting out the sources of distinct classes of effects on learning?

2. Steps toward an ecology of assessment information infrastructure

What specific problems are encountered when discontinuities in levels of linguistic abstraction are ignored in the design of information infrastructures? Each different paradigmatic approach to understanding the relation of data, theory, and instruments—modern positivist, postmodern antipositivist, and unmodern postpositivist [25]—creates its own kinds of problems in this regard. Philosophically modernist conceptions of science are positivist in the sense of prioritizing a focus on data as the ultimate criterion of objectivity. Postmodernism, in contrast, is sensitized by historical changes in what data count as worthy of attention and so is concerned with the role of theory in making data salient. Unmodern (also known as amodern) postpositivist perspectives [17, 26] assert that the debate between modern and postmodern focuses too exclusively on the mutual implication of theory and data, and so will remain unresolved as long as the roles of instruments and knowledge technologies are not taken into account. Instruments encapsulate what is learned from data and what can be explained by theory. That assemblage is then circulated within networks sharing common standards (vocabularies, units, grammars, etc.).

Unmodern philosophical perspectives and research in the history of science [5, 9-11, 15-19] focus on the collective cognition and team-based coordinations made possible when theory-data-instrument assemblages are expressed in a uniform language distributed throughout a community of practice. Metrology's concern with measuring instruments traceable to unit standards then becomes a matter of focal interest as a way in which everyday model-based reasoning has been extended productively into science [9]. Recent developments suggesting metrological paths forward for the constructs of psychology and the social sciences [20-24, 27-28] also extend everyday model-based reasoning [27] and open up new possibilities for enhanced innovation in education, health care and other fields. A significant problem that remains unaddressed is how varying levels of information complexity can be integrated into a new metrological culture encompassing all of the arts and sciences.

For instance, in applying educational measures for formative and accountability purposes, we face a problem of coherently coordinating and aligning representations of processes and outcomes across discontinuous levels of complexity. Taking steps toward an ecology of infrastructure alongside Bateson's [4] steps to an ecology of mind, Star and Ruhleder [5] point out:

> These discontinuities have the same conceptual importance for the relationship between information infrastructure and organizational transformation that Bateson's work on the double bind had for the psychology of schizophrenia. If we, in large-scale information systems implementation, design messaging systems blind to the discontinuous nature of the different levels of context, we end up with organizations which are split and confused, systems which are unused or circumvented, and a set of circumstances of our own creation which more deeply impress disparities on the organizational landscape.

In educational assessment, the usual approach to counting up correct answers and treating that score as a measure is known as True Score Theory (TST) [28]. TST works at the local level of simple denotation. Bateson's [4] example is "The cat is on the mat." TST says, "Your test score is x." Bateson observes that this kind of simple statement of fact is very different from the metalinguistic statement, "The word 'cat' has no fur." The object of reference has changed from a small mammal to a word. Similarly, in saying "Your score of x on the test means you're proficient," TST wants to shift the focus from recorded responses to a broader linguistic construction. But it does so in a limited way.

The difference is that a word like 'cat' has a place in the larger culture recognized by virtually everyone who can speak the language, whereas the meaning of the test score depends on the local context of particular questions asked. 'Cat' emerged from a self-organized historical process that was not directed by any one person or group, and that resulted in a commonly shared sign universally understood within a cultural context as representing a particular kind of animal. The TST score, in contrast, was determined by the intentions of someone who composed the questions and recorded the responses, not by a collective, undirected, self-organized process. Statistical models applied to these data take a modern positivist approach primarily interested in describing the facts of the situation, with little or no effort invested in developing construct theories or in calibrating interval-level measuring instruments capable of supporting more complex metalinguistic statements.

What possibilities exist in educational measurement for fostering the manifestation and observation of emergent, self-organizing phenomena that could be represented in signs referring to shared objects in universally accessible and shared symbols? Do not group level constructs emerge from the internal coherence of responses to test questions? Psychometric constructs are invariant in a way that is not under the control of intentions imposed from the outside by the person asking the questions and recording the responses (though the construct is given a medium for its self-organized expression). Might not there be a way to interpret and contextualize these constructs so as to bring them into language in meaningful and useful terms, in a way more akin to the way the word "cat" emerged than is obtained by TST?

Rasch's models for unidimensional measurement [28-31] are sensitive to the distinction between the local denotative and mid-range metalinguistic levels. These models enable the development of construct theories and the calibration of interval-level instruments. Stopping with analyses applying Rasch's models for measurement to given data sets, though, does not complete the process. To follow through, we need to distribute the concept throughout a network of shared signification by embodying it in words that consistently mean the same thing wherever they turn up. This is the task of metrological traceability. Coordinating and aligning meanings on this scale is incredibly difficult given the variation in the questions and answers that exists, and, especially, given the widespread lack of appreciation for the value of explanatory, predictive theory capable of efficiently removing the need for data-based measure estimation [32-33].

The difficulty becomes even greater at the third, metacommunicative level. This level's statements include "My telling you where to find the cat was friendly," or "Your instructional methods in the class with the mean test score that changed so much must have been superior." Now comments on associated aspects of the situation are being brought to bear and have to be evaluated. Maybe saying where to find the cat was not friendly but was motivated by a desire to have the animal removed from your garden. Perhaps the mean test score changed less because of instructional methods than because of the students' home environments, genetic propensities for learning, or some combination of these.

So, in the same way that a group-level effect occurs in the shift from denotative to metalinguistic communications, another one occurs at the metacommunicative level. Here we encounter the problem of the ecological fallacy, the problem of inferring individual-level characteristics from group-level statistics. We overcome one facet of this problem with Rasch's probabilistic models by testing the hypothesis of meaningful comparability at the group-level of coherent constructs, and by accepting individual departures from the model expectations in technologies like the kidmap [34]. But other factors affecting comparability remain unevaluated.

That is, the causal relationship between question and answer exists in a context in which the consistency of ability relative to difficulty may be stable across a wide range of different circumstances, such as nutrition, genetics, environmental quality, etc. Until we identify what these factors are and account for them in a metrological context in which we share common languages and can sort things out effectively in generalizable ways, we will continue to fail in our efforts to create institutions that systematically support and value human sensitivity and relationships. Models and methods effectively integrating the denotative, metalinguistic and metacommunicative levels [35] take these hierarchical effects into account in ways that do not penalize but reward enacting these qualities in individual and small group relationships in classrooms and clinics.

3. A reading measurement ecosystem

Reading measures are linked together in an ecosystem that has capitalized on the literacy form of life that consistently asserts itself across samples of students, texts, test items, time, and space [24]. Niches in this ecosystem span a wide range of classrooms, schools, homes, libraries, testing agencies, and book publishers. Reading test item difficulties have been shown to be remarkably stable over decades of use [36] and moreover can be predicted by an explanatory theory accounting for over 90 percent of the observed variance [24, 32]. More than 100 English language reading tests across the world measure in a common unit. Over 30 million student measures annually are interpreted relative to 250,000 book measures and 200 million article measures, where matching student and text measures predict a 75 percent comprehension rate. Books, articles, assessments, and students have been brought into a common frame of reference in a process now over 27 years old and still accelerating. Text complexity corresponds with reading learning progressions such that student measures enable the individualization of instruction.

Each of the three forms of coherence sought for educational assessment information infrastructures [13-15] is provided in this system. Student measures are tracked over time and across grade levels, instantiating developmental coherence. Teachers are able to compare learning outcomes across their own and each other's classes, realizing horizontal coherence. And in many locations, state end-of-year or graduation tests report in the common unit, providing parents, students, teachers, principals, librarians, researchers, and the public with the vertical coherence needed for connecting classroom formative assessments with accountability standards.

4. Implications

An obvious starting point for developing effective educational assessment information infrastructures and associated ecosystem communities of practice lies in the complexity of the partially connected relationships between stakeholder groups in technologically mediated networks of various kinds. Galison [25, pp. 844-845] wonders what new analogy might be capable of informing models of science that have emerged in the wake of postmodern deconstructions showing communities of theoreticians, experimentalists, and instrumentalists as disunified. Galison raises examples of amorphous crystals in electronics and laminated materials in structural engineering as instances in which certain amounts of disorder provide greater reliability and utility than could be obtained if the materials were perfectly and rigidly structured. Wittgenstein's metaphor of concepts as intertwined strands of fibre in a thread that is stronger than it would be if it were formed from only one single continuous fibre is plainly relevant, as Galison recognizes. But Galison is sensitive to the dynamic processes involved in social networks and seeks instead a nonmechanical way of coordinating the "different symbolic and material actions [through which] people create the binding culture of science." Berg and Timmermans [10, p. 56] independently concur, saying that the stability and reach of the medical decision networks they studied were "not due to more (precise) instructions: the protocol's logistics could thrive only by parasitically drawing upon its own disorder." The ways in which differing orders of things chaotically aggregate in harmonic and dissonant patterns is similar to the effects of stochastic resonance, which may provide an apt metaphor for psychological and social ecosystem processes [37].

Instead of demanding strict conformity with standards, then, it is likely more realistic and productive to think of standards as potential universals in partially interconnected, resonant, and multilevel traceability networks, bypassing the strictly local problem-solution dependency and the universal problem-solution independence at the same time [11, p. 229]. These "glocal" media, simultaneously local and global, are characterized by Ricoeur [38, p. 289] as potential or inchoate universals. Dewey [39, p. 215] similarly held that "The local is the ultimate universal, and as near an absolute as exists." The interconnections of metrological networks supporting local approximations and translations of standards is pointed to by Golinski [18, p. 35] as replacing the uniform universality assumed in modern science. And Haraway [16, pp. 439-440] suggests another account as to how locally embedded relationships offer an alternative to both relativism and transcendence.

The sustainability opportunities created within an ecologizing paradigm stem from the co-evolution of (a) concepts embodied in linguistic and measurement technologies and (b) the institutional rules, roles, and responsibilities within multilevel social, political, and economic ecologies [40]. The end results are systems of tools embodying individually unique problem-solution unities that are useful in negotiating local particularities while still recognizable as belonging to an identifiable general class. These results suggest potentially large payoffs of new analogies from existing online engineering models of global cooperation enabling intelligent metrology applications [41]. Perhaps caring for our measuring technologies in education and other fields the same way we care for our children will yet lead to creation of forms of social life sensitive to the values and experiences of those who inhabit them.

References

[1] Whitehead A N and Russell B 1910-1913 *Principia Mathematica, 3 vols* (Cambridge, England: Cambridge University Press)
[2] Wittgenstein L 1922 *Tractatus logico-philosophicus* (London: Harcourt Brace)
[3] Carnap R 2003 *The logical syntax of language* (New York: Harcourt Brace)
[4] Bateson G 1972 *Steps to an ecology of mind* (Chicago: University of Chicago Press)
[5] Star S L and Ruhleder K 1996 *Information Systems Research* **7** 111-34
[6] Scott J C 1998 *Seeing like a state: How certain schemes to improve the human condition have failed* (New Haven: Yale University Press)

[7] Susser M and Susser E 1996 *Am J Public Health* **86** 674-77
[8] Bizouarn P 2016 *Médicine/Sciences* **32** 500-505
[9] Nersessian N J 2012 *Mind, Culture, and Activity* **19** 222-39
[10] Berg M and Timmermans S 2000 *Configurations* **8** 31-61
[11] Latour B 2005 *Reassembling the social* (Oxford, England: Oxford University Press)
[12] Gorin J and Mislevy R 2013 *Inherent measurement challenges in the next generation science standards for both formative and summative assessment* (Princeton, NJ: ETS)
[13] Wilson M 2004 *Towards coherence between classroom assessment and accountability* (Chicago: University of Chicago Press)
[14] National Research Council 2006 *Systems for state science assessment* eds M R Wilson and M W Bertenthal (Washington, DC: The National Academies Press)
[15] Latour B 2011 *Breakthrough J* **2** 21-8
[16] Haraway D J 1996 *The disunity of science* eds P Galison and D J Stump (Stanford, California: Stanford University Press) pp 428-41
[17] Dewey J 2012 *Unmodern philosophy and modern philosophy* (Carbondale, Illinois: Southern Illinois University Press)
[18] Golinski J 2012 *Osiris* **27** 19-36
[19] Overton W F 2015 *Handbook of child psychology and developmental science, Volume I* eds W Overton and P C M Molenaar (Hoboken, NJ: Wiley) pp 9-62
[20] Mari L and Wilson M 2014 *Measurement* **51** 315-27
[21] Pendrill L and Fisher W P Jr 2015 *Measurement* **71** 46-55
[22] Pendrill L 2014 *NCSLi Measure: The Journal of Measurement Science* **9** 22-33
[23] Wilson M, Mari L, Maul A and Torres Irribara D 2015 *J Physics Conf Series* **588**(012034)
[24] Fisher W P Jr and Stenner A J 2016 *Measurement* **92** 489-96
[25] Galison P 1997 *Image and logic: A material culture of microphysics* (Chicago: University of Chicago Press)
[26] Latour B 1993 *We have never been modern* (Cambridge, Massachusetts: Harvard University Press)
[27] Fisher W P Jr 2010 *J Physics Conf Series* **238**
[28] Wilson M R 2013 *Measurement* **46** 3766-74
[29] Rasch G 1980 *Probabilistic models for some intelligence and attainment tests* (Chicago: University of Chicago Press)
[30] Andrich D 1978 *Psychometrika* **43** 561-73
[31] Wilson M 2005 *Constructing measures: An item response modeling approach* (Mahwah, New Jersey: Lawrence Erlbaum Associates)
[32] Stenner A J, Fisher W P Jr, Stone M H and Burdick D S 2013 *Frontiers in Psychology: Quantitat Psychol and Measurement* **4** 1-14
[33] DeBoeck P and Wilson M 2004 *Explanatory item response models: A generalized linear and nonlinear approach* (New York: Springer-Verlag)
[34] Mead R J 2009 *J Appl Measurement* **10** 208-24
[35] Beretvas N and Kamata A 2007 *Rasch measurement: Advanced and specialized applications* ed E V Smith Jr and R M Smith (Maple Grove, MN: JAM Press) pp 290-470
[36] He W and Kingsbury G G 2016 *J Physics Conf Series* **772** 012022
[37] Fisher W P Jr and Wilson M 2015 *Pensamiento Educativo: Revista de Investigacion Educacional Latinoamericana* **52** 55-78
[38] Ricoeur P 1992 *Oneself as another* (Chicago, Illinois: University of Chicago Press)
[39] Dewey J 1954 *The public and its problems* (Athens, Ohio: Swallow Press, Ohio University Press)
[40] Hutchins E 2014 *Philosophical Psychol* **27** 34-49
[41] Durakbasa N M, Bauer J M, Bas G and Riepl D 2015 *Internatl J Online Engineering* **11** 14-9

IOP Conf. Series: Journal of Physics: Conf. Series **1044** (2018) 012026 doi:10.1088/1742-6596/1044/1/012026

Education and training in coordinate metrology for industry towards digital manufacturing

L Rocha[1], E Savio[2], M Marxer[3], F Ferreira[4]

[1] CATIM Technological Centre for the Metal Working Industry, Porto, Portugal, luis.rocha@catim.pt

[2] Università di Padova, DII – Dipartimento di Ingegneria Industriale, Padova, Italy, enrico.savio@unipd.it

[3] NTB Interstate University of Applied Science, Buchs, Switzerland, michael.marxer@ntb.ch

[4] CATIM Technological Centre for the Metal Working Industry, Porto, Portugal, fernando.ferreira@catim.pt

Abstract. Today manufacturing industry is based on multiple suppliers located in different countries and continents, intensively using automation, data exchange and advanced manufacturing technologies embedded in the digital era, and a number of national and international initiatives (e.g. Industrie 4.0) are devoted to support this effort. New opportunities to advance manufacturing of high value products are based on innovating the measuring technologies and procedures for Geometrical Quality Control, in particular addressing the training needs since having well trained professionals is an important key of success for manufacturing companies in several sectors. The paper is summarising recent experience in the field, with quantification of learning success related to the implementation of remote workshops.

1. Introduction

Several European initiatives, like the well-known German initiative Industrie 4.0 [1,2], the Smart Industry in the Netherlands [3], Industrie du Futur in France [4] and the Portugal i4.0 Initiative [5], are supporting the Digitisation of Manufacturing. Due to the last evolutions and trends, across the industrialised countries, it´s imperative giving to the company's workforce the digital skills they need to fill the actual and future jobs. Digital technologies became the central role in creating value in the economy and bring several innovations to all aspects of development, production and related services. Europe's reindustrialisation will be possible if digitisation of products, processes and business models is mainstreamed across all industrial sectors. The European Union manufacturing sector accounts for 2 million companies and 33 million jobs; it is the source of 15% of EU GDP and 80% of exports. Digital Manufacturing (i.e. Industry 4.0, Smart Industry, etc.) is the strategy at EU level to support innovation processes [6]. Coordinate Metrology (including 3D digital measuring technologies) is by far the most important tool for these specialised activities. As widely reported, operators with their decisions are frequently the most relevant error sources in Coordinate Metrology operations, especially when dealing with new measuring technologies supporting Digital Manufacturing [7].

2. Education and Training in Coordinate Metrology

Highly qualified labour force remains a key source of future growth, and a particular focus is needed on the automotive and aerospace sectors [8,9,10]. A key enabler for innovation is the competent use of advanced equipment for the 3D measurement and digitisation of parts, as needed for advanced product/process engineering and quality control in customer-supplier business-to-business (B2B) interactions. Coordinate Metrology (including 3D digital measuring technologies) is by far the most important tool for these specialised activities, in particular Coordinate Measuring Machines (CMMs). As widely reported, operators with their decisions are frequently the most relevant error sources in Coordinate Metrology operations, especially when dealing with new measuring technologies supporting the Digital Manufacturing. Capacitate the industry employees is urgent, especially those who operating in SMEs that are newcomers on 3D measuring technologies. This target group benefits from getting access to advanced 3D measuring equipment and special training for support the new needs in digital manufacturing.

3. The CMTRAIN Education and Training Approach

After a project was carried out, supported by the European Union, the EUKOM project, based on the experience of several national training concepts for coordinate metrology [11], the project returned a European-wide harmonized concept for teaching metrological competence and machine-independent basic knowledge for coordinate metrology. The Association "CMTrain – Training for Coordinate Metrology e.V. (CMTrain) [12] was founded, for assure the sustainability of the project. Nowadays CMTrain manages training activities and the further development of new learning materials and contents.

The CMTrain training approach is based on the manufacturer-independent concept developed by the EUKOM project (Figure 1). The training program comprises at several hierarchical levels (levels 1 to 3); this includes hands-on training on real CMM's blended learning concept [13]. Level 1 is basic training, level 2 advanced training and level 3 training for specialists. Each training level is followed by further training and hands-on experience using CMM manufacturer's specific measuring devices.

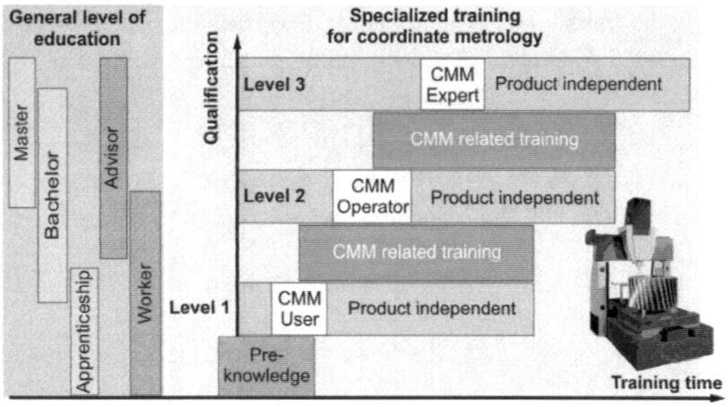

Figure 1. CMTrain learning concept

Each level of training consists of a combination of several forms of learning (blended learning). Participants work on e-competence and also with some of the experts during the contact hours. This form of learning is also used for supervised intensive hands-on training in small groups (4-5 people per group). Participants use a learning platform to develop their area of expertise. At the end of each level of training there is a practical and theoretical exam. The theoretical part of the exam is performed in a written form using a combination of multiple choice questions as well as open questions to be answered by the participants. Together with the evaluation, by the tutors, of the practical skills during the

IOP Conf. Series: Journal of Physics: Conf. Series **1044** (2018) 012026 doi:10.1088/1742-6596/1044/1/012026

workshops, a pass or fail evaluation is being held. It's necessary to reach a minimum score of 60%. If the candidate does not reach the required number of points, the test can be repeated within a certain period of time defined by the participant together with the tutor. This period is typically within four to six weeks after attending the first one. If this second attempt is not successful, the participant has to repeat the level to be able to attend the test once again [13].

In addition to hands-on training in small groups, practical e-training is also available. For this type of practical training, that feature was dominated as Remote Learning (RL) participants have access to a real CMM via the Internet (Figure 2). They have full access to the control and programming of the CMM and are in touch with a competent tutor who is on site where the CMM is located. Communication with this tutor also takes place via the Internet using information and communication technologies such as video and audio transmission.

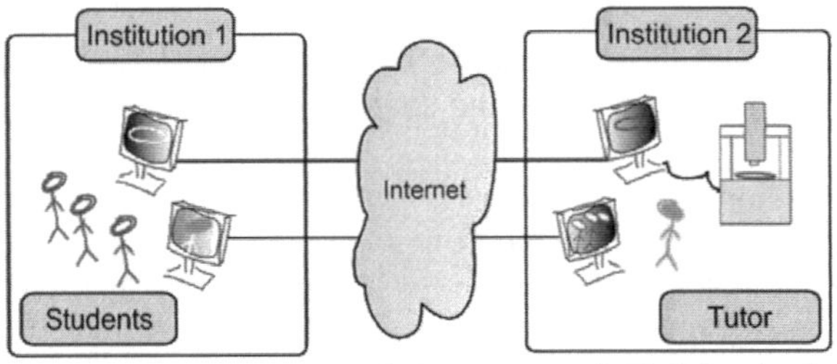

Figure. 2. Remote hands-on workshops on real CMMs via the Internet

This training approach is used for advanced training in industry and education organisations up to tertiary level [13]. For use in industrial enterprises, courses are offered as blended learning way in various languages and include a standardized final exam, which is internationally recognized.

The feedback of first remote workshop courses was registered based on three actions. The participants of each action have been divided into two groups. The participants in group 1 attended conventional workshop in a presence-based way, the participants of group 2 attended the workshop on a remote basis. The feedback of the pilot courses showed that the remote workshops were rated at least as good as the conventional workshops. This result was surprising and it showed that the approach is promising.

The comparison of the learning success in a written examination showed that the participants of the remote workshop had better results than the participants of the conventional workshop in two out of three actions (Figure 3).

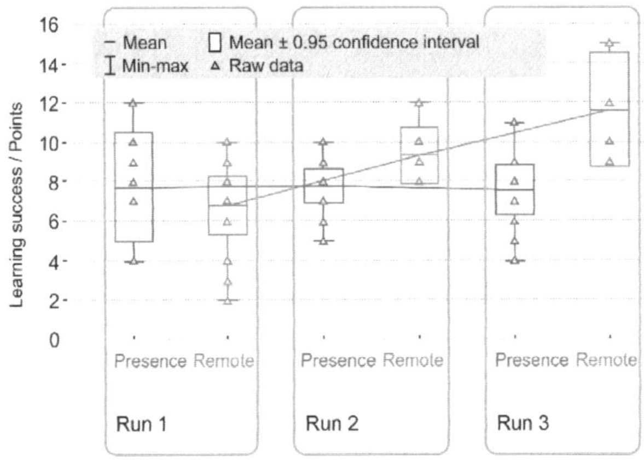

Figure. 3. Learning success comparison presence-based workshop to remote workshop (Marxer, 2012)

Following the promising results based on the first pilot tests comparing presence-based workshops and remote workshops (Figure 3), the remote workshop method was incorporated for further development within the ongoing EU project "European Training for Coordinate Metrology 4.0", currently under development and in an early stage.

The learning concept of CMTrain is used in universities as well, some packages are integrated into conventional lectures and the e-learning component supports self-study. This process is articulated with the Bologna Process initiated by the European Union. Up to now, this concept has been used in Bachelor and Master Courses in Germany, Switzerland, Italy and Poland. The existing training contents are available in several languages: English, German, Romanian, Polish and Portuguese. New versions in French and Italian are currently being developed.

4. Added value for companies

We need to take in account several factors that brings added value to companies, and trainees, by getting CMTrain courses as a perspective of sustainability (Figure 4) [13, 14, 15, 16]. We can identify environmental aspect, this training model enables participants to reduce travels (energy consumption), paper consumption for course materials is significantly reduced as information is made, distributed and available electronically, contributing to pollution reduction. In the economic direction, the CMTrain concept implies lower travel costs, lower course fees and less time out of the working place, such comprehensive training was not possible in the past for small and medium-sized companies as it meant that employees were not at their place of work.

Figure. 4. Sustainability aspects

Regarding the social issues, this training approach certification presented significantly increases equality of opportunity when it comes to accessing to knowledge, as for a company or learners, for example, giving the opportunity to small companies and emerging markets that do not have widespread modern infrastructure but that can now benefit from this training method [16].

5. Future developments

The distribution channels are evolving very fast due to the technology evolution (metrology and media distribution), opening new ways and opportunities to exploit these factors. Boundaries between technologies are becoming blurred, like normal smartphone is now a car navigation system, a health monitor or a training delivery device, at the same time, in any place. Consequently, we began several trials with complementary devices to receive the training contents, beyond the traditional desktop or laptop computer, such as tablets and smartphone, using the increasing presence of large internet bandwidth in and out of working places, to maximise the impact of training.

Other efforts are made for create new contents, to provide in a near future, through CMTrain courses, aligned with the paradigm of digital industry and digital manufacturing. Contents such as optical measuring systems for digitisation of surfaces, computed tomography for full digitisation of parts, point-cloud metrology, reverse engineering for digital modelling of products, data protocols and analysis among others contents.

6. Conclusions

After several editions of CM courses, during the last years, with the blended learning approach, we obtained positive evaluation, showing that the concept is adequate at industrial and university learning environments. For this reason, it is very important to propagate the concept of CMTrain internationally to give a wider access to this model of training to fill the present and future needs of the industrial companies in several sectors.

7. Acknowledgments

The authors wish to acknowledge the support of the members of CMTrain association and gratefully appreciate the feedback of the manufacturing companies, and university students that have participated in the field tests, and training editions, for their important contributions.

References

[1] Hermann M, Pentek T, and Otto B 2015 *Design principles for Industrie 4.0 scenarios: a literature review.* Technische Universität Dortmund.
[2] Zhou K, Taigang L, and Lifeng Z 2015 Industry 4.0: Towards future industrial opportunities and challenges." Fuzzy Systems and Knowledge Discovery (FSKD), 12th International Conference on. IEEE, Taiwan, 2147-2152.
[3] Netherland Smart Industry initiative, http://www.smartindustry.nl/eng/ (17.03.2017).
[4] http://industriedufutur.fim.net/ (18.03.2017).
[5] http://www.industria4-0.cotec.pt/ (20.03.2017)
[6] Background Paper on "Digitising European Industry", European Commission, DG Connect, June 2015
[7] Weckenmann, A.: Koordinatenmesstechnik: Flexible Strategien für funktions- und fertigungsgerechtes Prüfen. München, Wien: Hanser, 2011
[8] Hermawati S, Lawson G, D'Cruz M, Arlt F, Apold J, Andersson L, ... and Malmsköld L, 2015. *Understanding the complex needs of automotive training at final assembly lines. Applied ergonomics*, 46, 144-157.
[9] van der Heiden P, Pohl C, Mansor S B, and van Genderen J 2015. *The role of education and training in absorptive capacity of international technology transfer in the aerospace sector. Progress in Aerospace Sciences*, **76**, 42-54.
[10] Restructuring in Europe, European Commission, ISBN: 978-92-79-25056-9, May 2012.
[11] Keferstein C P, Marxer M.,Weckenmann A, Beetz S:2005 *EUKOM. Ein europäisch abgestimmtes, auf eLearning basierendes, neues Ausbildungskonzept für die Koordinatenmesstechnik; Innovative Entwicklungen im Fokus des Anwenders. Tagung Braunschweig 15. und 16.* S. 109-118. Düsseldorf: VDI Verlag, (VDI-Berichte 1914).
[12] http://www.cmtrain.org (15.03.2017)
[13] Marxer M, Keferstein C, Weckenmann A 2010 *Sustainable Manufacturing Using a Global Education Concept for Coordinate Metrology with a Blended Learning Approach, 8th Global Conference for Sustainable Manufacturing, Abu Dhabi*, November 2010.
[14] Marxer M, Rocha L, Araújo N, Götti R 2012 *Hard-Skills on the Cloud – Certification Skills by Blended Learning, Groningen, October 2012.*
[15] Marxer M, Keferstein C 2011 *Innovative Concept for Coordinate Metrology using a Blended Learning Approach, 15th International Congress of Metrology, Paris, October 2011.*
[16] Marxer M, Rocha L, Araújo N and Kuster R 2015 *CMtrain–Remote training in Coordinate Metrology. In ICEL2015-10th International Conference on e-Learning: ICEL 2015 (p. 201). Academic Conferences and publishing limited.*

IOP Conf. Series: Journal of Physics: Conf. Series **1044** (2018) 012027 doi:10.1088/1742-6596/1044/1/012027

Employment of a Monte Carlo method in the dynamic characterization of pressure transducers

F R F Theodoro[1], M L C C Reis[2], C d'A Souto[2]

[1] Instituto Tecnológico de Aeronáutica.[2] Instituto de Aeronáutica e Espaço. Brazil.

E-mail: flavio.theodoro@aedu.com

Abstract. Dynamic calibration is a multidisciplinary theme which is attracting interest from the metrological community. Methods and devices are being tested, such as Shock Tube and Drop-weigh systems. There is a great demand for research into this issue, in particular regarding uncertainty analysis in dynamic calibration. This paper deals with the use of a Monte Carlo method in the dynamic characterization of pressure transducers. We analyse the process to characterize the transducer Kistler 701A. The results showed that the method can be efficient for evaluating the uncertainties involving the dynamic characterization of piezoelectric transducers that cover frequency ranges from 1 to 60 kHz and a pressure range from 1 to 1000 kPa.

1. Introduction

We use the words "dynamic calibration" to define the process that characterizes a dynamic pressure transducer, determining some properties, such as natural frequency, damping, peak time, stabilization time and sensitivity [1].

In this calibration method we consider a theoretical signal generated by a device, such as a shock tube, as the standard. Based on the characteristics of this device, a known and repetitive signal is generated and this signal is compared with that captured by the transducer under calibration [1].

Due to the demand for dynamic measurement pressure in areas such as the aeronautics and automotive industries [2], research centres and large corporations have an interest in improving its measurement processes. Some examples are the National Physical Laboratory, NPL, in England; the Physikalisch-Technische Bundesanstalt (PTB), in Germany, Rolls-Royce, Volkswagen and Porsche [3].

To find a standardized method that serves for several pressure and frequency ranges is a challenge for the dynamic calibration of pressure transducers. Another challenge is to determinate the uncertainties of this process [4].

Therefore, this article deals with the use of a Monte Carlo Method to determine the uncertainties in the dynamic characterization of pressure transducers. The Monte Carlo Method is used to estimate the mean values of the experimental parameters and associated uncertainties. The properties described in this paper were peak time, natural frequency and damping ratio, all identified in time domain.

From the analysis of the results we see that a Monte Carlo method can be efficient for evaluating the uncertainties involving the dynamic characterization of the piezoelectric pressure transducers that cover frequency ranges from 1 to 60 kHz and pressure ranges from 1 to 1000 kPa [1]. The sensor employed in the test is a Kistler model 701A.

The article is the result of studies carried out by the Institute of Aeronautics and Space, IAE, Brazil.

2. Facility and Mathematical Modelling

2.1. Facility

The quality of the signal generated for calibration is a determining factor for the success of the dynamic calibration. Thus, a good laboratory structure can make all the difference. The facility used in the study is the Henry T. Nagamatzu Laboratory of the Aerodynamics and Hypersonic Division, of the Institute of Advanced Studies, IEAv, of Department of Aerospace Science and Technology, DCTA, Brazil [1].

A shock tube (figure 1) is a device that contains two compartments which are separated by a diaphragm. The first compartment is called "driver" where a gas (generally air) is stored with "P_1" pressure. The second compartment is the "driven" where the "P2" and "P2[1]" sensors are located. This compartment contains a "P_2" pressure.

When the diaphragm is broken by pressure difference between P_1 and P_2, it generates a shock wave that is used to calculate a step signal that will be employed in the characterization of the transducers [5].

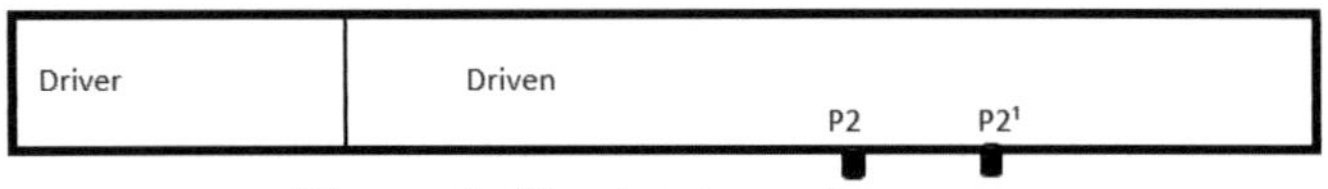

Figure 1. Shock tube and sensors.

2.2 Mathematical Modelling

The start of the dynamic characterization process of the sensor occurs immediately after the rupture of the diaphragm separating the driver and driven tubes. Thus, firstly we calculate the velocity of the front shock wave which passes by sensors P2 and P2[1] (see figure 1). Using a theoretical model (equation (1)) and based on the value of the velocity of the wave front (*L*/*t*) where *L* is the distance between the two sensors, P2 and P2[1] and *t* is the time of the passage of the front wave between them), we estimate the value of the pressure step signal, ΔP [3].

$$\Delta P = P_2 - P_1 = \frac{7}{6} P_1 [\left(\frac{\frac{L}{t}}{\gamma_1 R_1 T} \right)^2 - 1] \tag{1}$$

In equation (1) γ_1 is the ratio of specific heat, equal to 1.4 for the air considered as a perfect gas; R_1 is the perfect gas constant, equal to 287 J/(kg.K) for the air; and T_1 is the temperature in the driven section of the tube, expressed in kelvin [5].

We can represent the dynamic behaviour of the pressure sensor membrane by a mass-spring system with damping, a second order system. The Laplace Transform of the output function for this system can be the result of the multiplication between the Laplace Transform of the input function and the Laplace Transform of the differential equation which represents the mass-spring system behaviour.

Assuming that the input signal is a unit step and by applying the Laplace Inverse Transform to the output function, one obtains the temporal response of the sensor expressed by equation (2), estimating the damping ratio ζ and the natural frequency ω_n values. The gain *K* of the system can be determined in the steady state regime [6].

$$x(t) = K\left[1 - e^{-\zeta\omega_n t}\left(\cos \zeta\omega_n \sqrt{1-\zeta^2}\, t + \frac{\zeta}{\sqrt{1-\zeta^2}}\, sen\, \omega_n \sqrt{1-\zeta^2}\, t\right)\right] \tag{2}$$

The damping ratio, ζ, can be obtained experimentally employing equation (3) [6]. Where $x\ (t_{\max 1})$ and $x\ (t_{\max 2})$ are two consecutives peaks above the mean value of the steady signal, $x\ (t_\infty)$

$$\zeta = \left(\sqrt{1 + \left(\frac{2.728}{\left(\log_{10} \left(\frac{x(t_{max1}) - x(t_{\infty})}{x(t_{max2}) - x(t_{\infty})} \right) \right)} \right)^2} \right)^{-1} \tag{3}$$

In figure 2, we can see the points used in Equation (3) to calculate the damping ratio value.

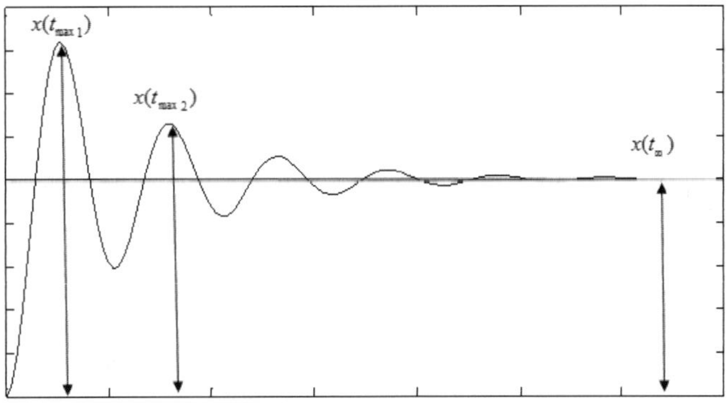

Figure 2. Two consecutives peaks and mean value of the steady signal.

The natural frequency, ω_n, can be determined by peak time, t_p, or by the time for the signal to achieve overshoot, $x\,(t_{max\,1})$, and by damping, according to equation (4) [6].

$$\omega_n = \frac{\pi}{t_p\,\sqrt{1-\zeta^2}} \tag{4}$$

3. Results and discussions

The Monte Carlo method is based on the Probability Distribution Functions, PDF, of the input quantities. These distributions are propagated M times, independently, according to each mathematical model of the measurement and the mean and standard deviations of the results are estimated.

Its main advantage is that it can be used to calculate uncertainties with asymmetric distribution. As such, the uncertainty is determined by the lower and upper limits of the PDF resulting from the measurement [7, 8].

Three tests were carried out on a shock tube using the same configuration to obtain the parameters to perform a Monte Carlo simulation. Mean and standard deviation values are seen in Table 1.

Table 1. Input experimental data.

Parameters	Mean values	Standard deviation	Units
T	469	5	µs
L	0.400	0.014	m
T_1	20.40	0.5	ºC
P_1	95.90	0.12	kPa
t_p	109×10^{-7}	12×10^{-7}	s
$x\,(t_{\infty})$	2.85	0.05	V
$x\,(t_{max\,1})$	4.09	0.16	V
$x\,(t_{max\,2})$	3.68	0.09	V

Using the values from Table 1, assuming a t-*Student* distribution, we performed 10^6 trials for each measurement model (equations (1), (3) and (4)) to estimate the mean values and uncertainties of the parameters ΔP, ζ, ω_n, for a 95% level of confidence, as shown in Table 2 [8].

Table 2 - Input quantities for propagation distributions to estimate ΔP, ζ and ω_n.

Output parameters	Input parameters	Mean values	Standard deviation	Uncertainties lower limits	Uncertainties upper limits	Output units
ΔP	t, L, T_1, P_1	5.8×10^2	0.5×10^2	4.8×10^2	6.8×10^2	kPa
ζ	$x(t_\infty), x(t_{max\,1}), x(t_{max\,2})$	0.58×10^{-1}	0.21×10^{-1}	0	1.1×10^{-1}	-
ω_n	t_p	290×10^4	3×10^4	25×10^4	36×10^4	rad/s

The histograms of figures 3 (a), (b) and (c) respectively represent the resulting PDF for outputs ΔP, ζ and ω_n considering the input values of Table 1 applied to Equations (1), (3) and (4).

In these histograms, the red line represents a normal distribution, according to mean values and standard deviation of the quantities. The blue bars represent the results of a Monte Carlo simulation considering the input parameters and model adopted.

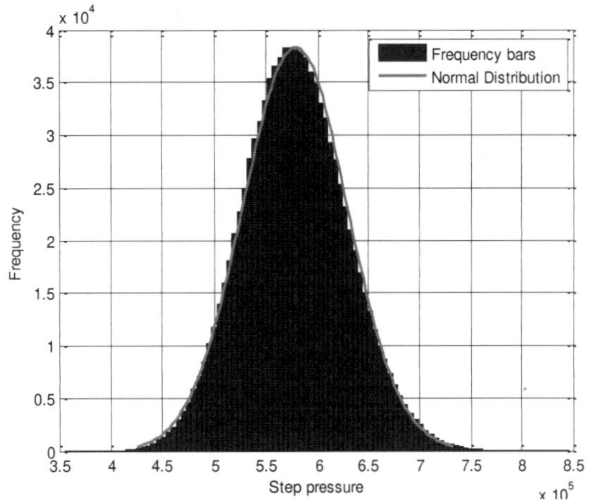

Figure 3 (a). Histogram of ΔP

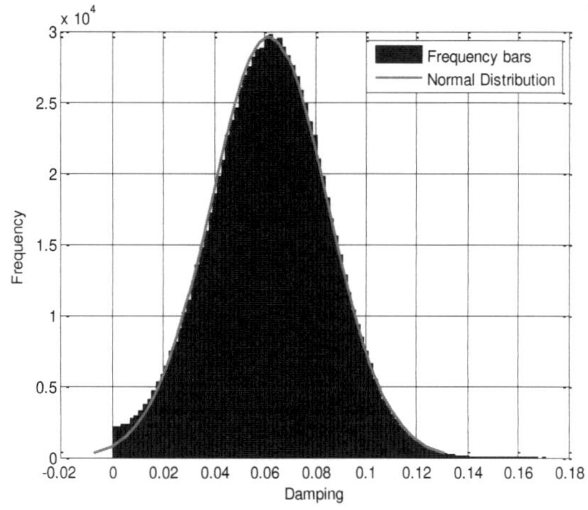

Figure 3 (b). Histogram of ζ

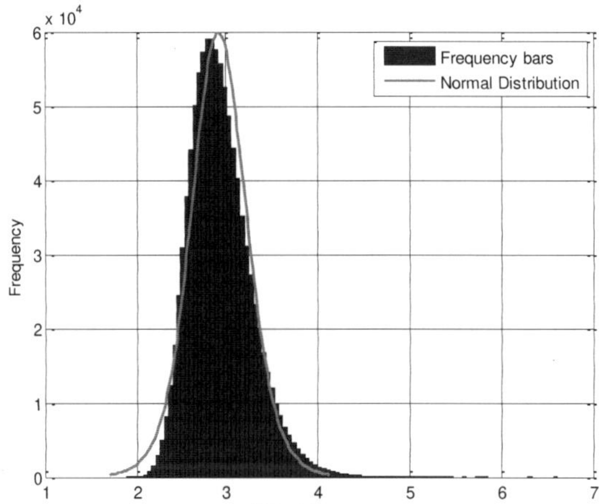

Figure 3 (c). Histogram of ω_n

In figures 3 (a), (b) and (c) we can see an acceptable agreement between the red line, which represents a normal distribution, and the blue area, which represents the experimental data. It can be seen that uncertainties values for damping and natural frequency are asymmetric, as was also observed in Table 2.

Comparing the uncertainties limits shown in columns five and six of the Table 2, it can be seen that uncertainties values of the ΔP quantities are symmetric as shown in figure 3 (a). Thus, when using a coverage factor k = 2 [9], we have a coverage interval for (the) ΔP value which is similar to (the) uncertainties limits obtained according to (a) Monte Carlo Method.

Concerning values of the uncertainties limits for ζ, and ω_n, when using a coverage factor k = 2 [9] we have a coverage interval different from that obtained according to (the) Monte Carlo Method, in other words, a Monte Carlo Method supplies asymmetric values in relation to (the) mean values, as seen in figures 3 (b) and 3 (c).

4. Conclusion

In this paper we discussed the use of a Monte Carlo method for estimating uncertainties in dynamic calibration of a pressure transducer. Firstly we considered the applications of the pressure sensors. Then we briefly the facility of the Institute of Advanced Studies, IEAv, in Brazil, and the mathematical modeling used in this study.

The Monte Carlo Method was used to determine the mean values and the uncertainties of the peak time, natural frequency and damping ratio parameters. This Monte Carlo method proved to be capable of evaluating the uncertainties involving the dynamic characterization of the piezoelectric pressure transducers that cover frequency ranges from 1 to 60 kHz and pressure range from 1 to 1000 kPa.

5. References

[1] Theodoro F R F, Reis M L C C, Souto C d' and Barros E, Measurement uncertainty of a pressure sensor submitted to a step input. 2016 *Measurement Vol.* ***88***. p. 238-247 doi:http://dx.doi.org/10.1016/j.measurement.2016.03.043.

[2] Hjelmgren J, Dynamic measurement of pressure-A literature survey. *SP Rapport 2002:34*. In: http://www.diva-portal.org/smash/get/diva2:962216/FULLTEXT01.pdf

[3] Matthews C, Pennecchi F, Eichstadt S, Malengo A, Esward T, Smith I, Elster C, Knott A, Arrhen A and Lakka A, Mathematical modelling to support traceable dynamic calibration of pressure sensors. 2014 *Metrologia* ***51***. p. 326–338 doi:10.1088/0026-1394/51/3/326.

[4] Diniz A C G C, Vianna J N S and Neves F J R, Dynamic calibration of the pressure sensors: method and means (In Portuguese) 2006. *Metrologia (SBM Congress)* September 01-05 Brazil

[5] Anderson J D 2003*Modern Compressible Flow with Historical Perspective*, third ed, (USA: Mc Graw Hill)

[6] ISA–37.16.01-2002. A guide for the dynamic calibration of pressure transducers. North Carolina: The Instrumentation Systems and Automation Society

[7] Couto P R G, Damasceno J C and Oliveira S P, Monte Carlo Simulations Applied to Uncertanty in Measurement. 2013. *INTECH* In: http://cdn.intechopen.com/pdfs-wm/43533.pdf

[8] BIPM/JCGM 101:2008 Evaluation of Measurement Data – Supplement 1 to the Guide to the expression of uncertainty of measurement - Propagation of distributions using a Monte Carlo method. Joint Committee for Guides in Metrology, BIPM

[9] BIPM/JCGM 100:2008 Evaluation of Measurement Data - Guide to the expression of uncertainty of measurement, Joint Committee for Guides in Metrology, BIPM

IOP Conf. Series: Journal of Physics: Conf. Series **1044** (2018) 012028 doi:10.1088/1742-6596/1044/1/012028

Energy Savings Resulting from Energy Management Program Using Measurement and Verification Procedure

R S Vieira[1]; R F Calili[1]; A S Aranha[2] and L F Fabbriani[3]
[1] Graduate Program in Metrology, Pontifical Catholic University of Rio de Janeiro Marquês de São Vicente Street, 225 - Gávea, 22451-900, Rio de Janeiro - RJ, Brazil
[2] Electrical Engineering Department, Pontifical Catholic University of Rio de Janeiro, Marquês de São Vicente Street, 225 - Gávea, 22451-900, Rio de Janeiro - RJ, Brazil
[3] Civil Engineering Department, Pontifical Catholic University of Rio de Janeiro, Marquês de São Vicente Street, 225 - Gávea, 22451-900, Rio de Janeiro - RJ, Brazil

rsantosvieira@hotmail.com

Abstract. Energy efficiency is one of the main solutions for problems such as the increase in energy consumption and greenhouse gas emissions. Energy savings are obtained by the increase of efficiency, a solution that reflects directly on the energy matrix, lowering the amount of gases emitted during energy generation. According to the available literature, there are many works that deal with existing equipment exchanges for more efficient ones. The aim of this article is to show the effectiveness of the energy efficiency program in saving energy, showing how the amount of energy reduced is proven and discussing the accuracy errors. The case study is a resort in Rio de Janeiro - Brazil. In order to prove the energy saved, measurement and verification models were built according to recommendations in international protocol (IPMVP) and standards (ISO 50001, GUM).

1. Introduction

Studies point out that the excessive emission of greenhouse gas – GHG can cause global warming, which is responsible for climate changes [1]. One of the major GHG emitter are coal-fired power plants, which are also one of the major sources of electric power in the world. This scenario is aggravated by an increasing demand for energy, due to changes in the consumption profile with the advent of new technologies. [1]

To make available more energy and, at the same time emitting less GHG, one can use alternatively renewable energy sources or simply make consumers more efficient. Thus, it is possible to save energy that can be used for other purposes, or even to withdraw the obligation of generating energy by the power plants that pollute the energy matrix of a country.

This article aims to detail, through a case study, the amount of energy savings achieved by energy management in the commercial sector. An energy efficiency work carried out by an Energy Service Company (ESCO) at a Resort located in state of Rio de Janeiro will be used for this study.

In order to prove the savings, data collected from the energy bills, energy performance improvement actions (EPIA) and diesel generator dispatch data were used in conjunction with International Performance Measurement and Verification Protocol (IPMVP)[2], ISO 50001[3], ISO 50002[4] and ISO 50015[5]. These documents include good practices in energy audits, measurement and verification (M&V) and establish minimum requirements of precision and confidence to measure the real savings obtained after the EPIA.

IOP Conf. Series: Journal of Physics: Conf. Series **1044** (2018) 012028 doi:10.1088/1742-6596/1044/1/012028

2. Case Study

This article is based on a case study, using real data and analyzing an energy efficiency work carried out by an ESCO at a resort in the state of Rio de Janeiro. For confidentiality reasons stated in the contract, names of both the parts are going to be omitted.

The resort has 5 power meters, one in each substation, with an area of 137.068 m². In the main substation named SEDE, which is the largest consumer, are connected restaurants, event venues and the hotel lobby. In the substation named Apartments – APTO, 150 hotel rooms are connected. This is the second largest consumer. CANTINA is the substation responsible for feeding the maintenance area, warehouse and employees' cafeteria. At last, the SPA substation feeds massage rooms, saunas and the administrator's residence.

The ESCO was hired by the customer in February 2016, to develop an energy audit at its facilities in order to find energy performance improvement actions (EPIA) that would reduce the hotel's energy consumption.

In the audit accomplished, following the precepts of ISO 50002 that guides the requirements for the performance of an energy audit, it was verified that the Energy Performance over the total annual cost of the use of energies was of R$ 1,056,572.43 (base year 2015). It was also estimated that a set of energy efficiency measures (EEM) would save around 9.27%, which corresponds to R$ 98,256.91 per year. This value was found prioritizing equipment exchanges that should be made by the client.

During the step of collecting data, it was noticed that the external temperature and the occupation of the hotel were independent variables that highly influenced the energy consumption. This influence was proved by the correlational method between these variables and the energy measured constantly in bills.

One year after the start of the energy management program, the ESCO noticed that because of alterations and cuts in the investments of new efficient equipments that were needed to reduce cost, there was a big risk that initial savings predictions wouldn't be reached. On the other hand, the energy bills showed different. During the monitoring and after the engineering technical team critical analysis, savings of R$ 51,315.16 were estimated until December of 2016.

Four possible EPIA responsible for the savings generated by the energy management program were pointed out:

- Changes in the operation of the diesel generator, monitoring the peak load, and reducing the usage of energy from the electricity distribution utility;
- Adjustment of the energy meters' clock, which was desynchronized with Brasilia time, charging as consequence a high tariff when it was still off-peak time;
- Adjustment in the capacitor banks.

From May 2016 to mid-October 2016, a nonconformity on the capacitor bank panels was corrected. These panels were installed without the ESCO's supervision and were poorly configured. Instead of correcting the inductive reactive energy they were producing at the time where the energy is usually capacitive, thus generating billing. The ESCO intervened in the month of October to correctly parameterize the bank of capacitors as to the time of entry and its capacitive load.

The resort's energy bills are charged according to the "Green Time" type of tariff. This means that there are two energy consumption tariffs divided into consumption at on-peak hours (from 6:00 pm to 9:00 pm) and consumption at the off-peak hours. There is also a charge for contract demand. The hotel also has generator sets to work as the power source at peak times. The savings over the contract demand optimization will not be accounted for in this study.

3. Proof of Energy Conservation

As a matter of proving the energy savings, the ISO 50006 and ISO 50015 standards were used along with the International Performance Measurement and Verification Protocol (IPMVP) and the Guide to Expression of Uncertainty in Measurements (GUM) [6]. This set of standards guides the rules for

IOP Conf. Series: Journal of Physics: Conf. Series **1044** (2018) 012028 doi:10.1088/1742-6596/1044/1/012028

Measurement and Verification (M&V), in order to prove the savings generated by the EPIA implemented.

According to the family of Standard ISO 50001 and the IPMVP there is no way to directly measure energy savings, for this it is necessary to create a mathematical model called the baseline, which reproduces the behavior of the energy consumption before the EPIA occurs in the post-period Implementation of the EPIA. And with this baseline and measurement values after EPIA, it is possible to estimate the energy savings.

For a good baseline, the IPMVP recommends that the coefficient of determination (R^2) should be at least 0.75, the coefficient of variation (CV) of 20% at most and the coefficients resulting from the linear regressions should have a t-statistic greater than 2. It is also specified that a minimum precision of 10% shall be used for a 95% confidence level, if sampling is required.

For the case study, although the savings are still not fully measured and the EPIA are not fully implemented, it is wanted to find energy savings that are consequences of low cost energy management actions. Therefore, Method C from IPMVP will be used to prove the economy. This method uses the distribution utility meter. Other methods use specific measurements for each equipment/systems (method A and B) or perform consumption simulations (method D).

For the application of method C there is the restriction that the total energy saved should be greater than 10%. If the total economy is less than 10%, it should be checked whether the EPIAs are well spread so that individualized measurement is not possible. As the hotel case study the EPIAs are largely in the awareness of people, it is not possible to carry out the individualized measures, thus being method C most appropriate to carry out the M&V process.

Since several baselines will be carried out, one for each item (on-peak and off-peak time consumption, demand, and generator usage), and since each energy efficiency action affects different items, in this work they will be unified, converting their values to a unit that is common to all variables: money. Thus, it will be easier to perceive the final savings. In this way, the measurement uncertainties will be combined to find general precision.

The variables used to build baselines for the commercial sector of hotel industry are commonly the temperature and occupation of the consumer unit. The temperature is collected from the official meteorological websites and the occupation can be taken directly with the hotel's manager. Later, the data is adjusted to match the billing days.

4. Results

As a result, 5 baselines models were built: equation (1) the consumption of electric energy in off-peak hours, equation (2) demand, equation (3) the consumption of reactive electric energy, equation (4) the consumption of electric energy in on-peak hours, and equation (5) consumption of liters of diesel in the generator used at peak hours. The baseline equations are shown in table 1, the statistics resulting from the regression process are shown in table 2, which can be seen in t-statics the parameters are significants. Energy consumption, ambient temperature and hotel occupancy data were used considering the horizon from 2015 to 2016. These data can be found in table 3.

Table 1 – Table of equations that model the consumption of energy utilities given by linear regression.

Consumption of energy utilities	Linear regression equation.	
Electric energy in off-peak hours [kW h]	$1456{,}88 \times temp_i + 1283{,}42 \times occup_i - 130{,}51$	(1)
Demand [R$/kW]	$1283{,}42 \times occup_i + 11{,}68$	(2)
Reactive electric energy [R$/(kV A h)]	$240{,}69 \times temp_i - 53{,}00 \times occup_i - 37{,}85$	(3)
Electric energy in on-peak hours [R$/kW h]	$100{,}45 \times occup_i - 491{,}61$	(4)
Liters of diesel [L]	$123{,}51 \times temp_i - 183{,}27$	(5)

IOP Conf. Series: Journal of Physics: Conf. Series **1044** (2018) 012028 doi:10.1088/1742-6596/1044/1/012028

Table 2 – Table of the statistical results of linear regression.

energy utilities	Baseline Period	T-statistic temperature coefficient	T-statistic occupation coefficient	R^2	CV	Standard error (R$)
Electric energy in off-peak hours [kW h]	From jan/15 to jun/16	2,58	4,23	0,99	12,4%	R$ 8.162,01
Demand [kW]	From jan/15 to jun/16	-	24,66	0,97	17,5%	R$ 329,70
Reactive electric energy [kV A h]	From jun/15 to abr/16	6,33	-2,63	0,97	20,3%	R$ 495,64
Electric energy in on-peak hours [kW h]	From jun/15 to abr/16	-	16,72	0,97	20,3%	R$ 1.048,78
Liters of diesel [L]	From jan/15 to abr/16	16,61	-	0,97	22,0%	R$ 754,26

Table 3 - Consumption data table of energy utilities and independent variables used for baseline and energy saving calculations.

	Electric energy in off-peak hours [kW h]	Demand [kW]	Reactive electric energy [kV A]	Electric energy in on-peak hours [kW h]	Liters of diesel [L]	Temperature [°C]	Occupation [%]
jan/15	166739	494	3954	11786	2609	36,3	86,8
fev/15	140133	524	4314	11347	2363	30,3	79,4
mar/15	135912	503	3991	11815	2861	31,2	55,4
abr/15	110355	404	4901	10568	2878	28,1	40,1
mai/15	99141	370	6062	9486	3227	29,0	48,2
jun/15	79107	281	5149	4427	3289	26,9	36,9
jul/15	72744	345	4995	5081	3331	26,8	42,7
ago/15	95970	307	4933	5964	3277	29,6	48,6
set/15	84546	274	4711	5262	2539	28,1	42,6
out/15	89355	327	4766	6394	2428	28,4	51,2
nov/15	113715	370	4505	6870	3301	33,2	58,9
dez/15	101598	378	3385	4216	3805	31,8	54,2
jan/16	157479	486	3870	5873	4502	33,3	73,8
fev/16	145698	482	3582	5782	4878	34,1	71,9
mar/16	115857	416	3157	5765	4541	28,8	57,5
abr/16	122430	407	4219	4934	4997	31,5	47,1
mai/16	106428	419	8993	3442	3898	25,9	40,5
jun/16	93954	310	14553	3166	3300	25,5	36,9
jul/16	89397	326	16925	4432	3221	28,7	45,3
ago/16	88725	290	4830	3677	3328	25,7	46,2
set/16	95571	311	4369	948	4135	28,8	47,4
out/16	90510	373	2222	1795	4381	25,5	51,0
nov/16	105168	383	1389	1143	5326	24,8	56,9
dez/16	113190	389	0	1114	5481	24,8	57,0

For the measurement uncertainty calculation, because of the diversity of measurement units, the standard errors found were converted to monetary value using the respective tariffs, which can be seen in table 4.

Table 4 - Energy utilities tariff table

Year	Electric energy in off-peak hours [R$/(kW h)]	Demand [R$/kW]	Reactive electric energy [R$/(kV A h)]	Electric energy in on-peak hours [R$/(kW h)]	Liters of diesel [R$/L]
2015	0,48	24,66	0,39	2,20	2,73
2016	0,53	26,58	0,39	2,42	3,02

The result of crossing the actual consumption energy measurements, and baseline in the period of economy are shown in figure 1. This is the graph of total monthly expenditure, in *Reais*, adding up all utility expenses in each month. Then, it is possible to identify: (a) the energy measurements used to calculate the baseline until the arrival of ESCO, (b) the baseline in the period of measurement of the economy, representing what would be the consumption without the application of the EPIA (c) energy measurements after the EPIA, *i.e.* after the entry of ESCO. Making the difference between (c) and (b) is found monthly energy savings.

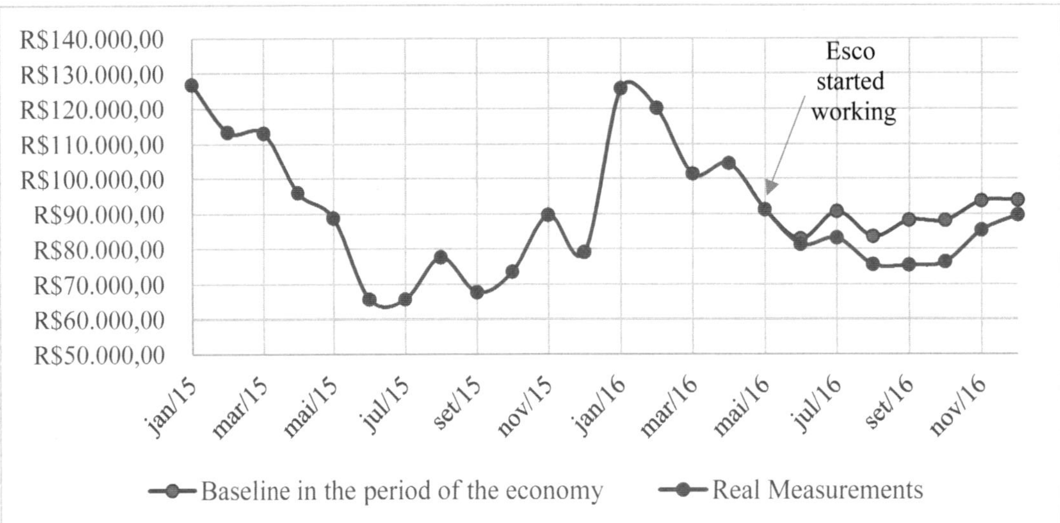

Figure 1 – Graph of the total consumption in *Reais* comparing the measured consumption and consumption after the EPIAs carried out by the ESCO energy management program.

Thus, to find the energy savings and the variation caused by the measurement uncertainty arising from the application of the energy management program carried out by ESCO considering a 95% confidence, which results in a total savings of R$ 54,123.89 accumulated in 7 months of operation of ESCO. These savings represent 7.60% of average in comparison to the monthly bills, with a relative precision of 17.6%. For this level of confidence, the absolute precision is R$ 9,532.64, meaning that the savings can vary between R$ 44,591.26 and R$ 63,656.53.

5. Conclusion

In this article, it was proved that programs of electric energy management can generate savings, measuring some actions of energy efficiency in management made with a relative precision of 17.6%.

The ISO 50001, 50002, 50004, 50006, 50015, PIMVP and GUM standards were used successfully to obtain the final savings.

It is necessary to deepen studies on this field, because energy efficiency and energy conservation gain more and more importance nowadays.

Instead of using IPMVP C option was used in this project, because of the amount of energy other IPMVP options should be used, which would give more detailed metering. The next step of this work will be usage of others metering options to improving the effectiveness of the energy savings and the proposed mathematical model.

Acknowledgments

The authors would like to thank the Brazilian agency CAPES for the financial support in the development of this research and Otimiza Sustentável by releasing data for this article.

References

[1] Nejat P, Jomehzadeh F, Taheri MM, Gohari M, Muhd MZ 2015 A global review of energy consumption, CO2 emissions and policy in the residential sector (with an overview of the top ten CO2 emitting countries). *Renew Sustain Energy Reviews* (Elsevier)
[2] EVO 2012 IPMVP – International Performance Measurement and Verification Protocol
[3] ISO 2011 Standard ISO 50001 - Energy Management System
[4] ISO 2014 Standard ISO 50002
[5] ISO 2014 Draft Standard ISO 50015
[6] Jcgm JCFGIM 2008 GUM - Evaluation of measurement data - Guide to the expression of uncertainty in measurement.

IOP Conf. Series: Journal of Physics: Conf. Series **1044** (2018) 012029 doi:10.1088/1742-6596/1044/1/012029

Estimated impact force for an electromagnetic hammer

L C Lemos Júnior, J J Silva and J S Rocha Neto, *IEEE senior member*

Federal University of Campina Grande
Department of Electrical Engineering
58109-970, Campina Grande-PB, Brazil

luiz.lemos@ee.ufcg.edu.br, jaidilson@dee.ufcg.edu.br, zesergio@dee.ufcg.edu.br

Abstract. This paper presents a force estimation of an electromagnetic hammer based on the principle of the magnetic force generated by the field of a solenoid that displaces its metal core to strike a structure under analysis. This technique known as the Hammer Impact Test is used as a method for vibration analysis. For the correct electronic system design a mathematical model for the hammer is needed to estimate impact force based on the input electrical current the hammer physical parameters.

1. Introduction

Vibration analysis is a method widely used in industry to detect failures in structures and preventive maintenance. Currently, the hammer impact test is used as a non-invasive-technique, it consists of striking a structure and observing its vibrational behavior, whether it conforms to standard norms or presents some atypical characteristic [1].

Impulse excitation is a technique used in conventional vibration tests to determine the dynamic behavior of structures. The impact of a hammer is used to generate the excitation by an impulse after striking the structure. The pulse, with specific amplitude and with a minimum possible duration, can excite all vibration modes of the structure in the determined frequency range [2][3].

However, for each specific application, it is necessary to adapt the hammer model, since it must vary the impact intensity on the structure to be analyzed. In order to determine the intensity of this force at the moment of impact, this paper brings an equation for the impact force by virtue of the electric input current, having as parameters the dimensions of the solenoid and test body for the construction of the excitation system.

This paper focuses only on estimating the impact force of the hammer so that its construction is not simply empirical, but based on a previous study of the estimated force required. The purpose is to allow the design to be based on the need of the intensity of the impact force varying the pattern for the construction of the excitation structure.

2. Principle of Operation

When the system is excited by the impact of a hammer, it vibrates while the energy is dissipated. In fact, the duration time of the free vibration depends on the physical characteristics of the system, specifically the damping coefficient [4]. Figure 1 shows the schematic diagram of the electromagnetic hammer.

IOP Conf. Series: Journal of Physics: Conf. Series **1044** (2018) 012029 doi:10.1088/1742-6596/1044/1/012029

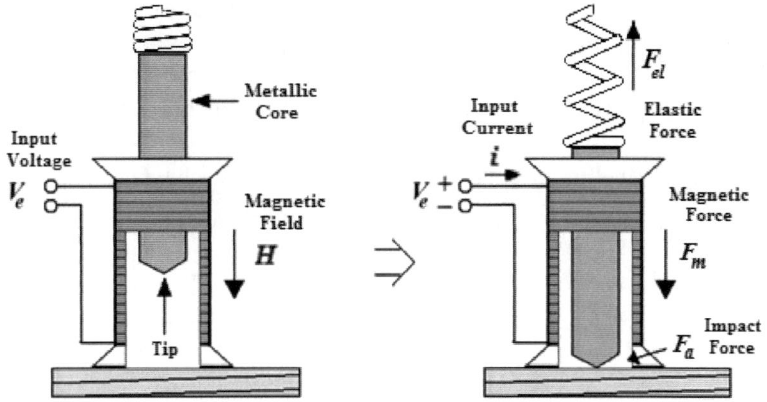

Figure 1. Structure of the electromagnetic hammer.

Where: V_e is voltage input; i is current input; H is magnetic field; F_{el} is elastic force; F_m is magnetic force; F_a is impact force.

2.1. Hammer impact test

The principle of the electromagnetic hammer is a mobile metallic core inserted inside a solenoid with one end attached to a spring. When the solenoid is triggered, it generates a magnetic force responsible for the metallic core displacement, which returns to its original position due to the elastic force exerted by the spring after the magnetic field cease [4][5].

2.2. Magnetic force

In general, it is widely observed the use of the magnetic field referenced to the interior or extremity of the solenoid, but in order to determine the impact force of the hammer on the tubular structure, it is necessary to first discover the influence of the produced field on the moving metallic core.

For a solenoid of length L formed by N turns with cylindrical ferromagnetic core of radius r, the interior magnetic field is given by the action of the field H along the longitudinal core axis as function of the variable distance x from extremities to center of the coil as shown in figure 2.

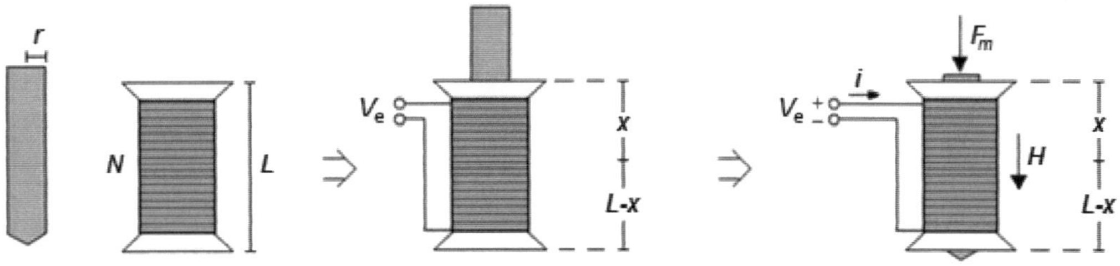

Figure 2. Structure of the electromagnetic hammer – core movement.

The magnetic field in the middle of solenoid is given by (1):

$$H = \frac{Ni}{2L}\left[\frac{x}{\left(r^2+x^2\right)^{1/2}} + \frac{L-x}{\left(r^2+(L-x)^2\right)^{1/2}}\right] \tag{1}$$

According to Zaro [7], the magnetic force work depends only on magnetic energy, so $U_m = F.x$. Thus the magnetic force measured in the x axis direction is given by the variation of the magnetic energy along x i.e., $F_x = \Delta U/\Delta x$. From this variation it is possible to find a relation for the magnetic force as function of the displacement of core from the magnetic energy, obtaining the magnetic force as function of the magnetic field H, the permeability of the nucleus μ_n and the permeability of the air μ_0 and the area of the cylinder A, in the extremities (F_{m_e}) and the center (F_{m_c}) respectively by (2) and (3):

$$F_{m_e} = \frac{N^2.i^2.A.X_m.\mu_0}{8.(r^2 + L^2)} \tag{2}$$

$$F_{m_c} = \frac{N^2.i^2.A.X_m.\mu_0}{2.(4r^2 + L^2)} \tag{3}$$

Where: N is the number of turns of the solenoid; i is the input current of the system controlling the resulting impact force; X_m is the magnetic susceptibility of the metal core; μ_0 is the magnetic permeability of free space; A is the cross sectional area of the magnetic core.

The behavior of the magnetic force must be considered to obtaining the impact force, since this occurs at the extremities of the coil, where the magnetic force is smaller, but is influenced by the center component which has the highest intensity.

2.3. *Impact force*

The impact force F_a is obtained by the resultant forces acting on the moving core of the hammer. The resultant is given by the action of the magnetic force, due to the magnetic field to which the core is inserted subtracted from the elastic force exerted by the spring in opposition to the movement. Once k is the elastic constant of the spring and α is acceleration of the system and is obtained from the beginning of the movement, the impact force is modeled by (4):

$$F_a = \frac{N^2.i^2.X_m.\mu_0.A}{8.(r^2 + L^2)} + m.\alpha - \frac{k.L}{2} \tag{4}$$

In the first moment, it is observed that the body leaves the rest position, which implies in the minimum elongation after the influence of the force weight, which is balanced with the elastic force. Note it is necessary to determine the acceleration α that causes a additional force (second Newton's Law), at that moment only the magnetic force acts on the moving core. That is, the body is at rest, therefore, $F_{el} = P$, so:

$$\alpha = \frac{N^2.i^2.X_m.\mu_0.A}{2m.(4r^2 + L^2)} \tag{5}$$

Thus the expression for the impact force estimate by:

$$F_a = \frac{N^2.i^2.X_m.\mu_0.A}{2} + \left[\frac{1}{4.(r^2 + L^2)} + \frac{1}{4r^2 + L^2}\right] - \frac{k.L}{2} \tag{6}$$

From (6) which represents the intensity of the impact force, the behavior of this force during the time was simulated in MATLAB, highlighting important parameters for characterization of the hammer operation as: electric current intensity, magnetic core permeability, number of turns And dimensions of the solenoid and mobile core. These results are discussed in the following section.

3. Results

The results obtained were analyzed by the variation of the input current of the hammer. The behavior of the impact force F_a can be observed in time and as function of the current input, as well as other determining factors such as number of solenoid turns and magnetic permeability of the core.

To obtain the simulated F_a results, the following values were adopted for the hammer description parameters for (6): $k = 15\ N/m$; $N = 300\ turns$; $L = 0.05\ m$; $r = 0.005\ m$; $\mu_r = 800$. These parameters are the same as the hammer built in the laboratory in order to compare the values obtained from the force estimation and the impact measurement.

The value of F_a will follow the input current waveform i. The current input signal used in this simulation was a 3.9 A amplitude pulse train and can be observed in figure 3.

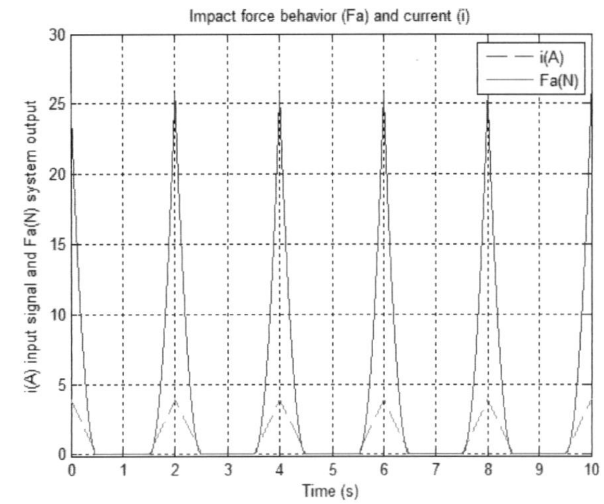

Figure 3. Impact force with input current of 3.9 A of amplitude in time.

However, its intensity can be altered by varying the other parameters of the system such as magnetic permeability. Note that by increasing the value of the current supplied to the hammer (figure 4), keeping the other parameters constant, there is a considerable increase for the impact force, especially if μ_r increases.

Although the behavior of the impact force depends only on the value of the input current supplied to the hammer, its intensity can be altered by varying the other parameters of the system such as magnetic permeability, number of turns and dimensions of the solenoid and ferromagnetic core.

The parameters used as reference for this model were based on the magnetic hammer already used in the [1], a compact device for the purpose of having a good performance and low power consumption for the construction of a portable scale detection system for field analysis.

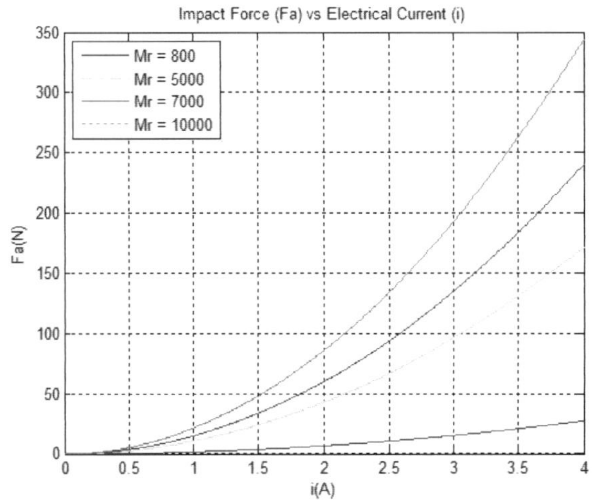

Figure 4. Impact strength vs. the input current for different values of μ_R.

To confirm the estimated values, the impact force of the hammer constructed with the same parameters of the simulation was measured. In the Vibration and Instrumentation Laboratory, with the aid of a PCB 086D05 load cell and an Agilent Dynamic Signal Analyzer 35670A (figure 5).

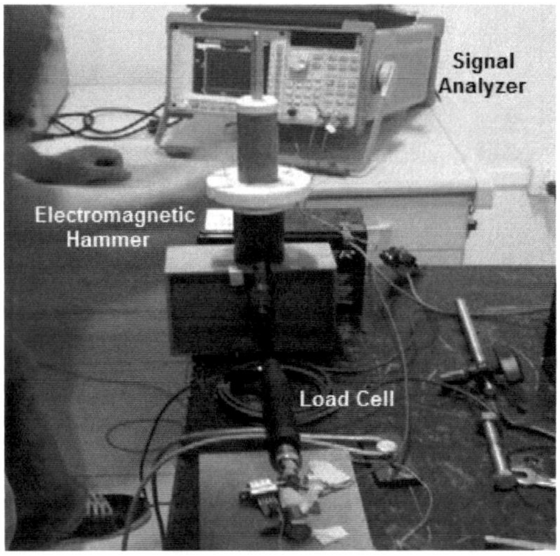

Figure 5. Assembly for measuring the impact force.

The experiment was performed to confirm the estimated value determined by (6). A load cell was used at the hammer impact point so that it was possible to capture the maximum force exerted by the force. A impulse excitation was applied in input and the measurement result was an inertness force peak of 25.78 N. In figure 6 it is possible to observe the peak value obtained with the impact of the hammer.

The value of the peak impact force obtained using the modeled hammer through (6) was 25.72 N for an input current of 3.9 A, it was the current consumed by the hammer during the measurement, considering 12 V DC power supply, while in the experimental measurement for the hammer constructed this value was 25.78 N. The error obtained was 0.233 %.

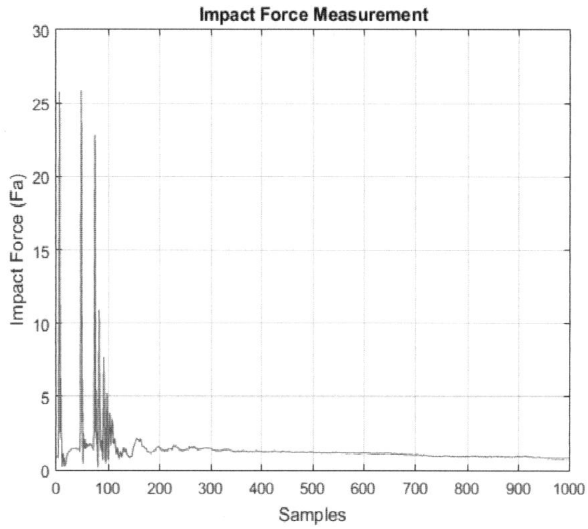

Figure 6. Impact force measurement.

Note that, although the maximum value of the impact force is close to the estimated one, there are some minor peaks present. These values should contain a secondary impact from the first strong impact hammer in the structure. The metallic core after collide the structure generates backward thrust, but the magnetic field, still active, pulls back the metal core again for another less intense impact. This undesired effect can be corrected by decreasing the duration of the impulse excitation.

4. Conclusions

The objective was reached since the results obtained were within expected, the obtained error was close to zero and the measured value was very close to the expected one. With these results, it is now possible to build an impact hammer by molding its dimensions according to the desired force instead to test many different methods of building a hammer in order to find the best, avoiding trial and error.

Acknowledgement

The authors would like to thanks CAPES and COPELE for financial support.

References

[1] Lemos Junior L C, Souza Neto J M R, Silva J J and Rocha Neto J S 2011 Fouling Detection using Hammer Impact Test and Wireless Comunication *International Instrumentation and Measurement Technology Conference (Binjiang, Hangzhou, China, May 2011)*
[2] Lai W P and Fang W 2002 Novel Bulk Acoustic Wave Hammer to Determinate the Dynamic Response of Microstructure Using Pulsed Broad Bandwidth Ultrasonic Transducers *(Elsevier – Sensors and Actuators)* p 96:43–52
[3] Mansoori G A 2001 Physicochemical Basis of Arterial Blockage/Fouling Prediction and Prevention *(University of Illinois at Chicago, USA. September, 2001)*
[4] Ahn S J and Jeong W B 2005 Improvement of Impulse Response Spectrum and its Application *(Journal of Sound and Vibration)* p 288:1223–1239
[5] Lohr K R and Rose J L 2002 Ultrasonic Guided Wave and Acoustic Impact Methods for Pipe Fouling Detection *(Journal of food engineering, March, 2002)*
[6] Tipler P A, Mosca G 2012 *Física Para Cientistas e Engenheiros, Volume 2* (LTC, São Paulo)
[7] Zaro M A, Blanco R L D, Thomas C A K 1989 Medição de força produzida pelo solenoide sobre o núcleo móvel em bobinas usadas nos sistemas de partida *(Revista de Ensino de Física) vol. 11* p.88-106.

IOP Conf. Series: Journal of Physics: Conf. Series **1044** (2018) 012030 doi:10.1088/1742-6596/1044/1/012030

Estimating the uncertainty of the specific surface area of Titanium Oxide Nanotubes calculated from nitrogen adsorption isotherms

E Ayala Blanco[1 2], F. Martínez Ortega[1], C E García Sánchez[2]

[1]Centro de Investigación en Catálisis (CICAT) –Escuela de Química, Universidad Industrial de Santander, Ciudad Universitaria, Cra. 27 #9, Bucaramanga, Santander, Colombia.

[2] Grupo de Investigación en Fluidos y Energía (GIFE) - Corporación Centro de Desarrollo Tecnológico del Gas, km 2 Vía Refugio, Parque Tecnológico UIS Guatiguará, Piedecuesta, Santander, Colombia.

eayala@cdtdegas.com

Abstract. The characterization of the textural properties of porous solids by isotherms of adsorption-desorption of nitrogen is a procedure commonly used for the study of the porosity of a solid, which is a great importance for meso and microporous solid materials. In the present work an analysis of the estimate the uncertainty associated to the calculation of the specific surface area was carried out. Here, the specific surface area of titanium oxide nanotubes was determined from N_2 adsorption isotherms data. The number of moles of gas adsorbed per gram of adsorbent were the input quantities with higher contribution to uncertainty of the surface area.

1. Introduction

An area of active research is the production of one-dimensional nanostructures, such as nanotubes, nanoribbons and nanorods, because they exhibit interesting optical, magnetic or chemical properties as a result of their small size. Some of the most relevant properties of the aforementioned nanostructures are the high mobility of electrons, high specific surface and quantum confinement effects. As a result, the exploration of new materials based on nanostructures made up of new materials, such as the inorganic nanotubes of metal sulfides and oxides, has grown. Titanium dioxide is one of the compounds used to create nanostructures used in photocatalysis, solar cells and medical devices, among others, due to its high chemical stability, low cost, high area-volume ratio and high surface activity **[1]**.

The nitrogen adsorption characterization is the most used method to evaluate the textural properties of the adsorbents. The specific surface area represents the surface per unit mass of a solid, which has an important effect on the physicochemical interactions between adsorbent and adsorbate. An appropriate determination of the surface area depends of the adequate measurement of N_2 adsorption-desorption isotherms at 77 K, and its correct physical interpretation. A good use of the measurements requires the estimation of the level of doubt or uncertainty that it possess. The process of uncertainty estimation also allows to obtain information about the sources that have higher contribution to the uncertainty of the result of a measurement. Taking into account the above considerations, the present work is focused on the analysis of the uncertainty of the specific surface area of adsorbent materials, determined from isotherms of nitrogen adsorption of samples nanotubes of TiO_2 (TiNTs) prepared by a hydrothermal method.

2. Materials and methods

Materials

Two samples of TiNTs were prepared using a hydrothermal process in an autoclave reactor under autogenous pressure. The temperatures of 383.15 K and 393.15 K were used for the hydrothermal process, anatase nanopowder (Sigma Aldrich) and a solution 10 M of NaOH were used as precursors [2].

Equipment

3FLEX Surface Characterization Analyzer Micrometrics was used for the adsorption measurements. It operates on the basis of a volumetric method, establishing the volume of N_2 adsorbed at the boiling temperature of the liquid nitrogen (77 K) as a function of the equilibrium pressure. The samples were degassed previously during 12 h at 383 K.

Mathematical model

The uncertainty estimation was performed using the procedure described in the Guide for Uncertainty Estimation (GUM method) [3]. The software Microsoft Excel was used to perform the calculations.

The specific surface area was determined using the BET method [4], which is based on the following Langmuir assumptions: (1) all adsorption centers are equivalent, (2) the adsorption capacity of a center is not affected by the occupation in the adjacent centers, and (3) each center can adsorb several layers of molecules. The measurement model of the BET method is represented in Figure 1.

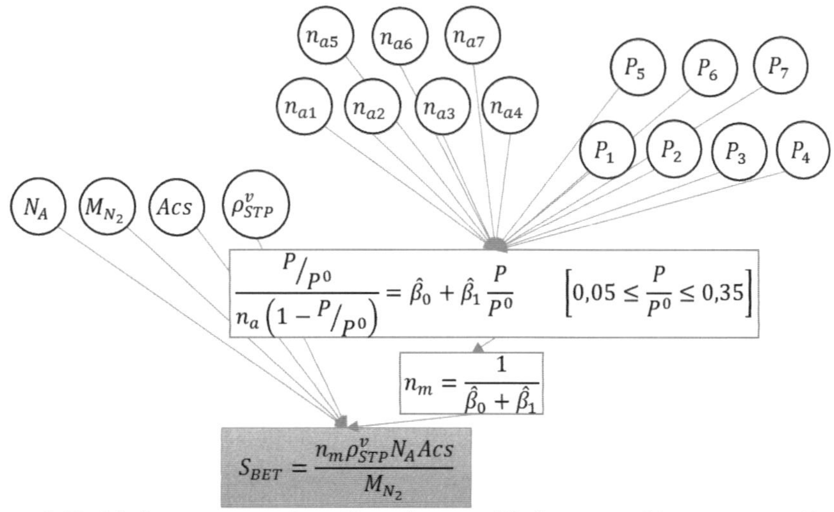

Figure 1. Model of measurement of specific surface area. The input quantities are presented in circles.

Where N_A is Avogadro number, M_{N_2} is the molar mass of nitrogen, Acs is the cross-sectional area of the nitrogen molecule, ρ_{STP}^{v} is the density of the nitrogen vapor at standard temperature and pressure, n_m is the number of moles of gas adsorbed per gram of the adsorbent, P^0 is the standard saturation pressure, P is the adsorptive pressure, and S_{BET} is the specific surface area of the adsorbent.

Applying the GUM method, the calculation of the surface area uncertainty is described by Equation (1).

$$u^2(S_{BET}) = \left(\frac{\partial S_{BET}}{N_A}\right)^2 u^2(N_A) + \left(\frac{\partial S_{BET}}{M_{N_2}}\right)^2 u^2(M_{N_2}) + \left(\frac{\partial S_{BET}}{Acs}\right)^2 u^2(Acs) + \left(\frac{\partial S_{BET}}{\rho_{STP}^{v}}\right)^2 u^2(\rho_{STP}^{v}) + \sum_{k=1}^{n}\left(\frac{\partial S_{BET}}{n_{a,k}}\right)^2 u^2(n_{a,k}) + \sum_{k=1}^{n}\left(\frac{\partial S_{BET}}{P_k}\right)^2 u^2(P_k) \qquad [Equation\ 1]$$

3. Results

3.1 Adsorption isotherms

The determination of the surface area of the evaluated samples was obtained through an adsorption analysis. In this technique, the solid material is contacted with a gas (N_2), and the pressure is gradually increased in the system until the solid surface reach its nitrogen saturation pressure. As the pressure increases, nitrogen molecules are adsorbed on the surface of the nanotubes to complete a monomolecular layer at a constant temperature. If the pressure continues increasing, a multi-layer adsorption and smaller pore filling are generated, resulting in capillary condensation. In the Figure 2 shows the adsorption isotherms obtained from the samples of TiNTs prepared. According to the classification of IUPAC, the isotherms correspond to a class IV, which is characteristic of the mesoporous adsorbents with weak affinity [5].

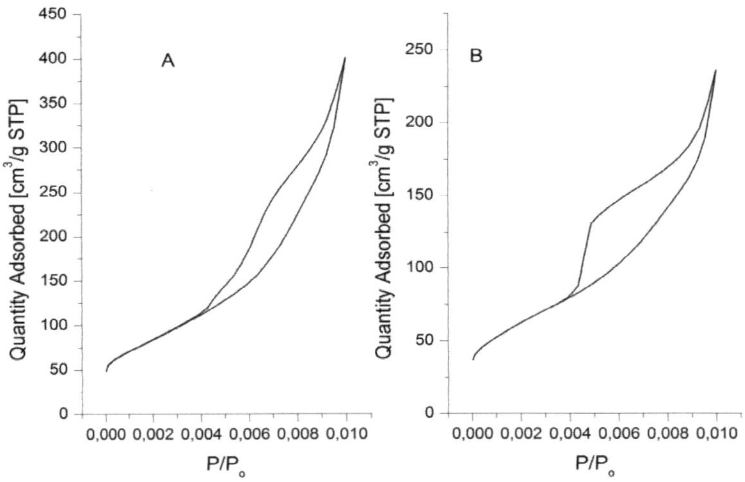

Figure 2. Adsorption isotherms of TiO_2 nanotubes, prepared by a hydrothermal process (A) TiNTs at 383.15 K and (B) TiNTs 393.15 K

The appearance of the hysteresis cycle (adsorption - desorption) is due to a process of capillary condensation in the pore of mesoporous materials [5].

3.2 Estimation of measurement uncertainty

Analysis of the uncertainty of properties measured at very small scale is important in determining whether the process of preparation of nanostructures is appropriate [6]. From the mathematical model proposed for the determination of the surface area indicated in figure 1, the sources of uncertainty associated with the measurement process were identified, and the quantification of each one of the uncertainties that came from the sources identified as shown In Table 1. The uncertainty values differ for each of the evaluated samples.

Table 1. Input quantities with their respective uncertainties

Sample 1 TiNTs				**Sample 2 TiNTs**			
Input quantities	**Uncertainty u(i)**	**Input quantities**	**Uncertainty u(i)**	**Input quantities**	**Uncertainty u(i)**	**Input quantities**	**Uncertainty u(i)**
P_1[mmHg]	0,0019	n_{a3}[mol]	1,89	P_1[mmHg]	0,0013	n_{a3}[mol]	1,37
P_2[mmHg]	0,0018	n_{a4}[mol]	2,00	P_2[mmHg]	0,0012	n_{a4}[mol]	1,45
P_3[mmHg]	0,0011	n_{a5}[mol]	3,48	P_3[mmHg]	0,00081	n_{a5}[mol]	2,51
P_4[mmHg]	0,00039	n_{a6}[mol]	7,52	P_4[mmHg]	0,000033	n_{a6}[mol]	5,50
P_5[mmHg]	0,0028	n_{a7}[mol]	12,50	P_5[mmHg]	0,0017	n_{a7}[mol]	9,29
P_6[mmHg]	0,0069	N_A[mol]	0,0000037	P_6[mmHg]	0,0049	N_A[mol]	0,0000027
P_7[mmHg]	0,013	M_{N_2}[g/mol]	3,062	P_7[mmHg]	0,010	M_{N_2}[g/mol]	2,26
n_{a1} [mol]	3,48	ρ_{STP}^{v}[m³/g]	0,0021	n_{a1} [mol]	2,55	ρ_{STP}^{v}[m³/g]	0,0016
n_{a2}[mol]	3,18	Acs[m²]	0,0049	n_{a2}[mol]	2,323	Acs[m²]	0,0036

The quantitative evaluation of the uncertainty expressed as the standard deviation of the probability distribution calculated for the measurand following the guidelines established in GUM expressed in equation (1) for the surface area, where $\frac{\partial S_{BET}}{N_A}, \frac{\partial S_{BET}}{M_{N_2}}, \frac{\partial S_{BET}}{Acs}, \frac{\partial S_{BET}}{\rho_{STP}^{v}}, \frac{\partial S_{BET}}{n_{a,k}}, \frac{\partial S_{BET}}{P_k}$ represent coefficients of sensitivity and $u(N_A)$, $u(M_{N_2})$, $u(Acs)$, $u(\rho_{STP}^{v})$, $u(n_{a,k})$, $u(P_k)$ the contributions of their uncertainty. The uncertainty values of the constants present in the mathematical model were taken from the literature reports and some values of uncertainties such as the Avogadro number were taken from the NIST Chemistry WebBook [7]. The probability distributions assigned to each of the input quantities were normal distributions, except for pressure, whose uncertainty was calculated from a combination of resolution and accuracy uncertainties simulated with rectangular distributions.The uncertainty values of the moles of gas adsorbed per gram of adsorbent at different pressures was attributed a 5%.

The sensitivity coefficients were determined by numerical derivation using central finite differences, which is a numerical analysis technique that allows us to calculate an approximation of the derivative of a function at a point, using some values of the function [8]. The formula of numerical differentiation by central finite differences is presented in equation 2 [9].

$$\left[\frac{\partial S_{BET}}{N_A}\right] = \frac{f(N_A + h) - f(N_A - h)}{2h} \qquad \textit{Equation 2}$$

The surface area and its estimated uncertainty were calculated for the two samples of TiNTs. The results, including the uncertainty budget, are presented in Tables 2 and 3.

Table 2, Uncertainty estimation Sample 1 TiNTs

Experimental conditions: TiNTs, 90 h, reaction at 383.15 K, pH = 7.0									
S_{BET}		306,24 m^2/g			u(S)		31,93 m^2/g		
Input quantity	P_1	P_2	P_3	P_4	P_5	P_6	P_7	n_{a1}	n_{a2}
Relative contribution[%]	$1,44*10^{-8}$	$1,22*10^{-8}$	$4,46*10^{-9}$	$5,72*10^{-10}$	$3,14*10^{-8}$	$1,81*10^{-7}$	$6,57*10^{-7}$	4,60	3,82
Input quantity	n_{a3}	n_{a4}	n_{a5}	n_{a6}	n_{a7}	N_A	Acs	M_{N_2}	ρ^v_{STP}
Relative contribution[%]	1,35	1,51	4,57	21,43	59,15	$5,11*10^{-14}$	3,55	$1,81*10^{-8}$	$9,08*10^{-8}$

Table 3. Uncertainty estimation Sample 2 TiNTs

Experimental conditions: TiNTs, 90 h, reaction at 393.15 K, pH = 7.0									
S_{BET}		226,08 m^2/g			u(S)		23,57 m^2/g		
Input quantity	P_1	P_2	P_3	P_4	P_5	P_6	P_7	n_{a1}	n_{a2}
Relative contribution[%]	$1,17*10^{-8}$	$9,9*10^{-9}$	$4,59*10^{-9}$	$7,53*10^{-12}$	$2,09*10^{-8}$	$1,67*10^{-7}$	$7,42*10^{-7}$	4,51	3,74
Input quantity	n_{a3}	n_{a4}	n_{a5}	n_{a6}	n_{a7}	N_A	Acs	M_{N_2}	ρ^v_{STP}
Relative contribution[%]	1,31	1,46	4,38	21,05	59,99	$5,11*10^{-14}$	3,55	$1,81*10^{-8}$	$9,08*10^{-8}$

The expanded uncertainty estimation is defined as half the length of a coverage interval containing a given percentage of the values reasonably attributable to the measurand with a coverage factor k = 2, that corresponds to a coverage level of about 95%. The expanded uncertainty of the surface area in the case studies was 31,93 m^2/g and 23,57 m^2/g, which represents about 10,43% and 10,42% of the measured surfaces. In tables 2 and 3 the relative contributions of each of the sources of uncertainty evaluated can be observed. For sample 1 and sample 2 the greatest contribution to the uncertainty was given by the moles of gas adsorbed per gram of adsorbent at different pressures. The difference in the contributions observed in each of the samples is mainly due to the fact that each pre-paved TiNT showed differences in the porosity of the material as observed in the adsorption isotherms indicated in figure 2, where perhaps this difference is associated with the temperature of TiNT preparation. The preparation and characterization of nanomaterials require advances in measurement and instrumentation capabilities to validate their properties, these measurements must be reliable and comparable at a global level. The correct identification of these sources for any measurement contributes to greater precision in decision making.

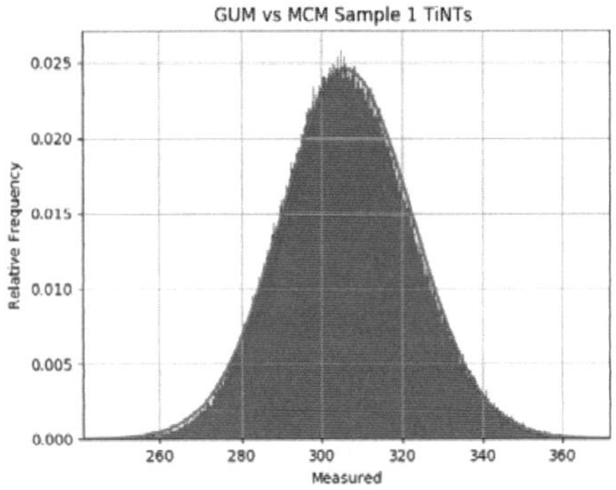

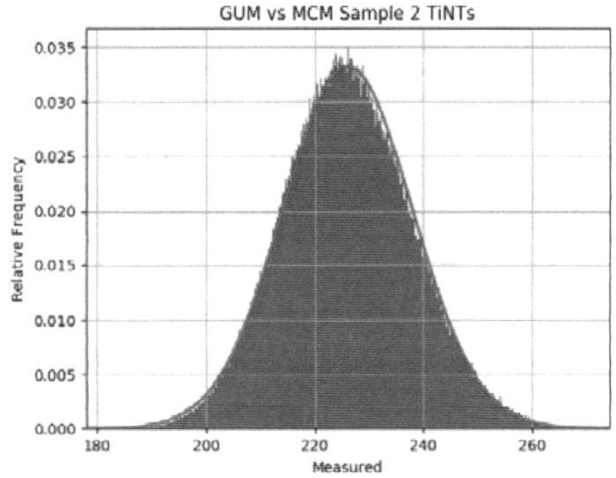

Figure 3. Uncertainty estimation comparison using the GUM and Monte Carlo

To evaluate the importance of correlation effects in the mathematical model, it was used the Monte Carlo method to estimate the uncertainty, and the results were compared with those obtained with the GUM method without correlations. Figure 3 shows the results of the comparison. It is concluded that the GUM method without correlations estimates adequately the uncertainty in the study case, so it is not necessary to include correlation terms in the GUM method in this case.

4. Conclusions

- It was possible to estimate the uncertainty of the surface area of two different types of TiNTs applying the GUM method. It is important to consider the uncertainty in all the measurement processes, because a correct uncertainty estimation is fundamental for proper decision-making and for guiding improvement actions.

- The input quantity with the higher contribution to the uncertainty of the surface area in our case was the number of moles of gas adsorbed per gram of adsorbent. These results allow to correctly orient the efforts to reduce the uncertainty of surface area measurement by adsorption-desorption isotherms. Since the amount of gas adsorbed by the sample depends on temperature, vapour pressure and the nature of the solid, the use of pressure meters with higher precision could decrease the uncertainty of the amount of gas adsorbed by mass unit of the adsorbent.

Acknowledgments

The authors are grateful to Centro de Investigación en Catálisis (CICAT), Internal project Dief 1868 from the Universidad Industrial de Santander and the Corporation Centro de Desarrollo Tecnológico del Gas, for their financial support to the present work.

References

[1] K. Lee, A. Mazare, y P. Schmuki, «One-Dimensional Titanium Dioxide Nanomaterials: Nanotubes», *Chem. Rev.*, vol. 114, n.º 19, pp. 9385-9454, oct. 2014.
[2] T. H. T. Vu *et al.*, «Synthesis of titanium dioxide nanotubes via one-step dynamic hydrothermal process», *J. Mater. Sci.*, vol. 49, n.º 16, pp. 5617-5625, ago. 2014.
[3] International Bureau of Weights and Measures, International Organization for Standardization. «Guide to the Expression of Uncertainty in Measurement». 1st ed. 1993. Genève, Switzerland: International Organization for Standardization
[4] M. F. De Lange, T. J. H. Vlugt, J. Gascon, y F. Kapteijn, «Adsorptive characterization of porous solids: Error analysis guides the way», *Microporous Mesoporous Mater.*, vol. 200, pp. 199-215, dic. 2014
[5] M. Thommes *et al.*, «Physisorption of gases, with special reference to the evaluation of surface area and pore size distribution (IUPAC Technical Report)», *Pure Appl. Chem.*, vol. 87, n.º 9-10, ene. 2015.
[6] Leach, R.; Chetwynd D.; Blunt, L.; Haycocks, J.; Harris, P.; Jackson, K.; Oldfield, S. and Reilly, S. (2006) Recent advances in traceable nanoscale dimension and force metrology in the UK. Meas. Sci. Technol. 17(3), 467–476.
[7] NIST Chemistry WebBook. Disponible en: http://webbook.nist.gov/chemistry/. Consultado: 6 de Febrero de 2017.
[8] Chapra, S., Canale, R. P., Numerical Methods for Engineers, McGraw-Hill Higher Education, Boston 2010.
[9] N. García, A. M. Baró, R. Miranda, H. Rohrer, Ch. Gerber, R. García Cantú and J. L. Peña, Metrologia, vol.21, p.135 (1985).
[10] The critical role of metrology in nanotechnology. M.L. Schattenburg, H.I. Smith, Proc. SPIE, Vol. 4608, p. 116, Workshop on Nanostructure Science, Metrology and Technology, Gaithersburg, Maryland, 5-7 September 2001.

IOP Conf. Series: Journal of Physics: Conf. Series **1044** (2018) 012031 doi:10.1088/1742-6596/1044/1/012031

Exact correction of image distortion in a filter wheel multispectral camera with focus adjustment

C. Zhang, R. Celestre, M. Rosenberger, G. Notni
Ilmenau University of Technology
Department of Quality Assurance and Industrial Image Processing
Gustav-Kirchhoff-Platz 2, 98693 Ilmenau, Germany

E-mail: chen.zhang@tu-ilmenau.de

Abstract. The correction of inter-channel image distortions in multispectral imaging is the key process for high-precision applications. In this work, image distortion in a special filter wheel camera with sensor actuation based focus adjustment is investigated. Based on the analysis of the optical path, error compensation models at different levels are derived and qualitatively compared with each other. The experimental evaluation of the models shows that the proposed extended radial-tangential model achieves the highest image registration accuracy, especially in the marginal image areas. Furthermore, the stability of these algorithms is evaluated with repeated measurements.

1. Introduction

In multispectral imaging, an accurate registration of the images acquired at different wavelengths is of great advantage to the spatial-spectral analysis of the multispectral image data. A typical multispectral imaging technique is the spectral scanning using a rotating filter wheel with narrow bandpass optical filters. In this technique, the optical aberration originates from the chromatic aberration in the lens, the different spectral refractive indices of the filter materials as well as the deviations in the mechanical filter system, e.g. the different thicknesses and tilt angles of the filters [1] [2]. The optical aberration could be divided into a longitudinal and a transversal part, resulting in an image blurring compared to a reference channel and an image displacement, respectively.

A conventional tool for image sharpness recovery is the image restoration method using point spread function (PSF). In [3] it is shown that a spatially varying free-form PSF model is most suitable for a filter wheel camera system. In [3] a PSF estimation method using a random noise target is proposed to calculate the relative PSFs relating to the reference channel, and in [4] the blind restoration method is adapted for the PSF characterization. Due to the block-wise PSF estimation used in these works the channel images must be geometrically aligned with each other before the PSF estimation. With a large spectral range, the blurring due to longitudinal aberration could be extremely significant and disturbs this geometric image alignment. Moreover, the performance of image restoration method is strongly limited with heavy blurring, and irregular distortions could be produced due to the anisotropic free-form PSF model. Therefore, mechanical focus correction methods could be more meaningful for high-precision measurements.

In a previous work of the authors [5], a focus adjustment via mounting glass correction plates at the filters is implemented. On this basis, a 2D polynomial function is estimated for the correction of image displacements. With this method, about 95% of the image distortion could be removed. However, this focus correction approach counts only for the special objective as well as working distance. In order to facilitate a higher flexibility concerning the change of objective and variation of working distance, an autofocus approach via linear sensor actuation is developed in a following work [6]. Nevertheless, the sensor actuation in this approach causes a variation of the distance between the objective and image sensor, leading to an alteration of the imaging characteristics. In this work, the image distortion in this circumstance and its compensation will be analyzed and experimentally investigated.

2. Theoretical analysis of the transversal aberration

In the focus correction, the channel of 550 nm is set as reference channel due to its high signal-to-noise ratio. In this channel, a perfect focusing is realized with manual lens adjustment. Afterwards, the sensor positions for the best focusing in the other channels are estimated through maximization of the graph Laplacian energy at a target object. In this work, the image distortion is defined as the relative image displacement in comparison to the reference channel and includes all the influences of filter aberration, lens chromatic aberration, lens distortion, sensor actuation as well as all the machining and mounting deviations in the mechanical guide.

2.1. Filter aberration under linear sensor actuation

In a previous work of the authors [7], the optical aberration caused by the filters with regard to the variation of the sensor-aperture-distance (focal length) is analysed. The relative filter aberration in comparison to the reference channel is dependent on the refractive indices of both filter materials n and n_{ref}, the filter normal vectors $\mathbf{n}$ and $\mathbf{n}_{\mathbf{ref}}$, the sensor positions d_s and $d_{s,ref}$ as well as a transversal shift of the sensor $\mathbf{S}_\mathbf{S}$ concerning the positioning errors of the mechanical guide. It is proved that the image point displacement $\mathbf{e}_{\mathbf{F,rel}}$ caused by the filter aberration and sensor actuation consists of a global part and a location dependent part, which could be modelled as a shift of the image principle point and an image zoom, respectively, under the paraxial approximation.

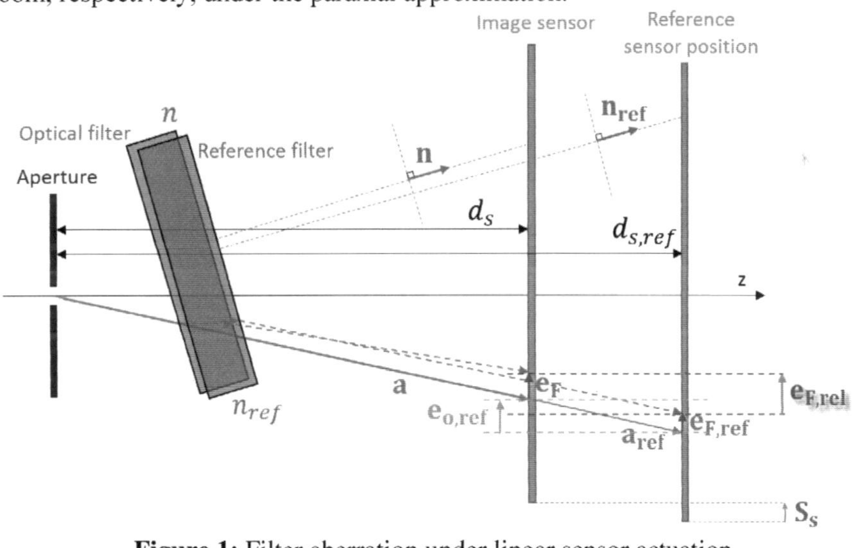

Figure 1: Filter aberration under linear sensor actuation.

2.2. Lens chromatic aberration and lens distortion

Under a perfect focus correction and the paraxial approximation, it is obvious that the linear sensor actuation for the compensation of the longitudinal aberration leads to a lateral image displacement $\mathbf{e}_{\mathbf{c,rel}}$ in figure 2 that corresponds to the shift $\mathbf{e}_{\mathbf{o,rel}}$ in figure 1. It indicates that an explicit modelling of the transversal lens chromatic aberration is not needed in this ideal case. Considering the risk of an imperfect focus correction or a large field of view that could disturb the paraxial approximation, the affine model [2] and radial-tangential model in [8] proposed for filter wheel cameras with fixed sensor position will be taken as reference models.

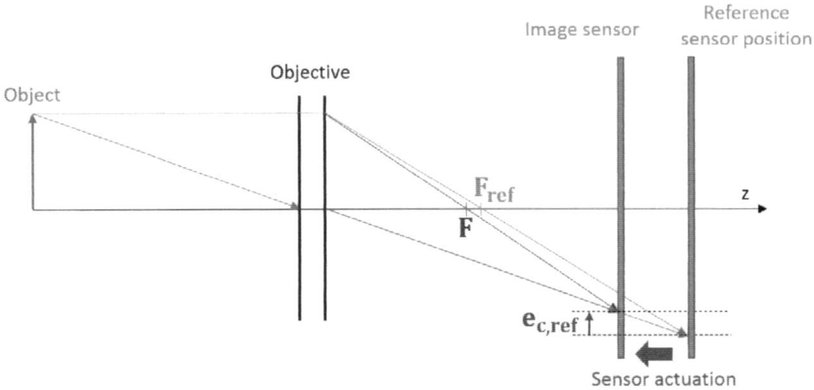

Figure 2: Transversal chromatic aberration in sensor actuation.

Next to the lens chromatic aberration, the channel dependence of the lens distortion could also take a considerable part in the total image distortion. The lens distortion is influenced by the wavelength-dependent refractive index of lens material [9] as well as the image sensor position. Furthermore, the shift of the image principle point, which is usually assumed as the distortion-free reference point, causes also an alteration of the lens distortion parameters relating to the local image coordinate system and could be several pixels due to the mechanics [6]. The channel dependence of lens distortion could be more remarkable in the image areas with severe distortions, mostly the marginal areas. For the requirement for a higher precision, the channel dependence of lens distortion could be also taken into consideration by the means of a channel-wise correction.

3. Proposed image correction models

3.1. Scaling-translation model

According to the analysis in section 2, image correction models at different levels can be derived. Assuming a perfect focus adjustment and ignoring the channel dependence of lens distortion, a simple scaling-translation model (ST model) can be used with parameters k_1 to k_3:

$$\begin{pmatrix} x_d \\ y_d \end{pmatrix} = \begin{pmatrix} k_1 x_n + k_2 \\ k_1 y_n + k_3 \end{pmatrix}, \quad (1)$$

where (x_d, y_d) and (x_n, y_n) denote the locations of distorted und undistorted image point referring to the image principle point that is simultaneously estimated in the model fitting.

3.2. Affine model

As a reference model, the combination of the ST model with the affine model for chromatic aberration in [2] could be used. An affine transformation consists of a rotation, a translation and an anisotropic scaling, and allows a finer modelling of the possible irregularity of distortions. Because that it contains already the ST model, this combination results in an affine image correction model in turn.

3.3. Channel-wise lens distortion correction

In order to deal with the channel dependence of lens distortion, a channel-wise correction is performed before the implementation of these both aforementioned models. For the correction, the lens distortion model in [10] is used:

$$\begin{pmatrix} x_d \\ y_d \end{pmatrix} = \begin{pmatrix} x_n(1+d_1 r^2 + d_2 r^4) + 2d_3 x_n y_n + d_4(r^2 + 2x_n^2) \\ y_n(1+d_1 r^2 + d_2 r^4) + d_3(r^2 + 2y_n^2) + 2d_4 x_n y_n \end{pmatrix}, \quad (2)$$

where r denotes the distance between the undistorted image point and the principle point, and d_i are the lens distortion coefficients.

3.4. Radial-tangential model

Under the fact that the polynomial terms in the radial-tangential chromatic aberration model in [8] are already contained in the lens distortion model in [10], an extended radial-tangential model (RT model) combing the filter aberration, lens chromatic aberration and lens distortion is proposed as follows:

$$\begin{pmatrix} x_d \\ y_d \end{pmatrix} = \begin{pmatrix} x_n(1+k_1+k_2r^2+k_3r^4)+2k_4x_ny_n+k_5(r^2+2x_n^2)+k_6 \\ y_n(1+k_1+k_2r^2+k_3r^4)+k_4(r^2+2y_n^2)+2k_5x_ny_n+k_7 \end{pmatrix}, \tag{3}$$

where k_i are the model parameters.

Due to the large number of model parameters, the implementation of the RT model succeeds in three steps. In the first step, the image translation and zoom parameters as well as the image principle point are estimated using the least squares method. Then, these parameters are fixed, and the parameters in the radial and tangential terms of higher degree are estimated. In the last step, all the model parameters are refined again through the Levenberg-Marquardt optimization.

4. Experimental results

Experimental investigations are performed at a 12-channel filter wheel camera. The spectral channels are between 400 nm and 950 nm in step of 50 nm und have a full width at half maximum (FWHW) of 50 nm. As target object, a dot grid glass target is placed orthogonal to the camera in a distance of ca. 400 mm and illuminated with uniform backlight. For the acquisition, the objective Carl Zeiss Tevidon 1.4/25 is used. In order to exclude the influences of vibration caused by the sensor actuation, a latency of 0.5 seconds between channel switch and image acquisition is specified.

For the extraction of the circular markers on the target, the two-step procedure in [11] is used. The markers are roughly detected based on Hough transform, and the centers of the dots are then located with higher accuracy using the search-line based edge point detection and circle fitting. These centers are used as controlling points for the model evaluation. Deviations between the dot centers detected in the corrected channel images and the reference image are identified as residual errors.

4.1. Model evaluation

In order to oppress stochastic errors, the model parameter estimation is performed with ten repeated full spectral acquisitions, whereby the dots are repetitively detected and their coordinates are averaged. From the raw image data, the mean and maximal errors in each channel are determined and averaged across all the channels. The cross-channel mean error is 2.958 pixels, and the maximal error among the channels is 8.326 pixels with an averaged maximal error of 4.67 pixels.

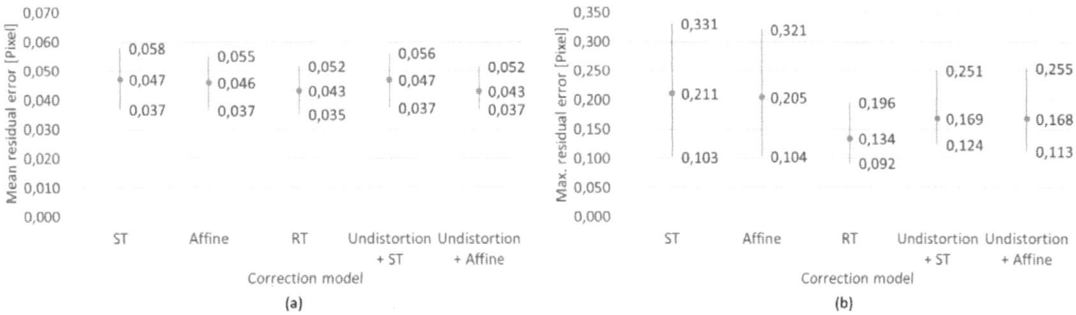

Figure 3: Model evaluation results: cross-channel mean (a) and maximal residuals (b).

Figure 3 shows the cross-channel mean and maximal residual errors in the corrected image data, with an indication of the maximal and minimal values among the channels. It can be seen that the cross-channel mean residual could be reduced to lower than 0.05 pixels with all the models. Among these models, the RT model exhibits a minor advantage at thousandth pixel level. This is yet negligible in practical applications.

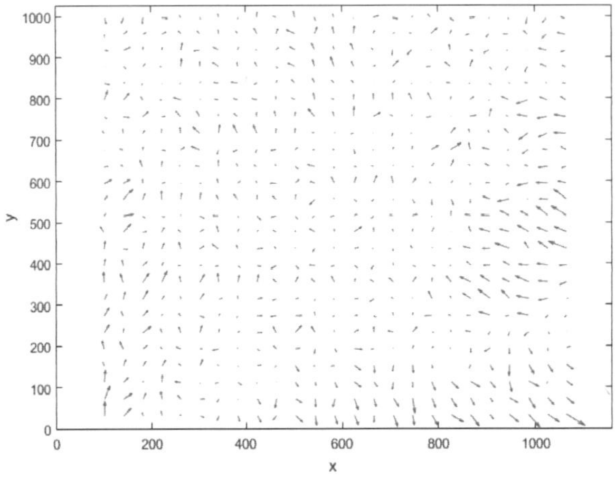

Figure 4. Residual errors with the extended RT model in 900 nm channel.

In comparison to the mean residual error, the maximum residual is much larger. As an example, the residual errors in the 900 nm channel at the RT model are presented in figure 4, in which the residual errors are drawn as vectors at the controlling points. It could be seen that the residuals in the marginal areas of the image are more significant, which correspond to areas with stronger lens distortion. By the means of explicit lens distortion correction, the cross-channel maximal residual error can be reduced from ca. 0.21 pixels to ca. 0.17 pixels at the ST and affine model. A more considerable advantage of the RT model is recognizable. With this model, a cross-channel maximal residual of 0.134 pixels is achieved, while the upper limit is lower than with other models.

4.2. Performance evaluation with repeated measurements

As a next step, the stability of the image correction methods is evaluated with repeated measurements. This evaluation is performed with ten further spectral acquisitions that are corrected using the model parameters estimated in section 4.1. At each spectral acquisition the cross-channel averaged mean and maximal residual are determined. Table 1 shows that the models exhibit almost the same performance relating to the cross-channel mean residual, with a minor general growth than the values in section 4.1.

Table 1. Mean and standard deviation of cross-channel mean residual.

	ST	Affine	RT	Undistortion + ST	Undistortion + Affine
Mean	0.063	0.061	0.060	0.063	0.059
Standard deviation	0.003	0.003	0.002	0.002	0.002

Table 2. Mean and standard deviation of cross-channel maximal residual.

	ST	Affine	RT	Undistortion + ST	Undistortion + Affine
Mean	0.225	0.211	0.173	0.190	0.183
Standard deviation	0.022	0.022	0.009	0.013	0.009

The evaluation concerning the cross-channel maximal residual reflects also the same trend in section 4.1, see table 2. Overall, a cross-channel mean residual of 0.060 ± 0.002 pixels and maximal residual of 0.173 ± 0.009 pixels are achieved with the extended RT model. These lower values could validate the advantages of the RT model over the other models, primarily in the marginal image areas.

5. Conclusion and future work

The experimental results show that the simple scaling-translation (ST) model can already compensate more than 98% of image distortion. It indicates the correctness of the theoretical analysis and a high-precision focus correction. The affine model remains nearly at the same level as the ST model, while these both models suffer from the irregular lens distortion in the marginal image areas. An explicit correction of lens distortion scores an improvement, but could produce additional process errors and is hence less accurate than the extended radial-tangential (RT) model. With a higher comprehensiveness, the RT model ensures a more accurate compensation of image distortion in the marginal image areas and hence a mitigation of artefacts in multispectral image processing that could be caused by inter-channel pixel displacements. So, the RT model could be implemented for high-precision geometric and color measurements as well as spatiospectral analysis, in which the entire area of the image sensor should be used.
The measurement uncertainty with this filter wheel camera arises from the illumination fluctuation and mechanical sensor actuation, whereas the last one concerning the accuracy and uncertainty of sensor repositioning could play a crucial role. In high-speed acquisitions, a higher sensor actuation frequency could cause a stronger vibration and lead to an increased measurement uncertainty. Evaluation of the error compensation algorithms in this case will be performed as a future work.

Acknowledgements

We thank the Federal Ministry of Education and Research for the strong support of this work. The work is related to the project *Qualimess next generation* (03IPT709X) and the program *Zwanzig20 – Partnership for Innovation* as part of the research alliance 3Dsensation.

References

[1] Brauers J and Aach T 2008 Longitudinal Aberrations Caused by Optical Filters and Their Compensation in Multispectral Imaging *Proc. of 2008 IEEE International Conference on Image Processing (ICIP2008)* 525-528
[2] Brauers J, Schulte N and Aach T 2008 Multispectral Filter-Wheel Cameras: Geometric Distortion Model and Compensation Algorithms *IEEE Transactions on Image Processing* **17** 2368-2380
[3] Brauers J, Seiler C and Aach T 2010 Direct PSF Estimation Using a Random Noise Target *Proc. SPIE 7537, Digital Photography VI*, 75370B
[4] Klein J 2015 *Multispectral imaging: aberrations and acquisitions from different viewing positions* (RWTH Aachen University, PhD thesis)
[5] Rosenberger M and Linss G 2015 Multispectral Image Correction for Geometric Measurement *Journal of Physics: Conference Series 588* 012037
[6] Rosenberger M, Preissler M, Fuetterer R, Zhang C, Celestre R and Notni G 2016 Development and characterization of a high speed linear-moving-stage for multispectral measurements *Journal of Physics: Conference Series 772* 012054
[7] Zhang C, Rosenberger M, Breitbarth A and Notni G 2016 A novel 3D multispectral vision system based on filter wheel cameras *Proc. of 2016 IEEE International Conference on Imaging Systems and Techniques (IST 2016)* 267-272
[8] Klein J, Brauers J and Aach T 2010 Spatial and Spectral Analysis and Modeling of Transversal Chromatic Aberrations and their Compensation *Proc. of IS&Ts 5th European Conference on Color in Graphics, Imaging, and Vision (CGIV)* 516-522
[9] Haferkorn H 1986 *Bewertung optischer Systeme* (Berlin: Deutscher Verlag der Wissenschaften)
[10] Braeuer-Burchardt C 2005 A new methodology for determination and correction of lens distortion in 3D measuring systems using fringe projection *Pattern Recognition* **3663** 200-207
[11] Breitbarth A, Correns M, Zimmermann Z, Zhang C, Rosenberger M, Schambach J and Notni G 2017 Verification of real sensor motion for a high-dynamic 3D measurement inspection system *Proc. SPIE 10329, Optical Measurement Systems for Industrial Inspection X*, 10329-22

IOP Conf. Series: Journal of Physics: Conf. Series **1044** (2018) 012032 doi:10.1088/1742-6596/1044/1/012032

Exposing all the decay paths of ^{152}Eu and their probabilities

Alfredo L Ferreira Filho[1,2,3]; P A L da Cruz[1]; J U Delgado[1]; Ricardo T Lopes[2]

[1] Laboratório Nacional de Metrologia das Radiações Ionizantes (LNMRI/IRD/CNEN), Av. Salvador Allende, s/n, Recreio, CEP 22780-160, Rio de Janeiro, Brazil;
[2] Laboratório de Instrumentação Nuclear (LIN/PEN/COPPE/UFRJ), Caixa Postal 68509, CEP 21945-970, Rio de Janeiro, Brazil.

alfredo@ird.gov.br

Abstract. Checking the ^{152}Eu 13-year very complex decay scheme, the aim of this paper is the identification, followed by a graphical presentation of all its decay paths and their probabilities, based on the information about particles, emitted energies and probabilities. The decay path followed in the course of the decay process is defined by the cascade transitions from the parent ground state to the daughter ground state. They are 74 paths to the ^{152}Gd ground state and 285 to ^{152}Sm, totalizing 359 cascades. As the decay path is seen as a time-series, the probability to follow this path, calculated here by an excel spreadsheet, can simply be seen as a product of transition probabilities. All such data is necessary for the accurate primary standardization of ^{152}Eu solution through methods as the CIEMAT/NIST for liquid scintillation counting or even the sum peak method, an absolute measurement technique used for activity standardization by gamma-ray spectrometry techniques. The idea is to expose and highlight the difficulties arising from the complexity of such decay, meeting the needs of a didactic presentation to the user community not specifically available in the literature.

1. Introduction

For many cascade photon emitting nuclides, more than one mode of decay is possible. The ^{152}Eu has two branches of decay, which are feeding several excited states in ^{152}Sm and ^{152}Gd, promptly de-excited via γ-ray transitions (photon emission + conversion electron emission). Direct decay of parent to the ground state of daughter does not exist. The ^{152}Sm branch has nineteen excited levels in addition to the basic level, as the ^{152}Gd has fifteen. It disintegrates into ^{152}Sm by electron capture (72,1%) and a negligible amount of positron emission (0,027%), and into ^{152}Gd disintegrates by beta minus emission (27,9%) [1]. About 132 γ-ray transitions are known to follow the decay, with intensities above 10^{-5} photons per parent decay. Due to the very complex decay scheme of ^{152}Eu, the activity determination was attended by difficulty. The particular interest in this radionuclide, with a half-life of 13.522 years, is explained by its wide use for energy and efficiency calibration of solid-state γ-ray detectors since its energy range of the emitted photons is broad, that is, 40 - 1769 keV, making this radionuclide suitable for the aforementioned purpose.

[3] Corresponding author

IOP Conf. Series: Journal of Physics: Conf. Series **1044** (2018) 012032 doi:10.1088/1742-6596/1044/1/012032

2. Decay paths and transitions probabilities

The main features of the complex decay scheme of ^{152}Eu are shown in figure 1. The ground state of the parent nucleus is presented as an excited level of the daughter nucleus, labelled by n. Note that, in this modified scheme, n is equal to 20 in the ^{152}Sm branch and 16 in the ^{153}Gd branch. The ground state of the daughter nucleus is labelled by 0. Between these two ground states, there are $n-1$ daughter nucleus excited levels themselves.

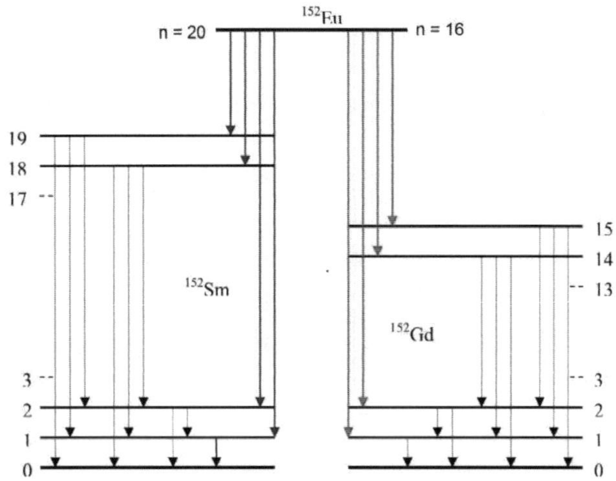

Figure 1. ^{152}Eu decay scheme consisting of $n-1$ excited levels.

The decay path, followed in the course of the decay process, is defined by the cascade transitions from the ground state of the parent nucleus to the ground state of the daughter nucleus. Transition from ^{152}Eu to the ^{152}Sm or ^{152}Gd ground state can occur through alternative paths. Figure 2 shows a datasheet with the decay paths and, on the bottom of each cascade transition line, the corresponding normalized transition probability for each step from the ^{152}Eu to the ^{152}Gd ground state, in 2, 3, 4, 5 and 6 steps. Representing an Excel spreadsheet, the figure also shows (at the end of each probabilities line) the probability of each path (corresponding to the product of the normalized transitions probabilities). Data details may be seen more clearly by zooming-in on the figure in the on-line version of the journal.

Corresponding datasheet for ^{152}Sm is too extensive for publishing now, in this paper; hence, it can be obtained in electronic form from the corresponding author, together with the MS Excel spreadsheet used in calculations. Using filters to narrow down data in the Excel spreadsheet, the figures 3 and 4 show the decay paths for the transition from ground state of ^{152}Eu to the ground state of ^{152}Sm, by way of example, only in 2 and 3 steps, respectively. The number of decay paths of ^{152}Eu to ^{152}Sm and ^{152}Gd are summarized in table 1. As direct decay of parent to the ground state of daughter does not exist, there isn't decay path in 1 step. The probability of transition from ^{152}Eu to ^{152}Sm by positron emission was ignored.

The necessary nuclear data (as the transition probabilities, $P_{\gamma\text{-ce}}$) to calculate the normalized probability ($x_{i,j}$) of each γ-ray transition $\gamma_{i,j}$ were obtained from table 2.4 of reference [1]. The $P_{\gamma\text{-ce}}$ probability of each γ-ray transition (photon emission+conversion electron emission) is divided by the sum of the probabilities $P_{\gamma\text{-ce}}$ of all the transitions of this given level to all others, obtaining the corresponding normalized probability, displayed in the bottom of each cascade transition line, as mentioned.

IOP Conf. Series: Journal of Physics: Conf. Series **1044** (2018) 012032 doi:10.1088/1742-6596/1044/1/012032

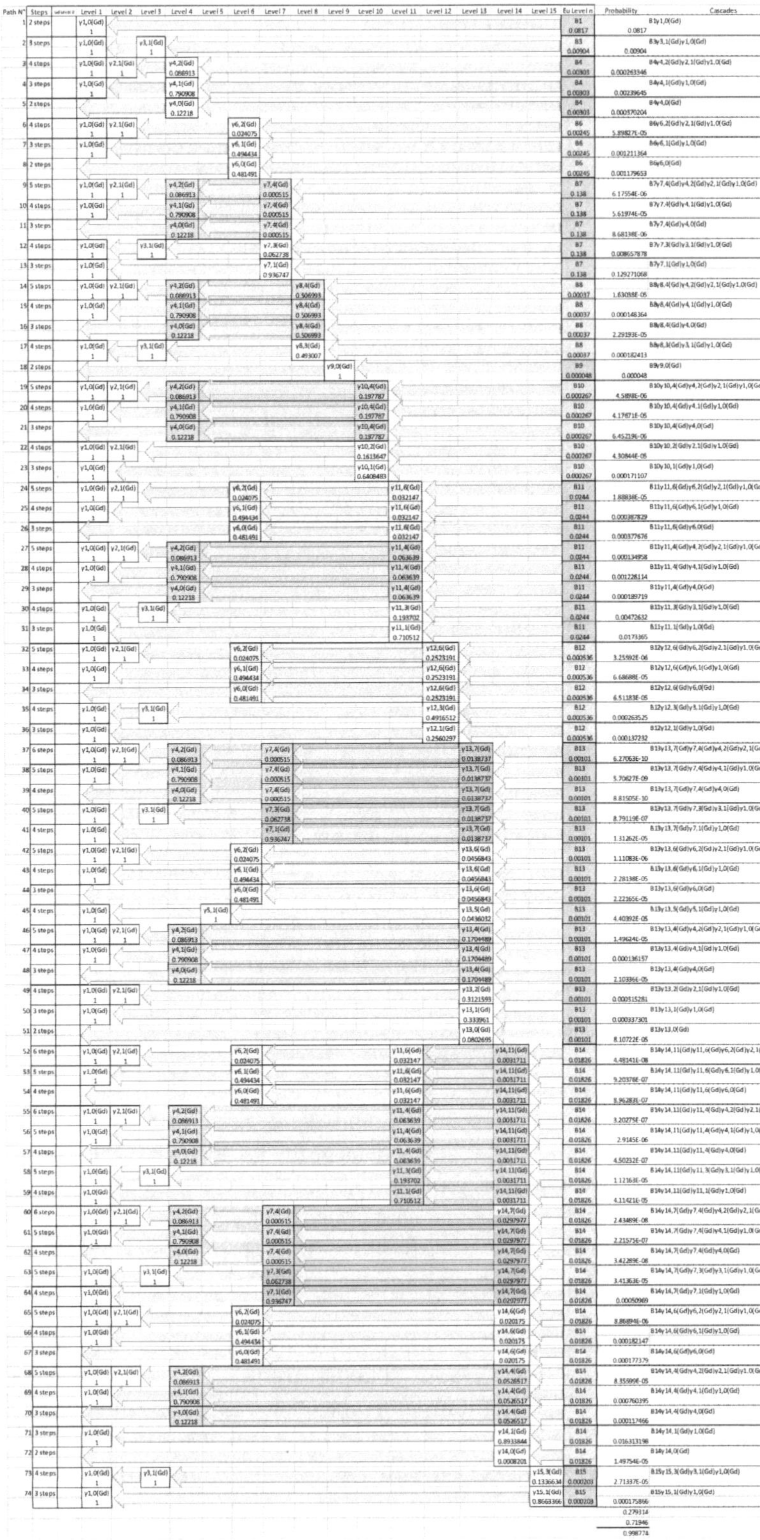

Figure 2. Decay paths and corresponding probabilities from ^{152}Eu to the ^{152}Gd ground state (from right to left) in 2, 3, 4, 5 and 6 steps.

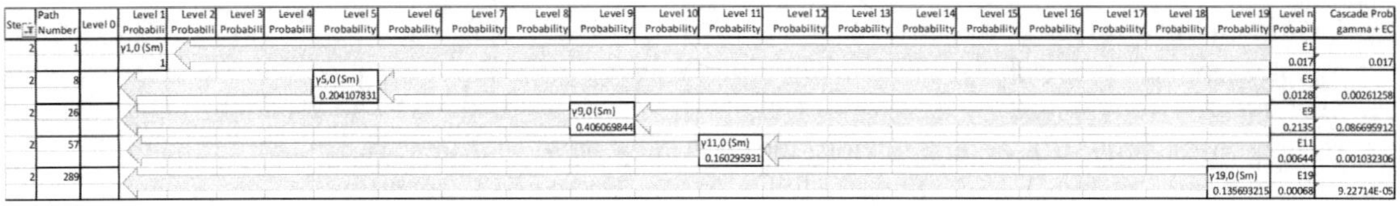

Figure 3: Decay paths for the transition from ^{152}Eu to the ground state of ^{152}Sm in 2 steps

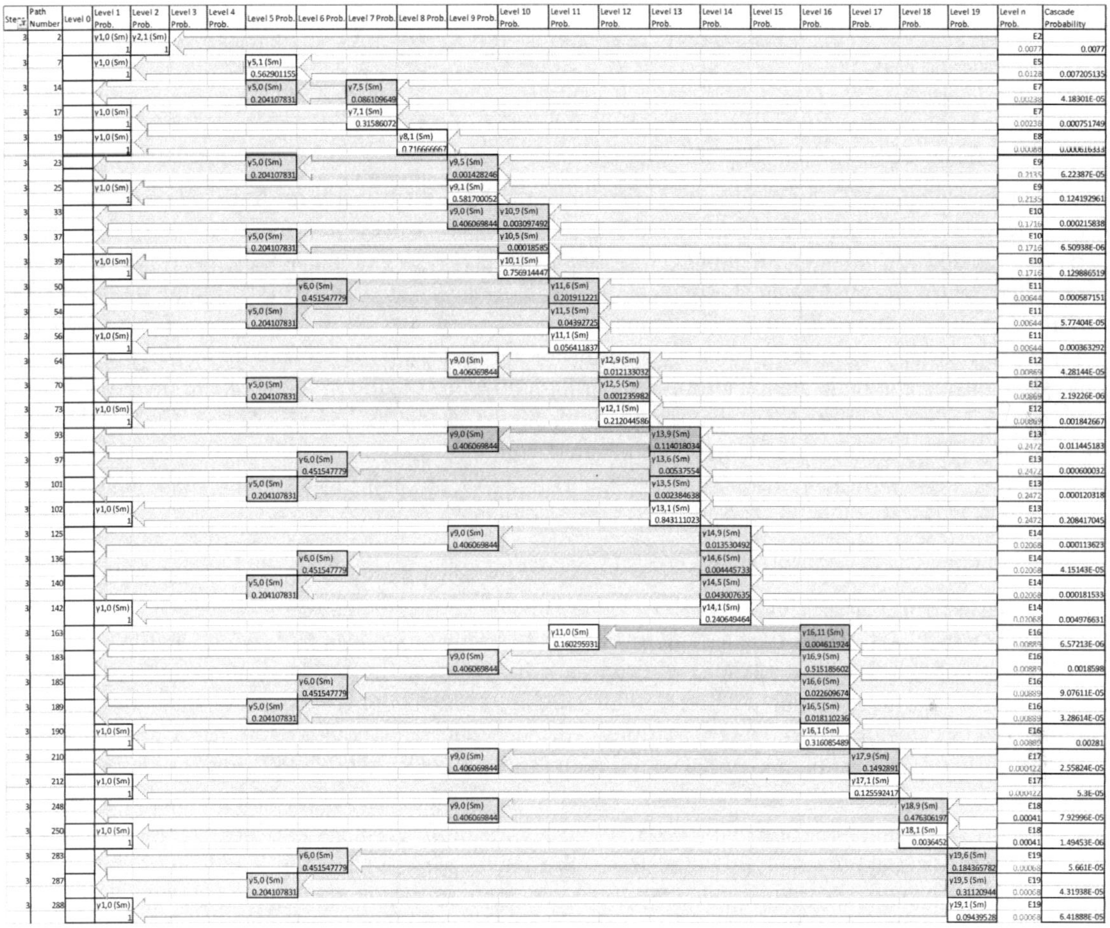

Figure 4: Decay paths for the transition from ^{152}Eu to the ground state of ^{152}Sm in 3 steps.

Table 1. Number of decay paths for the transition from the ground state of ^{152}Eu to the ground states of ^{152}Sm and ^{152}Gd in different number of steps.

Number of steps	**1**	**2**	**3**	**4**	**5**	**6**	**7**	**8**	**Total**
Number of paths (^{152}Sm)	0	5	36	74	85	62	21	2	**285**
Number of paths (^{152}Gd)	0	6	20	27	17	4	0	0	**74**

3. Using a matrix formalism

A possible approach to check the identification of all decay paths and their probabilities is based on a matrix formalism developed by Semkov et al. [2], Novković et al. [3] and Nađđerđ et al. [4] for complex decay schemes. In order to determine the decay paths and their respective probabilities, are introduced square matrices, as **X** (strictly lower-triangular matrix), equation 1. A square matrix is called strictly lower triangular if all the entries on and above the main diagonal are zero.

$$\mathbf{X} = \begin{bmatrix} 0 & 0 & 0 & \cdots & 0 \\ x_{10} & 0 & 0 & \cdots & 0 \\ x_{20} & x_{21} & 0 & \cdots & 0 \\ \vdots & \vdots & \vdots & \ddots & \vdots \\ x_{n0} & x_{n1} & \cdots & x_{nn-1} & 0 \end{bmatrix}, \qquad (1)$$

The matrix element $x_{i,j}$ represents transition or normalized probability for transition from level i to level j, recalling that the ground state of the parent nucleus is presented as an excited level of the daughter nucleus. The ^{152}Eu daughter nucleus ^{152}Sm has 19 excited levels, and the transition probability matrix formed using the nuclear decay data has 21 rows and 21 columns, including the ground states of ^{152}Eu and ^{152}Sm. The transition probability matrix of ^{152}Gd has 17 rows and 17 columns. As each element of this matrix determines the transition probability between two levels, if this matrix is denoted by **X**, then the transition probabilities from ^{152}Eu ground state to the ground state of ^{152}Sm in 1, 2, 3, 4, 5, and 6 steps are given by matrix elements $[\mathbf{X}]_{20,0}$, $[\mathbf{X}^2]_{20,0}$, $[\mathbf{X}^3]_{20,0}$, $[\mathbf{X}^4]_{20,0}$, $[\mathbf{X}^5]_{20,0}$, and $[\mathbf{X}^6]_{20,0}$. These are elements in the first column (or column number 0 in our notation) of the 20st row (row 20) of these matrices, respectively. All possible decay paths were identified by calculating $[\mathbf{Y}^k]_{20,0}$ (k = 1, 2, 3, 4, 5, 6, 7, 8) for ^{152}Sm and $[\mathbf{X}^k]_{16,0}$ (k = 1, 2, 3, 4, 5, 6) for ^{152}Gd. See Novković et al. [5].

4. Discussion and Conclusions

As the decay path, defined by the cascade transitions from the ground state of the parent nucleus to the ground state of the daughter nucleus, is seen as a time-series, the probability to follow this path, calculated here by an excel spreadsheet and checked by a matrix formalism, can simply be seen as a product of normalized transition probabilities. All such data are required for the accurate primary standardization of ^{152}Eu solution [6] through methods as the CIEMAT/NIST for liquid scintillation counting or even the sum peak method, an absolute measurement technique used for activity standardization by gamma-ray spectrometry techniques. For the CIEMAT/NIST method, a procedure was devised that combined experimental data (information about the spectrometer, liquid scintillator and radionuclide) with theoretical calculations that depend on the radionuclide to be standardized. In fact, the identification of all possible decay paths and of their probabilities is required for this purpose. The results constitute, for example, input data for the freeware program CN2003 (E. Günther, PTB/Germany), or CN2004, a useful computer code to calculate the LSC efficiency of nuclides with a more complex decay scheme, by obtaining the coefficients of a fit function of the nuclide efficiency vs. efficiency of the tracer ^{3}H (CIEMAT/NIST method). See Cruz *et al.* [7].

Acknowledgements

The authors are grateful to Dr. Marie-Christine Lepy for valuable discussions on the decay of atomic states and Dušan Novković and Aleksandar Kandic from Vinča Institute of Nuclear Sciences for providing tips about the matrix method.

References

[1] Laboratoire National Henri Becquerel, Internet Information, Updated: 24th March 2014 http://www.nucleide.org/DDEP_WG/Nuclides/Eu-152_tables.pdf

[2] Semkov T M, Mehmood G, Parekh P P and Virgil M 1990 Coincidence summing in gamma-ray spectroscopy, *Nuclear Instruments and Methods* A **290**, p. 437

[3] Dušan Novković, Aleksandar Kandić, M. Đurašević, Ivana Vukanac, Zoran Milošević, Laslo Nađđerđ. Coincidence summing of X-ray and γ-rays in γ-ray spectrometry. Nucl. Instrum. Methods Phys. Res. Sect. A, 578 (2007), pp. 207–217

[4] Laslo J. Nađđerđ, Dragana J. Jordanov, Miloš. D. Davidović. A new matrix method for calculating coincidence summing effects for gamma spectroscopy. Nuclear Instrum. Methods A, 662 (1) (2012), pp. 21–25

[5] Dušan Novković, Mirjana Đurašević, Aleksandar Kandić, Ivana Vukanac, Bojan Šešlak and Zoran Milošević. January 2016. Coincidence summing corrections for point and volume ^{152}Eu sources, *Applied Radiation and Isotopes* **Volume 107**, Pages 138–144

[6] Stefaan Pommé, Methods for primary standardization of activity, (2007) Metrologia 44 S17

[7] Paulo A. L. da Cruz, Carlos J. da Silva, Akira Iwahara, Jamir S. Loureiro, Antônio E. de Oliveira, Luiz Tauhata and Ricardo T. Lopes, 2016, TDCR and CIEMAT/NIST Liquid Scintillation Methods applied to the Radionuclide Metrology, *Journal of Physics: Conference Series*, Volume 733, Number 1, 012099.

[8] Gunther E 2003 Program CN2003: A program to calculate the LC efficiency of a nuclide vs. efficiency the tracer H-3 (CIEMAT/NIST). PTB/Germany.

IOP Conf. Series: Journal of Physics: Conf. Series **1044** (2018) 012033 doi:10.1088/1742-6596/1044/1/012033

Implementing a proficiency test provider for sphygmomanometers in Brazil

B A Rodrigues Filho[1], R F Farias[1]and W Anjos[2]
[1]National Institute of Metrology, Quality and Technology – INMETRO
[2]Weights and Measures Institute of São Paulo – IPEM/SP
E-mail: bafilho@inmetro.gov.br

Abstract. Self-verification represents a level of control in legal metrology where manufacturers are responsible for the conformity assessment of measuring instruments. To solve the independence and impartiality concerns, accreditation is introduced as a tool to monitor the competence of laboratories providing self-verification tests. Especially in health measurements, an inaccurate measurement may lead to misdiagnoses. To support the traceability of accredited laboratories, a proficiency test due to interlaboratory comparison is necessary in order to assess the conformity of measurements of manufacturers of sphygmomanometers for blood pressure measurement. Then, this study presents the implementing process of a proficiency test provider laboratory for sphygmomanometers based on legal metrology requirements in order to fulfill the necessity of an interlaboratory comparison among manufacturers that assess the conformity of sphygmomanometers. We also show a comparison of the measurement tests for three distinct manufacturers in order to identify divergences of processes.

1. Introduction and motivation

In order to provide accuracy and precision for measurements that impact the society, an unbroken calibration chain is required. Then, metrological traceability due to a documented unbroken chain of calibrations to a reference is necessary to provide confidence to the results, the measurements [1].

Accreditation is a consolidate methodology applied to disseminate traceability, applying formal, international well-known and standardized procedures, evaluating the competence, infrastructure, methods, standards, training, independence, impartiality, proficiency and several others aspects of accredited laboratories [2-3].

In this context, an interlaboratory comparison is a tool applied in order to check the comparisons of measurement uncertainties among the parties and it is widely used in accreditation due to proficiency tests to validate the competence of the accredited bodies. Consequently, deviations provide feedback to the laboratories to correct eventual problems in their processes.

However, accreditation in legal metrology is a relatively new area and publications regarding it are still sparse [4].

Although the working bodies in legal metrology use standards calibrated in accredited laboratories, accreditation for legal metrology tests is still incipient. There are some critical points for manufacturers that run self-verification tests that can undermine the interlaboratory comparison, such as:

---

[1] bafilho@inmetro.gov.br

- The tests required by legal metrology requirements also demand administrative tests, sealing marks checking, constructive features, and usage conditions since the instrument is used *in* the field and it is exposed to several environmental aspects such as temperature, mechanical impacts, and others;
- There is a resistance of manufacturers participating in interlaboratory comparison due to commercial issues.

Sphygmomanometers are generally under the metrology regulation as they are widely used in diagnoses in medicine, as an example, diagnoses and treatment of hypertension. Then, an accurate measurement is vital to an adequate medical treatment since a deviation in the measurement might either induce an unnecessary treatment or provide treatment to an inexistent disease. Consequently, the instruments must be accurate and precise [5].

Small systematic deviations in blood pressure measurements may cause large variations in the proportion of false positive diagnosis of hypertension. A deviation of + 3 mmHg increases in 68% the false positives of patients with blood pressure superior to 90 mmHg, and a deviation of – 3 mmHg, 44% of hypertension patients would not be diagnosed. A deviation of +1 mmHg would decrease it to 20% [6]. Furthermore, in 2015, in the United Kingdom, 28% of the mechanical sphygmomanometers and 42% of the mercury-based ones presented measurement errors of + 4 mmHg [7].

In Brazil, in 2015, the self-verification process was responsible for 45% (2,668,086) of sphygmomanometers tests and it is estimated that in 2017, approximately 100% of the instruments might have their conformity assessed due to self-verification processes.

Then, to support the traceability and conformity for all aspects involving legal metrology, the present project proposes the implementation of a third part independent facility to provide the interlaboratory comparison based on tests required by legal metrology requirements, focusing on sphygmomanometer accredited laboratories for self-verification.

2. Methodology

A partnership between the National Institute of Metrology, Quality and Technology – INMETRO, Brazilian National Metrology Institute (NMI) and the Weights and Measurement Institute of São Paulo – IPEM/SP enables the implementation of the interlaboratory comparison incorporating the already existing infrastructure of an ISO/IEC 17025:2005 accredited laboratory. The State Institute is a public body and it has an INMETRO delegation for legal metrology activities, including sphygmomanometers verification and inspection. Therefore, the project uses both accredited laboratory and legal metrology expertise of the Pressure Laboratory of the State Institute in order to conduct the present project, which will play a role as a proficiency test provider.

As an OIML member, Brazil adopts the procedures for sphygmomanometer tests based on the International Recommendation R 16 -1 and 2 for Non-Invasive Sphygmomanometers. Additionally, the laboratory conducts tests using an Onneken calibrator OM-HM 621-S having a measuring interval from 0.0 to 375.0 mmHg and resolution of 0.1 mmHg, adequate for sphygmomanometer tests. Moreover, the traceability is assured by periodic comparisons of the Onneken calibrator to a dead weight tester, traced to the national pressure standard in Brazil.

Figure 1 shows the steps to be followed in order to implement the interlaboratory comparison laboratory for sphygmomanometer tests based on legal metrology requirements.

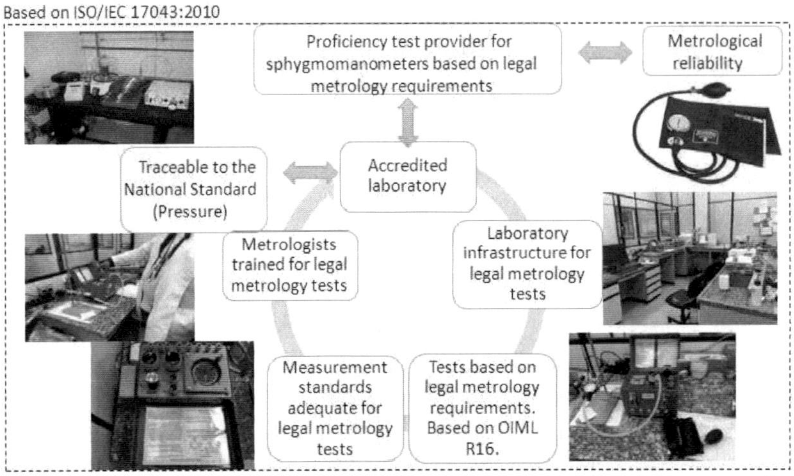

Figure 1. Stages required in order to implement the laboratory for comparisons for sphygmomanometer tests.

3. Results of a comparison among manufacturers

Since Brazil is an OIML Member, the tests for sphygmomanometer are based on the OIML Recommendation R 16, which defines the maximum permissible errors (MPE) for initial verification measurement tests in $\pm$ 3 mmHg [8-9].

Then, in order to illustrate the importance of a proficiency test provider in legal metrology, we collected a sample of results of tests conducted by three already working laboratories of distinct manufacturers. The samples were collected according to the technical standard ABNT NBR Sampling Procedures for Inspection by Attributes [10]. The samples were also collected considering similar sphygmomanometers (automated) and the same period of time, representing the production during September/2016.

The results of the pressure test for three different manufacturers are shown in figure 2 and 3 for two distinct points (P) of the measurement test, P = 80 mmHg and P = 160 mmHg.

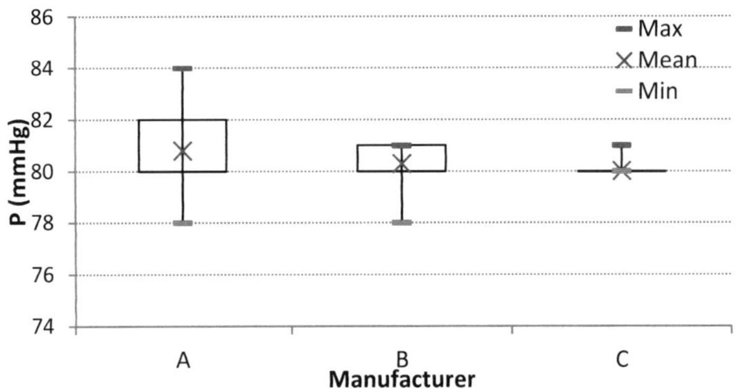

Figure 2. Comparison among three different manufacturers for pressure test (P = 80 mmHg).

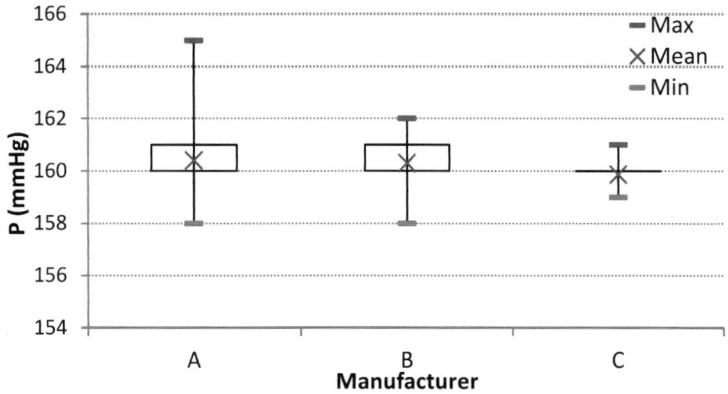

Figure 3. Comparison among three different manufacturers for pressure test (P = 160 mmHg).

Figure 2 and 3 show that the mean for measurement tests are according to the MPE enforced by the national regulation. However, it is possible to note that the maximum values presented by the manufacturer A for both 80 mmHg and 160 mmHg surpass the MPE. It is also possible to conclude that manufacturer C presents the minimum variation between maximum and minimum values for the tests, demonstrating more precision when compared to A and B.

Figures 4 and 5 show the confidence interval for the mean of the points P = 80 mmHg and P = 160 mmHg for the three analyzed manufacturers.

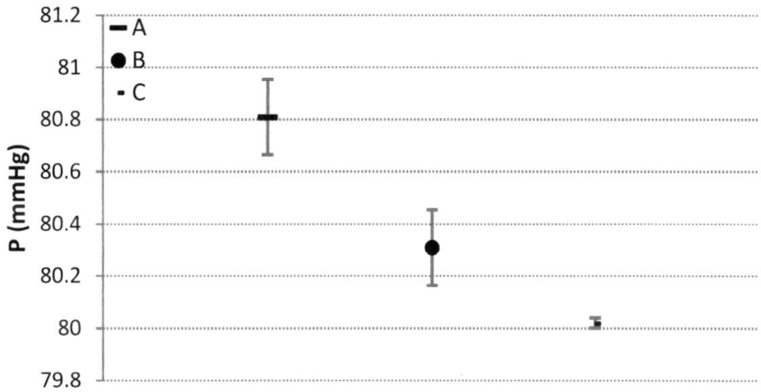

Figure 4. The confidence interval for the mean of the pressure test (P = 80 mmHg) for the tested manufacturers.

Based on the confidence interval for the mean, it is possible to conclude that all manufacturers are according to MPE, considering an interval of confidence of 95.45%.

We also conducted a Tukey test in order to determine the independence of the results to analyze the difference of significance of the results produced by different manufacturers. Table 1 and 2 show the results for the test comparing A vs B, A vs C and B vs C for the points P = 80 mmHg and P = 160 mmHg [11].

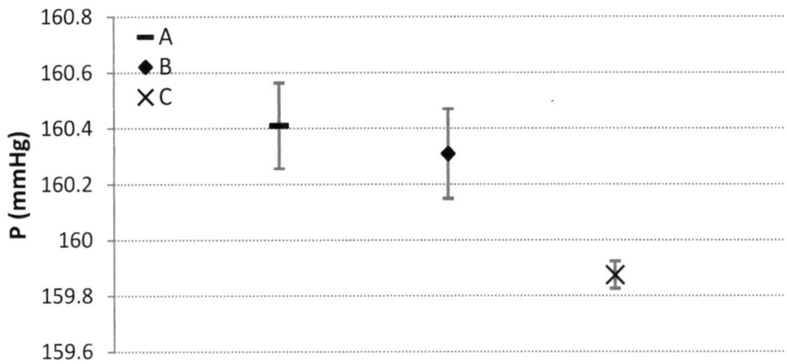

Figure 5. The confidence interval for the mean of the pressure test (P = 160 mmHg) for the tested manufacturers.

Table 1. Tukey Test for the samples for the pressure test (P = 80 mmHg).

Treatments pair	Tukey HSD p-value	Tukey HSD inference
A vs B	0.001005	** p<0.01
A vs C	0.001005	** p<0.01
B vs C	0.002322	** p<0.01

Table 2. Tukey Test for the samples for the pressure test (P = 160 mmHg).

Treatments pair	Tukey HSD p-value	Tukey HSD inference
A vs B	0.53693	insignificant
A vs C	0.00101	** p<0.01
B vs C	0.00101	** p<0.01

Although A, B and C are within the MPE (± 3 mmHg), the Tukey test for P = 160 mmHg, shows that the results of manufacturer C are different when compared to A and B for a confidence interval of 95.45%. For P = 80 mmHg, differently from the previous test, A, B and C present different results for the same confidence interval.

4. Discussion

As the methodology applied to the tests follows the international recommendation, OIML R 16, and based on figure 4 and 5, it is possible to conclude that all manufacturers are according to the MPE within a confidence interval of 95.45%. Additionally, according to the Tukey test, considering both P = 80 mmHg and P = 160 mmHg, A, B and C produce independent results.

We also note a significant difference when comparing C to A and B, which results are more precise and accurate.

Additionally, based on figures 4 and 5, to P = 80 mmHg, the results of A, B and C are within ± 1 mmHg considering the confidence interval, and ± 0.6 mmHg from the reference value to P =160 mmHg.

Although the selected samples represent similar sphygmomanometers, the devices present minor differences as they are produced by distinct manufacturers, consequently, it is not possible to conclude if the divergence in precision can be explained by quality or by the competence of the producer.

It is worth noting the importance of a proficiency provider in order to confirm the results produced by the laboratories. Based on the results of A, B and C, it would also be possible to decrease the MPE for initial verifications improving the accuracy, consequently impacting the misdiagnoses.

As a proficiency test provider, the IPEM/SP's laboratory will give a panorama of the sphygmomanometer manufacturers' uncertainties allowing the authorities to survey deviations processes, providing a market surveillance tool improving the efficiency in legal metrology control.

Moreover, the IPEM/SP's laboratory acting as proficiency test provider will also be an important independent and impartial body allowing comparisons in legal metrology level for the manufacturers, preventing commercial disputes.

Although manufacturer's laboratories are accredited for legal metrology tests, once a deviation in a measurement is identified in a proficiency test, the authorities will be able to suspend authorizations for self-verification, eliminating bad-practices of the market. It will also be a tool to ensure penalties in legal metrology level. Consequently, it will provide harmonization and standardization for self-verification bodies and initial verification, permitting fair competition among competitors.

Consequently, once the market is properly surveyed, we expect to guarantee blood pressure measurements of devices in initial verification level, increasing the reliability of the legal metrological control once approximately 100% of the devices should be self-verified in 2017.

5. Conclusion

A proficient test provider due to an interlaboratory comparison of measuring instruments under the legal metrology control, conducted by IPEM/SP' laboratory will allow authorities ensuring conformity in blood pressure measurement field. It will also permit authorities to identify and enforce legal actions to laboratories which practices are not in conformity with legal metrology requirements.

In addition, the implementation of a proficiency testing provider in legal metrology also allows evaluating whether laboratories are able to detect nonconforming instruments through interlaboratory comparisons that use as reference sphygmomanometers purposely altered.

The comparison among different manufacturers also presented the importance of a reference, in order to check the accuracy and precision of producers.

The IPEM/SP's laboratory will also permit decrease the MPE, therefore improving the accuracy of blood pressure leading to a decrease of misdiagnoses, and consequently, the costs involving in treatments as well as the health issues involved in either non-necessary and mistreatments.

Acknowledgments

The authors acknowledge FAPESP proc. nº 2017/50173-0 for the financial support.

References

[1] International Vocabulary of Metrology—basic and general concepts and associated terms. JCGM 200:2012.

[2] Lagauterie, G.; Pecchioli, G. Accreditation and legal metrology in France. OIML Bulletin, v. LII, n. 2, p. 19–29, Apr. 2011.

[3] **ILAC**, 2015. Available in: http://ilac.org/about-ilac/ . Accessed in: 14 Oct. 2016

[4] Rodrigues Filho, B. A.; Gonçalves, R. F. Legal metrology, the economy and society: A systematic literature review. Measurement, v. 69, p. 155–163, Jun. 2015.

[5] Serafim, T. S. et al. Avaliação das condições de uso de esfigmomanômetros em serviços hospitalares. Acta Paulista de Enfermagem, v. 25, n. 6, p. 940–946, 2012.

[6] Turner, M. J.; Baker, A. B.; Kam, P. C. Effects of systematic errors in blood pressure measurements. Blood Pressure Monitoring, v. 9, n. 5, p. 249–253, Oct 2004.

[7] Hathway, P. et al. Comparative study of two ambulatory blood pressure monitors that use differing methods of calibration to determine central aortic blood pressure from the brachial waveform. Journal of Hypertension, v. 33, p. e181–e182, Jun. 2015.

[8] **OIML**. R 16-1 Non-invasive mechanical sphygmomanometers, 2002. Available in: https://www.oiml.org/en/files/pdf_r/r016-p-e02.pdf. Accessed in 04/04/2017.

[9] **OIML**. R 16-2 Non-invasive automated sphygmomanometers, 2002. Available in: https://www.oiml.org/en/files/pdf_r/r016-p-e02.pdf. Accessed in 04/04/2017.

[10] **ABNT NBR 5426**. Technical Standard - Sampling procedures for inspection by attributes, 63 p., 1985.

[11] Tukey test available in: http://astatsa.com/OneWay_Anova_with_TukeyHSD/ .

IOP Conf. Series: Journal of Physics: Conf. Series **1044** (2018) 012034 doi:10.1088/1742-6596/1044/1/012034

Influence of the light spectral distribution used in the radiometers calibration

J C Moraes and L O Guerra

Allergisa Pesquisa Dermato-Cosmética Ltda – GRUPO INVESTIGA, 452. Dr. Romeu Tórtima Avenue, Postcode: 13084-791, Campinas, São Paulo (SP), Brazil.

E-mail: juliana.moraes@grupoinvestiga.com.br

Abstract. The use of radiometers is common as a form of control in tests that engage exposure to natural or artificial electromagnetic radiation. The photodiode is a very much used device, capable of converting the incident radiation into electrical current, which is transformed in irradiance by a duly calibrated meter. The calibration of radiometers is made by comparing it with a spectroradiometer and a calibration factor is obtained. Considering that the photodiodes responsiveness is not constant for all wavelengths, the spectral distribution of the light source is a great influence factor in the calibration. Thus, the objective of this work was to evaluate the variation in the calibration factor of the radiometers resulting from the spectral distribution variation of the light source used. The results showed a variation of up to 16% in the calibration factor, reinforcing the importance of calibrating the radiometers with the same light source type used daily, in order to assure the accuracy of the measurements made with this type of equipment.

1. Introduction

The use of radiometers is common as a form of control in tests that engage exposure to natural or artificial electromagnetic radiation. There are several types of radiometers available in the market, with different principles of working. The photodiode is the most used device, capable of converting the incident radiation into electrical current, which is transformed in irradiance by a duly calibrated meter. Associated to optical filters that filter part of the radiation, more specific sensors are obtained for specific spectral ranges.

Radiometers manufacturers usually provide a spectral response curve, better known as responsiveness, which represents the sensor response over its spectral range. The responsiveness of the photodiodes is not the same for all wavelengths, meaning, it is more sensitive for certain wavelengths and less sensitive for other ones, it can even record different irradiance values for different wavelengths, even when they have the same intensity.

Thus, as other pieces of equipment, it is necessary to calibrate it with defined intervals, in order to assure the accuracy of the radiometers. The calibration of the radiometers is made by following the ASTM G130 standard, comparing its response with the response of a spectroradiometer, which measures the spectral irradiance, obtaining a calibration factor to correct the radiometer reading.

Considering that the spectral responsiveness of the photodiodes is not constant, the distribution of the light source is a factor with great influence in the calibration and measurements made with this type of equipment. Therefore, it is necessary to calibrate it with the same light source measured daily with the radiometer.

IOP Conf. Series: Journal of Physics: Conf. Series **1044** (2018) 012034 doi:10.1088/1742-6596/1044/1/012034

2. Objective

Evaluating the variation on the calibration factor of the radiometers resulting from the spectral distribution variation of the light source used, to reinforce the necessity of calibration with the same light source used daily.

3. Methods

The calibration factor of the radiometer was obtained according to ASTM G-130 standard. Different light sources were used to calibrate the radiometer and the variation in the calibration factor was analyzed.

3.1. Equipment used

Two UVA radiometers manufactured by Solar Light, model PMA 2110, with datalogger PMA 2100.

Spectroradiometer manufactured by Optronic OL756, calibrated against a spectral irradiance standard with NIST traceability, following the ASTM G-138 standard, with specified bandwidth of 0.8nm, wavelength interval of 1nm, integrating sphere of 5cm diameter, with 5mm-opening and optical fiber.

The light sources used in the radiometer calibration were: xenon, xenon + UV filter, xenon + UVA filter, sunlight and mercury.

3.2. Radiometer Calibration

The radiometer calibration was performed according to ASTM G-130 standard. The radiometer was positioned in a way its entrance port was at the same level as the input of the integrating sphere relative to the source of light used, with ±1mm of tolerance.

After the light source was stabilized, the radiometer reading was recorded ($n \geq 5$) and then the integration sphere was positioned on its place. Three ($n=3$) measurements of spectral irradiance were taken, using the reading range of the radiometer – (320-400) nm – in 1nm intervals.

After the measurements with the spectroradiometer were completed, there was a 10-minute wait before repeating this procedure, until a total of 3 replicates were obtained.

The spectral irradiance in the range of (320-400)nm was integrated using the trapezium method and converted to the reading unit of the radiometer (mW/cm^2). The integrated irradiance was divided by the average irradiance obtained by the radiometer, and an F calibration factor was obtained (Equation *1*).

$$Calibration\,Factor\,(F) = \frac{Integrated\,Irradiance\,(\text{Spectroradiometer})}{Obtained\,Irradiance\,(Radiometer)} \quad (1)$$

The calibration factor can be used to correct the spectroradiometer reading to obtain the corrected value, corresponding to the real value obtained by the standard (Spectroradiometer).

The calibration was repeated by using 5 types of light sources:

- Xenon + UV filter
- Xenon + UVA filter
- Sunlight
- Mercury

4. Results

Figure 1 shows the responsiveness of the UVA radiometer used, as declared by the manufacturer Solar Light. It is noticeable that the radiometer is more sensitive between 360 nm and 370 nm and the responsiveness is reduced at the extremes of the spectral range analyzed.

Figure 2 shows the spectral distribution of different light sources used for the calibration of the radiometer.

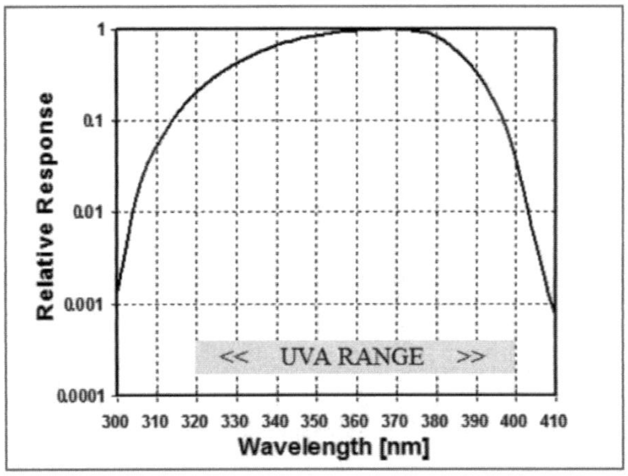

Figure 1. Responsiveness of the UVA radiometer used, as declared by the manufacturer Solar Light.

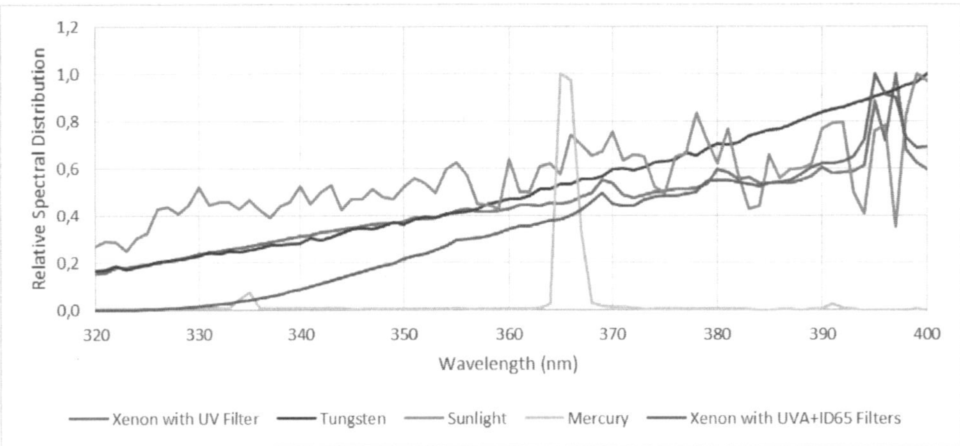

Figure 2. Spectral distribution of the light sources used.

A great difference between the spectral distributions is observed. The mercury lamps contain emission peaks that are characteristic of the atomic emission of mercury, the reason why it is called spectral source. As for the sunlight and the tungsten lamp, they are continuous light sources, characterized by emitting continuous radiation throughout the wavelengths, similar to the emission of a blackbody.

The xenon lamp has its spectral distribution modified according to the combination of filters used. It is observed that the UVA+ID65 filters block a part of the UVA radiation between 320 nm to 370 nm when compared to the UV filter.

Table 1 shows the results obtained when comparing the calibration using the tungsten lamp and the xenon lamp + UV filter, both continuous light sources, but with slight differences in the spectral distribution between 360 nm and 400 nm.

Table 1. Results obtained in the UVA radiometer calibration with 1.10 scale for the tungsten and xenon lamps with UV filter.

	Xenon with UV Filter	Tungsten
Irradiance - Radiometer (mW/cm^2)	8.19	0.10
Irradiance - Spectroradiometer (mW/cm^2)	7.61	0.09
Calibration Factor (F) (s/u)	0.93	0.89

A difference of 3.8 % in the calibration factor of the radiometer when calibrated with the tungsten lamp was observed in relation to the calibration carried with the xenon lamp with UV filter.

As the relative intensity of the tungsten lamp in the range of the radiometer greater responsiveness is similar to that of the xenon lamp, the calibration factors resulted in very close values.

Table 2 shows the results obtained when comparing the calibration using the mercury lamp, the xenon + UVA + ID65 filter and the xenon lamp + UV filter.

Table 2. Results obtained in the UVA radiometer calibration with 0.96 scale for the tungsten and xenon lamps with UV filter.

	Mercury	Xenon with UVA+ID65 filter	Xenon with UV Filter
Irradiance - Radiometer (mW/cm^2)	1.94	4.62	6.54
Irradiance - Spectroradiometer (mW/cm^2)	1.35	4.77	6.56
Calibration factor (F) s/u	0.70	1.03	1.00

A difference of 30.5 % in the calibration factor of the radiometer when calibrated with the mercury lamp and a difference of 3.1 % when calibrated with the xenon lamp with UVA+ID65 filter were observed, both in relation to the calibration carried with the xenon lamp with UV filter.

Table 3 shows the results obtained when comparing the calibration using sunlight and xenon lamp + UV filter.

Table 3. Results obtained in the UVA radiometer calibration with 1.10 scale for the tungsten and xenon lamps with UV filter.

	Sunlight	Xenon with UV Filter
Irradiance - Radiometer (mW/cm^2)	0.87	1.37
Irradiance - Spectroradiometer (mW/cm^2)	5.62	7.61
Calibration factor (F) s/u	6.43	5.55

A difference of 16.0 % in the calibration factor of the radiometer was observed when calibrated with sunlight, in relation to the calibration carried with the xenon lamp with UV filter.

5. Conclusion

The international standard ASTM G-130 which describes the method of broadband radiometer calibration using a reference spectroradiometer requires that the light source used is the same as the one used daily with the radiometer, and even specifies methods of assurance that the positioning is also the same.

Taking into consideration the 30 % difference found in the calibration factors, the importance of calibrating the radiometers with the same light source type used daily is reinforced, in order to assure the accuracy of the measurements taken with this type of equipment.

References

[1] ASTM-G138 2012 *Standard Test Method for Calibration of a Spectroradiometer Using a Standard Source of Irradiance 2012*
[2] ASTM-G130 2012 *Calibration of Narrow- and Broad-Band Ultraviolet Radiometers Using a Spectroradiometer 2012*
[3] CIE 53 1982 *Methods of characterizing the performance of radiometers and photometers.*
[4] NIST Special Publication 250-37 1997 *Photometric Calibrations*

IOP Conf. Series: Journal of Physics: Conf. Series **1044** (2018) 012035 doi:10.1088/1742-6596/1044/1/012035

Inline process monitoring method for geometrical characteristics in additive manufacturing

M Preissler, J Broghammer, M Rosenberger, G Notni

Technische Universität Ilmenau, Faculty of Mechanical Engineering, Group of Quality Assurance and Industrial Image Processing

E-mail: marc.preissler@tu-ilmenau.de

Abstract. This method provides a suitable solution for monitoring an additive manufacturing process. The implementation is evaluated in a manufacturing machine of fused filament fabrication. The main parts of the work are the developing and integration of an adaptable 3D hardware platform, the pre-processing of captured layer point clouds and the developing of object point cloud stack. The result is sufficient to estimate the manufacture quality and complies requirements for an inline process control method.

1. Introduction

The area of additive manufacturing (AM) is increasing in almost every sector of industry and is getting more and more important as an alternative for traditional factoring methods like milling. The advantages begins with the process of developing in Computer Aided Design (CAD). The user can create more and more complex geometrical structures without to generate a blueprint. The transformation in machine readable code follows after mechanical developing in CAD. The so-called *slicing process* is usually preceded and manufacturing parameter are to be set. The parameter setting allows an advantage for AM at once. The knowledge of the requirements for the developed part and adjusting of the right parameters for the AM process might be a cost effective manufacturing process in comparison to traditional manufactured parts. The manufacturing time is certainly one of the most important factors in costing, particularly for single pieces. The initial costs are a very high and getting lower with a higher number of manufactured parts. [1]

New and more effective AM technologies are still in developing and might be able to reduce the manufacturing time. But to reduce manufacturing time is just one way for increasing the cost effectivity in AM. The increasing of the manufacturing quality is simultaneously desirable. These include higher geometric accuracy, convenient surface and the conformance for mechanical requirements of the manufactured pieces. An approach for controlling the manufacturing process could be a possibility for monitoring the AM process, what is called inline process monitoring.

Inline process monitoring and also control are one of the evolving areas in AM. It could be the next evolution for this manufacturing technologies. Monitoring in manufacturing processes is useable for visibility of the manufacturing quality and responsible for traceability at once.

Process parameters can be monitoring, but also the manufacturing results can be compared with expected values. These values can describe geometrical and mechanical specifications for the manufactured part. Inline process monitoring is only capturing the manufacturing process. Inline process control has the claim to manipulate instantly process parameters for constant quality of the manufacturing results in a quality control loop (Figure 1). Inline process metrology describes the measurement of the workpiece inside the manufacturing machine and defines the process control loop within the time of the manufacturing process [2].

The requirements for manufacturing parts are getting more and more complex. Lightweight, composite materials and effective stiffness are fields with high research potential for manufacturer. New potential is available through the width accessibility of additive manufacturing methods. These methods

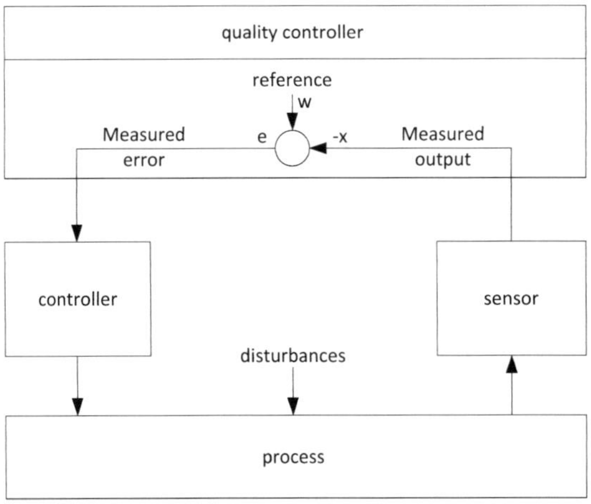

Figure 1. Quality control loop

scan be differentiate in two major summarized groups based on their usage. Firstly the classical additive manufacturing processes, which includes Stereolithography (STL) and Selective Laser Sintering (SLS). These systems are well-established in development departments of large companies and service companies. But the purchasing and using of these machines is still relatively expensive in comparing to other AM technologies. The second group of generative manufacturing processes is constituted by fused filament fabricationModelling, Multi Jet Modelling and Binder Jetting. The primary use of the second group converts more and more to engineering and design companies for manufacturing objects directly from CAD. Meanwhile additive manufacturing systems on the market are slightly more expensive than laser printers a few years ago. These are usually based on the fused filament fabrication (FFF) method. Meltable synthetic materials are liquefied in a heating nozzle and coated in layer construction equivalent of CAD Data. [3] The low costs results of the cheap raw material and the simple mechanical structure of the overall system. Furthermore the large variety in material of available filaments for an FFF manufacturing system requires adapted machine parameters for an acceptable manufacturing result. The parameters are normally only adjustable with the help of iterative manufacturing process steps or the machine operator has high expertise in handling with different filament. The number of possible process failures is very diverse and can be reasonable in different causes. A great support is the knowledge of all relevant process parameters for better understanding of the AM process and should content of this work.

An inline process monitoring method could be implemented in each of these additive manufacturing groups. This paper presents an inline process monitoring method for a fused filament fabrication manufacturing machine.

2. Potential process failures in fused filament fabrication

The fused filament fabrication process is one of the cheapest AM method for initial costs and thus currently most popular for consumers. The manufacturing process works additive, what means the objects structure is manufactured in layers. The raw material for manufacturing is mostly plastic wire on a coil (Filament) or granulate and is melting in a nozzle, which builds the separate layers. The subsequent cool down process of the melted material is very sensitive and is influencing through a lots of parameters of the surrounding [4] [5]. Also changing of filament from another supplier can cause in some process failures, because different chemical additives are being used. The additives are necessary to influence the filament color and should make it easier to handle the material for the manufacturing process.

Example effects for process failures in fused filament fabrication are showing Table 1, if the manufacturing process isn't under stable conditions or the machine parameters settings is wrong.

Table 1. Examples for Process Failures in fused filament fabrication

Failure description		Potential reasons
Not sticking to platform		• manufacturing speed too fast • wrong platform- or nozzle temperature • trouble with platform surface • build platform is not level or wrong distance to noozle
Warping	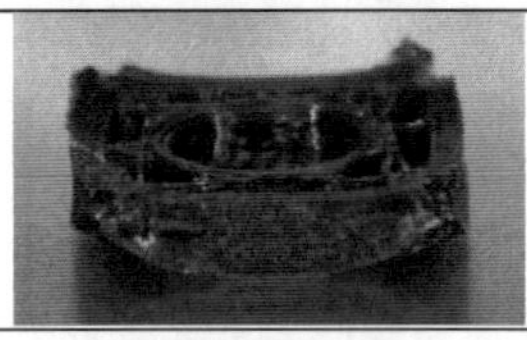	• wrong platform temperature • fan cooling • trouble with platform surface
Blobs and Zits		• nozzle temperature too high • wrong retraction parameter • too many extrusion of filament
Weak Infill	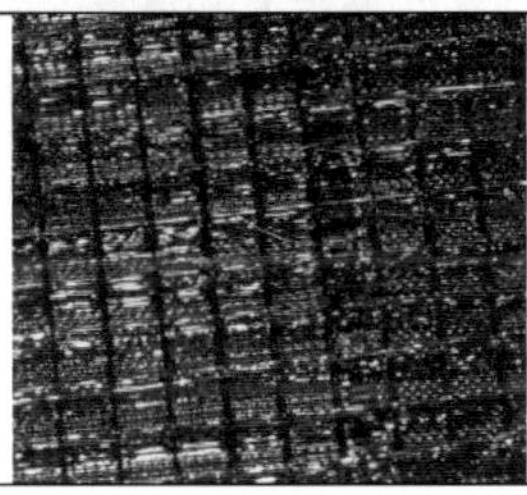	• manufacturing speed too high • wrong infill extrusion

The manufacturing process in FFF starts with the first layer and is very important for the following layers. The parallel and correct platform levelling to the nozzle is a basic requirement for stable process conditions and has immense affect for layer sticking. If these conditions aren't achieve, the first layer doesn't stick to the platform and some warping effects can follow. In worse case the object starts to lift completely and damages the manufacturing machine.

The correct adjustment of the machine parameters is also important for mechanical properties to get correct surfaces of the manufactured object. Also the variable filament diameter has influence for the material extrusion and consequently more or less material extrudes for layer building. The affects can be blobs and zits on the surface, which impairs the next layer and more manufacturing failures can follow. The fluctuating material extrusion has also influence the infill, which is most responsibly for mechanical properties and stiffness.

The variety of potential process failures in Additive Manufacturing is wide and is almost impossible to give a complete overview in this paper.

3. Experimental design

The aim for this inline process monitoring method is to get dimensional information of the manufactured object on the platform within the process time. A capturing directly after process finishing is possible with conformable sensor hardware equipment. But therefor the construction of the manufacturing machine needs a rebuilding process and that isn't claim of this work.

For this purpose an adaptable 3D - inline process platform for evaluating additive manufacturing processes in fused filament fabrication has been developed. The self-made and adaptable hardware platform for the inline process monitoring [Figure 2] is able to output 3 - dimensional information about the manufactured object during process time. It is positioned over the additive manufacturing machine in bird view to capture the information from layer to layer. Other places for positioning are not acceptable, because a modification of the additive manufacturing machine is necessary and that could decrease the efficiency of the manufacturing process. Furthermore the self-made hardware platform should also usable for other AM machines and inline process monitoring tasks.

The hardware platform has the structure of stereoscopic camera systems and the necessary projector for fringe projection is placed between both. The GigE Vision cameras and C-mount lenses are replaceable to vary the field of view. At once the field of view defines the accuracy of the measurement system and furthermore the point cloud resolution is being affected. The experimental design for this paper composes of a Sony IMX249 2.4 MP image sensor and lens with a focal length of 25 mm. This setup offers to capture the maximum available workspace in the manufacturing machine of 220 x 220mm. The maximum depth in focus depends on the selected lens and the selected aperture. But the depth of focus has to be a minimum of the chosen layer thickness. Common fused filament fabrication machines are able to manufacture a layer thickness of 50 - 400 µm.

A special feature of the hardware platform is the possibility to change and adjust the camera angle and base, which has influence to the output result in form of 3D informations. All necessary translational and rotational degrees of freedom are adjustable. A standard, but high light intensity output, DLP projector is located between both cameras and displays the pattern projection on object. The object distance is reasonable in the minimum projector work distance of 550 mm.

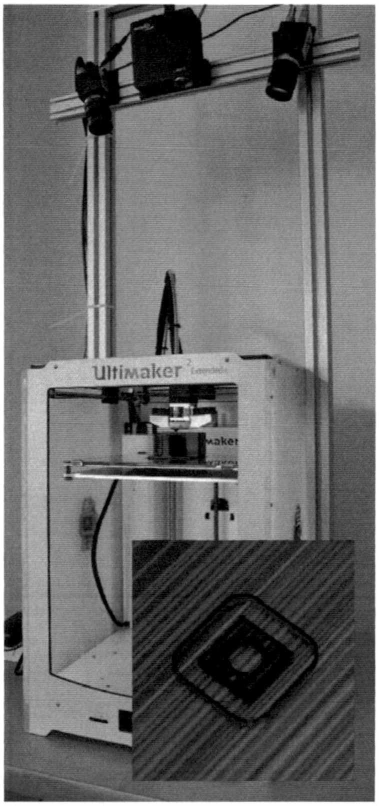

Figure 2. Experimental design

The developed 3D Sensor system is using the fringe projection method [7] [8] and is currently assembled in an Ultimaker 2 extended+, but adaptable in other additive manufacturing systems with requirement to have an free field of view to the manufacturing process on the upper side. A 3D scan in a bird eyes view arrangement is captured after every finished manufactured layer and is the basis for the subsequent image processing. The projector - and camera control, calibration, image acquisition and processing is managed by MathWorks software Matlab. The work is realized by the Image Acquisition Toolbox and the custom stlTools Toolbox.

4. Measurement method

Aim of this work is to get information about the manufactured object structure for a following evaluating purpose. The object information are captured with the hardware platform from the previous chapter. The

CAD file, which is basis for manufacturing process, describes the object in his ideal form and is used for the slicing process. The challenge is the object capturing without losing of geometrical information and afterwards a comparing with the CAD file. Extra time and object handling should be avoided for the capturing. Therefor the idea of inline process monitoring has advantages and should has no influence to the manufacturing process. Another way is to capture the object after finished manufacturing process. Thus it's certainly possible to integrate sensors, but needs mostly rebuilding steps in the AM machine for lossless information capturing. But the aim of this proposal is to get a high system adaptivity for possible subsequent evaluation in other additive manufacturing machines.

The inline process monitoring begins with capturing first layer after manufacturing. The machining head is moving to an outer place to have a clear field of view to the object from bird view. If the calibration process has been performed before [8], the sensor of the hardware platform captures the necessary 2D images of the first manufactured layer and creates a point cloud of the layer, which is called single point cloud in this work. The image data processing for creating the point cloud is performing on a connected PC, which can also control the manufacturing machine. A possible controlling would be a machine stop, if the manufacturing process goes in wrong way. Afterwards the single point cloud is prepared for analysis. The area where the manufactured layer is just placed will be interesting for analysis. The rest of the point cloud around the object is uninteresting and is erased in the single point cloud. Furthermore pseudo points are removed, which constitutes in noise effects. This steps has effect in reducing of the necessary data storage and consequently the subsequent data processing. But the captured platform ground of the AM machine is the virtual reference layer for all subsequent processing steps and is saved in background. The next manufactured layer is following and the layer capturing process starts again. The preprocessed single point clouds might be used for analyses of the manufacturing process, but isn't content of this work. After finishing of the manufacturing process all individuals point clouds are saved on the connected PC and prepared for the compounding of all layers to the final result.

After manufacturing is finished the number of single point clouds are same like the number of layers. The assembly of single point clouds to the object point cloud is necessary and is doing subsequently. One Advantage is the position of the layer to 3D capturing system, because the manufactured layer is always in the same z position. This results in moving of platform in z direction and the fix position of machine head. The compounding of the layers starts at the first and the next layer is orientated to the previous layer. Therefor apriori knowledge is used to position the layers to each other. The z movement of the build platform is constant and very accurate. The predefined layer height in slicing process is also used for point cloud assembly. The overlapping points from one layer to the other is be filtered to reduce data storage.

In summary the object point cloud is created and can be used for analysis. The STL model, what describes the manufactured object, is also be used for result analysis. The STL model is virtually positioned in the object point cloud and geometrical differences are coloured [Figure 3]. For this analysis a user input is required to give a tolerance.

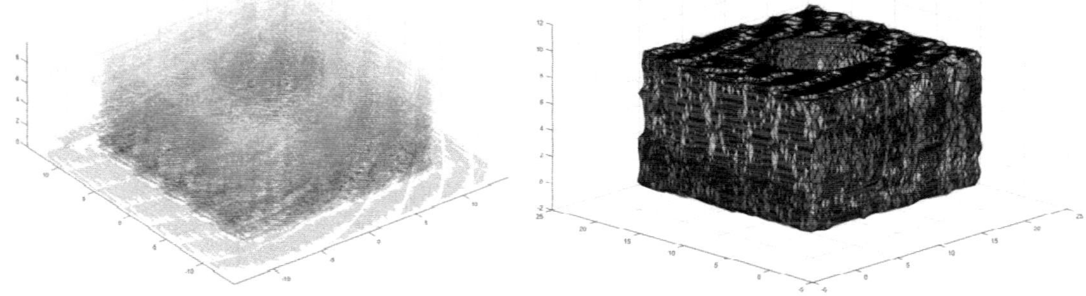

Figure 3. Assembled point cloud and final result

5. Results

The results are achieved with a triangulation angle of 28.7° and a baseline of 271 mm. Camera image size is 1920 x 1200 px and captures in combination with projector an area of 230 x 200 mm. A 20 x 20 x 10 mm cuboid has a density of several million points. The identified measurement stability for 3d reconstruction is 15 μm for white coloured ABS filament and current used algorithm. Admittedly it depends on the measurement field.

6. Conclusion

The investigation of the adaptable 3D sensor system in additive manufacturing system shows the possibility to monitoring the manufacturing process from start to end. The accuracy and number of point clouds are sufficient to give reproductive results for evaluating. A consequent next step is to close the quality control loop and manipulate the machine parameters for increasing manufacturing quality. Furthermore the single points cloud can be used for analysis and can declare more about the manufactured object.

7. Acknowledgment

We thank the federal ministry of education and research for the strong support of this work. The work is related to the project *Qualimess next generation* (03IPT709X).

References

[1] Lindemann C, Jahnke U, Moi M, Koch R 2012 Analyzing product lifecycle costs for a better understanding of cost drivers in additive manufacturing *23th Annual International Solid Freeform Fabrication Symposium–An Additive Manufacturing Conference (Austin Texas USA) 6th-8th August*

[2] Weckenmann A 1999 Koordinatenmesstechnik: Flexible Meßstrategien für Maß, Form und Lage 2nd edition Carl Hanser Verlag GmbH Co KG (München) 224

[3] Crump S S 1992 Apparatus and method for creating three-dimensional objects *U.S. Patent 5121329*

[4] Sun Q, Rizvi G M, Bellehumeur C T, Gu P 2008 Effect of processing conditions on the bonding quality of FDM polymer filaments. *Rapid Prototyping Journal*, *14*(2), 72-80

[5] Rankouhi B, Javadpour S, Delfanian F and Letcher T 2016 Failure analysis and mechanical characterization of 3D printed ABS with respect to layer thickness and orientation. *Journal of Failure Analysis and Prevention*, *16*(3), 467-481

[6] Heist S, Kühmstedt P, Tünnermann A and Notni G 2015 Theoretical considerations on aperiodic sinusoidal fringes in comparison to phase-shifted sinusoidal fringes for high-speed three-dimensional shape measurement. *Applied optics*, *54*(35), 10541-10551

[7] Heist S, Kühmstedt P, Tünnermann A and Notni G 2016 Experimental comparison of aperiodic sinusoidal fringes and phase-shifted sinusoidal fringes for high-speed three-dimensional shape measurement *Optical Engineering*. 55(2), 024105-024105

[8] Zhang Z 2000 A flexible new technique for camera calibration. *IEEE Transactions on pattern analysis and machine intelligence*, *22*(11), 1330-1334

Inmetro hydrostatic weighing system – determination of solids volume

Santos Junior J J P[1,2], Pereira, RG[2], Rosendahl, M[1,2,3], Santo Filho D M[1] , Gouveia J M G[1], Barbosa, A P F[1], Castro, C S C[1], Farias, M H[1]

1- National Institute of Metrology, Standardization and Industrial Quality - Inmetro
2- Fluminense Federal University – PGMEC-UFF
3- Estácio de Sá University – UNESA
4- State University of Rio de Janeiro - UERJ

Abstract. This article presents the hydrostatic weighing system of the National Institute of Metrology, Quality and Technology (Inmetro) Fluid Laboratory (Laflu) under the Fluid Dynamics Metrology Division (Dinam). The hydrostatic weighing system allows determining the volume of solids and the density of liquids and solids. Objective is to determine the volume of three spheres using the hydrostatic weighing method. Results are presented to the spheres of identification codes EF-001, EF-006 and EF-008, to one ring with code AN-017 and two sinkers with codes SK-001 and SK-002. The traceability of the method is also presented. This study evaluates the contribution of the input quantities to the uncertainty of the volume of solids and the impact of the influence of the mass of the solid on the value of the expanded uncertainty of the specific mass of the liquid.

1. Introduction

The Fluids Laboratory (Laflu) of the Division of Metrology in Fluid Dynamics (Dinam) of the National Institute of Metrology, Standardization and Industrial Quality (Inmetro) is responsible for providing traceability to density measurements in Brazil [1]. Density is a physical property defined as the ratio of mass to volume at a given temperature.

Figure 1- Inmetro hydrostatic weighing system and detail of the sphere.

To measure the density, there are many methods such as, for example, hydrometers [Salvatore], digital density meters [Tati 1 e 2], pycnometers and hydrostatic weighing [2]. After the implantation and qualification of the Laflu hydrostatic weighing system, the traceability happened by this primary method, composed by a mass comparator with device for weighing underneath, thermostatic bath and silicon sphere immersed in a liquid of stable characteristics, as defined in the diagram as shown by the picture in figure 1. The traceability of the method is presented by the diagram, figure 2.

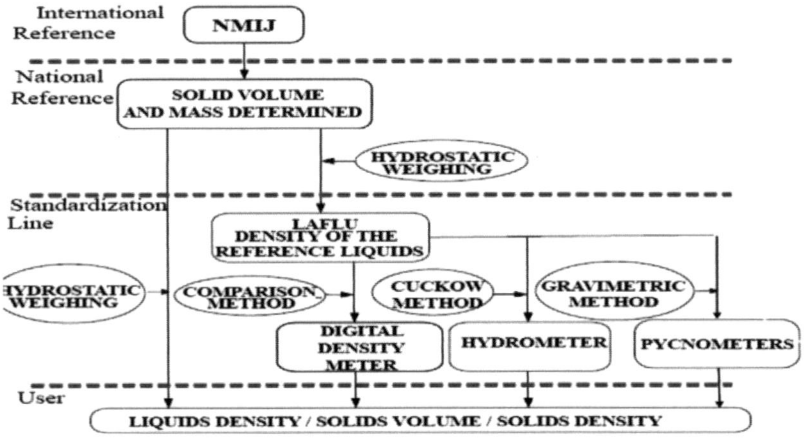

Figure 2 - Laflu Traceability diagrams

2. Objectives

Volume results of three spheres are obtained using the hydrostatic weighing method. Results are presented to the spheres of identification codes EF-001, EF-006 and EF-008, to one ring with code AN-017 and two sinkers with codes SK-001 and SK-002.

3. Primary System

Measurement procedure to determine the volume using the hydrostatic weighing method of solids adopt the following steps:

3.1- Measurement of the density of the reference liquid:

The n-Dodecane P.a. is the reference liquid. To determine the density of the liquid, a thermostatic bath with capacity to hold 8 L beaker and with a temperature gradient of about 0.005 °C were used. Also two temperature sensors, each of 0.001 °C uncertainty, were placed on the apparatus. Besides the cited equipment and sensors, in order to determine the environmental conditions the experimental apparatus contains a comparator scale with a 0.0001 g resolution, a thermohygrometer and a digital barometer. Accessories composing the apparatus are a set of class E2 weights, a solid (silicon sphere) with mass of 1000.63279 g ± 0.00056 g and volume of 429.62429 cm³ ± 0.00014 cm³. To determine the density of the liquid, equation 1 [1] is used:

$$\rho_L = \frac{M_s - M_{sL}}{V_s \cdot 1+\beta_s \cdot T_L - T_r} \cdot 1+\beta_L \cdot T_L - T_r \qquad (1)$$

Where, in equation [1], $\rho_L(T_r)$ is the density of the liquid at the reference temperature (g/cm³), M_S is the mass of the silicon sphere (g); M_{SL} is the mass of the silicon sphere immersed in the liquid (g); V_S is the volume of the silicon sphere (cm^3); β_S is the coefficient of thermal expansion of the solid; β_L is the coefficient of thermal expansion of the liquid; T_r is the reference temperature (in the case 20 °C); T_L is the temperature of the liquid (°C). The value obtained for density of n-Dodecane P.a. at 20 °C and pressure of 1013 hPa was (0.748842 ± 0.000005) g/cm^3.

3.2- Solid Characteristics

The solids used have the characteristics shown in table 1.

Table 1: Sinkers versus solids used as standards (Solid Characteristics) [3-5]

Solids	Code Identification	Material	Coef. thermal expansion (°C^{-1})	Mass (g)
Ring	AN-017	silicon	0,0000077	201,27489 ± 0,00038
Sphere	EF-001	glass	0,0000099	4,96226 ± 0,00016
Sphere	EF-006	silicon	0,0000077	1000,625303 ± 0,00056
Sphere	EF-008	silicon	0,0000077	500,33550 ± 0,00072
Sinker	SK-001	fused silica	0,0000016	321,54341 ± 0,00020
Sinker	SK-002	fused silica	0,0000016	24,78417 ± 0,00069

Figure 3 – Solids.

3.3 - Volume of Solids

The apparatus used to measure the density of the reference liquid are the same for volume measurements. The figures 4 and 5 presents the schematics of the assemblies adopted for the balls EF-001, EF-006, EF-008, AN-017 and SK-001.

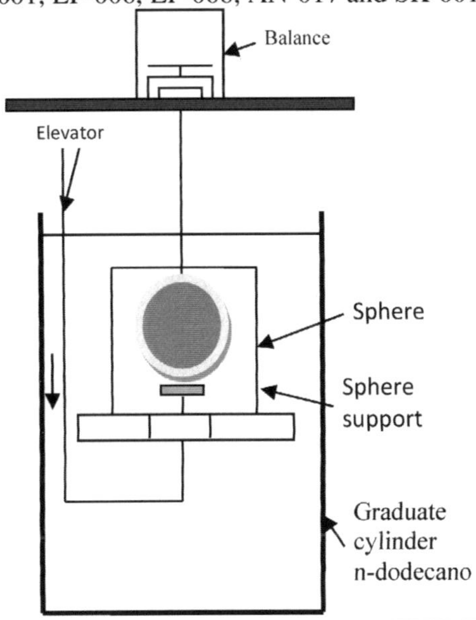

Figure 4 – Solid Lifted

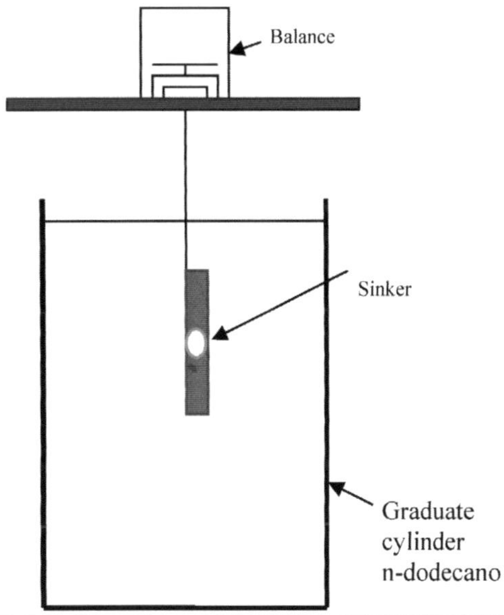

Figure 5 – Hanging solid

IOP Conf. Series: Journal of Physics: Conf. Series **1044** (2018) 012036 doi:10.1088/1742-6596/1044/1/012036

From the schemes it can me observed that the solid is supported in a suitable support for the sizing, figure 4. The assembly considering the hanging solid, for the case with the sinker SK-002, figure 5. To determine the solids volume, the following equation was used [2].

In order to determine the mass of solids immersed in the liquid (M_{O2}), the cyclic loading process ABA and BAB [5] is used, where the parameter B corresponds to the mass of the standard set of weights used in the weighing and the parameter A corresponds to the mass of the solid inside the liquid. The solid volume is determined by equation (2).

$$V_O = \frac{M_{O1} - M_{O2} \cdot 1 + \beta_o \cdot T_r - T_L}{\rho_L - \rho_a} \quad (2)$$

where M_{O1} is the mass of solid (g); M_{O2} is the mass of solid immersed in the liquid (g); ρ_L (20 °C) is density of the reference liquid at 20 °C, T_r is the reference temperature; T_a is the ambient temperature; ρ_a is the density of the air (g/cm³) [6]; T_L is the temperature of the liquid (°C); βo is the coefficient of thermal expansion of the solid.

4. Results

The results of the volume measurements of the solids are stated in Table 2. Experiments were conducted at the temperature of 20.00 ºC and pressure of 1013 hPa.

Table 2 - Results

Solids	Nominal	Volume (cm³)	Relative expanded uncertainty (%)
EF-006	1000	429,6199	0,0008
EF-008	500	214,8209	0,0008
SK-001	321	145,9681	0,0007
AN-017	201	86,4195	0,0014
SK-002	25	10,0092	0,0133
EF-001	5	2,0792	0,0140

The contributions of the input quantities to the solids with mass between 200 g and 1000 g are shown in the figure 6. In this case the analysis was performed for the solids coded EF-006, EF-008, SK-001 and AN-017.

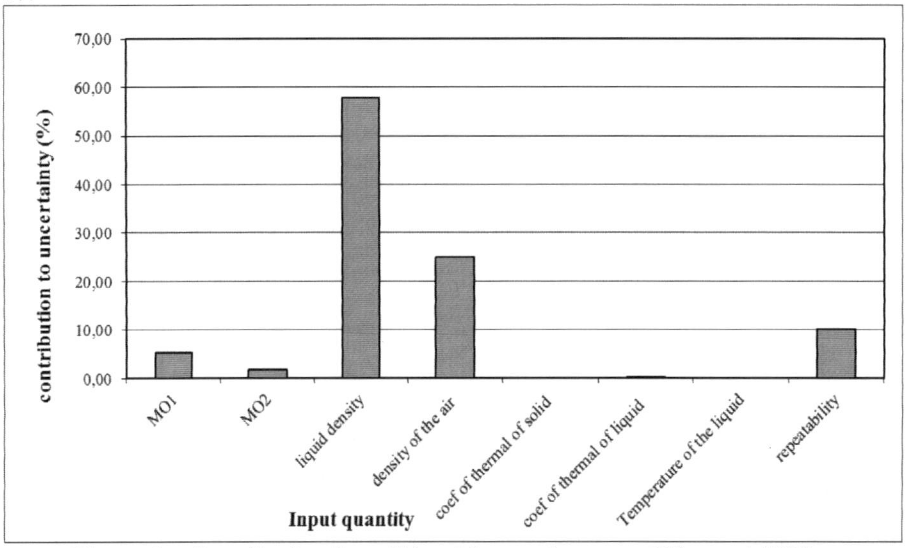

Figure 6 – Contribution for solids with mass between 200 g and 1000 g

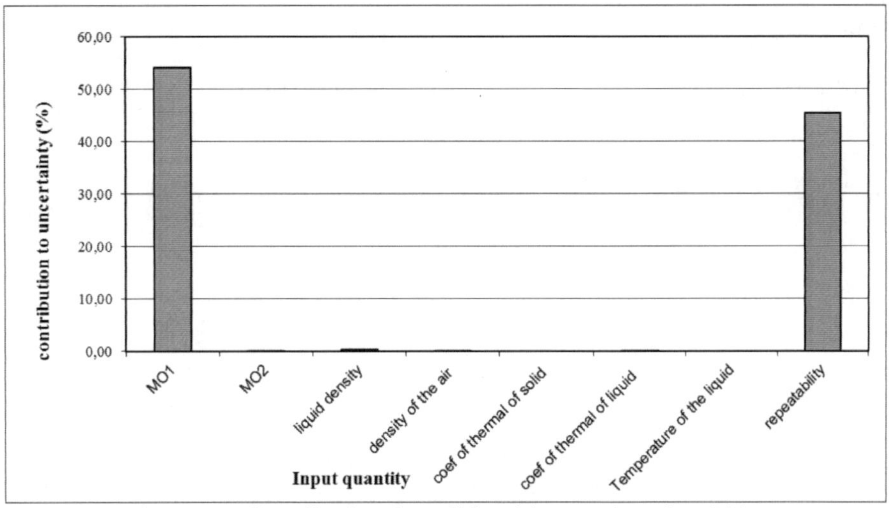

Figure 7 – Contribution for solids with mass less than 200 g

The contributions of the input quantities to the solids with mass less than 200 g are shown in figure 7. In this case the analysis was performed for the solids of code EF-001 and SK-002.

5- Conclusion

From the results obtained in Table 2, it can be stated that solids with nominal mass equal to or above 200 g can be used to obtain the density of liquids with a relative expanded uncertainty in the order of 0.0006% by the method of the hydrostatic weighing, since the expanded uncertainty of the volume of these solids is in the order of 0.001%, as shown in figure 8.

In solids with mass greater than or equal to 200 g, the quantities that contribute most to volume uncertainty are the specific mass of the reference liquid and the repeatability of the process.

We can reduce the uncertainties of these parameters by better control of the temperature gradient inside the liquid and the improvement of the bench automation system.

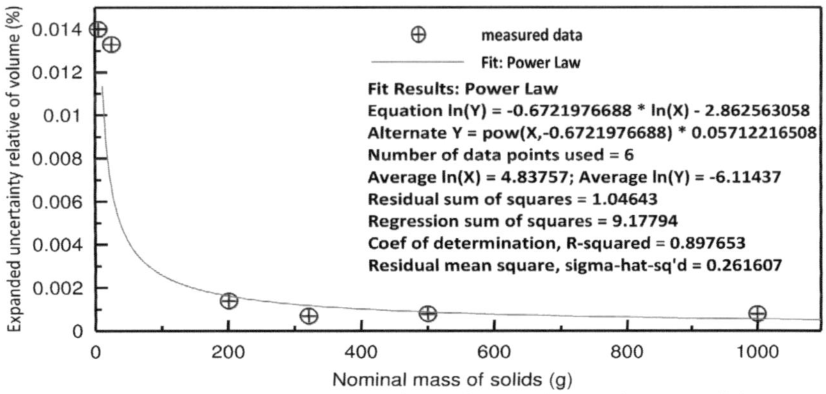

Figure 8 – Expanded uncertainty of the volume of these solids

In the case of solids with a mass of less than 200 g, the quantities that contribute most to the volume uncertainty are the mass of the solid, the specific mass of the liquid and the repeatability of the process. For these solids, even reducing the uncertainties of influence quantities, it will provide a high uncertainty for the volume due mainly to external factors such as small disturbances in the liquid during the measurement that affected the repeatability of the measurement.

Acknowledgements

The author R.G. Pereira is grateful to the National Research Council of Brazil, CNPq, Authors Santos Junior J.J.P. and M. Rosendahl are grateful to PRONAMETRO and UNESA for the financial support.

References

[1] DOS SANTOS JÚNIOR, José Júlio Pinheiro et al. Utilização do método da pesagem hidrostática para determinação massa específica de líquidos. 2008.
[2] BECERRA SANTIAGO, Luis Omar; PEZET SANDOVAL, Felix. Determinación de la densidad en sólidos y líquidos. Centro Nacional de Metrología.(México).
[3] OIML R 111-1 Edition 2004 (E), Weights of classes E1, E2, F1, F2, M1, M1-2, M2, M2-3 and M3 – Part 1: Metrological and technical requirements.
[4] EA-4/02, 1ª Edição Brasileira: Expressão da Incerteza de Medição na Calibração.
[5] JCGM 100:2008 GUM 1995 with minor corrections - Evaluation of measurement data — Guide to the expression of uncertainty in measurement
[6] A. Picard, R. S. Davis, M. Gläser, K. Fujii, Revised formula for the density of moist air (CIPM-2007). Metrologia 45(2008), p. 149-155.
[7] MAZIOLI, Tatiane dos Santos et al. Procedimento do laboratório de fluídos do Inmetro para limpeza de densímetros digitais. 2011.
[8] MAZIOLI, T. S. Medição de Massa Específica de Líquidos Utilizando o Densimetro Digital, monografia do Curso Técnico em Metrologia. Colégio Estadual Círculo Operário/Inmetro. Duque de Caxias, 2008.
[9] LOREFICE, Salvatore; MALENGO, Andrea. Calibration of hydrometers. Measurement Science and Technology, v. 17, n. 10, p. 2560, 2006.
[10] JÚNIOR, JJP Santos; SANTO FILHO, D. M. E. Estudo da contribuição da massa específica da água pura na calibração do volume de um recipiente. Tc, v. 3, n. 3, p. 2.

IOP Conf. Series: Journal of Physics: Conf. Series **1044** (2018) 012037 doi:10.1088/1742-6596/1044/1/012037

Instrumentation for respiratory flow estimation using tracheal sounds analysis: Design and evaluation in measurements of respiratory cycle periods and airflow amplitude

L L Gomes[1], N V Oliveira[1], L M Tauil[1], R A Mattos[1] and P L Melo[1,2]
[1]Biomedical Instrumentation Laboratory, Institute of Biology and Faculty of Engineering, State University of Rio de Janeiro, Brazil
[2]Laboratory of Clinical and Experimental Research in Vascular Biology, State University of Rio de Janeiro, Brazil

E-mail: plopes@uerj.br

Abstract. The increase in the incidence of respiratory diseases resulted in an increased interest in the improvement of the instruments used in research and diagnosis of respiratory dysfunction. We developed a non-invasive instrument able to estimate ventilation parameters by means of tracheal sounds analysis (TSA). The development of the hardware and software used in the instrument is described, as well as initial in vivo results obtained in a normal subjects. These tests showed that the TSA was able to provide adequate airflow estimative. Moreover, it was found that accurate respiratory periods can be obtained from tracheal sounds. These results were in close agreement with the physiology, confirming the high scientific and clinical potential of this system.

1. Introduction

There is a consensus in the scientific community that the costs associated with respiratory diseases are increasing and are likely to grow exponentially in the next generations [1]. This resulted in increased interest in the improvement of the methods used in research and diagnosis of respiratory dysfunction.

The Phonospirometry appears to be a very promising method in this regard. This method is based on the estimate of the ventilation parameters by means of measuring breathing sounds in the trachea [2]. From the patient's point of view, obtaining tracheal sounds signal is relatively easy, since the signal is taken non-invasively. These sounds are generated by turbulence mainly within the large and medium airways, being influenced by the flow rate and the dimensions of the airways. As a result, for a given position of the subject and the microphone, the sound amplitude is proportional to the flow velocity, which suggests the possibility of obtaining flow estimates from these measurements. This is currently a topic of great interest because this method can provide a simple alternative to respiratory analysis [3-5]. The areas of interest are the use of tracheal sounds as indicators in the upper airway obstruction and as a source for the qualitative and quantitative evaluation of airflow. In this context, the tracheal sounds analysis (TSA) was applied in measuring the periods related to the different stages of the respiratory cycle [6], to monitor respiration [7], as well as in the detection of the onset of respiratory cycles of inhalation and exhalation phases [8]. Previous studies have reported a connection between the signal in

[2] To whom any correspondence should be addressed.

IOP Conf. Series: Journal of Physics: Conf. Series **1044** (2018) 012037 doi:10.1088/1742-6596/1044/1/012037

the trachea, the underlying geometry of the airways and air flow [9]. Based on this principle, Yonemaru *et al.* [10] have developed a new method for the detection of stenosis in the trachea, while Shirazi *et al.*, [4] used this principle to detect abnormal swallowing involving silent aspiration. In sleep medicine, intensity changes of tracheal sounds were used in the identification of apnea events [11]. Kraman *et al.* pointed out that although the TSA is promising as a simple and non-invasive method of assessing airway, more studies are needed to correlate specific anatomical features with the sound characteristics [12]. In this context, this paper describes the design and performance analysis of an instrument capable of evaluating respiratory parameters through TSA.

2. Materials and Methods

2.1. Hardware design

Figure 1 describes a simplified block diagram of the instrument. Tracheal sounds were measured using a Sony microphone (ECM-77B). The balanced outputs are initially processed using an Instrumentation Amplifier (INA118, Texas Instruments Incorporated, Texas, USA) and then analogically high-pass filtered (0.05 Hz, first order) to remove off-set and low-pass filtered (Butterworth, 8th order, 5000 Hz) to reduce external noise. The resulting signal was measured at a sampling rate (fs) of 40 kHz *[13]* by a data acquisition module (NI 6003, National Instruments, Austin, Texas, USA), which presents a resolution of 16 bits, eight channels and fs maximum of 100 kHz. To simplify the practical use of the system, the hardware platform used was a touch screen notebook (Dell Inspiron series 137000, Intel, Core i5, 4G of RAM, hard disk of 456 GB).

The airflow obtained by TSA was compared with a reference airflow measurement system (Figure 1). It is composed by a fleisch pneumotachometer (PNT) coupled to a differential pressure transducer (176PC, Honeywell Inc. USA), whose signal is amplified by an instrumentation amplifier (INA118, Texas Instruments Incorporated, Texas, USA) and subsequently processed by a low-pass filter (Butterworth, 4th order, 10 Hz) and adapted to the data acquisition system.

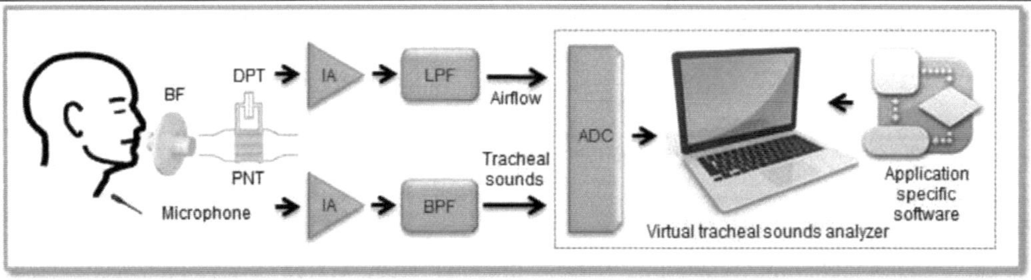

Figure 1: Simplified block diagram of the instrument. BF: biological filter; DPT: differential pressure transducer, PNT: pneumothacometer, IA: instrumentation amplifier, LPF: low pass filter, BPF: band-pass filter, ADC: analog-to-digital converter.

2.2. Software design

The software was developed in LabVIEW 2012 environment (National Instruments, Austin, TX). It is composed by two main modules; the first contains the data acquisition subroutines through the USB port, allowing the visualization of PNT and tracheal sounds signals and a visual quality control analysis. At the end of each exam the system allows researchers to save the data in an ASCII file.

The second module performs the tracheal sounds signal processing. A simplified flowchart describing the dedicated software and typical signals obtained in the signal processing chain are presented in Figure 2A. It begins with a band-pass filtering (300 Hz to 1 kHz, Butterworth, 5th order) to minimize cardiac effects and to remove noise from the external environment. The signal is then manually segmented to start before the beginning of an inspiration (Figure 2B graphs 1 and 2). The automatic identification of the beginning and end of inspiration and expiration, which allows the estimation of the inspiratory, expiratory and respiratory time, was obtained using the minimum values of Shannon

Entropy [14]. This parameter was calculated at intervals of 20 ms (800 samples) and the resulting signal was low-pass filtered (1 Hz, first order, red points in graph 2B graph 4). As can be seen in Figure 2B graph 3, the respiratory signal obtained from the tracheal sounds is rectified containing only positive values. To perform the reconstruction of the respiratory signal, the segment of the tracheal sounds signal associated with the expiratory phase was multiplied by -1. A typical result of this process is described in Figure 2B graph 5. The final results may be saved in an ASCII file for further clinical/research use. The software also includes the analysis of the spectrogram (or sonogram) of the tracheal sounds.

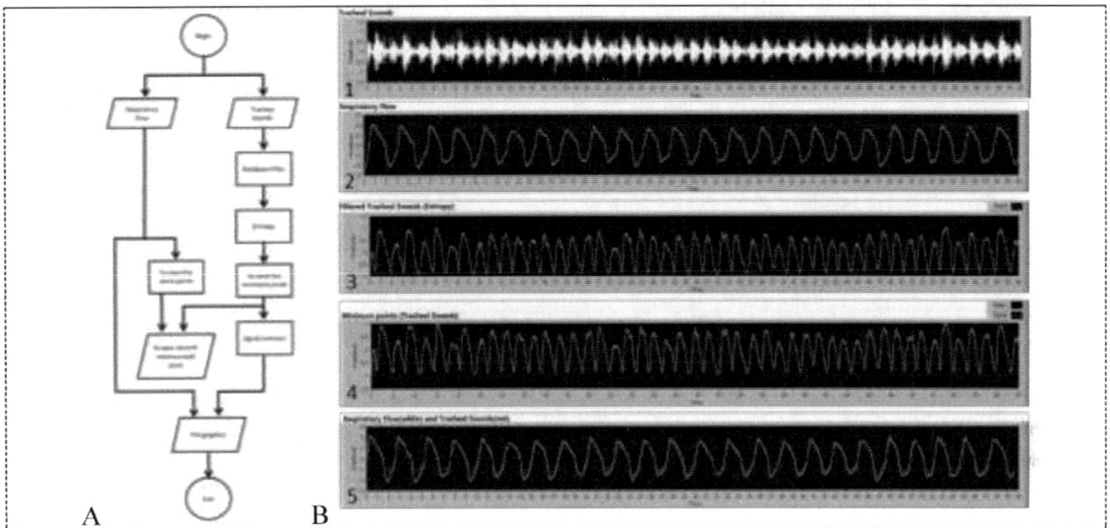

Figure 2: Simplified flowchart describing the dedicated software (A) and typical signals obtained in the signal processing (B).

2.3. *In vivo tests in healthy individuals.* The ability of the system to identify the duration of the respiratory cycle phases and airflow amplitude was evaluated through a comparative analysis with the results obtained in the reference system (PNT). It was performed analyzing 5 young healthy individuals that do not present history of respiratory diseases or tobacco use (3 male, 24.6 years, 172 cm, 68.7 kg). During the exams, the individuals remained seated, with the head in a neutral position.

2.4. *Statistical analysis.* The comparison between the results was performed by correlation analysis using ORIGIN 8.0. Values of $p<0.05$ were considered significant. The RMS error was also calculated.

3. Results

3.1 System development

Figure 3 illustrates the developed system, its typical use and the front panel of the program used in data acquisition. The supporting elements used during the exams, including the biological filter and the microphone are also described.

3.2. System evaluation

3.2.1 Respiratory cycle time in normal individuals. Figure 4 describes the comparative analysis of the PNT and TSA results. It were observed RMS errors of 0.19 s, 0.27 s and 0.03 s in the estimation of expiratory, inspiratory and total respiratory cycle time, respectively.

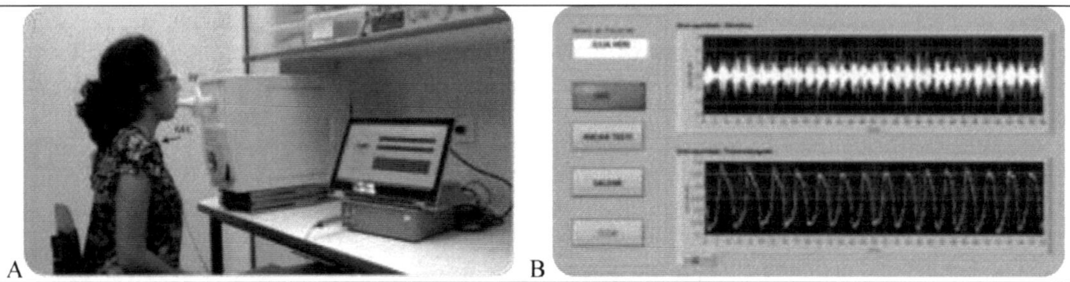

Figure 3: Picture describing the system and its use (A) and frontal panel of the data acquisition module subsystem (B). MIC = microphone; BF = bacterial filter.

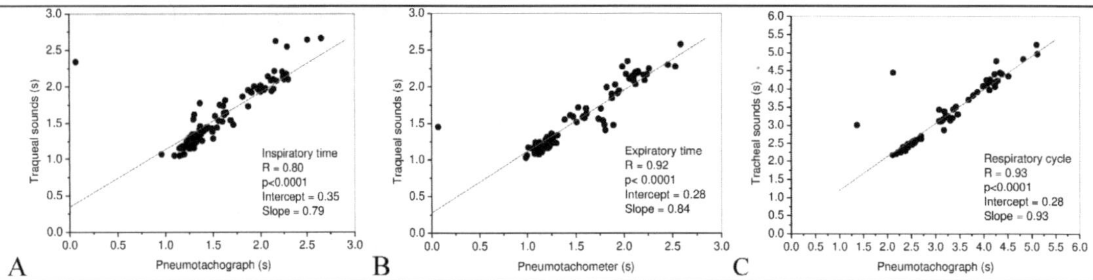

Figure 4: Associations of inspiratory (A), expiratory (B) and respiratory cycle (C) times measured using the pneumotachometer and tracheal sound analysis in five normal subjects during spontaneous ventilation.

3.2.2 Airflow amplitude in normal individuals. Figure 5A shows a typical result observed in the studied individuals. The correlation coefficient between the PNT and TSA was 0.95±0.02 (range 0.93-0.98) and significant (p<0.0001) in all studied individuals. Figure 5B present a typical spectrogram obtained in a segment of 20 s of tracheal sounds obtained during spontaneous ventilation.

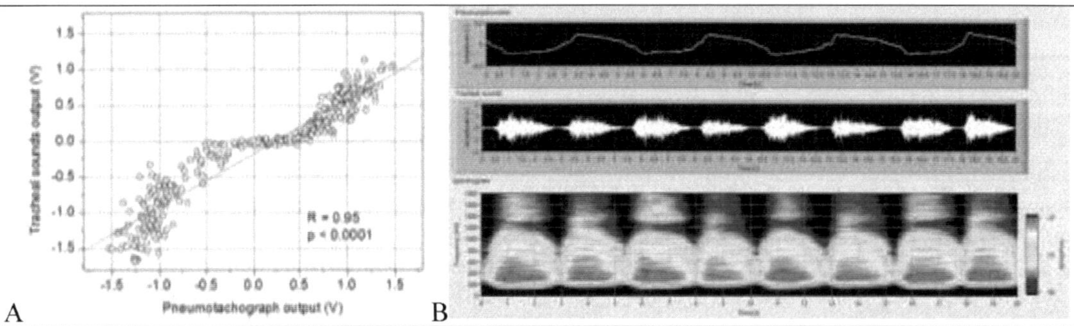

Figure 5: Typical correlation of the airflow measured using the pneumotachometer and tracheal sound analysis observed during spontaneous ventilation (A). Tracheal sound signal spectrogram observed during spontaneous ventilation (B; low) along with the tracheal sound signal (B; middle) and the corresponding typical flow (B; top graph).

4 Discussion

In vivo tests. The morphology of the tracheal sounds signal described in Figure 2B1 is coherent with that presented in a recent study of Elwali and Moussavi [15] and in the studies of Sierra et al. [16] and Que et al. [2], in which the tracheal sounds morphology of normal awake subjects were investigated. In Figure 2B we can also observe that the estimation of airflow from tracheal breath sounds entropy resulted in a signal that follows the airflow variation very closely. These results confirm previous findings [14]. They are also consistent with the study of Yu et al [17] in which the tracheal breath sounds

entropy was used to detect abnormal reductions in airflow during sleep. Both these earlier results and ours confirmed the high potential of the TSA in physiological and clinical studies.
Respiratory cycle time in normal individuals. Sohrabi et al. [18] recently suggested that TSA was adequate to track the different phases of the breathing cycle. The results presented in Figure 4 clearly provide additional support to this hypothesis. The strong linear correlations observed in Figure 4 are also consistent with previous works investigating normal awake subjects [16]. The observed RMS errors were small, and probably associated with the presence of laminar flow and reduced breathing sounds at low airflow values. Comparing with the mean values of the inspiratory, expiratory and total respiratory cycle periods, these errors were < 18%, < 13% and < 1%, respectively. These values represent a reasonable accuracy, and may be considered appropriate to clinical applications. However, it is important to point out that some measurements were observed far from the regression line (Figure 4: one in the inspiratory, one in the expiratory and two in the respiratory cycle times). The precise definition of the nature of these errors is a work in progress.
Airflow amplitude in normal individuals. It was observed that the physiological tracheal sounds signal was absent during low airflow in the respiratory cycle (Figure 5A, near ±0.5V of the PNT output). This is consistent with the study developed by Sohrabi et al. [18], and can be explained considering that below this level there are no breathing sounds due to the presence of laminar flow. The very high correlations observed between the actual measured flow and the acoustically determined flow was also consistent with the cited work. The obtained results corroborate the hypothesis that the band pass filtered tracheal sounds entropy provides adequate estimation of airflow [14] and provides further experimental evidence of the relevance of TSA ability to provide effective noninvasive airflow quantitative measurements.

The spectrogram (Figure 5B, bottom) describes the amplitude spectrum for each time segment of the tracheal sounds signal (Figure 5B, middle). In the reference respiratory signal (Figure 5B, top) the positive values refer to inspiration and the negative values refer to expiration airflow. The spectrogram showed that, in the studied subject, the tracheal sound energy is concentrated in frequencies below 700 Hz and that the energy drops off between 700 and 1500 Hz. It was also observed that there is not much difference comparing the higher frequency components in the inspiratory and expiratory segments of the tracheal sounds. Another interesting finding in the spectrogram is the presence of a silent period separating the inspiratory and expiratory segments, which is consistent with the tracheal sound waveform (Figure 5B, middle). These results give additional support to the behaviour observed recently by Bersain et al. [19]. In agreement with the mentioned work, and with the physiological phenomena, this finding clearly describes the cessation of turbulence. This is characteristic of the of low airflow velocities events in the beginning of inspiration and expiration.

5 Conclusion & future work

A novel system for the acquisition and processing of tracheal sounds signals was developed in the current study. The ability of the system to evaluate respiratory cycle time and amplitude was validated by a study in normal subjects. The proposed system may be useful in a wide spectrum of clinical and research studies of subjects with respiratory diseases, enabling simple and noninvasive exams and improving the assistance offered to these patients.

Based on these promising results, future plans include the development of a portable version to help home monitoring and telemedicine applications [20]. They also include studies in well-defined groups of patients with respiratory diseases in order to evaluate the clinical contribution of the proposed instrument, and to integrate the TSA instrumentation described in the present work with machine learning algorithms [21], contributing to improve the diagnostic, understanding and management of respiratory diseases.

Acknowledgments
This study was supported by the Brazilian Council for Scientific and Technological Development (CNPq) and Rio de Janeiro State Research Supporting Foundation (FAPERJ).

References

[1] Anonymous 2013 The juggernaut of respiratory diseases gains momentum *The Lancet Respiratory Medicine* **1**

[2] Que C L, Kolmaga C, Durand L G, Kelly S M and Macklem P T 2002 Phonospirometry for noninvasive measurement of ventilation: methodology and preliminary results *Journal of applied physiology* **93** 1515-26

[3] Kulkas A, Huupponen E, Virkkala J, Saastamoinen A, Rauhala E, Tenhunen M and Himanen S L 2010 Tracheal sound parameters of respiratory cycle phases show differences between flow-limited and normal breathing during sleep *Physiological Measurement* **31** 427-38

[4] Sarraf Shirazi S and Moussavi Z 2012 Silent aspiration detection by breath and swallowing sound analysis *Conf Proc IEEE Eng Med Biol Soc* **2012** 2599-602

[5] Yadollahi A, Montazeri A, Azarbarzin A and Moussavi Z 2013 Respiratory Flow-Sound Relationship During Both Wakefulness and Sleep and Its Variation in Relation to Sleep Apnea *Annals of Biomedical Engineering* **41** 537-46

[6] Hult P, Wranne B and Ask P 2000 A bioacoustic method for timing of the different phases of the breathing cycle and monitoring of breathing frequency *Medical Engineering & Physics* **22** 425-33

[7] Hult P, Fj¨allbrant T, Wranne B, Engdahl O and Ask P 2004 An improved bioacoustic method formonitoring respiration. In: *Technol Health Care,* pp 323–32

[8] Moussavi Z K, Leopando M T, Pasterkamp H and Rempel G 2000 Computerised acoustical respiratory phase detection without airflow measurement *Medical & Biological Engineering & Computing* **38** 198-203

[9] Harper V P, Pasterkamp H, Kiyokawa H and Wodicka G R 2003 Modeling and measurement of flow effects on tracheal sounds *Ieee Transactions on Biomedical Engineering* **50** 1-10

[10] Yonemaru M, Kikuchi K, Mori M, Kawai A, Abe T, Kawashiro T, Ishihara T and Yokoyama T 1993 Detection of tracheal stenosis by frequency analysis of tracheal sounds *Journal of applied physiology* **75** 605-12

[11] Yadollahi A, Montazeri A, Azarbarzin A and Moussavi Z 2013 Respiratory flow-sound relationship during both wakefulness and sleep and its variation in relation to sleep apnea *Annals of biomedical engineering* **41** 537-46

[12] Kraman S S, Pasterkamp H, Kompis M, Takase M and Wodicka G R 1998 Effects of breathing pathways on tracheal sound spectral features *Respiration Physiology* **111** 295-300

[13] Huq S and Moussavi Z 2012 Acoustic breath-phase detection using tracheal breath sounds *Medical & Biological Engineering & Computing* **50** 297-308

[14] Yadollahi A and Moussavi Z M 2006 A robust method for estimating respiratory flow using tracheal sounds entropy *IEEE transactions on bio-medical engineering* **53** 662-8

[15] Elwali A and Moussavi Z 2016 Obstructive Sleep Apnea Screening and Airway Structure Characterization During Wakefulness Using Tracheal Breathing Sounds *Annals of biomedical engineering*

[16] Sierra G, Telfort V, Popov B, Durand L G, Agarwal R and Lanzo V 2004 Monitoring respiratory rate based on tracheal sounds. First experiences *Conference proceedings : ... Annual International Conference of the IEEE Engineering in Medicine and Biology Society. IEEE Engineering in Medicine and Biology Society. Conference* **1** 317-20

[17] Yu L, Ting C K, Hill B E, Orr J A, Brewer L M, Johnson K B, Egan T D and Westenskow D R 2013 Using the entropy of tracheal sounds to detect apnea during sedation in healthy nonobese volunteers *Anesthesiology* **118** 1341-9

[18] Sohrabi K A, Basu D, Schudt F, Scholtes M, Seifert O, Koehler U and Gross V 2012 Quantification of nasal respiratory flow by tracheal sound analysis *Biomed Tech* **57** 733-5

[19] Reyes B A, Reljin N and Chon K H 2014 Tracheal sounds acquisition using smartphones *Sensors* **14** 13830-50

[20] da Silva Junior E P, Esteves G P, Dames K K and Melo P L 2011 A telemedicine instrument for Internet-based home monitoring of thoracoabdominal motion in patients with respiratory diseases *The Review of scientific instruments* **82** 014301

[21] Amaral J L, Lopes A J, Faria A C and Melo P L 2015 Machine learning algorithms and forced oscillation measurements to categorise the airway obstruction severity in chronic obstructive pulmonary disease *Computer methods and programs in biomedicine* **118** 186-97

IOP Conf. Series: Journal of Physics: Conf. Series **1044** (2018) 012038 doi:10.1088/1742-6596/1044/1/012038

Instrumentation for the analysis of changes in the knee joint of patients with rheumatoid arthritis: focus on low-frequency vibrations

A C D Faria[1], G R C Pinheiro[2], J Neri[1] and P L Melo[1,3,*]
[1]Biomedical Instrumentation Laboratory, Institute of Biology and Faculty of Engineering, State University of Rio de Janeiro, Brazil
[2] Department of Rheumatology, Pedro Ernesto University Hospital, State University of Rio de Janeiro, Brazil
[3] Laboratory of Clinical and Experimental Research in Vascular Biology, State University of Rio de Janeiro, Brazil

E-mail: plopes@uerj.br

Abstract. One of the most common occurrences in patients with rheumatoid arthritis (RA) is the presence of problems in the knee joint. The methods used in the diagnosis of this damage, however, are expensive and/or invasive. This work describes the design and application of a system able to non-invasively evaluate changes in knee joint using accelerometry. Three accelerometers mounted perpendicularly were used, providing signals to a portable instrument. In vivo tests investigating the knee joint vibration power between 2 and 10 Hz showed increased values in RA patients ($p<0.006$). The developed system could be a useful tool for the evaluation of changes in the knee joint introduced by RA.

1. Introduction

The rheumatoid arthritis (RA) is an autoimmune disease of unknown etiology, characterized by peripheral polyarthritis, symmetrical, which leads to deformity and a destruction of the joints [1]. In the United States there are two million adults with RA [1]. It is estimated that RA reaches approximately 0.5% of the Brazilian population [2]. In this disease, the main changes are related with articular problems [1]. The presence of knee joint abnormalities in these patients is one of the most common occurrences, being its detection often based on the simple report of pain and stiffness in the articulation [3]. Among the techniques used to assess the knee joint, the most popular methods of evaluation of joint cartilage damage are arthroscopy and arthrography. These methods, however, are of high price, invasive and may result in infection or further damage to the cartilage [3]. The evaluation of the degree of passive and active movement, ultrasonography, scintigraphy, radiography and magnetic resonance imaging are non-invasive techniques able to evaluate knee joint abnormalities. However, these techniques have disadvantages including inability to radiograph the joint at all degrees of flexion, presence of radiation, high cost, extensive time of examination, and limited ability to delimit the extent of joint cartilage destruction. The auscultation of the knee has been suggested as a diagnostic method, but the ambient noise and the lack of sensitivity of the microphone to the low

[*] To whom any correspondence should be addressed.

frequencies emitted by the joint limit the effectiveness of this method [4]. Therefore it was observed the need for the development of new non-invasive methods for evaluating knee disorders [3].

The measurement of the vibration produced during knee movement using accelerometry was proposed by Reddy and contributors [4] as a non-invasive method for the analysis of pathologies of the knee. This movement in normal individuals results in an acceleration pattern that is changed in the presence of RA [4]. The technique is minimally intrusive, requiring only the positioning of the transducers on the surface of the knee on analysis, contributing to the diagnosis and follow-up of patients. In addition, this method provides new parameters for analysis that may contribute to deepen the knowledge about the pathophysiology of joint diseases. Previous studies used high frequencies in the evaluation of the oscillations produced as a result of knee disorders [3, 5], typically above 10 Hz. However, there are evidences indicating that this signal is predominantly low-frequency in nature and that most of the signal power fall below 10 Hz [6]. Previous works also point out the importance of subsonic range present in the signal [7, 8]. In this context, the objectives of this study were: (1) to describe the development of a portable system suitable for performing knee accelerometry exams and (2) Investigate the potential of using this methodology in patients with RA, with emphasis on low frequency analysis.

2. Materials and Methods

2.1. System development

2.1.1. Hardware design. Figure 1 presents a simplified block diagram of the instrument. It basically consists of a set of transducers, an analogic signal processing system, including amplifiers and filters, an analogue-to-digital conversion stage and a personal computer running a program responsible for the acquisition of the signals, analysis, presentation and storage of results. The transducers consist of three piezoelectric accelerometers mounted perpendicularly (low power, 3-axis model ADXL335; Analog Devices Inc. EUA). This transducer measures static and dynamic acceleration in a band comprised between ±3.0 g and has a maximum operating frequency of 550 Hz [9]. The transducers are powered by means of a reference voltage of 3.3 V obtained from the integrated LM 2671 (National Semiconductor Inc.) [10]. The signals from the three accelerometers are initially amplified and then processed by analogic low-pass filters (Butterworth, 4th order, 500 Hz) to reduce interference and avoid spectral leakage [3]. This initial stage is powered by two batteries (9 V). Low energy consumption operational amplifiers (LMC 6484, National semiconductor) [11] were used to increase the system autonomy. The acceleration signals were connected to a data acquisition module (NI 6008, National Instruments, Austin, Texas) with a resolution of 12 bits, four channels and maximum sampling frequency (fs) of 10 kHz. In the present project the fs used was 3.0 kHz [3]. To simplify clinical evaluations, the hardware platform used was a small netbook (LG X120, LG Electronics). An important additional advantage resulting from the use of this subsystem is to allow the complete isolation of the patient in relation to the electrical network.

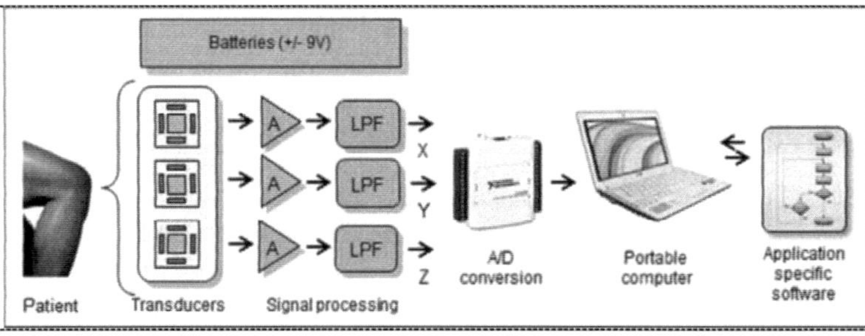

Figure 1: Simplified block diagram of the instrument. A = amplifier and LPF = low-pass filter.

2.1.2. Software design. The program was developed in a LabVIEW® environment, containing two main modules. The first, whose front panel is shown in figure 2, contains the acquisition subroutines via the USB port and allows the presentation of the vector sum of the acceleration signal. The acquisition is carried out over 10 s, during movement of the patient's leg, whose frequency must be between 0.4 Hz and 0.5 Hz for the test to be accepted [3, 4]. The program also has routines that allow the user to store patient data, including biometric features, data for contact, and disease history. At the end of the test, the system offers the possibility to save an ASCII file with the acceleration signals over time. The values of the periods and frequencies associated with each cycle of knee movement are also saved.

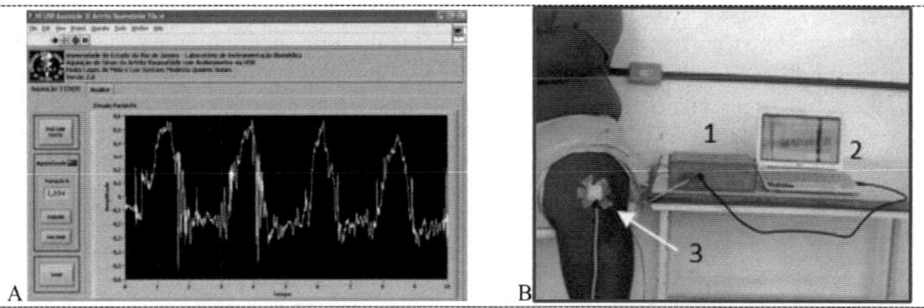

Figure 2: Front panel of the program developed for data acquisition (A). Picture illustrating the developed system and its application (B) describing the analog signal processing (1); the netbook used for data acquisition (2) and the positioning of the transducers (3).

The second module provides a quantitative analysis of the measured accelerations. Initially, the signal is normalized dividing the acceleration values by the maximum value of the basic knee movement signal [3], which was obtained by means of a low pass filter (Butterworth, 4th order, 2 Hz). Subsequently, the length of the signal to be analysed is chosen to include complete cycles of movement. This segmentation is done manually using cursors to identify the start and end points of the segment to be analysed. The signal is processed by a high-pass filter (Butterworth, 4th order, 2 Hz) and, using the routines available in LabVIEW, the program estimates the RMS power spectrum of the signal. Based on this spectrum, it estimates the average power between 2 and 10 Hz (Pm). The software also allows the evaluation of the short-time spectra. This is very useful for characterization of the frequency distribution of knee vibration signal segments in the transient time spans. The spectrograms were obtained by computing the short-time Fourier transform and the Hanning window in the length of 3000 samples, with the overlapping segment length of 2700 samples.

2.2. *In vivo tests in healthy individuals and patients with rheumatoid arthritis*

The present study was approved by the Ethics Committee of the Pedro Ernesto University Hospital (HUPE). The protocol obeys the guidelines of Declaration of Helsinki and the post-informed consent of all volunteers was obtained. A comparative analysis was carried out among the results obtained in seven healthy individuals (23.7±3.8 years, 72.6±12.9 kg, 177.3±7.5 cm) and six patients with RA (52.7±9.8 years, 78.6±15.2 kg, 168.0±4.5 cm). The transducers were adapted to the patella of the volunteers with a tape of medical use. Measurements were obtained with the volunteers by cyclically moving the 90° flexion leg to full extension. The flexion-extension procedure was repeated at a uniform rate of approximately 0.5 Hz [3], with the aid of a physiotherapist who guided the volunteer in relation to periods of extension and flexion. The volunteers were analysed wearing loose clothing or shorts and were invited to sit so that they could perform full flexion and extension without feet touching the floor (figure 2B). The RA is a disease that results in bilateral attack [12]. Thus, the

IOP Conf. Series: Journal of Physics: Conf. Series **1044** (2018) 012038 doi:10.1088/1742-6596/1044/1/012038

analyses were performed on both knees of all volunteers. Three measurements were performed on each knee, completing a total of 42 normal exams and 36 exams in patients with RA. Results are presented as mean ± standard deviation. The comparison between the results was performed using the STATISTICA version 5.0 program by means of the Mann-Whitney test.

3. Results

3.1. System development

Figure 2B illustrates the developed system and its use. The current consumed by the circuit is 7.9 mA, which results in an autonomy of 45.7 hours. In this project, the limitation of autonomy refers to the autonomy presented by the netbook (3 hours). Considering the average time of an evaluation (10 min), observed during the examinations carried out in the present study, this autonomy allows the accomplishment of approximately 18 exams. Figure 3 shows typical signals obtained in patients and normal volunteers. Figure 4 describes representative high pass-filtered signal courses and associated spectrograms registered for a control individual and a patient.

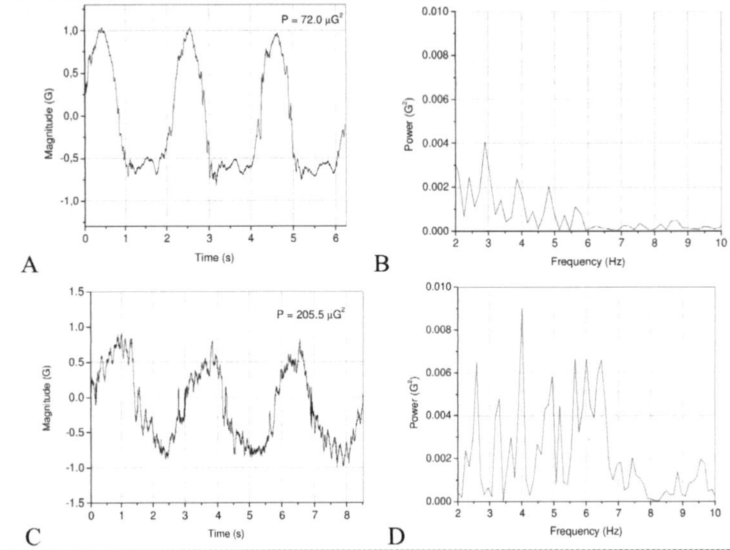

Figure 3: Typical signals obtained in normal individuals (A) and RA patients (C), and their respective power spectral densities (B) and (D). Note the presence of high frequency vibrations in the signal associated with the patient.

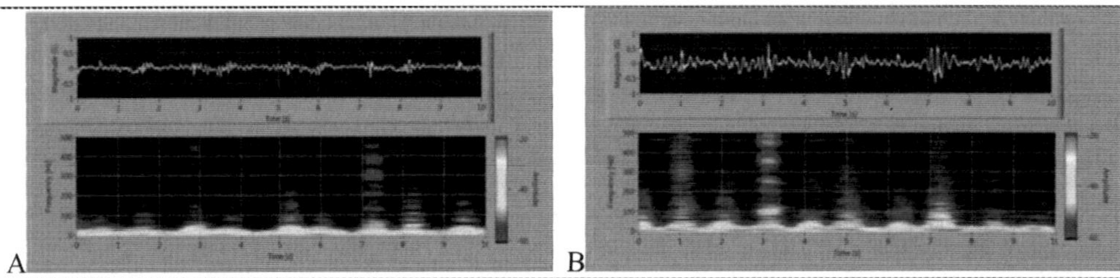

Figure 4: Frequency characteristics of the high-pass filtered knee vibration signal (top) examined by a short-time Fourier transform analysis (bottom). Subject representative for normal persons (A) and patients (B). Note the higher amplitudes in the low frequency vibrations in the patient.

On average, were observed significantly higher values of Pm ($p<0.006$) in patients with RA (figure 5A). Considering the mean values obtained, including the two knees in each individual studied (figure 5B), it was observed that 4 of the 6 patients studied were above the values observed in normal individuals.

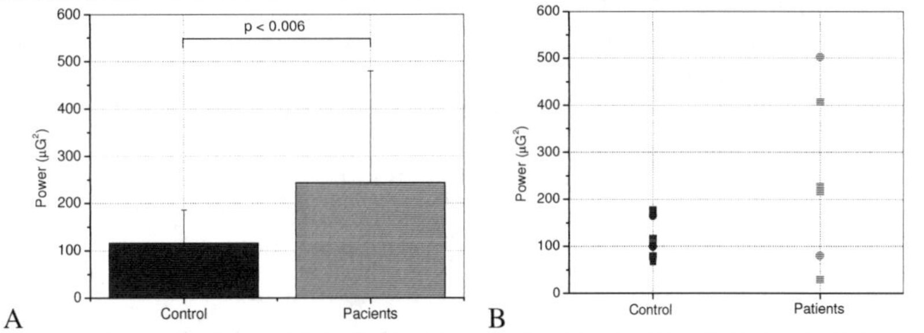

Figure 5: Power obtained in the two groups studied in average terms by group (A) and considering the average values obtained in each individual including the two knees (B).

4. Discussion

The system proposed in the present work is suitable for outpatient use, being small (figure 2B). It was observed that the control program is easy to use by biomedical professionals (Figures 2A). The instrument allows the accomplishment of 18 exams with complete isolation of the patient in relation to the electrical network, which makes the system intrinsically safe. An important advantage refers to allowing a quantitative characterization of the process. The Pm calculated by the instrument presents a high potential to contribute as a useful objective clinical parameter, both for the diagnosis and for the monitoring of knee injuries.

The signals in the time domain show an approximately sinusoidal, low-frequency component associated with the leg movement of the volunteers (Figure 3A,C). It may also be noted that higher frequencies components overlapping on these signals. These components present higher amplitude values in the patients with RA, which allows differentiating the groups (Figure 5). The amplitude of vibrations in normal individuals (figure 3A) is reduced because the surfaces of the knee joints are smooth and integral. In RA patients (figure 3C), on the other hand, there is an increase in friction between the joint surfaces due to the extensive erosion typical of this disease [4]. These principles are clearly described in the results presented in figures 3 and 4. These results are also consistent with those described previously by Reddy et al. [3, 4], which, however, estimate the mean power in a frequency range between 100 and 500 Hz. On the other hand, in the present study, no differences were observed between normal individuals and RA patients in this frequency range. The visual analysis of the results (figure 3) indicates that differences between normal individuals and RA patients can be observed at lower frequencies, which was confirmed by the spectral analysis of the signals (Figure 3B,D). Based on these practical results, it was studied the Pm in the range where the spectral analyses were more selective (2 and 10 Hz). The high-pass filtered signals described in figure 4 eliminate the low-frequency component associated with leg movement (approximately 0.5 Hz). These signals were consistent with the recent work of Yang et al. [13] and clearly showed increased spectral amplitudes in the 2-10 Hz frequency in the presence of frictional events (figure 4). Considering the studied groups, this increased amplitudes resulted in significantly higher values of power in the patients (figure 5A).

It is important to point out that the present study investigated patients in the early stages of RA. In this case there is only inflammation in the synovial tissue; the joint surfaces have not yet been eroded, still encountering the smoothing close to normal. This results in small friction between the surfaces [4]. This may explain the low Pm values observed in two patients described in Figure 5B. One of these patients was in the initial phase of treatment because it was a case of a recent diagnosis of RA. The other case was a patient with no joint pain. Thus, these results are consistent with the clinical evaluations and the pathophysiology of RA and indicate that the proposed system may be useful in discriminating between the early and late phases of the disease. Studies in a larger number of patients are needed to confirm this hypothesis.

5. Conclusion & future work

A new system was developed for non-invasive analysis of knee disorders using accelerometry. The main characteristic of this system is the low frequency analysis. The application of the system in normal individuals and patients with RA showed results in close agreement with the literature, as well as with the physiological principles involved. The system has a high potential to contribute to the improvement of the analysis of alterations resulting from RA or associated to rehabilitation procedures. The next phases of this project include its use in the evaluation of patients with different degrees of joint impairment and the evaluation of the performance of medications in patients attended at our University Hospital.

Acknowledgments
This study was supported by the Brazilian Council for Scientific and Technological Development (CNPq) and Rio de Janeiro State Research Supporting Foundation (FAPERJ).

References

[1] Brown K K 2007 Rheumatoid lung disease *Proc Am Thorac Soc* **443-448**
[2] Senna E R, De Barros, A.L., Silva, E.O., Costa, I.F., Pereira, L.V., Ciconelli, R.M., Ferraz, M.B. 2004 Prevalence of rheumatic diseases in Brazil: a study using the COPCORD approach. *J Rheumatol* **31(3)** 594-7
[3] Reddy N P, Rothschild B M, E. V and A. J 2001 Noninvasive Measurement of Acceleration at the Knee Joint in Patients with Rheumatoid Arthritis and Spondyloarthropathy of the Knee *Ann Biomed Eng* **29** 1106-11
[4] Reddy N P, Rothschild B M, Mandal M and Gupta V 1995 Noninvasive acceleration measurements to characterize knee arthritis and chondromalacia *Ann Biomed Eng* **23** 78-84
[5] Baczkowicz D, Majorczyk E and Krecisz K 2015 Age-related impairment of quality of joint motion in vibroarthrographic signal analysis *BioMed research international* **2015** 591707
[6] Tavathia S, Rangayyan R M, Frank C B, Bell G D, Ladly K O and Zhang Y T 1992 Analysis of knee vibration signals using linear prediction *IEEE transactions on bio-medical engineering* **39** 959-70
[7] Mollan R A, McCullagh G C and Wilson R I 1982 A critical appraisal of auscultation of human joints *Clinical orthopaedics and related research* 231-7
[8] Kernohan W G, Beverland D E, McCoy G F, Shaw S N, Wallace R G, McCullagh G C and Mollan R A 1986 The diagnostic potential of vibration arthrography *Clinical orthopaedics and related research* 106-12
[9] ADXL335 2009 Data sheet In: *Analog Devices, Inc.* , (http://www.analog.com/en/sensors/inertial-sensors/adxl335/products/product.html.
[10] LM2671 2007 Data sheet In: National Semiconductor http://www.national.com/pf/LM/LM2671.html.)
[11] LM6484 2000 Data sheet. In: National Semiconductor (http://www.national.com/mpf/LM/LMC6484.html.
[12] Bértolo M B, Brenol C V, Schainberg C G, Neubarth F, Lima F A C and Laurindo I M e a 2007 Atualização do consenso brasileiro no diagnóstico e tratamento da artrite reumatóide *Rev Bras Reumatol* **47** 151-9
[13] Yang S, Cai S, Zheng F, Wu Y, Liu K, Wu M, Zou Q and Chen J 2014 Representation of fluctuation features in pathological knee joint vibroarthrographic signals using kernel density modeling method *Medical engineering & physics* **36** 1305-11

IOP Conf. Series: Journal of Physics: Conf. Series **1044** (2018) 012039 doi:10.1088/1742-6596/1044/1/012039

Investigations on infrared-channel-image quality improvements for multispectral imaging

M. Rosenberger, C. Zhang, R. Celestre, G. Notni
Ilmenau University of Technology
Department for Quality Assurance and Industrial Image Processing
Gustav-Kirchhoff-Platz 2, 98693 Ilmenau, Germany

E-mail: maik.rosenberger@tu-ilmenau.de

Abstract. The improvement of image quality in the infrared range on silicon-based sensors is one major topics using these long wavelength channels for geometric measurements. The reason behind the bad quality of infrared images in comparison with the visible sampling range is explained by the wavelength response dependency of the silicon. Photons are able to either to pass the sensitive range of the pixel or can tunnel to the neighbor pixel. This effect leads to blurred images, which will not only increase the uncertainty of measurement, but also the aesthetical quality of the image. In this paper, methods to improve the image quality using blind convolution as well as a special infrared focusing to improve the sharpness will be presented.

1. Introduction

Multispectral and hyperspectral imaging technologies allow detecting several features in industrial inspections tasks using the different spectral channels [1] its applications range from cultural heritage, arts, remote sensing, environmental monitoring, medicine, biology, food quality control, military applications, factory automation and manufacturing [2]–[7]. Besides an accurate registration of images acquired at different wavelengths for the spatial-spectral utilization of multispectral data [8], the consideration of the semiconductor characteristics is a big issue. Therefore, in [9] a method for the characterization of pixel inhomogeneity's caused by the manufacturing process and read out process is proposed. These investigations show how the specific data can be processed to compensate the inhomogeneity's as function of the current integration time. A fact which is actually not discussed in these works, is the crosstalk characteristics between the pixels in dependence of the wavelength. The material characteristic plays an important role. The characteristics of penetration depth versus absorption of photons for different materials are given in [10]. It is shown in [10] and [11] that in the near infrared region (NIR) the generation of electrons will happen in an range of tens to hundreds of microns inside the semiconductor. Therewith it is possible that the electrons flow into neighbor pixels, which is presented in [12]. Finally, these effects in the modular transfer function of the image sensor and the images have a blurred appearance as well as blurred information across edges, which account for the deterioration of the accuracy of geometrical measurements in the NIR regime: blurred images have less sharp edges (see Figure 2) and edge detection based algorithms rely on precise estimation of edge's position. Unfortunately, this effect has a statistical behavior so a discrete model based deconvolution of the blurred image is not possible. In this paper, an approach to improve this characteristic using a statistical method based on a combined image restauration algorithm is part of the discussion. Other sources of instabilities and how to tackle them are discussed in [1],[8]-[9],[13].

IOP Conf. Series: Journal of Physics: Conf. Series **1044** (2018) 012039 doi:10.1088/1742-6596/1044/1/012039

2. Analysis of the blurred images in the infrared region

The blurred images caused by the electron generation of the incoming photons and the flow inside the neighbour pixel regions inside the deep regions of the semiconductor are hardly measurable without extensive setups. Nevertheless, for practical experiments the estimation of the sharpness on real pictures can be evaluated. For the investigations in the infrared regions a multispectral filter-wheel camera was used to generate the image data in the VIS-NIR region from 650-950nm. To minimize the defocus caused by the optics in the different spectral channels, a refocus for every wavelength has to be made inside the multispectral imager [1]. Therefore, a high precision sensor actuation inside the multispectral camera is used [13].

2.1. Autofocus calculation

As mentioned in the introduction, the multispectral imager offers the possibility to correct the longitudinal chromatic aberration using a linear moving stage to set the image sensor in the correct position. To ensure a certain repeatability and to ensure comparability between the different infrared channels, Laplace approach for the calculation of the best focus position was applied. The Laplace approach uses the Laplace energy as given in equation 1:

$$M_{Lap}(I) = \sum_{(x,y)} \left(\frac{\partial^2 I(x,y)}{\partial x^2} + \frac{\partial^2 I(x,y)}{\partial y^2}\right)^2. \quad (1)$$

A linear moving stage inside were actuated and the M_{Lap} were calculated for each position. The optimal focus position is the sensor position with the highest Laplace energy (Figure 1 c). After the focus procedure the Sensor was driven to the optimal position. The different focus positions along the optical axis can be stored inside the camera [1].

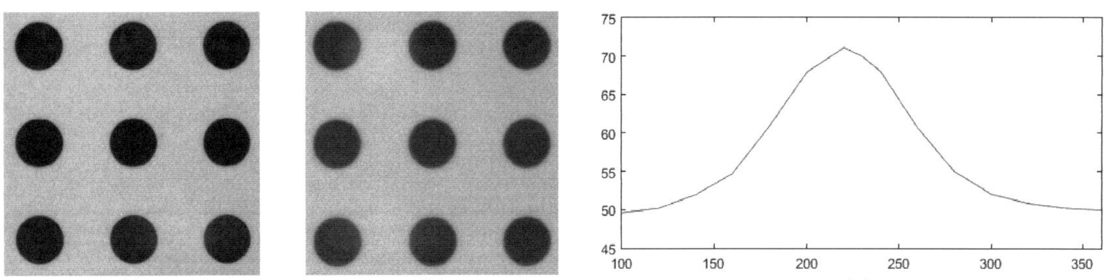

(a) **(b)** **(c)**

Figure 1: Focused images using Laplace method at (a) 550nm and (b) 900nm, (c) Laplace energy versus sensor position.

The depicted images (Figure 1) show as an example the difference between an NIR-image and a VIS image. Next step is to find a method to get the blured information in an evaluable value.

2.2. Approach for the evaluation of the image sharpness

After the images were refocused for each wavelength, the evaluation of sharpness was processed using the edge transition of a chrome glass target depicted in Figure 1. Therefore, search lines along the radii will be fitted on the outer contour of the circles. The result of this method delivers a grey value transition for each line (normalized grey value (blue) Figure 2), which must be processed.

Using a normalization method, the edge transition was converted to a normalized region between zero and one. With these values a best fit using a gaussian fitting method was applied to get a continuous edge transition description. Hence a Heaviside function (Figure 2 black) will be fitted over the real edge transition (Figure 2 fitted gaussian error function (red)) and the grey values were normalized to that function (min-max grey value). After that, an iterative algorithm tries to get the same area on the left and right side of the Heaviside function shifting itself along a virtual x-axis. If the areas are equal, the width is calculated by the pixel amounts, which are involved getting the areas equal (this is also known as integral criteria of the photometric middle). The following illustration shows the result of applying the proposed method for the evaluation of the wavelength dependent edge transition enlargement. As expected the need of the proposed image correction is confirmed.

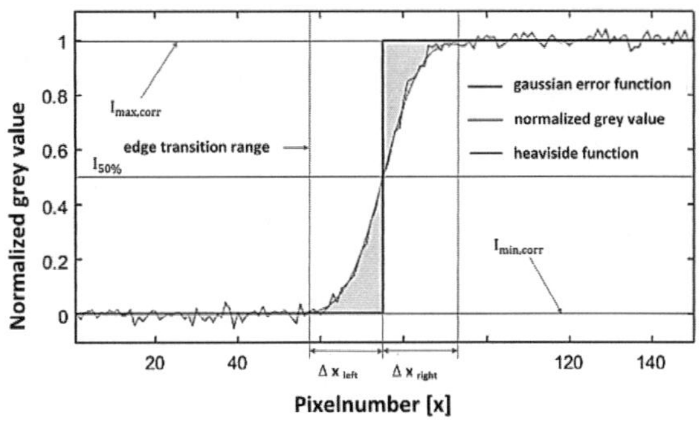

Figure 2: Edge transition width estimation, method for calculating the image blur.

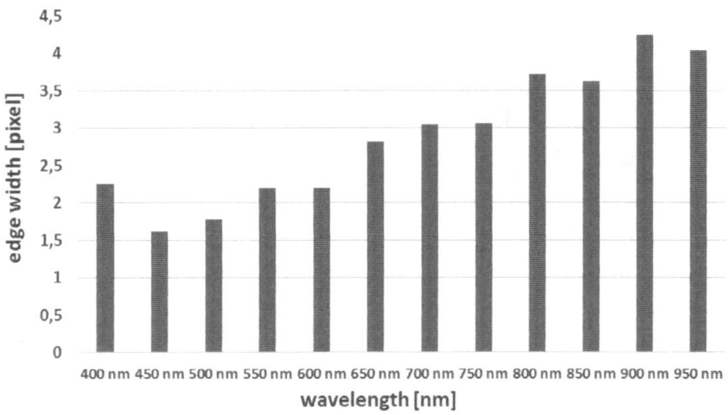

Figure 3: Characteristic edge width behaviour starting in VIS up to the NIR wavelength range, where edge width becomes large due to cross-talk between pixels and the transparency of the semicobdctor for higher wavelengths. The seaminly blured image found at 400nm can be explained by two factors: low photon count on the UV region (light source depentent), which makes necessary higher integration times and with that, higher noise content; and the fact that the Si sensor is less sensitive to this region of the spectrum [1].

3. Proposed image correction models using an estimation of the point spread function

3.1. Direct point spread function measurement

Figure 6 shows the values representing the edge width in the NIR channels. As a comparison, the width in the visible wavelength region is less than 2.5 pixels. To improve the system behaviour in the NIR range, discussed in section two, a deconvolution method was chosen to improve performance in the NIR. First of all, a method to estimate the point spread function is necessary. To overcome this problem two methods were tested. The direct PSF-estimation, which is based on the calculation out of a white noise pattern, does not perform accurate. The reasons behind that can be studied in the quality of the white noise target.

3.2. Point spread function estimation using blind deconvolution

The second approach uses a blind deconvolution models using the expectation-maximization optimization (EM) [14] and the maximum – a – *postori* estimation (MAP) [15]. While the MAP method calculating the PSF and the corrected image simultaneously, the EM method calculates the PSF and the image correction separately. This leads to lower number of parameters, which must be calculated during for the EM PSF estimation. Furthermore, the image restauration algorithm can be chosen independently from the PSF estimation method. After a few investigations in the 900nm infrared channel the EM method delivers more reasonable PSF-functions. The generated PSF using MAP delivers approximately the Dirac function as result. With this behaviour, a performant image restauration cannot be achieved. Furthermore, the lower number of parameters of the EM method leads to a better function performance which leads to the final decision for the EM-processing method.

3.3. Spatial PSF Processing using EM method

The PSF has a spatial dependency. This leads to the necessity of calculating the PSF per the described method within a defined image-block region [16]. With the spatial information of the local PSF distribution inside the sub image blocks, the image is convoluted using the MAP optimization.

Taking into consideration the spatial variation of the PSF, the acquired calibration image divided into several sub-image blocks. By means of the blind deconvolution, a PSF is calculated for each sub-image block (as shown in Figure 4).

Under the assumption that the PSF is piecewise continuous and a soft function, that means, it varies slowly along the x and y coordinates, an algorithm for the stitching and merging of neighbouring PSFs is applied. Figure 4 illustrates how the algorithm works:

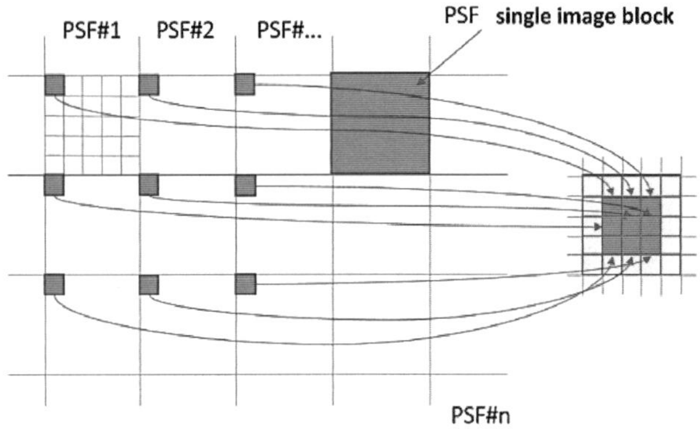

Figure 4: Schematic showing the PSF calculation.

The pixels of the estimated PSFn for different picture blocks are transformed in new images in the following manner. The pixels at the same position within each PSF are arranged to a new block, as shown in Figure 4, where each pixel from the right handed image comes from the calculated PSF for the most upper-left pixel. Each one of these blocks is then filtered with a low-pass filter (Equation 2 shows the core of the filter). This filter reduces the stochastic error between neighbouring blocks.

$$H = \begin{pmatrix} 1 & 2 & 1 \\ 2 & 4 & 2 \\ 1 & 2 & 1 \end{pmatrix}. \tag{2}$$

Filtering ensures a certain uniformity between adjacent PSFs. After this filtering, the pixels are reset to their original positions in the PSFn. Subsequently, a threshold-based filtering of the PSF is performed by setting the values which lie below a specific location-dependent threshold value to zero. Thereby, the noise which is close to the limit of a PSF and and must have a low value is reduced. The thresholding windows are defined as follows:

$$T(n) \equiv \begin{cases} 1 & 0 \leq |n| \leq \alpha\frac{N}{2} \\ 0.5\left[1+\cos\left(\pi\frac{n-\alpha\frac{N}{2}}{2(1-\alpha)\frac{N}{2}}\right)\right] & \alpha\frac{N}{2} \leq |n| \leq \alpha\frac{N}{2} \end{cases}, \tag{3}$$

where N is the size of the window. This Turkey-window is chosen because the width of the top of the window can be chosen by setting the parameter α.
The threshold operation is then performed by:

$$P(x,y) = \begin{cases} 0 & \tilde{P}(x,y) \leq \tilde{P}_{max} \cdot T \\ \check{P}(x,y) & \text{otherwise} \end{cases}, \quad (4)$$

where $\tilde{P}_{max}$ is the maximum value of the unfiltered PSF $\tilde{P}(x,y)$of each block, and $P(x,y)$ are the filtered values. Finally, the PSFn are normalized so that the sum of all PSF values is equal to 1.

3.4. Image restauration using spatial PSF estimation

Regarding the location-dependent PSF estimation, each image is also split into several image blocks and each image block is restored with the corresponding PSF. There are already a variety of methods for image restoration, for example, the Richardson-Lucy method and the Wiener filtering. These methods can allow a certain increase in the sharpness of the image, but cause annular artifacts in the vicinity of the edges. In order to suppress these effects, the method used in [17] is used in this work, which is based on MAP optimization. For the subsequent investigations, the original code is used by the author.

4. Experimental results and Discussion

In order to evaluate the improvement of the sharpness of the restored images, edge detection on five search lines was performed both on the original and restored images. The positions of the search beams are indicated in Figure 5 (the search beam for each edge detection is shown in yellow – right handed image). Point 1 is at the center of the image, and points 2 to 5 are at the four corners of the calibration target.
Figure 6 shows the edges of the original images and the restored images. It shows that the edge width can be significantly reduced by image restoration. The average edge width is reduced from 4.15 pixels to 2.49 pixels, which means that an increased image quality, i.e. sharpness, can be detected.

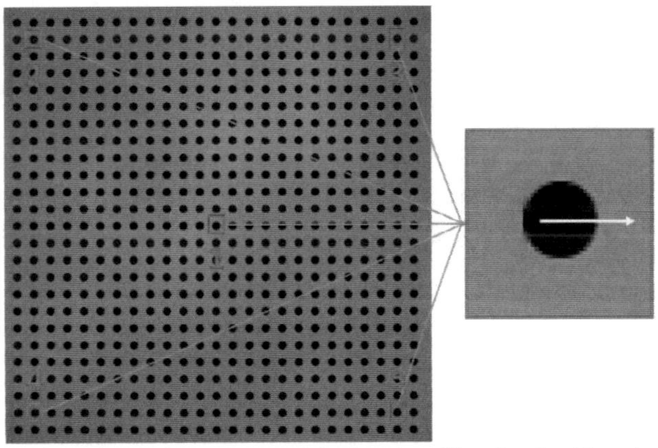

Figure 5: Shows the postion of five observation points used for benchmarking the algorithm. At each point of the calibration target, a search line was placed to measure the edge width for both original and corrected images.

The experimental results demonstrate that the image restauration works in a proper way. An improvement of more than approx. two pixel was achieved in the best case and approx. one and a half pixel in the worst case.

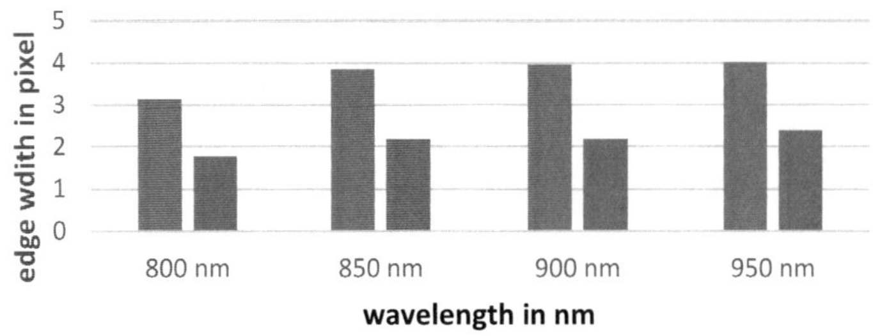

Figure 6: Experimental results using the image restauration algorithm in the NIR original value (blue), corrected value (red).

These results lead to better system performance for measurement tasks, which will be realized with NIR-imaging. Nevertheless, the error on the target depends additionally on the optics that will be used for the application.

5. Acknowledgements

We thank the federal ministry of education and research for the strong support of this work. The work is related to the project *Qualimess next generation* (03IPT709X).

References

[1] M. Rosenberger and R. Celestre, "Smart multispectral imager for industrial applications," *2016 IEEE International Conference on Imaging Systems and Techniques (IST)*, Chania, 2016, pp. 7-12. doi: 10.1109/IST.2016.7738189

[2] Y. Garini, I. T. Young, and G. McNamara, "Spectral imaging: Principles and applications," Cytom. Part A, vol. 69A, no. 8, pp. 735–747, Aug. 2006.

[3] G. A. Shaw and H. K. Burke, "Spectral Imaging for Remote Sensing," Lincoln Lab. J., vol. 14, no. 1, pp. 3–28, 2013.

[4] J. P. Robinson, Z. Darzynkiewicz, R. Hoffman, J. P. Nolan, P. S. Rabinovitch, and S. Watkins, Eds., Current Protocols in Cytometry. Hoboken, NJ, USA: John Wiley & Sons, Inc., 2001.

[5] S. Baronti, A. Casini, F. Lotti, and S. Porcinai, "Multispectral imaging system for the mapping of pigments in works of art by use of principal-component analysis," Appl. Opt., vol. 37, no. 8, pp. 1299–1309, 1998.

[6] R. Hedjam and M. Cheriet, "Historical document image restoration using multispectral imaging system," Pattern Recognit., vol. 46, no. 8, pp. 2297–2312, 2013.

[7] M. Kubik, "Chapter 5 Hyperspectral Imaging: A New Technique for the Non-Invasive Study of Artworks," in Physical Techniques in the study of Art, Archaeology and Cultural Heritage, vol. 2, D. Creagh and D. Bradley, Eds. Elsevier, 2007, pp. 199–259.

[8] M. Rosenberger and G. Linss, "Multispectral Image Correction for Geometric Measurement", in *Journal of Physics: Conference Series 588 (2015)*, p. 012037, 2015.

[9] R. Celestre, M. Rosenberger and G. Notni, "A novel algorithm for bad pixel detection and correction to improve quality and stability of geometric measurements", in *Journal of Physics: Conference Series 772 (2016)*, p. 012002, 2016.

[10] Zhou Fang, Ce Zhou Zhao, "Recent Progress in Silicon Photonics: A Review," *ISRN Optics*, vol. 2012, Article ID 428690, 27 pages, 2012. doi:10.5402/2012/428690

[11] J. B. Lincelles, O. Marcelot, P. Magnan, O. Saint-Pé and M. Bréart de Boisanger, "Enhanced Near-Infrared Response CMOS Image Sensors Using High-Resistivity Substrate: Photodiodes Design Impact on Performances," in *IEEE Transactions on Electron Devices*, vol. 63, no. 1, pp. 120-127, Jan. 2016

[12] G. Alper, "CCD vs. CMOS – Modulation Transfer Function (MTF) in NIR", *in Beyond the datasheet, Adimec blog,* 04/2015 http://info.adimec.com/blogposts/ccd-vs.-cmos-modulation-transfer-function-mtf-in-nir

[13] M Rosenberger, M Preissler, R Fütterer, C Zhang, R Celestre and G Notni, "Multispectral Image Correction for Geometric Measurement", in *Journal of Physics: Conference Series Conference Series 772 (2016) 012054*

[14] Levin, Anat; Weiss, Yair; Durand, Fredo; Freemann, William T.: "Efficient marginal likelihood optimization in blind deconvolution" *Computer Vision and Pattern Recognition (CVPR)*, 2011 IEEE Conference on, Seiten 2657-2664, ISBN 978-1-4577-0394-2.3

[15] Kotera, Jan; Šroubek, Filip; Milanfar, Peyman:"Blind deconvolution using alternating maximum a posteriori estimation with heavy-tailed priors", Computer Analysis of Images and Patterns, Vol. 8048, pages 59-66, 2013.

[16] Brauers, Johannes; Seiler, Claude; Aach, Til: "Direct PSF Estimation Using a Random Noise Target", *Proceedings of SPIE 7537*, Digital Photography VI, 75370B, 2010.

[17] Levin, Anat; Weiss, Yair; Durand, Fredo; Freemann, William T.: " Efficient marginal likelihood optimization in blind deconvolution ", *Lecture notes, The Weizmann Institute of Science, 2010*

IOP Conf. Series: Journal of Physics: Conf. Series **1044** (2018) 012040 doi:10.1088/1742-6596/1044/1/012040

Irritation-free pattern projection system for accurate 3D face and body scans

G Notni[1,2], C Bräuer-Burchardt[2], S Heist[2], A Brahm[2], P Kühmstedt[2]
[1]Technical University Ilmenau, Ehrenbergstraße 29, D-98693 Ilmenau, Germany
[2]Fraunhofer IOF Jena, Albert-Einstein-Str. 7, D-07745 Jena, Germany

E-mail: gunther.notni@iof.fraunhofer.de

Abstract. Three dimensional scanning of human bodies or parts of person become more and more a matter of interest especially in medical applications. In this paper a new 3D scanning system is introduced which is absolutely irritation-free based on the structured light illumination by aperiodic sinusoidal fringe patterns in the near infrared range. Hence, it is particularly suitable for continuing human face scanning. The system is based on optical stereo capturing and structured light illumination using the GOBO principle. The measurement principle of the system is explained and examples of face and body scans are given. The measurement accuracy was evaluated by experiments and will be presented and discussed as well as an outlook to future work.

1. Introduction

Three dimensional acquisition of the human face and body is a growing task in medicine, animation, virtual life, and computer games and entertainment. Gestures and facial expressions can be used for communication tasks or machine control. Although gestures and facial expressions can be detected in 2D image sequences yet, 3D data may provide more detailed information which for example can help to increase the power of gesture and expression analysis systems.

Optical scanners based on structured light projection technique are increasingly used for gesture and facial expression detection and recording [1, 2]. This technique provides the advantage of relative high spatial resolution and measurement accuracy. Recently high-speed camera technique provides 3D frame rates up to 1 kHz [3, 4]. The basis for many different fields of application of 3D person scanning is the technological development of hardware and software components for suitable 3D scanning systems. The scanning technology may depend on the fact, whether static or moving objects has to be scanned.

The 3D scanning of moving persons was strongly shifted into the focus of recent research. The 3D scanning of static objects or non-moving persons has its history mainly in medical applications such as computer-assisted tomography (CAT) [5] and nuclear magnetic resonance imaging (NMRI) [6].

An important requirement to 3D scanning systems for a permanent acquisition of human faces is the property of irritation freedom, i.e. the scanning process must not disturb the scanned person. This could be achieved using passive multi-camera systems. However, structured light illumination provides a very higher degree of object resolution and measurement accuracy. The approach for irritation free scanning using structured light illumination is using wavelengths at near infrared (NIR).

[1] To whom any correspondence should be addressed.

IOP Conf. Series: Journal of Physics: Conf. Series **1044** (2018) 012040 doi:10.1088/1742-6596/1044/1/012040

2. The pattern projection system

Fringe projection profilometry is one of the established methods of structured light projection (see e.g. [7]), but other structured illumination patterns (binary dot patterns, speckle patterns, aperiodic sinusoidal patterns, see [8]) are also suitable for this technique. This measurement principle works well using wavelengths over 780 nm. However, illumination sources should be selected accordingly or spectral filters must be used in the case of projected pattern generation. The recording process by one or more cameras should be optimized according to the used wavelengths.

The pattern projection system is based on the measurement principle of stereo observation and structured light illumination. It consists of a projection and an observation unit. The observation is done by a stereo-camera arrangement with blocking filters for visible light frequencies. The projection unit projects structured light patterns onto the measurement object which help to identify corresponding points in both cameras.

From the several possibilities for the used pattern for the structured illumination we applied aperiodic sinusoidal fringe patterns [3, 4]. The generation of the pattern sequence (n patterns) was realized by a so-called GOBO projector, that performs the projection by a rotating wheel with radially arranged fringes, illuminated by a high power LED lamp (see figure 1).

The GOBO principle is particularly known in the theatre-industry and means to place a partly transparent slide with a certain pattern is placed before a light source. The name GOBO is derived from "GOes Before Optics", describing that the pattern-depending slide object is positioned between the light source and the lenses. Usually, a GOBO projector contains a light source (e.g. a gas-filled tube or halogen lamp), a light collector (ellipsoidal reflector or condenser), a slide (e.g., made of steel or glass), and an imaging optical system (with fixed focal length or zoom lens) as optical components. For more details concerning the GOBO principle see [3].

The high power LED at 850 nm emission wavelength was used and homogenized with a light funnel made of mirrors with a protected gold coating. The aperiodic sinusoidal patterns were generated using a glass GOBO wheel (according to Heist et al. [3]) on which the pattern was made of a chrome coating. Thus, only the near-infrared radiation is projected into the measurement plane for an irritation-free measurement. Two CMOS cameras, each with a maximum resolution of 4 MPx were arranged in stereo-vision setup with a baseline of approximately 300 mm. For the working frame rate of 360 Hz, a resolution of only 1Mpx (in the image centre) is used. Exposure time is typically 2.7 ms.

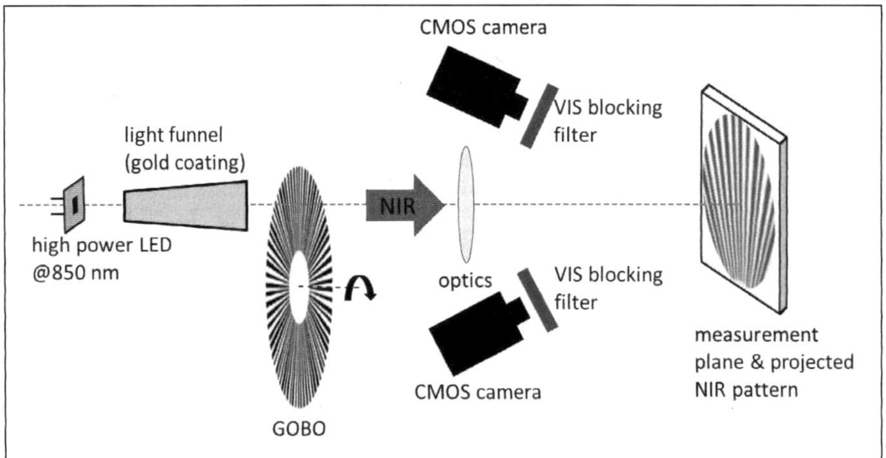

Figure 1. Principle of the irritation-free pattern projection system

To avoid overexposure and disruptive effects from the visible wavelength range, both cameras were equipped with an additional band-pass filter with cut-on and –off wavelengths of 830 nm and 870 nm, respectively. The working distance of the system can be scaled from 0.5 m to 1.5 m in which the measurement area varied from 0.3 x 0.3 m^2 to about 1.0 x 1.0 m^2, depending on the optics. The image acquisition was done by a frame grabber via Camera Link connection.

In order to get colour information additionally to the 3D data a colour camera is included in the system. The additional colour camera also has a resolution of 4MPx and can capture images with frame rates up to 90 Hz. Colour data are recorded with a resolution of 4 MPx with lower frame rate than the measurement cameras and transferred via USB 3.0.

The connection between the calculated 3D data and the colour information is realized by a 2D-3D mapping. The corresponding transformation matrix is obtained by inclusion of the colour camera into the calibration process. The complete 3D scanning system is shown in figure 2.

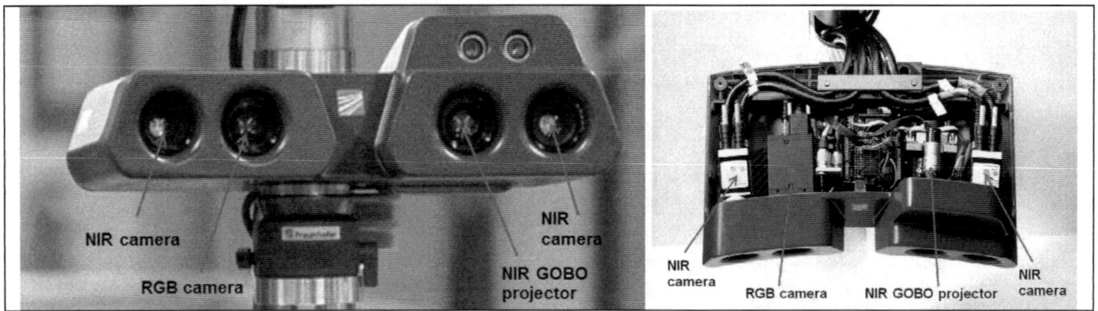

Figure 2. Complete Scanner system, front view(left and view from above (right)

The recorded data are processed to 3D data very quickly using a GPU-processing procedure (NVIDIA Geforce 1070 GTX) and the resulting images are displayed on a monitor in real-time. The latency from image recording to 3D point-calculation is only a few more than 100 ms. Furthermore, displaying the data on the screen needs some more time, depending on the hardware properties.

3. Measurement system evaluation

In order to evaluate the accuracy of the obtained 3D data the 3D scanner was used to perform measurements with certain specimens with known geometric properties. A flatness standard made of ceramic with a size of 300 x 50 x 40 mm³, a ball bar with length between the sphere centre-points of about 200 mm and a sphere with a diameter of 50.1 mm were used as specimens (see figure 3).

In order to quantify the measurement accuracy the characteristic parameters flatness measurement error ΔF, length measurement error ΔL, and probing error P_s were determined similar to the VDI/VDE guidelines [9] concerning optical measuring systems. The measurement volume was set to a size of approximately 0.5 x 0.5 x 0.5 m³. Sequences of either six or ten aperiodic sinusoidal fringe patterns were used. For the determination of each of the characteristic parameters at least six measurements were performed. The characteristic parameters were determined similar to the VDI/VDE guidelines with the supplement of complete noise removal at the determination of the flatness parameter ΔF by subtraction of six times standard deviation σ in order to exclude the influence of noise.

The parameters were defined as follows. Flatness measurement error is the maximum value of flatness deviation ΔF_i at n certain positions i of the flatness standard in the measurement volume.

$$\Delta F = \max_{i=1,\ldots,n} \{\Delta F_i\} \quad \text{with} \quad \Delta F_i = d_i^{\max} - d_i^{\min} - 6 \cdot \sigma_i \tag{1}$$

Here, d_i^{max} is the highest positive and d_i^{min} is the highest negative Euclidean distance of the 3D points from a fitted plane and σ_i are the standard deviations of the Euclidean distances (noise).

Length measurement error is defined as the span of the length measurements of the ball bar:

$$\Delta L = L_{\max} - L_{\min}, \tag{2}$$

where L_{max} is the largest and L_{min} is the smallest of the measured distances between the centre-points of the spheres of the ball bar.

Probing error P_s is defined as the span of the radius measurements of the sphere:

$$P_s = R_{\max} - R_{\min}, \tag{3}$$

where R_{max} is the largest and R_{min} is the smallest of the measured radii of the sphere.

The measurements were performed using both six-image and ten-image sequences. The following characteristic parameters were obtained: ΔF = 0.6 mm, ΔL = 2.0 mm, and P_s = 0.25 mm for the six-image and ΔF = 0.6 mm, ΔL = 1.3 mm, and P_s = 0.08 mm for the ten-image sequences.

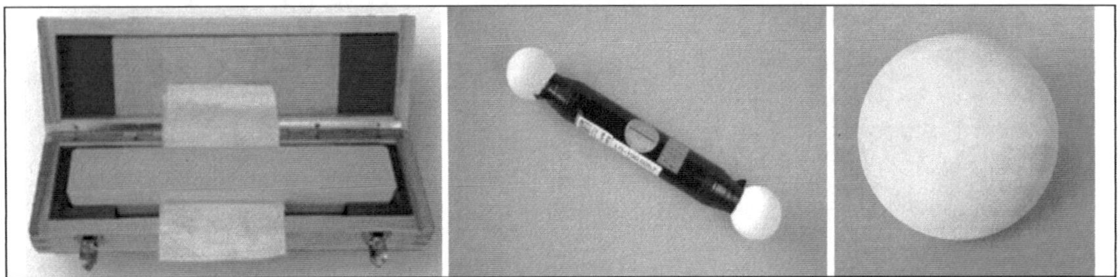

Figure 3. Specimens with known calibrated dimensions for evaluation measurements: flatness standard (left), ball bar (middle), and sphere (right)

Additionally, the noise of the 3D measurement points was defined as the standard deviation σ of the Euclidean distances of the calculated 3D points from a fitted plane at the measurements of the flatness standard. The mean noise value was σ = 0.11 mm (n=6) and σ = 0.09 mm (n=10). These values showing the good measurement accuracy compared to 3D scanning systems in the visible light range and/or single shot systems.

4. Measurement examples

Figures 4 and 5 show some examples of the measurements performed by the developed system. The first example shows a female face showing some emotional expression. A subsequently applied software module can detect certain facial points. Figure 4 shows a false colour representation of the 3D data and figure 5 shows the colour mapping of the data obtained by the colour camera onto the 3D data together with certain detected facial points. The colour mapping onto the 3D data allows a representation from an arbitrarily chosen viewing direction onto the face.

The 3D models obtained of these measurements can be used for detection of emotions based on the 3D face data.

The scanner was also used for quantification of lip-movements in order to support lip-reading [2]. Figure 6 shows an example of the determined trajectories in the three different spatial directions. Note, that lip movement in z-direction cannot be captured using only 2D image sequences.

It could be stated, that using 3D information about 25% more information can be obtained for recognition of the speech than using only 2D image sequences. More detailed information is given by the work of Brahm et al. [2].

5. Discussion and outlook

The measurement results show that the developed scanning system can acquire robust 3D face images of moving persons, obtained with a very short latency time of little more than 100 ms by processing of image data of 2 x 1 MPx recorded at 360 Hz. Consequently, a 3D rate of more than 30 Hz is obtained and displayed on the monitor.

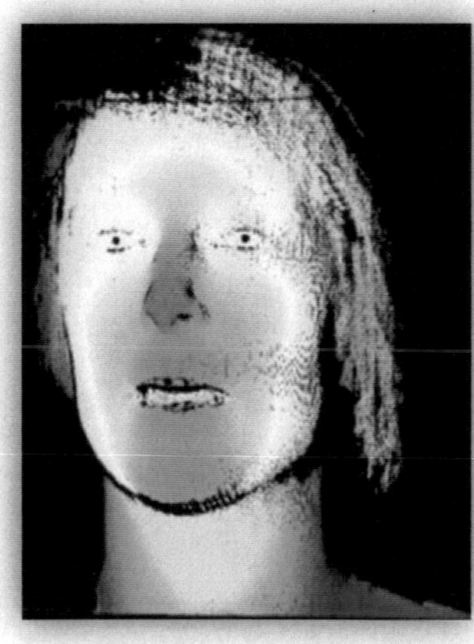

Figure 4. False colour representation of the calculated 3D data

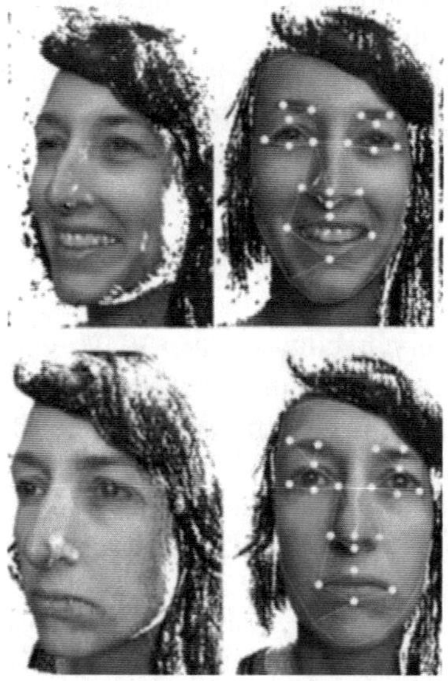

Figure 5. Colour mapping onto the 3D data and recognized facial points

The projection system is suitable for fast irritation-free 3D detection of moving persons, the acquisition of faces, and body parts with very short latency time. As the accuracy measurements show the system is not designed for high precision measurements of rigid body parts in the lower µm-range. In order to set the focus to highest measurement accuracy a pattern sequence with n≥10 can be used (see e.g. [4]).

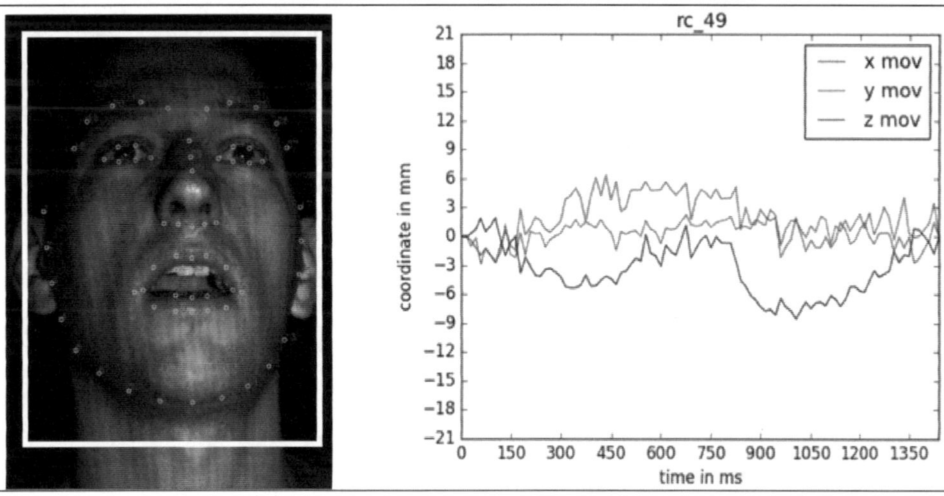

Figure 6. Application of the scanner for lip-reading. Detected facial points (left), detected lip-movement in x-,y-, and z-direction (right)

However, if maximum available 3D rate is not necessary and the actual movements are slow, noise can be reduced and robustness can be improved by using a higher number of images per sequence.

Here, a compromise should be found depending on the requirement of the measurement task and the power features of the scanning system.

Future work will be addressed to the further development of 3D processing algorithms, e.g. in order to increase robustness, suppress noise, and decrease the number of outliers (namely false positive detected 3D points). This may for example include optimization of the used aperiodic sinusoidal fringe pattern. For more details concerning these patterns see the work of Heist at al. [4]. Additionally, an appropriate analysis of the origins of outlier production should be performed initially.

Future work will also include the interpretation of facial expressions and body movements like gestures.

6. Conclusion

A new irritation-free pattern projection system for accurate 3D face and body scans was presented in this paper. The suitability of this 3D scanning system for acquisition of the expressions of facial emotions and the support of automated lip-reading could be shown.

The high possible 3D frame rate of 36 Hz with high resolution ($\sigma \approx 100\mu m$) makes the systems convenient for high-speed and high resolution 3D-face- and body-scans, e.g. at the perpetual 3D acquisition of body parts of moving humans.

References

[1] Beumier C 1006 3D face recognition IEEE 2006 pp 369-374

[2] Brahm A, Ramm R, Heist S, Rulff C, Kühmstedt P and Notni G 2017 Fast 3D NIR systems for facial measurement and lip-reading Proc of SPIE Vol. **10220** pp 102200P-1-9

[3] Heist S, Lutzke P, Schmidt I, Dietrich P, Kühmstedt P and Notni G 2016 High-speed three-dimensional shape measurement using GOBO projection Optics and Lasers In Engineering **87** pp. 90–96

[4] Heist S, Mann A, Kühmstedt P and Notni G 2014 Array-projection of aperiodic sinusoidal fringes for high-speed three-dimensional shape measurement Opt. Eng. **53** (11) pp 112208-1-12

[5] Brooks R A and Di Chiro G 1976 Principles of computer assisted tomography (CAT) in radiographic and radioisotopic imaging Physics in Medicine and Biology **21**, No 5 pp 689-732

[6] Rothwell. W P 1985 Nuclear magnetic resonance imaging Applied Optics Vol. **24**(23) pp 3958-3968

[7] Salvi J, Fernandez S, Pribanic T and Llado X 2010 A state of the art in structured light patterns for surface profilometry Pattern Recognition **43** pp 2666–2680

[8] Zuo C, Chen Q, Gu G, Feng S, Feng F, Li R and Shen G 2013 High-speed three-dimensional shape measurement for dynamic scenes using bi-frequency tripolar pulse-width-modulation fringe projection Optics and Lasers in Engineering **51** pp 953–960

[9] VDI/VDE 2634 2008 Optical 3D-measuring systems VDI/VDE guidelines, Part 2

IOP Conf. Series: Journal of Physics: Conf. Series **1044** (2018) 012041 doi:10.1088/1742-6596/1044/1/012041

Measurement system for parameters used in determining and monitoring the crops vegetation status

M C Luculescu[1], L Cristea, S C Zamfira and A Boer
29 Eroilor Blvd, Transilvania University of Brasov, 500036, Romania

E-mail: lucmar@unitbv.ro

Abstract. The paper presents a data acquisition system designed for measuring parameters involved in determining and monitoring the crops vegetation status, such as: plant height, temperature on plant leaves, environment temperature and humidity and spectral response at the incident sunlight. The system is mounted on a tractor and the measurements are made "on the go". Using the spectral response for different wavelengths, vegetation indices can be computed. They can offer important information about the health status of the crop, moisture stress, pest and diseases, necessary nutrients and so on. All data are correlated with the GPS coordinates in order to generate favourability and risk maps using a special designed software that imports data, processes them in order to be proper stored in the database, computes different vegetation indices and display corresponding maps.

1. Introduction

Vegetation status depends on different type of factors: climate characterized by solar radiation, precipitation, wind etc.; chemical nutrients as N-K-P Nitrogen-Phosphorous-Potassium complexes; soil chemistry; growing cycles; pests; diseases and so on. If we are referring at crops status, them health directly influences the productivity that is why we are interested in measuring and monitoring the entire development cycle.

The health vegetation status can be visually observed, but many times it is too late to intervene if something goes wrong. It is very important to early find out if the health status is threatened. We need for this a "special eye" that sees more than the biological one. Smart sensors are the solution for doing this.

Information that can be used for determining the health status can be measured using two categories of sensors: with contact and without contact [1].

Remote sensing is now one of the top subjects in investigating large areas vegetation status [2]. Satellites, aircrafts, Unmanned Aerial Vehicles (UAV) are used for taking photos with different resolutions, in different wavelength domains (VIS – visible, NIR – Near InfraRed, SWIR – Short Wave InfraRed and so on) [3-5].

When the solar radiation meets a surface, it can be absorbed, reflected or transmitted at a certain level. Each surface will be affected differently by sunlight, regarding the three actions mentioned before. In this way, each material will have its own footprint associated with the reflectance spectrum. This principle allows to identify vegetation areas and to determine the health or development of the

[1] Corresponding author.

vegetation. The leaf reflectance depends on the inside pigments, structure of the cells, content of water and so on and all these factors can be identified in different areas of the spectrum [6].

Depending on the number of wavelengths the reflectance is measured at, we discuss about multispectral or hyperspectral sensors. These values are used to compute the so called *vegetation indices*, very useful data for precision agriculture, that offer information about the health status of a crop, moisture stress, pest and diseases, necessary nutrients etc. [7].

2. Hardware structure

For measuring parameters involved in determining and monitoring the crop vegetation status a data acquisition system was designed and developed. It contains sensors for measuring plant height, temperature on plant leaves, environment temperature and humidity and spectral response at the incident sunlight – figure 1.

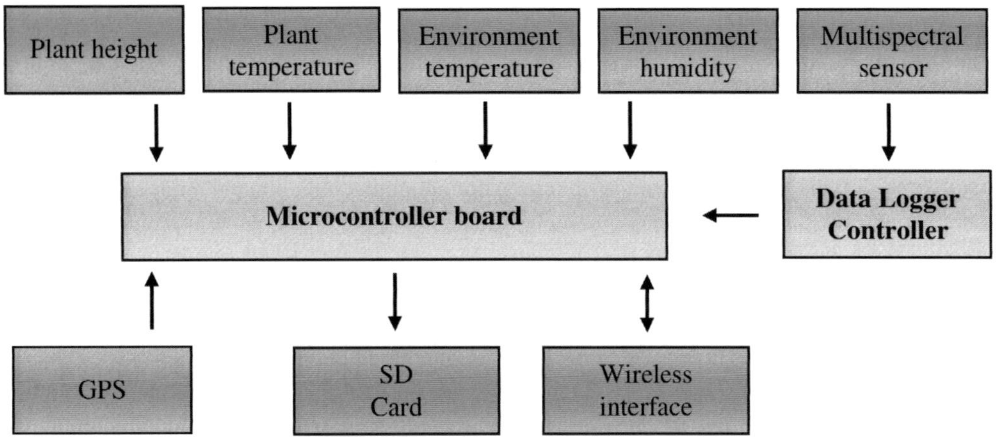

Figure 1. Hardware structure of the data acquisition system.

For plant height sensor an Ultrasound sensor MaxSonar® from MaxBotix™ was used, the temperature of plants is measured using an infrared sensor. Multispectral sensor is a Cropscan with it specific DLC – Data Logger Controller. Data from DLC and GPS are received by the microcontroller board through serial interfaces, while the wireless one is an Xbee module that communicates with a server. Data can also be stored on a SD Card.

3. Software aspects

The software consists in two programs: one stored in the microcontroller memory that manages the data acquisition process from sensors, stores data on the SD card or transmit them to the server; the other one stored on server receives data or import them from the SD card, processes data in order to store them in a database with a specific structure, computes vegetation indices, and displays georeferenced data on different type of maps.

3.1. Messages structure

In the data acquisition process, values are written as messages to a file on the SD card or are transmitted to the server. The message structure for each sensor respects the following format:

<Sensor_ID>;<Header>;<Data>;<Footer>

where

- *<Sensor_ID>* is the ID defined for the infrared plant temperature sensor (IRT), plant height ultrasonic sensor (SNR), Cropscan multispectral sensor (CRP) and GPS sensor (GPS);

- *<Header>* includes date and time using *YYYYMMDDhhmmssxx* format, where we denoted *YYYY* - year, *MM* - month, *DD* - day, *hh* – hour, *mm* - minute, *ss* - second, *xx* - hundredth of second;
- *<Data>* represent the acquired values, specific to the sensor type. Multiple values are comma separated *[val₁],[val₂], ..., [valₙ]*;
- *<Footer>* contains a similar structure as the header.

Headers contain the exact moments of the acquisition start, while footer contain the moments when the acquisition stops. Information are obtained from a Real Time Clock module (RTC) included in the system. Headers and footers are automatically added by software and are used later in the synchronization and interpolation process by the server routines. They are needed because sensors have different sampling rates, while we need the shortest total acquisition time.

3.2. Acquired and processed data

A part of a sample input file containing acquired data is presented in figure 2.

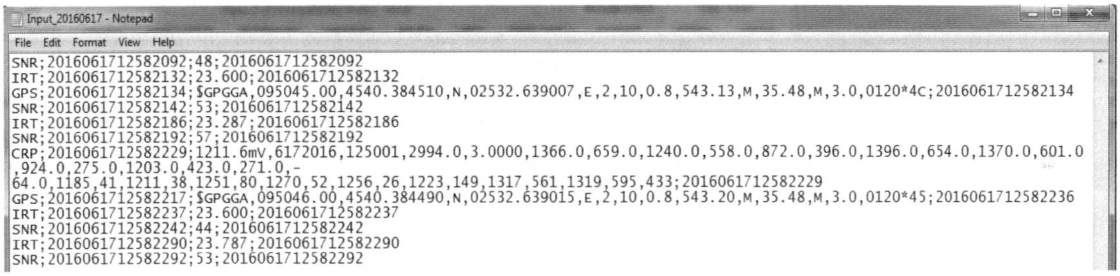

```
SNR;2016061712582092;48;2016061712582092
IRT;2016061712582132;23.600;2016061712582132
GPS;2016061712582134;$GPGGA,095045.00,4540.384510,N,02532.639007,E,2,10,0.8,543.13,M,35.48,M,3.0,0120*4C;2016061712582134
SNR;2016061712582142;53;2016061712582142
IRT;2016061712582186;23.287;2016061712582186
SNR;2016061712582192;57;2016061712582192
CRP;2016061712582229;1211.6mV,6172016,125001,2994.0,3.0000,1366.0,659.0,1240.0,558.0,872.0,396.0,1396.0,654.0,1370.0,601.0
,924.0,275.0,1203.0,423.0,271.0,-
64.0,1185,41,1211,38,1251,80,1270,52,1256,26,1223,149,1317,561,1319,595,433;2016061712582229
GPS;2016061712582217;$GPGGA,095046.00,4540.384490,N,02532.639015,E,2,10,0.8,543.20,M,35.48,M,3.0,0120*45;2016061712582236
IRT;2016061712582237;23.600;2016061712582237
SNR;2016061712582242;44;2016061712582242
IRT;2016061712582290;23.787;2016061712582290
SNR;2016061712582292;53;2016061712582292
```

Figure 2. Acquired data structured by messages.

It can be observed the structure discussed in previous section.

The program on the server processes the input data taking into account the headers information so that to be stored in a database. Each record will contain the following fields:

- Time – obtained from headers processing;
- Latitude, Longitude – obtained from GPS sensor;
- Snr - plant height obtained from the ultrasound sensor;
- Irt – plant temperature from the infrared sensor;
- Reflectance values computed from data obtained from the sixteen channels of the Cropscan sensor;
- Some computed vegetation indices

A part of some records from the database is presented in figure 3.

Time	Lat	Long	Snr	Irt	860nm	900nm	970nm	1050nm	1100nm	1200nm	1300nm	1500nm
2016061712582134	45.6730752	25.5439835	52.20	23.5884	45.1530	42.7053	42.0555	44.3609	44.1991	38.0858	39.8193	20.0301
2016061712582217	45.6730748	25.5439836	55.50	23.4773	45.1530	42.7053	42.0555	44.3609	44.1991	38.0858	39.8193	20.0301
2016061712582418	45.6730695	25.5439917	50.45	24.2993	45.3540	42.6324	43.1029	44.0128	44.7576	37.8093	40.8021	22.3852
2016061712582517	45.6730648	25.5439970	59.68	23.9483	43.3304	41.0243	42.1357	42.8380	43.3739	37.2594	41.0963	22.8140
2016061712582617	45.6730598	25.5440042	62.80	23.8189	40.9219	38.9013	39.7525	41.1131	41.6826	36.1153	39.8089	22.4072
2016061712582817	45.6730503	25.5440178	60.24	24.3342	41.6824	38.8160	39.8823	40.8907	42.5549	36.4478	39.8760	21.7904

Figure 3. Processed data are stored in a database.

Based on these information, the software can display georeferenced data on maps. Each field with data from sensors can be selected to be displayed on a map and also computed vegetation indices can be displayed.

4. Results

Data acquisition system was tested on a potato experimental field divided in areas that contains two variety of potato (S1 and S2), irrigated or not (B2 and B1), and having different amounts of fertilizer (D1, D2 and D3), presented in figure 4.

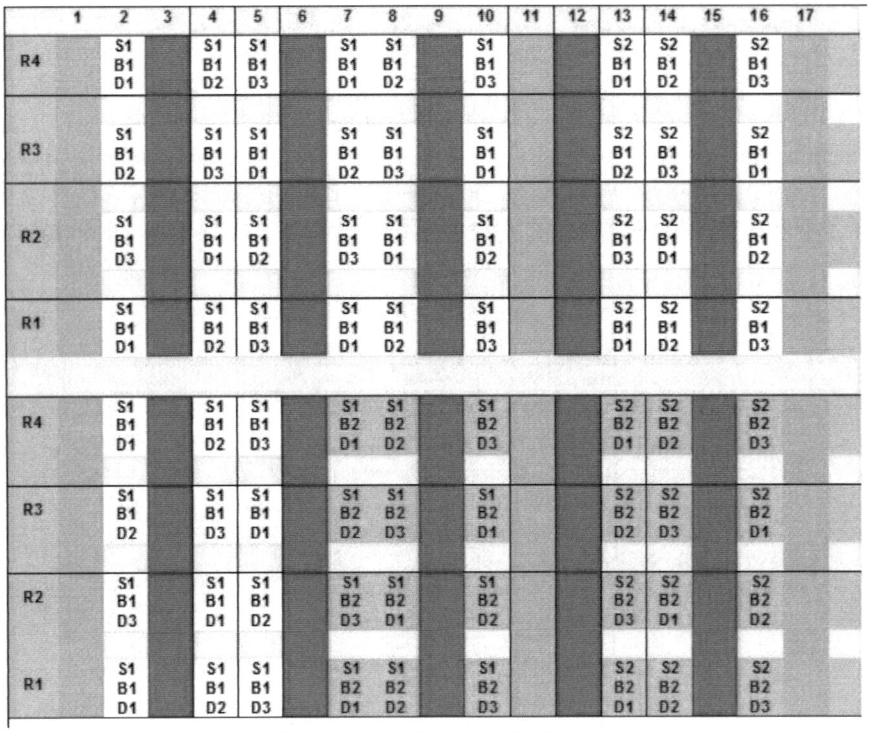

Figure 4. Experimental potato field structure.

The field was prepared by the researchers from the National Institute of Research and Development for Potato and Sugar Beet Brasov Romania so that to create different conditions for the vegetation status and to identify and monitor it using vegetation indices.

Using different quantities of fertilizer, it will be emphasized the need of nutrients. Irrigating only some areas, the plant water stress can be observed.

With red colour are represented technological paths (columns 3, 6, 9, 12, 15) followed by the tractor in the data acquisition process. With R1 to R4 are denoted the parameters combination rows. The same group of combinations were arranged in the lower and upper parts of the experimental field. Columns 1, 11 and 17 (in green colour) are used for potato varieties separation. In the lower part of the experimental field some areas are irrigated (blue colour – columns 7, 8, 10, 13, 14, 16).

In the beginning the data acquisition system was tested in a preliminary version without environment temperature and humidity sensors and also without the wireless interface. Two persons made static measurements, meaning they went to the experimental field with the equipment and they acquired data step by step from different points, situated at different distances in different areas. Data were stored on the SD card and after the entire acquisition process was finished, they were imported in the server.

Images in figure 5 present the experimental potato field (a), different areas marking (b), preliminary measurement tests (c) and an aerial view of the field obtained by a drone from 30 m altitude representing the NDVI - Normalized Difference Vegetation Index distribution (d). The acquired data were processed (synchronized, linear interpolated, computed reflectance and some vegetation indices) and stored in the database.

a. b. c. d.

Figure 5. Experimental potato field and preliminary measurement tests.

The multispectral sensor can detect the reflectance for sixteen wavelengths. O part of them are presented in table 1. Due to the paper space limit, reflectance values for the other 9 wavelengths and vegetation indices are not presented in another table.

Table 1. Georeferenced data obtained in the measurement process.

Time (yyyymmddhhmmssxx)	Latitude	Longitude	Plant height (cm)	Plant temp (°C)	Reflectance (%)						
					860nm	900nm	970nm	1050nm	1100nm	1200nm	1300nm
2016061712583017	45.6730442	25.5440271	46.00	23.4651	40.4044	38.0281	37.2654	38.7700	40.8212	34.0318	37.4393
2016061712583117	45.6730399	25.5440334	42.24	23.1708	51.5724	48.3095	47.0795	48.2698	50.7474	39.9054	44.1948
2016061712583217	45.6730345	25.5440389	50.28	23.4383	57.8486	54.6086	51.8646	53.7519	56.2666	42.7050	47.5778
2016061712583316	45.6730305	25.5440450	45.37	23.7791	57.8714	55.1065	50.6479	53.7455	56.3452	41.8552	46.7587
2016061712583517	45.6730240	25.5440553	44.76	23.2186	54.8101	51.3251	47.3942	50.9378	52.5533	40.1102	43.0353
2016061712583717	45.6730168	25.5440638	47.86	24.6922	40.5008	39.8610	37.3583	40.8112	40.3825	32.9648	36.6603
2016061712583817	45.6730130	25.5440678	49.60	22.6867	45.7191	44.3938	42.6143	45.1241	45.4190	37.0123	40.8111
2016061712583917	45.6730099	25.5440729	47.28	22.7116	53.3697	50.6017	48.8237	50.7524	52.4826	41.6550	45.8864
2016061712584017	45.6730063	25.5440779	74.90	22.9726	55.3507	52.2715	49.8117	51.8137	54.6981	42.0745	46.7753

The software on the server can compute the most used vegetation indices (NDVI - Normalized Difference Vegetation Index that measure the plant "greenness" or photosynthetic activity; LAI - Leaf Area Index; MCARI - Modified Chlorophyll Absorption Ratio Index; WBI - Water Band Index, WMI -Water Moisture Index etc.) and generates maps for spatial distribution of them.

Data were also analysed using ArcGIS software. Figure 6 presents the NDVI in the points corresponding to the acquired data (a) and the interpolated image of NDVI on the interest area (b).

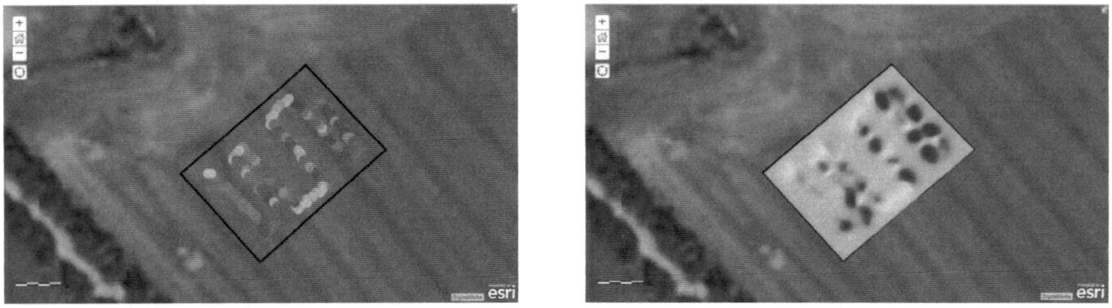

a. b.

Figure 6. Acquired data analysed with ArcGIS software.

5. Conclusions

A data acquisition system was designed and developed for measuring parameters involved in determining and monitoring the crops vegetation status. Data regarding plant height, temperature on plant leaves, environment temperature and humidity and spectral response at the incident sunlight are acquired and processed. Different vegetation indices are computed by a special designed software, offering information about the health status of the crop, moisture stress, pest and diseases, necessary nutrients and so on. Data are correlated with the GPS coordinates so that favourability and risk maps to be displayed. Tests were made on an experimental field with potato crop. Data were analysed also with ArcGIS software and the results confirmed the correlation between vegetation indices and crop vegetation status.

References

[1] Ruiz-Altisent M, Ruiz-Garcia L, Moreda G P, Renfu Lu, Hernandez-Sanchez N, Correa E C, Diezma B, Nicolaï B, García-Ramos J 2010 Sensors for product characterization and quality of specialty crops - A review, *Comput Electron Agr* **74** (2) 176-194.

[2] Gouache D. 2016 Remote sensing for crop cultivation: From research to industry *Sensing & Measurement* SPIE Newsroom

[3] Geipel J, Link J, Wirwahn J A and Claupein W 2016 A Programmable Aerial Multispectral Camera System for In-Season Crop Biomass and Nitrogen Content Estimation *Agriculture*, 6, 4.

[4] Hunt E R, Daughtry C S T, Mirsky S B and Hively W D 2013 Remote sensing with unmanned aircraft systems for precision agriculture applications. *Proc. 2nd Intl. Conf. on Agro-Geoinformatics* (*Fairfax, Virginia, 12-16 August 2013*) pp131-134

[5] Hunt E R et al 2005 High-Resolution Multispectral Digital Photography using Unmanned Airborne Vehicles, *20th Biennial Workshop on Aerial Photography, Ideography, and High Resolution Digital Imagery for Resource Assessment*, (4-6 October)

[6] Ashraf M A, Maah M J and Yusoff I 2001 *Introduction to Remote Sensing of Biomass*. In: Biomass and Remote Sensing of Biomass (InTech), chapter 8 pp 129-170.

[7] Huete A R, Liu H Q, Batchily K and van Leeuwen W 1997 A Comparison of Vegetation Indices Over a Global Set of TM Images for EOS-MODIS, *Remote Sens. Environ.* **59**(3) 440-451

Acknowledgments

This paper was realized within the Partnerships Programme in priority domains-PN-II, which runs with the financial support of MEN-UEFISCDI, Project no. PN-II-PT-PCCA-2013-4-1629 and it was also supported by a grant of the Romanian National Authority for Scientific Research and Innovation, CNCS/CCCDI – UEFISCDI, project number PN-III-P2-2.1-BG-2016-0132, within PNCDI III.

IOP Conf. Series: Journal of Physics: Conf. Series **1044** (2018) 012042 doi:10.1088/1742-6596/1044/1/012042

Measuring the performance of Science and Technology Parks: a proposal of a multidimensional model

R M Lyra[1] and M F L Almeida[1]
[1]Pontifical Catholic University of Rio de Janeiro - PUC-Rio, Postgraduate Programme in Metrology, Rua Marquês de São Vicente, 225 - 22451-900, Rio de Janeiro - RJ, Brazil

E-mail: fatima.ludovico@puc-rio.br

Abstract - This paper aims to propose a multidimensional model for monitoring and evaluating (ME) the performance of similar groups of Science and Technology Parks (STPs), instead of having a generic ME system. The development of this model considered STPs' business models and value propositions, their objectives and goals, development stage, stakeholder's commitment, and respective legal and governance structures. Firstly, we put into perspective the complexity and multidimensionality of measuring and evaluating performance of STPs. In sequence, we review the literature associated with STPs' performance assessment, covering the period of 1996-2016 and focusing on the dimensions/variables and methods that have been considered in previous works. Then, as a core part of this paper, we present the measurement framework itself describing the building blocks it is composed. Aiming to demonstrate the applicability of this model in the context of different science and technology parks in Brazil, an empirical study, focusing on the planning phase of the model, was carried out during 2016 with 26 participants of the 1st Course for Managers of Parks and Environments of Innovation, offered by the Brazilian Association of Science Parks and Business Incubators (Anprotec). The main contributions of the research are a flexible model for measuring and evaluating the performance of technological parks and sets of indicators and metrics associated with different types of parks.

1. Introduction

Science and Technology Parks (STPs) play a crucial role in creating an eco-system to promote innovation, new businesses at the local level, transfer of technologies, cooperation with the industry, impacting the development of knowledge economy. However, STPs' performance is a particularly problematical measure given the different objectives of the various types of science parks and incubators. There is a need to consider the interaction between objectives and the nature of performance. Additionally, there is a need to undertake further explication on the nature of governance and incentives for science parks and incubators [1].

In fact, a STP model can be considered a complex interaction of multiple factors of value, which enhance the park's capacity to achieve their goals and objectives. The main factors include governance and organization; knowledge creation, due to close links to universities, research institutes, public and private firms; value added services; attractive environment for tenant companies; creation of start-ups, spin-offs from universities or independent inventors, or also from resident companies; territorial influence in regional development public policies; and networking [2; 3; 4].

These aspects reinforce the need for a structural contingency perspective that associates the different types of STPs with diverse institutional circumstances and roles [5]. Thus, we argue that a conceptual model which generates measurement frameworks customized to groups of STPs – with common goals and characteristics – should be in best interest of managers and policy makers. The previous discussions lead to the following central research question, which will guide the whole development of an empirical study covering 26 STPs in Brazil:

"How to measure the performance of Science and Technology Parks (STPs), considering their business models and value propositions, objectives and goals, development stage, stakeholders' commitment, and respective governance structures?".

As an attempt to meet this challenge, this paper proposes a multidimensional measurement framework for monitoring and evaluating (ME) the performance of Science and Technology Parks' similar groups, instead of having a generic system.

2. Theorethical background

A systematic literature search was performed on peer-reviewed articles that were published between January 1996 and July 2016. Table 1 summarizes the measurement models and systems reviewed, highlighting the methods and the analytical grids adopted in these previous works [6; 7; 8; 9; 10; 11].

Table 1: Measurement models and systems for STPs' performance assessment

Authors	**Methodological approach**	**Evaluation method**	**Dimensions/ Variables**
Guy (1996)	*Ex-ante, mid-term, ex-post.*	Case study. Definition of an operational agenda for performance improvement.	Resources; inputs; structure variables; process variables; output variables; and impact variables characterizing the effect outputs have on the broader environment.
Hogan (1996)	*Ex-ante, mid-term, ex-post.*	Written survey and face-to-face interviews. Statistical analysis of data. KSIs metrics.	Dimensions: inputs; structures; processes; outputs; impacts. Set of 35 indicators considering key success indicators (KSIs).
Staton (1996)	*Ex-ante, mid-term, ex-post.*	Goal oriented planning including the logic model of performance assessment.	Set of input, output, and impact indicators.
UKSPA/Angle Technology (2003)	*Ex-ante, mid-term, ex-post.*	Survey and statistical analysis of data. Indicators' metrics.	Economic performance of tenant firms. Innovation and technology commercialization performance of their tenant firms.
Bigliardi *et al.* (2006)	*Ex-ante, mid-term, ex-post.*	Case study. Indicator metrics.	Alignment with STP mission; major stakeholders' commitment; regional economic conditions; legal forms; nature of the scientific competence base available within research centers; and STP's life-cycle stages
Monck & Peters (2009)	*Ex ante, mid-term, ex-post.*	Survey and statistical analysis of data. Indicators' metrics.	Two levels of impact evaluation: (i) direct effects; (ii) indirect and strategic added value (SAV) effects. (i) inputs; activities; gross outputs; outcomes; impacts; (ii) gross outputs; outcomes; impacts.

Monck & Peters (2009)	*Ex-ante, mid-term, ex-post.*	Survey and statistical analysis of data. Indicators' metrics.	Two levels of impact evaluation: (i) direct effects; (ii) indirect and strategic added value (SAV) effects. (i) inputs; activities; gross outputs; outcomes; impacts; (ii) gross outputs; outcomes; impacts.
Dabrowska (2011)	Balanced Scorecard	KPIs' metrics associated to BSC dimensions.	Model for 3G STPS comprising five dimensions: commercial; shareholder's perspective; owner's perspective; brand and reputation; and internal business processes.
Nosratabadi, Pourdarab & Abbasian (2011)	*Ex-ante, mid-term, ex-post.*	Survey and combined multi-criteria decision methods with fuzzy logic.	Constraints; Degree of integration with national or global markets; professional qualifications; managerial capabilities; stakeholder's interests; size; technological capabilities; maturity of the business; venture capital.
Hemati & Mardami (2012)	Balanced Scorecard	KPIs' metrics associated to BSC dimensions.	Four BSC perspectives: finance, customer, internal process and employee learning and growth.
Andreevna (2013)	Balanced Scorecard	KPIs' metrics associated to BSC dimensions.	Four BSC perspectives: finance, customer, internal process and employee learning and growth. Indicators are associated with the above dimensions, considering two roles of a given STP: (i) as the center of an innovative cluster; and (ii) as a commercial organization.
Wang *et al.* (2014)	Balanced Scorecard	Case study. KPIs' metrics associated to BSC dimensions.	Four BSC perspectives: finance, customers, internal processes and organizational learning and growth.
Kbar & Ally (2015)	*Ex-ante, mid-term, ex-post.*	Survey and combined multi-criteria decision methods. Case study for validation.	Alignment with the 4th generation STPs. Dimensions: R&D; business; management; and infrastructure. The optimum performance value of STPs can be estimated.
CMI (2015)	Business model generation (BMG)	Participative workshop. KPIs' metrics associated to BMG blocks.	The BMG is a strategic management that is a build-up of nine blocks: customer segments, customer relationships, key partners, key resources, key activities, channels, revenue stream, cost structure, and value proposition.
Ferrara, Lamperti & Mavilia (2016)	*Ex-ante, mid-term, ex-post.*	Survey and Choquet integral based Multi-Attribute Value Theory to elicit stakeholders' preferences on different dimensions of STPs' performances.	The index aggregates eight indicators associated with two dimensions: (i) innovation; and (ii) entrepreneurship. For innovation the indicators refer to research centers; patents; scientific network; projects. And for the second dimension: growth; employees; entropy; and geoconsistency.

Sources: [6; 7; 8; 9; 10; 11].

As can be observed in table 1, several practitioners and academicians and have made significant efforts to overcome the challenge to measure and assess the performance of STPs, by using different methodological approaches, but with emphasis on surveys and Balanced Scorecard methodologies, with the support of multiple criteria decision-making (MCDM) methods. In fact, STPs have specific goals and objectives, according to their missions, and can be in different stages of development. Besides, their activities can impact multiple stakeholders – from internal and external researchers to local government agencies, entrepreneurs, firms, and universities.

IOP Conf. Series: Journal of Physics: Conf. Series **1044** (2018) 012042 doi:10.1088/1742-6596/1044/1/012042

3. Methodology

The research methodology comprised: (i) literature and documental review on central themes and subthemes from the perspective of measuring the performance of groups of STPs, considering their business models, objectives and goals, maturity levels, and stakeholder's commitment; legal and governance structures; (ii) preliminary conceptual framework as a basis for the development of an empirical study in Brazil; (iii) development of this empirical study, with participation of representatives of 26 Brazilian STPs; (iv) discussion of results from the empirical study and formulation of recommendations for diffusing the model among other STPs around the world.

4. Results

The methods and analytical grids identified during the literature review on models for measuring the performance of STPs embodied the basis for designing a multidimensional framework for monitoring and evaluating the performance of STPs' similar groups, considering their business models and value propositions, objectives and goals, development stage, stakeholder's commitment, and legal and governance structures.

In STPs' context this type of ME system allows managers to modify and adjust both the theory of change [12] and the implementation processes to the achievement of desired objectives and outcomes. That was the central idea of the empirical study involving 26 Brazilian STPs, which participated in an Executive Program for STP Managers sponsored by the Anprotec.

The process flow diagram of the model in three phases is presented in Figure 1.

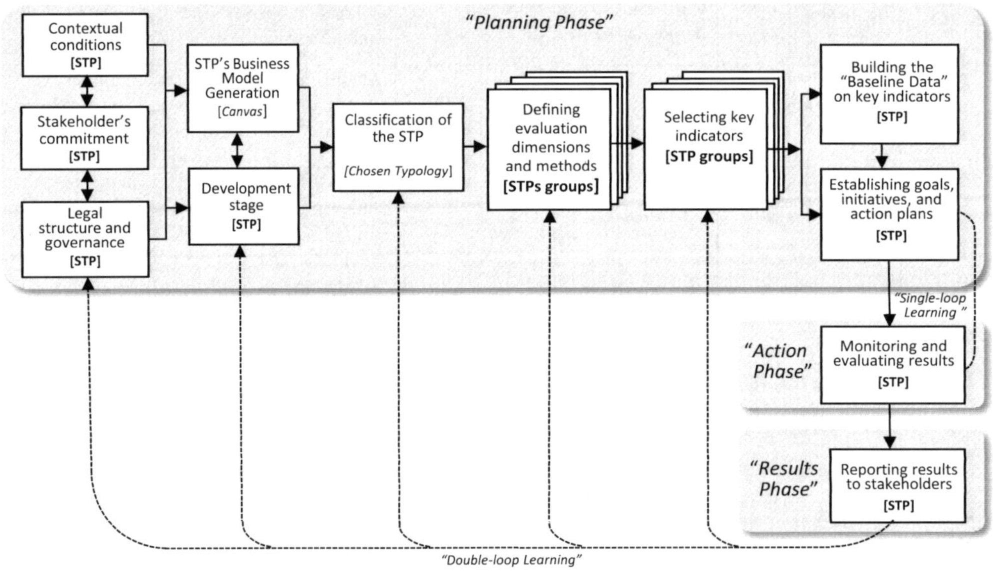

Figure 1: Proposed model for measuring and evaluating STPs' performance

The multidimensional model encompasses ten main blocks: (i) analysis of the contextual conditions of the focused STP; (ii) analysis of its stakeholders' commitment; (iii) analysis of its legal structure and governance; (iv) design of the STP's business model (Canvas); (v) classification of the STP in function of its development stage; (vi) classification of the STP according to a chosen STP's typology; (vii) defining evaluation dimensions and methods for STPs groups; (viii) selecting key indicators for STP groups; (ix)

IOP Conf. Series: Journal of Physics: Conf. Series **1044** (2018) 012042 doi:10.1088/1742-6596/1044/1/012042

building performance baselines on key indicators; and (x) establishing goals, initiatives, and action plans for the focused STP. Due to space limitations, we will not be able to give a detailed description of each block. Nevertheless, the whole model is already published in a MSc. dissertation, on which this article was based [13].

By way of illustration, we present in table 2 a subset of indicators associated to one of the strategic objectives of the Technology Park of the Lutheran University of Brazil (Ulbratech), located in Canoas, Rio Grande do Sul, Brazil. The general manager of this STP participated in the Executive Program for STP Managers sponsored by the Anprotec and Ulbratech was one of five chosen STPs for the multi-case study developed during this Program.

Table 2: Proposition of indicators and metrics for monitoring and evaluating the performance of the Ulbratech: short term (2 years)

Objective hierarchy	Objectives	Goals	Indicators	Metrics
Strategic Objective	***1.* To provide a favorable environment for innovation involving cooperation between the Brazilian Lutheran University (Ulbra) and resident companies of Ulbratech**			
Short- term objectives (2 years)	1.1. To deploy creative and collaborative environments to stimulate entrepreneurship and innovation	1.1a. Coworking spaces for collaborative activities and business incubation implemented (1 year).	Area of *coworking* spaces for collaborative activities and business incubation	Area of *coworking* spaces (m^2) Occupancy rate of coworking spaces (%) Degree of satisfaction of the users of the *coworking* spaces (Likert scale)
		1.1b. Generation of new jointly R&D projects and new businesses (2 years).	Growth rate of new jointly R&D and business projects	Number of pre-incubation candidates / period. Number of new jointly R&D projects / period. Number of business projects incubated / period
		1.1c. Quarterly workshops on business modeling, intellectual property and technology transfer and other strategic issues	Workshops on strategic issues for Ulbratech residents	Number of participants per workshop / number of places per workshop Number of workshops held per year / number of planned workshops
	1.2. To revitalize common areas within the area of Ulbratech	1.2a. Revitalized common areas for creating new coworking spaces (2 years)	Index of revitalization of common areas	Revitalized areas/period (m^2) Occupancy rate of coworking spaces (%) Degree of satisfaction of the users of the coworking spaces (Likert scale)
	1.3. To improve visual communication of Ulbratech's external and internal areas	1.3a. Projects of visual communication and signaling of Ulbratech's external and internal areas concluded (1 year)	Visual communication and signaling of external and internal areas	Degree of accomplishment of the projects of visual communication and signaling of Ulbratech's external and internal areas
	1.4. To organize cultural events to be held in the Ulbratech's Central Building	1.4a. Monthly cultural opening events held in Ulbratech's Central Building	Ulbratech Alive!	Number of cultural opening events held / period. Number of participants/cultural opening event. Degree of satisfaction of participants/event (Likert scale).

Due to space limitations, we will not be able to present the Ulbratech's objectives, goals, indicators, and metrics for medium- and long- term (5 years). The Ulbratech's complete table, as well as similar tables of performance indicators and metrics related to the other four STPs are already published in [13].

5. Conclusion

In this paper, an attempt was made to present a flexible and adaptive model, according to one of the STPs' taxonomies reported in the literature [14]. The applicability of this model for monitoring and evaluating the performance of similar groups of STPs, could be demonstrated through an empirical study involving 26 Brazilian STPs' managers, which participated in an Executive Program for STP Managers sponsored by the Anprotec. The proposed model distinguishes from the existing systems reported in the literature, because it will allow decision-makers to measure and compare the performance of STPs belonging to the same group and then prioritize actions which will add the most value to their business and programs, individually. It considers the common characteristics of these STPs' groups concerning business models and value propositions, strategic objectives and goals, development stage, stakeholder's commitment, and legal and governance structures.

6. References

[1] Phan P H, Siegel D S and Wright M 2005 Science parks and incubators: observations, synthesis, and future research. *Journal of Business Venturing* **20** 165–182

[2] Allen J 2007 *Third Generation Science Parks*. Manchester Science Parks Ltd

[3] Wasim M U 2014 Factors for science park planning. *WTR* **3** 97-108

[4] Nauwelaers C, Kleibrink, A and Stancova, K 2014 *The Role of Science Parks in Smart Specialisation Strategies.* Report EUR 26701. European Commission Joint Research Centre Institute for Prospective Technological Studies. Seville, Spain

[5] Donaldson L 2001. *The Contingency Theory of Organizations*. Thousand Oaks, CA: Sage Publications

[6] Dabrowska J 2011 *Measuring the success of science parks: performance monitoring and evaluation.* In: Proceedings of XXVIII IASP Conference on Science and Technology Parks, Kopenhagen, Denmark, 2011

[7] Hemati M, Mardami M 2012 Designing a performance appraisal system based on balanced scorecard for improving productivity: case study in Semnan Technology and Science Park. *Management Science Letters,* **2** 1619–1630

[8] Andreevna M A 2013 The balanced scorecard for estimation of science and technology parks. *World Applied Sciences Journal* **25**(5) 720–727

[9] Wang G, Wan J and Zhao L 2014. Strategy map for Chinese science parks with KPIs of BSC *Journal of Science and Technology Policy Management* 5 (2) 82 – 105

[10] Ferrara M, Lamperti F and Mavilia R 2016. Looking for best performers: a pilot study towards the evaluation of science parks. *Scientometrics* **106** (2) 717–750

[11] Albahari A, Catalano G and Landoni P 2013 Evaluation of national science park systems: a theoretical framework and its application to the Italian and Spanish systems *Technology Analysis & Strategic Management* **25**(5) 599-614

[12] Kusek J Z and Rist R C 2004. *Ten steps to a results-based monitoring and evaluation system.* World Bank, Washington, DC.

[13] Lyra, R M 2017 *Monitoring and evaluating the performance of science and technology parks: proposal of a conceptual adaptive model.* Rio de Janeiro, 171 p. MSc. Dissertation – Postgraduate Programme in Metrology. Pontifical Catholic University of Rio de Janeiro

[14] Associação Nacional de Entidades Promotoras de Empreendimentos Inovadores. Anprotec 2008 *Parques Tecnológicos no Brasil: Estudos, Análise e Proposições*. Anprotec, Brasília, Brazil.

IOP Conf. Series: Journal of Physics: Conf. Series **1044** (2018) 012043 doi:10.1088/1742-6596/1044/1/012043

Measuring the students' level of knowledge on the basis of other students' subjective opinions

I U Zulkarnay[1], **A I Dashkina**[2] **and D A Tarkhov**[2,3,4]

[1]Bashkir State University, Ufa, 450076, ul.Z.Validi, 32, Russian Federation
[2]Peter the Great St. Petersburg Polytechnic University, 195251, St.Petersburg, Polytechnicheskaya, 29, Russian Federation

[3]E-mail: dtarkhov@gmail.com

Abstract: This paper develops a new methodology of measuring the students' level of knowledge on the basis of other students' subjective opinions. It is done a research into the correlation between the subjective and objective indicators of the educational process. One of the key issues behind controlling the state of social processes is the fact that it is difficult to measure them objectively. In this case it is much easier to obtain subjective indicators. It implies assessing the values we are interested in by people. If we manage to measure the objective indicators with an acceptable degree of accuracy on the basis of the subjective ones, it will be much easier to study sociodynamic phenomena and processes. The methodology includes building a neural network function. The method allows the teacher to evaluate the progress made by a group of learners in a less time-consuming way.

1. Problem

When it comes to controlling and assessing social processes, one of the main problems is obtaining the objective indices for them. It is much easier to get subjective indices, people assessing the values we are interested in. If we manage to identify the objective indices on the basis of the subjective ones with adequate precision, it will be much easier to study sociodynamic phenomena and processes. Our research focuses on studying the correlation between the objective and subjective indices in the educational process.

There is a large body of literature on the measurement theory and a metrological perspective and on how they are applied in measuring social processes. Thus, Mari, L., & Wilson, M. give comparative discussion of deterministic and probabilistic measurement models on the example of social measurement in terms of the International Vocabulary of Metrology, giving an introduction to Rasch models in a metrological context they analyze measurement validity [1]. Maul, A., Wilson, M., & Torres Irribarra, D. try to formulate a philosophically coherent account of how measurement works in both the physical and human sciences. They argue that "…measurement is concerned with gaining knowledge about the real world, that it is an empirical process, and that it is shaped by the pragmatic concerns of the agents involved" [2]. Pendrill, L. studies how the metrology of human-based observations can be tackled with a measurement system analysis approach, focusing on the side where Man acts as a measurement instrument [3]. He discusses the problem of decision risks and suggests that through the Rasch invariant measure approach it is possible to provide a proper treatment of ordinal data. In this

[4] To whom any correspondence should be addressed.

IOP Conf. Series: Journal of Physics: Conf. Series **1044** (2018) 012043 doi:10.1088/1742-6596/1044/1/012043

case Rash models are similar to neural network ones, which provides a basis for successful application of neural networks in studying sociodynamic phenomena.

Our current research described in this article focuses on studying the correlation between the objective and subjective indices in the educational process.

2. Research Methodology

We measured how efficiently the students learned professionally orientated English depending on how the teaching process was organized. Our pedagogical experiment is based on the application of the hierarchical teaching system and mentorship, which prove to be much more effective than a conventional system of organizing the learning process, especially if the learners in the group have different levels of knowledge. In the hierarchical teaching system the knowledge is transferred not only from a teacher to learners, but also from one student to another, the system can be referred to as a hierarchical one.

The experiment was conducted at St. Petersburg Polytechnic University. It involved three groups of third-year students specializing in economics and management. At the end of the experiment we prepared a final test for the two experimental groups and the control one in order to check their progress. The test included multiple choice lexical tasks based on the limited amount of professional vocabulary studied in the course of the experiment. The students were also supposed to summarize a text based on the same lexical material.

The students in group 2 were with a high level of knowledge, whereas the learners in group 1 were with a low level of foreign-language competence. Mentorship meant that the students from group 2 helped the students from group 1 in their learning process.

The students from group 2 worked in the capacity of mentors in collaboration with the students from group 1. As group 2 outnumbered group 1 (20 and 7 students respectively), there were more mentors than mentees. The students from all the groups did the reading comprehension exercises, summarized the texts and discussed the cases referring to the texts. While doing the assignments, the mentors from group 2 helped the mentees from group 1 and the teacher checked the results of their work in class.

The students in control group 3 worked individually, without collaborating with each other, which means that the students with a high level of knowledge did not help the ones with a low level of foreign-language competence. They also summarized and discussed the texts, and did reading comprehension exercises.

At the beginning of the experiment we ran an input test aimed at checking the students' knowledge before the series of the experimental classes. At the end of this series of classes we ran a final test in order to assess objectively how the knowledge had changed.

We ran the same final test for the two experimental groups and the control one. Each of the three groups which took part in the experiment was given the multiple choice lexical tasks based on the vocabulary studied in the course of the experiment. The students were also supposed to summarize a text based on the lexical material studied in the course of the experiment.

Apart from the above mentioned testing, the students from group 2 assessed the knowledge of the students from group 1 on a five-point scale. The latter, in their turn, assessed the knowledge of the students from group 2. Besides, all the students' knowledge was assessed by the teacher.

3. The results of the experiments

The final test enabled us to determine how well they had acquired the vocabulary and how efficiently they summarize professionally-orientated texts. When we evaluated the results of summarizing the text, we considered the following criteria: rendering the ideas, grammar, vocabulary, sentence structure and independent judgment. The maximum score for each criterion was 5. The final test also included 10 multiple-choice sentences (one point per sentence). The maximum score in the assignment with true or false statements was 5.

Apart from that, we asked the students from group 1 to assess the progress of their mentors from group 2 in professionally-orientated English (table 2). The mentors (the students from group 2), in their turn, had to assess their mentees from group 1 (table 1). We wanted to discern the correlation between the students' subjective peer assessment and the objective assessment based on the results of the tests run by the teacher. The results of assessing the knowledge in group 3 we see in table 3.

Table 1. The results of assessing the knowledge in group 1 (The values are in the form of fractions: above the slash (/) – the results of the final test, below the slash (/) – the assessment by the students from group 2).

№	Overall result of the input test	Summa-rizing	Grammar	Vocabulary	Sentence structure	Opinions and judgments	Filling out the gaps	True or false statements	Overall result: final test / stud.'s assesm.
1	21	4/5	4/3	4/3	4/2	3/4	10	5	34/17
2	26	5/5	5/4	54	5/3	4/5	10	5	39/21
3	22	4/3	4/3	4/3	4/5	3/4	9	5	33/18
4	22	4/3	4/3	4/3	4/1	3/3	8	3	30/13
5	23	5/3	3/2	4/4	4/4	4/3	10	4	34/16
6	19	5/5	4/3	4/5	4/3	4/5	10	5	36/21
7	22	3/3	3/2	4/1	4/3	3/2	8	4	29/11

Table 2. The results of assessing the knowledge in group 2 (The values are in the form of fractions: above the slash (/) – the results of the final test, below the slash (/) – the assessment by the students from group 1).

№	Overall result of the input test	Summa-rizing	Grammar	Vocabulary	Sentence structure	Opinions and judgments	Filling out the gaps	True or false statements	Overall final test / stud.'s assesm.
1	36	5/5	5/5	5/4	4/4	5/4	10	5	39/22
2	33	4/4	5/5	5/4	5/3	5/5	9	5	38/21
3	38	5/5	4/5	5/5	5/5	5/5	10	5	39/25
4	39	5/5	5/5	5/5	4/5	5/5	10	5	39/25
5	35	5/4	5/5	5/5	4/4	5/5	10	5	39/23
6	32	5/4	4/4	5/5	4/5	5/4	10	5	38/22
7	37	5/5	4/5	5/5	5/5	5/5	9	5	39/25
8	37	5/5	5/5	5/5	5/5	5/5	9	5	39/25
9	33	5/5	4/5	5/5	5/4	5/5	10	5	39/24
10	40	5/4	5/5	5/5	5/5	5/4	10	5	40/23
11	39	5/4	5/5	5/5	5/5	5/4	10	5	40/23
12	39	5/5	5/5	5/5	5/5	5/5	10	5	40/25
13	30	4/4	4/4	5/4	4/4	5/4	9	5	36/20
14	37	5/4	4/5	5/5	5/4	5/4	10	5	39/22
15	35	4/4	5/5	5/5	5/5	4/5	10	5	38/24
16	32	5/5	4/5	5/5	5/4	5/5	10	5	39/24
17	35	5/5	4/5	5/5	4/4	5/5	10	5	38/24
18	35	4/4	5/5	5/5	4/5	5/5	10	5	38/24
19	32	5/4	4/5	4/5	4/4	5/4	10	5	37/22
20	28	4/4	5/5	5/4	4/4	4/4	9	5	36/21

Table 3. The results of assessing the knowledge in group 3.

№	Overall result of the input test	Summa-rizing	Grammar	Vocabulary	Sentence structure	Opinions and judgments	Filling out the gaps	True or false statements	Overall result of the final test
1	28	4	3	4	3	1	8	4	27
2	8	1	1	1	1	0	3	3	10
3	7	1	1	1	0	0	2	1	6
4	30	4	4	4	3	3	8	4	30
5	37	4	5	4	5	5	10	5	38
6	9	0	1	1	1	0	2	3	8
7	30	3	4	4	4	4	6	5	30
8	22	2	4	5	4	1	5	4	25
9	23	2	3	2	3	1	5	4	20
10	20	3	3	4	3	2	3	2	20
11	26	4	5	5	4	2	4	4	28

4. Analysis

Tables 1-3 let us build all figures below to see the progress of students. In figure 1, figure 2 and figure 3 there are the results (which are total for all the criteria) for both the input and the final tests. For illustrative purposes we arranged the results in ascending order and smoothed them with the neural network.

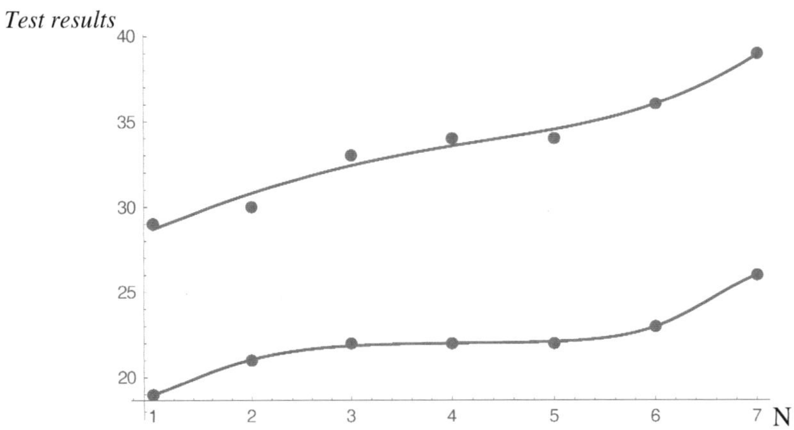

Figure 1. The results of the input and final tests (lower and upper ones respectively) for experimental group 1, N – students in the scaled sample.

It is obvious from figure 1 that the progress in the experimental group is substantial: the results of the final test are much better than those of the input one. It indicates the efficiency of collaborative learning.

Figure 2 illustrates that the students in group 2 also made a remarkable progress due to their teaching activities. However, the students whose scores in the input test were high did not show considerable improvement in the final test, which is illustrated with the graphs converging in one point. The results of the final test are approaching the saturation, which means that the students with mastery of English should have been given more challenging assignments.

In figure 3 it is quite obvious that there was very little difference in the results of the input and the final test, which indicates that the traditional teaching and learning process (without students' collaboration and with the focus on learning the language instead of concentrating on learning the professionally-orientated content), was not very efficient.

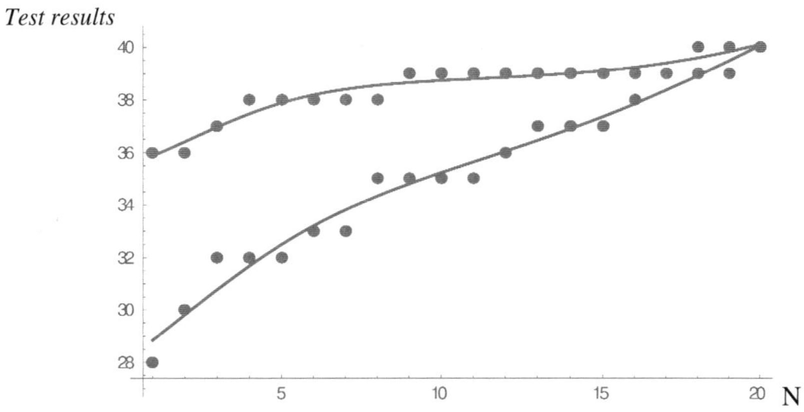

Figure 2. The results of the input and final tests (lower and upper ones respectively) for experimental group 2, N – students in the scaled sample.

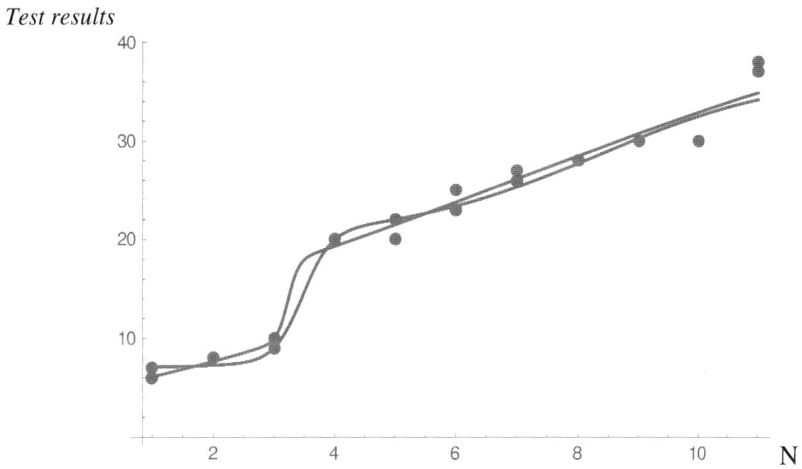

Figure 3. The results of the input and final tests (lower and upper ones respectively) for group 3, N – students in the scaled sample.

The results that have been obtained illustrate that it is possible to develop a reliable method of evaluating objective indices using subjective data. However, developing such a method and explaining the reasoning behind it requires further research.

This method can be integrated into the educational process because it allows the teacher to evaluate the progress made by a group of learners in a less time-consuming way. Instead of doing a test, the students will just fill out a short evaluation table with the results of each other's peer assessment. As the result of that tests can be done less frequently.

The teacher needs to assess the level of students' achievement and performance in order to provide operational management of educational process (the pace of learning, choosing the check points, determining the level of complexity of the current tests, etc.). Interviewing a few students may appear to be insufficient, whereas interviewing or testing a great number of students is time-consuming. For this reason we suggest that peer assessment by the students collaborating on the basis of the hierarchical system should be used for continuous monitoring. The results of verifying these assumptions on the basis of the data from tables 1 and 2 are considered further.

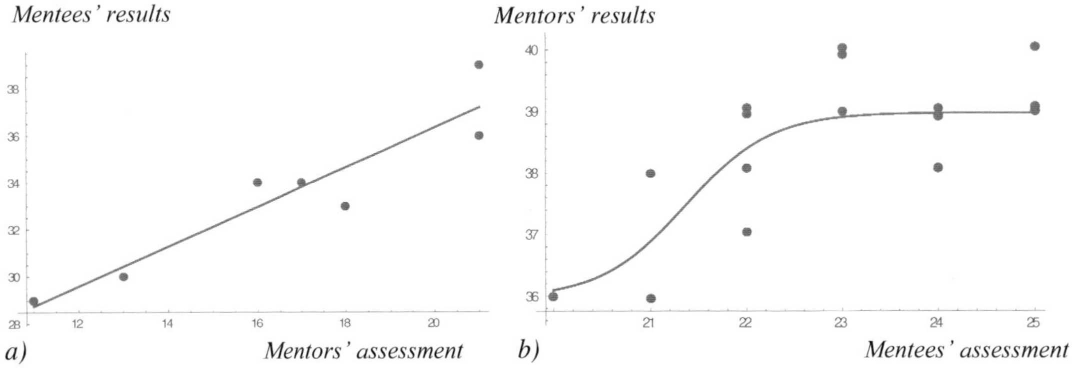

Figure 4. a) The combined results of the test in the mentees' group as the function of the mentors' assessment; b) the combined results of the test in the mentors' group as the function of the mentees' assessment.

The first relationship (figure 4a) appeared to be linear. The root-mean square error for it accounted for 1.12, the evaluation of the root-mean deviation of the sample was 3.41. If we take into account the error which resulted from the discreteness of assessment and its subjective nature, the result is quite acceptable.

When we considered the second relationship (figure 4b), we doubted whether it would be possible to obtain any meaningful results because we did not expect the students with a rudimentary level of foreign language knowledge to be able to assess the advanced students. However, we managed to come to some conclusions. In order to achieve that, the sampling was smoothed on the basis of the method of least squares through neural network function. The resulting relationship was the following (formula 1):

$$y = 38.5 + 1.05 Tanh(0.01x - 2.32) + 1.51 Tanh(1.13x - 24.2) \quad (1)$$

Neural networks are a standard tool for finding the implicit relationships in the data [6]. The root-mean square error for formula 1 accounted for 0.754, the evaluation of the root-mean deviation of the sample was 1.15. If we take into account the error which resulted from the discreteness of assessment and its subjective nature, the result is quite acceptable.

This formula refers only to the qualitative relationship, but not the quantitative one, and it gives a vivid illustration of the data. This relationship is similar to the relationship that results from applying Rasch methodology and approaches the minimum value at the beginning of the interval and the maximum one at its end. Here it is essential to mention that the mentees' results cannot exceed 40, so the students from group 2 with the highest level of knowledge had approximately the same results of the final test, because their score was close to the maximum possible value. However, their mentees' assessments differed from each other. The mentors who were interested in their mentees' results got the highest scores, whereas the mentors who were unwilling to explain the material got the lowest ratings.

The graph on figure 4b can be divided into two parts. The first part indicates that the subjective and the objective evaluations tend to change in one direction. The second part of the graph is virtually horizontal, which means that the objective evaluations were practically similar, whereas the subjective assessments were different. The similarity of the objective evaluations (i.e. the results of the final test) can be explained by the fact that most of the students got a maximum or nearly maximum score, which means that the test does not allow distinguishing between their levels of competence. According to the teacher, the real level of the students' foreign language proficiency does not always correspond to the results of the peer assessment. The students who were interested in their partners' progress were given higher scores by their mentees.

Further we describe the correlations between the individual students' results of the final tests and the assessment of their performance by the students who cooperated with them. The criteria of assessment (summarizing the plot, grammar, vocabulary acquisition, etc.) were considered separately. It is impossible to establish a clear correlation for most of the criteria. The most interesting results are considered below:

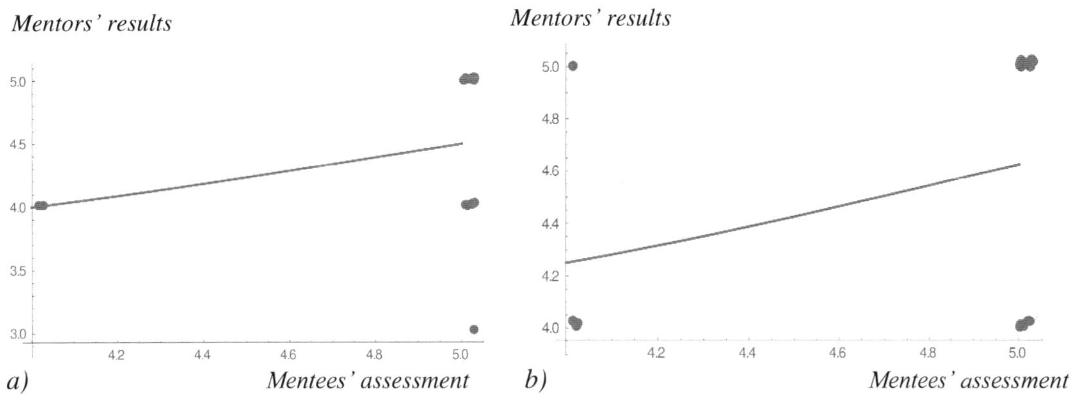

Figure 5. The mentors' results as the function of their assessment by: a) the mentees; b) grammar vocabulary.

Figure 5 illustrates that there was a certain degree of correlation between the actual results of the test done by the mentors and the subjective assessment of the same results by the mentees. It is obvious that the mentees had difficulty assessing the mentors' performance. On the other hand, it is easier to assess vocabulary and grammar in quantitative terms than other aspects of learning a foreign language.

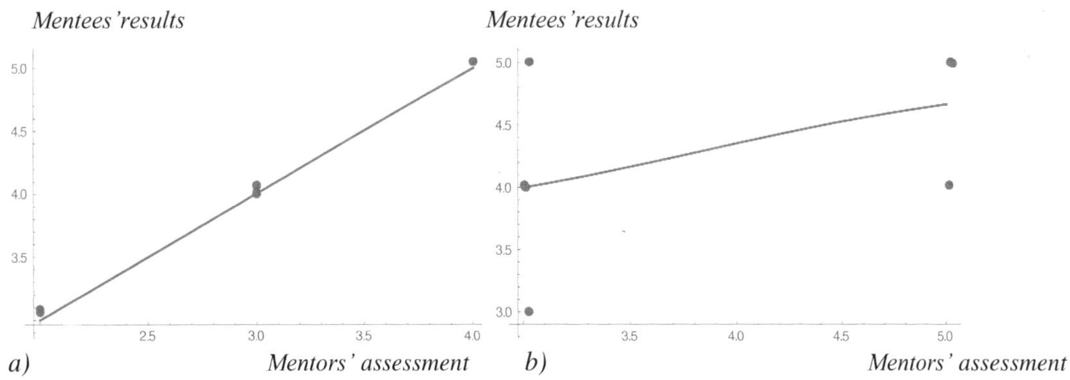

Figure 6. The mentees' results as the function of their assessment by: a) the mentors; b) grammar summarizing.

The graph on the left (figure 6a) describes the performance in grammar and shows a clear correlation between the mentees' objective results of the test and the mentors' subjective assessment. Such a high degree of consistency can be explained by the fact that the assessment was made by the students who have a mastery of English and who can make judgements about the mentees' level of knowledge.

Moreover, as we have already mentioned, the students' achievements in grammar can easily be assessed in quantitative terms. The graph on the right (figure 6b) shows the correlation between the mentees' progress in summarizing texts and the mentors' assessment of this progress. The correlation is not as clear as in grammar, because it is difficult to describe summarizing in quantitative terms. However, the mentors managed to make expert assessment of the mentees' progress.

The results given above allow us to come to two main conclusions: Mentors make more objective evaluations of the mentees, whereas the mentees' evaluations are less accurate. The results are more reliable when we deal with clearer criteria such as grammar.

5. Conclusions

The results that we have obtained reveal that it is difficult to develop a reliable method of assessing the objective indices on the basis of the subjective data. In order to succeed in applying such methods a researcher needs to define the desirable result. It is also important to determine whether it is possible to achieve the result on the basis of the data that are readily available or the data that can be obtained. The easiest way is to make conclusions about quality on the basis of the objective data. For instance, it is easy to come to the general qualitative conclusion about the efficiency of learning in different groups and the effectiveness of the teaching methods on the basis of figures 1-3.

Figure 4 illustrates that quite distinct qualitative conclusions can be made on the basis of the subjective data. Figures 5 and 6 indicate that it is difficult to conduct more detailed studies. More accurate results can be obtained when we deal with more formalized data (such as the level of proficiency in grammar). If the data are less formalized, it is more difficult to assess them because assessing the results of the tests by the teacher can be different from the results of the peer assessment (in this case – by other students).

This method can be applied in higher education, especially in open-source learning, in which students peer assessment is essential. However, developing such a method and explaining the reasoning behind it requires further research, for example designing models using the techniques [4, 5]. Neural networks were also used in measuring the efficiency of applying a hierarchical teaching system based on mentorship in several groups of students [7].

This method can be integrated into the educational process because it allows the teacher to evaluate the progress made by a group of learners in a less time-consuming way. Instead of doing a test, the students will just fill out a short evaluation table with the results of each other's peer assessment. As the result of that the tests can be run less frequently.

References

[1] Mari L and Wilson M 2014 May An introduction to the Rasch measurement approach for metrologists *Measurement* **51** 315-27

[2] Maul A, Wilson M and Torres I D 2013 On the conceptual foundations of psychological measurement *Journal of Physics Conference Series* **459(1)**

[3] Pendrill L 2014 December Man as a measurement instrument [Special Feature] *NCSLi Measure: The Journal of Measurement Science* **9(4)** 22-33

[4] Blagoveshchenskaya E A, Dashkina A I, Lazovskaya T V, Ryabukhina V V and Tarkhov D A 2016 Neural Network Methods for Construction of Sociodynamic Models Hierarchy *Lecture Notes in Computer Science (Springer International Publishing Switzerland)* ed L Cheng et al pp 513–520

[5] Tarkhov D A, Shanshin I K and Shakhanov D O 2016 The reversed problem of migration streams modeling *Journal of Physics: Conference Series* **772**

[6] Haykin S 1999 *Neural Networks: A Comprehensive Foundation* (Prentice-Hall Inc) p 199

[7] Dashkina A and Tarkhov D 2017 The application of neural network modeling in organizing a hierarchical teaching system based on mentorship *Complex Systems: Theory and Applications* ed R Martinez (New-York: Nova Publishers) chapter 4 pp 95-116

IOP Conf. Series: Journal of Physics: Conf. Series **1044** (2018) 012044 doi:10.1088/1742-6596/1044/1/012044

Method based on computational intelligence techniques for localization of firearms projectiles inserted into the human body, by high sensitivity magnetic measurements

J D T Jimenez and E C Silva
Department of Electrical Engineering, Pontifical Catholic University of Rio de Janeiro, PUC-Rio, Rio de Janeiro, RJ, Brazil

E-mail: ing.juandario2@gmail.com, edusilva@ele.puc-rio.br

Abstract. In modern society, one of the most frequent clinical cases involves location and extraction of firearms projectiles, usually made of lead, a non-ferromagnetic material. The development of a technique that allows the precise location of these projectiles will aid their surgical removal, which has a great relevance because it contributes directly to the increase of the survival rate of wounded patients. This work presents and discusses a new approach, based on computational intelligence techniques, aiming at locating firearm projectiles inserted into the human body, by processing the information contained in magnetic field maps. The projectile is modelled by a sphere with radius a, located on a search space contained in a xy plane, that is situated at a distance h from the sensor, along the z axis. The x and y positions of the projectile were estimated by a windowing algorithm. In turn, the distance h between the sphere and the sensor is inferred by neural network, and the radius of the sphere a is estimated by a genetic algorithm. The obtained results indicate that the developed technique was capable to estimate all parameters with high accuracy and precision.

1. Introduction

The penetration of metallic foreign bodies in human tissues is rather frequent, such objects might cause discomfort or be lodged close to organs at risk of perforation, often requiring surgical removal. Currently, the location of these foreign bodies is performed by conventional radiological techniques of imaging that have several drawbacks, such as the patients and staff exposure to ionizing radiation and the need for insertion of multiple radiopaque markers in different anatomical planes. Furthermore, the positioning information obtained by those methods is limited and imprecise [1]–[3]. These positioning parameters are essential for proper planning surgical procedures, impacting directly on the success rate levels and on the overall time required for surgical removal of the foreign body.

Previous works, highlighted in literature [1]-[3], developed methods for locating ferromagnetic foreign bodies inserted into the human body, by high sensitivity magnetic measurements. The results obtained by those methods have showed that they are considerably more precise than the conventional locating techniques. Moreover, the application of those methods led to the increasing of the success rates and the decreasing of the time spent on surgical removal procedures. However, one of the most frequent clinical cases in modern society involves locating and extracting firearms projectile fragments, which are usually made of lead, a non-ferromagnetic material. Despite the lead does not have remanent field ($\mu_r \approx 1.0$), it is a good conductor of electricity ($\sigma \approx 4.55$ x 106 S / m). Thus, as proposed by an initial theoretical analysis discussed in [6], the location of metallic non-ferromagnetic bodies can be performed by applying an alternating primary magnetic field over the object, in order to induce eddy currents [4], [5]. This work focuses on the development of a new method, based on computational intelligence techniques, to process the information contained in magnetic field maps, aiming at locating firearm projectiles inserted into the human body.

2. Methodology

The proposed location technique models the projectile by a sphere with radius a, located on a search space contained in a xy plane, that is situated at a distance h from the sensor, along the z axis. The *primary alternating magnetic field*, applied over the search space, is generated by a solenoid. If there is a firearm projectile in this region, parasitic currents will be induced in the projectile. As shown in figure 1, those currents will produce a *secondary magnetic field* that is measured by a gradiometric configuration of high-sensitivity magnetic sensors, in order to improve the signal-to-noise ratio.

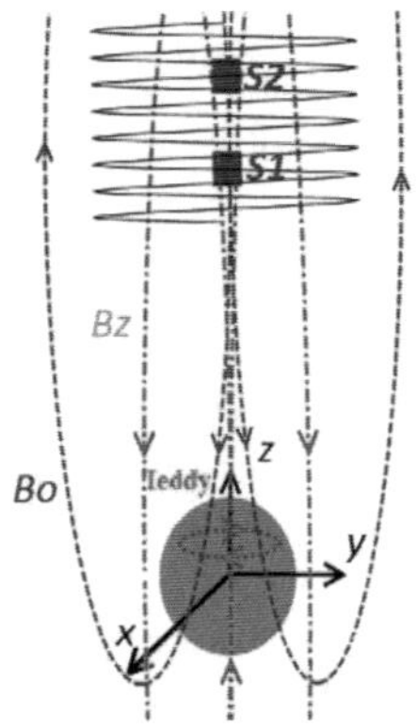

Figure 1. Measurement arrangement proposed in this work. The solenoid generates a time-varying primary magnetic flux density B_o (blue dashed line), which is applied to the search space in order to induce also time-varying parasitic currents (I_{eddy}) in the spherical foreign body (red dashed line). In turn, such currents produce a secondary magnetic flux density B_z (green dash line). Thus, the result of the gradiometric (differential) reading by the sensor elements S_1 and S_2 is composed mainly by the gradient of the secondary magnetic flux density [8].

An algorithm of location *(Position determination module)* was developed to determine the position of the projectile (x_b and y_b) in the xy plane. This algorithm was based on a windowing technique, that calculates successive averages within windows of decreasing sizes. In turn, the determination of the distance h between the geometrical centre of the sphere and the sensor S_1 is inferred by a neural network *(Depth determination module)*, and the radius of the sphere a is estimated by a genetic algorithm *(Radius determination module)*.

2.1. Primary magnetic flux density

The solenoid was mathematically modelled, in a simplified way, as being composed by n turns, parallel to each other, each of them with the same total length l. In this way, it is possible to use the Biot-Savart Law [7] to determine the components of the primary magnetic flux density $B_{ox}(\vec{r})$, $B_{oy}(\vec{r})$ and $B_{oz}(\vec{r})$ generated by the solenoid at any point $\vec{r}$ of the space. Assuming that the solenoid is excited by a current I and considering that the length l of the n turns can be discretized in an integer number of segments of size Δl, it is possible to define:

$$B_{ox}\left(\vec{r}\right)=\sum_{i=1}^{n}\sum_{j=1}^{l/\Delta l}\left(\frac{\mu_0 I}{4\pi}\frac{l_{yji}.r_{zji}}{r_{ji}^{3}}\right). \tag{1}$$

$$B_{oy}\left(\vec{r}\right)=-\sum_{i=1}^{n}\sum_{j=1}^{l/\Delta l}\left(\frac{\mu_0 I}{4\pi}\frac{l_{xji}.r_{zji}}{r_{ji}^{3}}\right). \tag{2}$$

$$B_{oz}\left(\vec{r}\right)=\sum_{i=1}^{n}\sum_{j=1}^{l/\Delta l}\left(\frac{\mu_0 I}{4\pi}\left(\frac{l_{xji}.r_{xji}}{r_{ji}^{3}}-\frac{l_{yji}.r_{xji}}{r_{ji}^{3}}\right)\right). \tag{3}$$

In order to model the problem as realistically as possible, the real parameters of the excitation solenoid, implemented in [8], were considered: radius of 0,01 m, 40 turns, total length of 0, 07 m and the thickness of the used wire of 0.00175 m. In addition, the solenoid was excited by a sinusoidal current with 2 mA amplitude and 8 kHz frequency.

2.2. Gradient of the secondary magnetic flux density

The system implemented in [8] used Giant Magnetoimpedance sensors to perform the readings of the gradient of the secondary magnetic flux density. Those sensors have only one axis of sensitivity, which was aligned to measure the z component of the secondary magnetic field. In turn, this component is affected by the three components of the primary magnetic flux density: $B_{ox}(\vec{r})$, $B_{oy}(\vec{r})$ e $B_{oz}(\vec{r})$. From the mathematical foundation explained in [4-8] and knowing that the proposed measurement structure is based on gradiometric (differential) readings between two magnetic sensors S_1 and S_2, separated by a distance l_s, along the z axis, and assuming that both sensors are subjected to the same primary magnetic flux density, the gradient of the secondary magnetic flux density (ΔB_z) between them can be expressed according to the equation (4).

$$\Delta B_z = B_z(x,y,h) - B_z(x,y,h+l_s) = \frac{3}{2}B_{ox}a^3V(a,f_o)(x_b-x_s)\left(\frac{h}{\left((x_b-x_s)^2+(y_b-y_s)^2+h^2\right)^{5/2}} - \frac{h+l_s}{\left((x_b-x_s)^2+(y_b-y_s)^2+(h+l_s)^2\right)^{5/2}}\right) \tag{4}$$

$$+\frac{3}{2}B_{oy}a^3V(a,f_o)(y_b-y_s)\left(\frac{h}{\left((x_b-x_s)^2+(y_b-y_s)^2+h^2\right)^{5/2}} - \frac{h+l_s}{\left((x_b-x_s)^2+(y_b-y_s)^2+(h+l_s)^2\right)^{5/2}}\right)$$

$$+\frac{B_{oz}}{2}a^3V(a,f_o)\left(\frac{2h^2-(x_b-x_s)^2-(y_b-y_s)^2}{\left((x_b-x_s)^2+(y_b-y_s)^2+h^2\right)^{5/2}} - \frac{2(h-l_s)^2-(x_b-x_s)^2-(y_b-y_s)^2}{\left((x_b-x_s)^2+(y_b-y_s)^2+(h+l_s)^2\right)^{5/2}}\right).$$

The factor $V(a, f_o)$ is defined by:

$$V(a,f_o) = a^3\left[\frac{(2\mu_r+1)-(2\mu_r+\nu^2(a,f_o)+1)\frac{\tanh\nu(a,f_o)}{\nu(a,f_o)}}{(\mu_r-1)-(\mu_r-\nu^2(a,f_o)+1)\frac{\tanh\nu(a,f_o)}{\nu(a,f_o)}}\right], \tag{5}$$

in turn

$$\nu = (1+j)\frac{a}{\delta} \tag{6}$$

$$\delta = \sqrt{\frac{1}{f_0}\frac{1}{\pi\mu_r\mu_0\sigma}} \tag{7}$$

where: B_o is expressed in teslas, x_b and x_s are the positions of the sphere and sensor respectively, with respect to the x axis, y_b and y_s are the positions of the sphere and sensor respectively, with respect to the y axis, f_0 is the frequency of the primary magnetic flux density, σ is the electrical conductivity of the sphere and μ_r is its relative magnetic permeability. In particular, assuming that the sphere is made of lead, the following average values can be used: $\sigma = 4.55 \cdot 10^6$ *S/m*, $\mu_r = (1.7 \cdot 10^6)$ and $\mu = \mu_r \cdot \mu_o = 1.26 \cdot 10^6 Tm/A$.

Using the equation (4), the gradient of the secondary magnetic flux density ΔB_z can be estimated at any position. Consequently, in order to make the model compatible with the system developed in [8], l_s was set to 0.07 m.

2.3. Position determination module (x_b, y_b)

The matrix containing the measurements of the gradient of the secondary magnetic flux density is processed by this module, in order to estimate the position of the sphere (x_b, y_b) in the xy plane. The processing methodology adopted by the developed localization algorithm can be subdivided into the following steps: **First**: Define a mask of dimension *n* x *n*, and make a sweep with this mask by all possible positions of the measurement matrix. In each position evaluated, the average value of the measurements contained in the region delimited by the respective mask is calculated and stored. At the end of the scan, it is identified in which position of the measurement matrix the mask obtained the highest average value. In sequence, a submatrix M_1 is created containing the values and positions of the mask that returned the highest average value. **Second**: A new mask with dimension (*n*-1) x (*n*-1) is defined, which is used to make a scan inside the region delimited by submatrix M_1, obtained in the first step. In an analogous way, the process performed in the previous step is repeated for submatrix M_1, returning a submatrix M_2 with the values and positions of the mask that returned the highest average value. **Third**: This process proceeds so that the mask size is reduced at each new step, until a scan mask of size 1 x 1 is reached. The position of the 1 x 1 mask correspondent to the highest magnetic flux gradient is considered as the best estimative for the position (x_b, y_b) of the centre of the conducting sphere.

2.4. Depth determination module (h)

This module aims to infer the depth *h* of the conducting sphere in relation to the sensor S_1, located at the bottom of the excitation solenoid. The module employs a Perceptron Multilayer Neural Network designed to infer *h*, from a set of input parameters extracted from the measurement matrix of the gradient of the secondary magnetic flux, in the neighbourhood of the position (x_b, y_b) of the sphere, estimated by the position determination module. In particular, three parameters of interest are extracted:

2.4.1 Area under the curve (A). Initially, the values contained in the measurement matrix are normalized with respect to the maximum value. Then, *y* is fixed at y_b and *x* is shifted around x_b, in order to identify the left (x_{besq}) and right (x_{bdir}) points of *x* corresponding to the normalized value of 0.8. Next, a polynomial fitting is adjusted to the data set and the area under the curve delimited by this polynomial is then calculated, between x_{bleft} and x_{bright}.

2.4.2 Average (md). This parameter is obtained by calculating the mean of the ΔB_z values contained in a sub-matrix of the measurement matrix. The sub-matrix contains only the normalized values of ΔB_z greater than 0.5.

2.4.3 Standard deviation (dp). This parameter is obtained by calculating the standard deviation of the values of ΔB_z in relation to *md*, taking into account the values contained in the same sub-matrix used to infer the attribute md.

2.5. Radius determining module (a)

This module uses the parameters estimated by the previous modules – sphere position (x_b, y_b) and depth (*h*) – in order to estimate the value of the radius *a* of the sphere, using the mathematical approach presented in equation (4) and an optimization technique based on genetic algorithms (GA). The optimization process was implemented by a genetic algorithm that seeks to identify the value of the radius (*a*) that minimizes the error between simulated matrices of the gradient of the secondary magnetic flux density, called ΔB_{ztGA}, and the measured matrix of the gradient of the secondary magnetic flux density, ΔB_{zt}. The GA evolves through successive generations repeating this process, until a certain stopping criterion is reached. Equation (8) presents the mathematical definition of the evaluation function employed by the developed GA, assuming that both matrices have dimension *n* x *n*, and considering that *i* indicates the row and *j* the column of the matrix.

$$e = \frac{1}{n^2} \sum_{i=1}^{i=n} \sum_{j=1}^{j=n} \left| \frac{B_{ztGAij} - B_{ztij}}{B_{ztij}} \right| \tag{8}$$

3. Results

The performance of the modules for determination of position and radius were tested for 6860 different reference matrices. On the other hand, the depth module was tested for 2899 different reference matrices. These matrices were generated for different combinations of the variables of interest: position (x_b, y_b), depth h and radius a. Random magnetic noise matrices were added to each primary magnetic flux density matrix, associated with a combination of x_b, y_b, h and a. The random magnetic noise matrices were generated with four different intensities (e_r = 0%, 1%, 3% and 5%) of the maximum value contained in the respective primary magnetic flux density matrix.

The table 1 indicates the lower and upper limits of the frequency distribution of the errors associated to the estimative of the variables x_b, y_b, h and a; called E_{xb}, E_{yb}, E_h and E_a, respectively. It is also presented their respective mean values, standard deviations and range of errors, for a confidence level of 99% [9-10].

Table 1. Parameters extracted from frequency distribution of the errors associated to the estimative of x_b, y_b, h and a.

	Minimum (mm)	*Maximum(mm)*	*Mean (mm)*	*Standard Deviation (mm)*	*Range of error (mm)*
E_{xb}	-5	5	-0,06	0,964	-2,54≤ E_{xb}≤2,42
E_{yb}	-5	5	-0,06	0,987	-2,60≤ E_{yb}≤2,48
E_h	-2,383	2,574	0,008	0,267	-0,68≤ E_h≤0,69
E_a	0,027	0,033	0,0003	0,002	-0,005≤ E_a≤0,006

The 6860 cases, to which the errors E_{xb}, E_{yb} and E_a were calculated, were divided into 28 subsets corresponding to specific combinations of h and e_r. On the other hand, the 2899 cases, to which E_h was calculated, were divided into 13 subsets according to h. The average of the absolute error of each subset was calculated and the results are presented in figures 2 to 5.

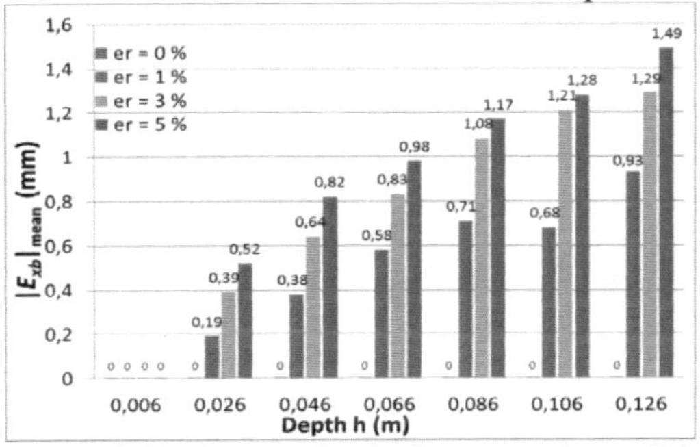

Figure 2. $|E_{xb}|_{mean}$ vs depth, for e_r equals to: 0%, 1%, 3%, 5%.

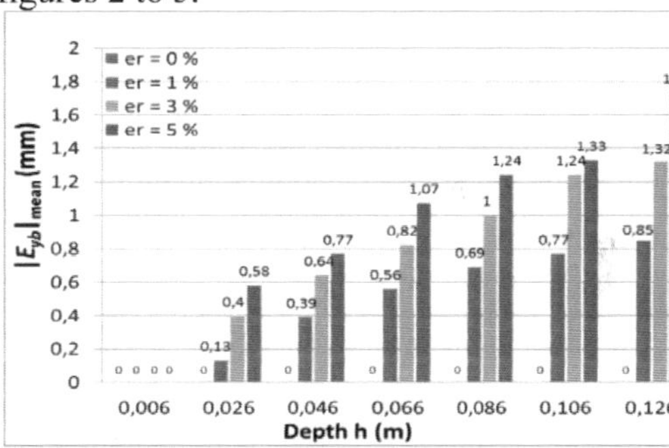

Figure 3. $|E_{yb}|_{mean}$ vs depth, for e_r equals to: 0%, 1%, 3%, 5%.

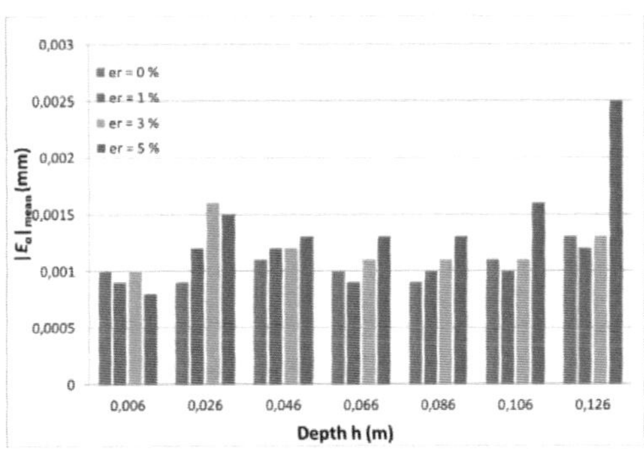

Figure 4. $|E_a|_{mean}$ vs depth, for e_r equals to: 0%, 1%, 3%, 5%.

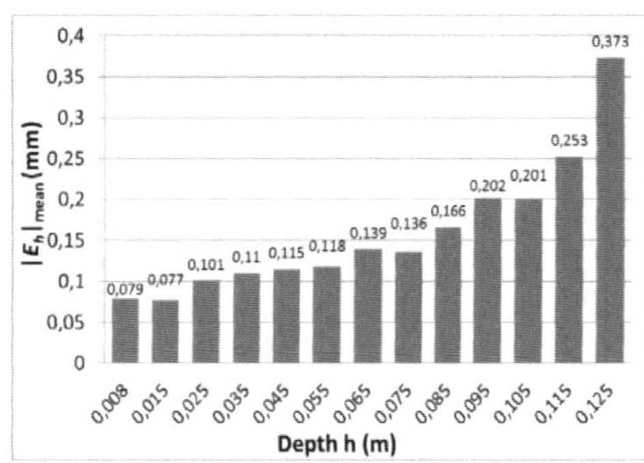

Figure 5. $|E_h|_{mean}$ vs depth, for $0\% < e_r < 5\%$.

According to the results shown in table 1, it is noticed that the developed method is an effective tool to estimate all of the desired parameters. By observing figures 2 and 3, it can be seen that the errors $|E_{xb}|_{\mathrm{mean}}$ and $|E_{yb}|_{\mathrm{mean}}$ tend to increase with depth h. In turn, by comparing the cases with the same h, it is verified that the errors $|E_{xb}|_{\mathrm{mean}}$, and $|E_{yb}|_{\mathrm{mean}}$ also tend to increase with the noise intensity e_r. In figure 4, it can be seen that there is a certain tendency to increase the error $|E_a|_{\mathrm{mean}}$ with the depth h and the noise intensity e_r, however this dependence is not as strong as that observed for the errors associated with the parameters x_b and y_b. In turn, figure 5 allows to observe that the error $|E_h|_{\mathrm{mean}}$ tends to increase with depth. Consequently, in general, as expected, the accuracy of the method worsens with increments of h and e_r.

4. Conclusions

The localization technique herein developed is based on the use of an alternating primary magnetic field source to induce parasitic currents in the fragments of firearm projectiles. In turn, these currents induce a secondary magnetic field, which can be measured by high resolution magnetic sensors. The measurements of these secondary magnetic fields can be processed in order to infer characteristics of its source. The present work focuses on the development of processing algorithms to determine the position (x_b, y_b), depth (h) and radius (a) of fragments of firearm projectiles, inserted into the human body, from the information contained in gradient maps of the secondary magnetic flux density.

The localization technique proposed here aims to aid the surgical extraction processes of non-magnetic metallic foreign bodies, such as lead projectiles, inserted in the human body. The proposed method does not subject the patient and medical team to ionizing radiation. It is imperative that the developed technique presents high precision and high repeatability in the determination of the parameters associated with the foreign body to be detected, in order to minimize the risks inherent to its surgical removal. The obtained results have showed that the proposed method is suitable for the intended application. The range of errors associated to the estimative of all variables were always satisfactorily small. Furthermore, the proposed technique presented high accuracy and repeatability.

Acknowledgments

The authors thank CAPES and CNPq for financial support to the development of this work.

References

[1] Monteiro E C, Kelber C, Lima E A, Ribeiro P C, Bruno A C, Barbosa C H, Ribeiro E P and Boechat P 1996 Magnetic Measurement Techniques for Locating Foreign Bodies in Humans *Biomag'96* (New York: Springer) pp 28-31

[2] Monteiro E C, Barbosa C H, Lima E A, Ribeiro P C and Boechat P 2000 Locating steel needles in the human body using a SQUID magnetometer *Phys. Med. Biol.* **45** 2389

[3] Barbosa C H, Monteiro E C, Lima E A, Santos S F, Cavalcanti E G and Ribeiro P C 2001 Improvement of a technique for localization of steel needles in humans using a SQUID magnetometer *IEEE Trans. Appiled Supercond.* **11**677–680

[4] William E and Smythe R 1968 *Static and Dynamic Electricity* (New York: McGraw-Hill)

[5] Silvester P P and Omeragic D 1996 Sensitivity maps for metal detector design *IEEE Trans. Geosci. Remote Sens.* **34** 788–792

[6] Barbosa C H 2004 Localization of firearm projectiles in the human body using a superconducting quantum interference device magnetometer: A theoretical study *Rev. Sci. Instrum.* **75** 2098–2106

[7] Justin A 2015 Biot-Savart Law *International Journal of Research* **2** 2348–6848.

[8] Fortaleza L G S 2016 Non-ferromagnetic metallic foreign body detection by eddy currents *Master Thesis* Programa de Pós-graduação em Metrologia, Pontifícia Universidade Católica do Rio de Janeiro (PUC-Rio)

[9] Lifshits M A 1995 *Gaussian Random Functions* (Netherlands: Springer)

[10] Dodge Y 2003 *The Oxford Dictionary of Statistical Terms* (Oxford: Oxford University Press)

Method of Monte Carlo simulation for the analysis of uncertainty for ultrasonic time-of-flight

F Santos, J Villanueva, R Gouveia and J. Silva
Department of Electrical Engineering, Federal University of Paraiba, JP 58051-900, Brazil
E-mail: felipeaugusto.santos@cear.ufpb.br, jmauricio@cear.ufpb.br, rafael.gouveia@cear.ufpb.br and joao.junior@cear.ufpb.br

Abstract. The measurement of wind speed may be performed using ultrasonic transducers based on the principle of measuring the time-of-flight. However, due to the propagation of the ultrasonic signal in the medium of measurement, noises are added which can be modelled by different natures that affect the measurement accuracy of transit time and therefore the speed of the wind. In this paper, the method of Monte Carlo Simulation was used for the analysis of uncertainties in the estimation of the ultrasonic transit time when considering different noise distributions of probability density functions such as Gaussian, Triangular and Uniform.

1. Introduction

The principle of measuring the wind speed using ultrasonic transducers is based on the time-of-flight estimation, which is the travel time of the ultrasonic wave measured as the time elapsed from the emission to detection (considering a pair of ultrasonic transmitter and receiver). The estimation of the transit time named ToF (Time of Flight) presents an increased uncertainty of the estimation when the measurement medium is influenced by noise, which could come from several forms such as electric, additive and multiplicative noises, thermal and more. Therefore, the ToF uncertainty is directly propagated to the wind speed measurement. This way, in the design of measurement systems using ultrasonic transducers, it becomes relevant to analyse the propagation of uncertainties in the estimative and measurement process of ToF.

The measurement uncertainty can be evaluated and expressed in several ways; two of them are the analysis by the Guide to the Expression of Uncertainty in Measurement (GUM) and the Method of Monte Carlo Simulation (MCS) [1] [2] [3]. GUM is utilized to determine a mathematical expression of the uncertainties in a process of measurement. However, it establishes some limitations where the uncertainties must be from random sources and known models of distribution. Moreover, the systematic uncertainties must be eliminated or not even exist. On the other hand, the method of Monte Carlo Simulation (MCS) allows us to build a probability density function (pdf) acquiring the statistical parameters of the variables studied [4]. In addition, the variables could have different distribution and not necessarily just Gaussian.

Given this scenario, this article aims to build probabilistic models associated with the estimation of the ultrasonic ToF considering distinct noise types present in the medium of measurement, with known distributions such as: Gaussian, Uniform and Triangular. For this purpose, the method of Monte Carlo simulation (MCS) will be used, performing under a computational model built to model the measurement process of wind speed using ultrasonic transducers.

2. Measurement of wind speed using ultrasonic transducers

Figure 1 illustrates the configuration for estimating the wind velocity, using a pair of ultrasonic transmitter - receiver. The transmitter transducer operates at a frequency of 40 kHz and emits a mechanical wave that passes through the medium and interacts with the air flow (wind). The medium introduces a random additive noise and attenuates the signal due to its impedance. When the receiving transducer detects the ultrasonic signal, it performs a domain change of mechanical energy into electrical energy. Usually, an amplification is required at the receiving stage [5] [6].

From the configuration of measurement are related the process variables that define the measurement of wind speed by:

$$\vartheta = \frac{1}{\cos\theta}\left(\frac{L}{ToF} - C\right) \quad (1)$$

Where: L is the distance between the ultrasonic transducers, θ is the alignment angle between the ultrasonic transducers and the wind direction, ToF is the ultrasonic transit time and C is the speed of sound. The speed of sound in dry air is influenced by temperature variations, which is expressed by:

$$C = 20.074\sqrt{273.15 + T} \quad (2)$$

Where, T is the medium temperature of measurement in °C.

Figure 2 illustrates the electrical signals in the transmitter (TX) and receiver (RX) ultrasonic transducer. The mathematical model of the electrical signal in TX is expressed by:

$$v_{TX}(t) = A_o \sin(\omega \times t) \quad (3)$$

Where: A_o is the generator amplitude, ω is the generator frequency, t is the time and $s(t)$ is the transmitted signal.

The mathematical model of the received ultrasonic signal can be built considering the attenuation of the medium into the signal, a delay due to the interaction with the medium (ToF), a phase difference due to the impedance of the medium and an additive noise (η). So, the modeling is given by:

$$v_{RX}(t) = A\sin(\omega(t - ToF) + \phi) + \eta \quad (4)$$

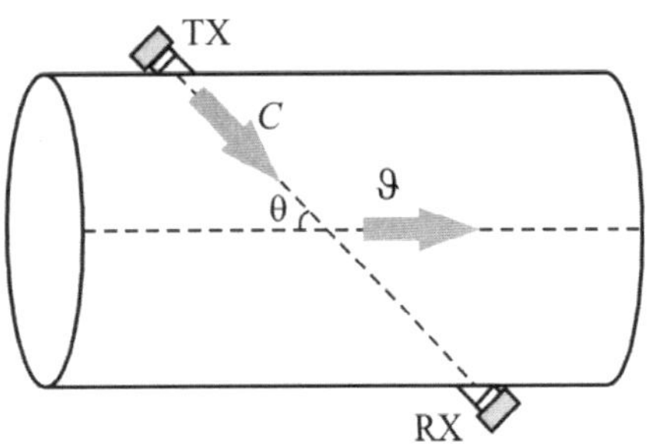

Figure 1. Measurement Configuration.

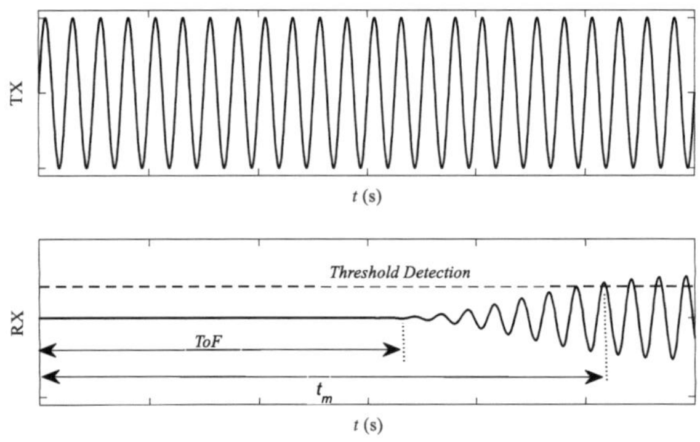

Figure 2. TX and RX signals.

From the received ultrasonic signal, the Threshold Detection method can be utilized for the ToF estimation. However, the estimation by this method can exhibit some changes in its probability density function (pdf) due to the model parameterization and the noise kinds present in the measurement medium, as illustrated in figure 3. ToF's pdf analysis can be made using the method of Monte Carlo Simulation (MCS) which is utilized in the evaluation of propagation of uncertainties as indicated in the supplement 1 to the GUM, where it demonstrates some procedures to find possible results associated to a distribution along with its respective probabilities of occurrence and its statistical moments [3]. To the pdf determination of ToF were utilized 3 kinds of noise distributions: Gaussian, Uniform and Triangular.

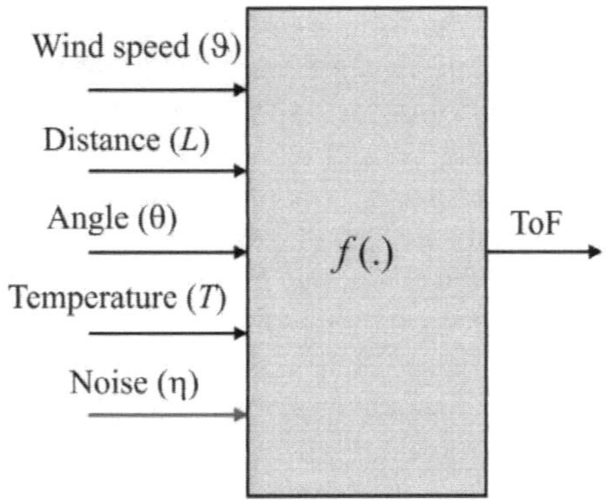

Figure 3. Model Parametrization.

In this model, the output ToF is the main variable to be estimated while the other independent variables are the inputs of the model parametrization, such as wind speed (ϑ), distance between the ultrasonic transducers (L), angle between the ultrasonic transducers and the wind direction (θ), medium temperature of measurement (T) and an additive noise (η). This model parametrization allows us to determine the combined uncertainties propagation by the following expression:

$$ToF = f(L, C, \vartheta, \theta) + \eta \tag{5}$$

where:

$$f = \frac{L}{C + \vartheta \cos\theta} \tag{6}$$

3. Monte Carlo simulation to estimate the uncertainties of the ToF

The method of Monte Carlo Simulation is defined as a class of statistical methods based on random sampling mass (random number) to obtain probability distributions of the variables of the problem, thus obtaining information about the future performance of systems or processes.

From the metrological point of view, the Monte Carlo simulation is a tool used for evaluation and propagation of uncertainties, which is indicated in the GUM Supplement 1, providing a range of possible outcomes associated with a distribution of their respective probabilities occurrence, providing useful information for the determination of statistical moments, such as mean, variance, standard deviation, kurtosis, and skewness. Figure 4 illustrates the flowchart for the estimating of ToF based on Monte Carlo simulation using different probability density functions (PDFs) of the additive noise [3].

This procedure has as inputs the mathematical model of the problem, which in this case was represented by two electrical signals associated with ultrasonic transducers transmitter (TX) and receiver (RX). Also, in the same level, it has the characterization of the influence of the uncertainties associated with an additive noise that is produced by means which propagates the ultrasonic waves. It is noticed that the received signal presents an additive noise that could have a probability density function (PDF) known that may be Gaussian, triangular or uniform shape. Additionally, we use an M number of simulations that for this work was 10000.

At the stage of processing, it was set up a random number generator capable of generating the distribution types Gaussian, Triangular and Uniform with known mean value and standard deviation, which are related to the properties of additive noise present in the propagation medium. To implement the method of Monte Carlo simulation, *M* sets of random numbers (simulations) were generated, where each set represents a signal of additive noise u_A.

The evaluation of measurement uncertainty using the MCS technique is performed in two phases. The first is to establish the model for the measurement, and subsequently, the second involves the evaluation of that model. The differences between more traditional methods and MCS are related to the several types of information describing the input parameters and the way this information is processed to obtain the measurement uncertainty.

Based on this procedure, we can show a sequence of actions necessary to evaluate the measurement uncertainty using the MCS, according to the Supplement Guide 1 to the Expression of Uncertainty in Measurement, entitled "Numerical Methods for the Propagation of Distributions." According to the GUM Supplement 1, from *M* simulations by the Monte Carlo method, PDFs were constructed of ultrasonic transit time (TOF). From these PDF forms were observed certain characteristics and extracted fundamental statistical parameters such as mean value (best estimate) and standard deviation (standard uncertainty).

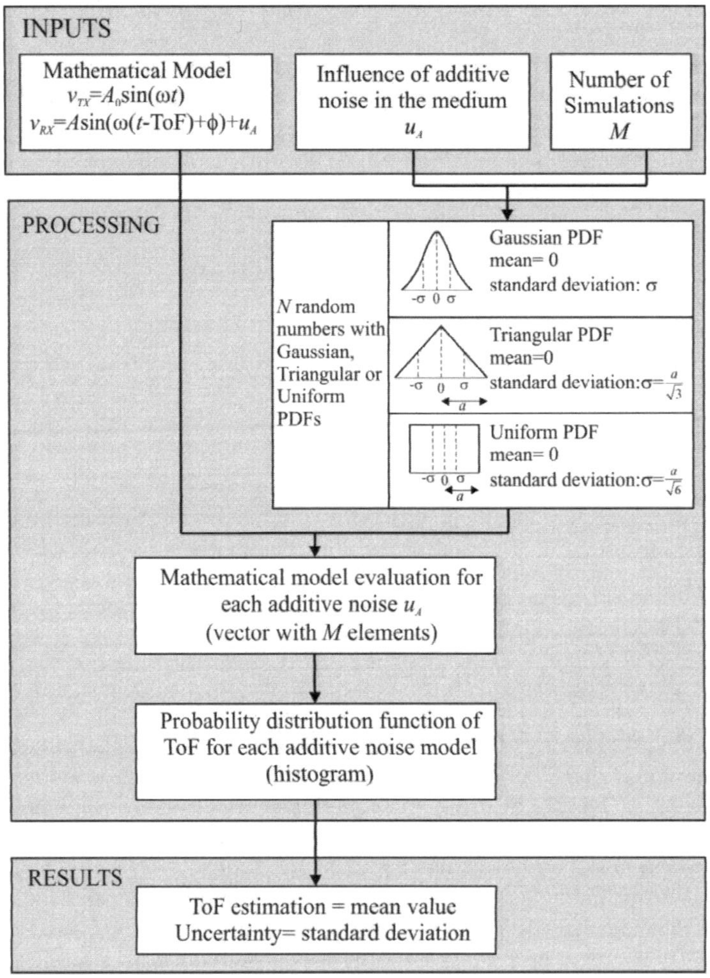

Figure 4. Procedure for the estimation of the PDFs of ToF for different PDF of the additive noise.

4. Evaluation of uncertainties of ToF using Monte Carlo Method

The system for measuring wind speed (figure 1) has been configured for a wind speed of 10 m/s, distance between the transducers $L = 0.1$ m, temperature $T = 20$ ° C, alignment angle $\theta = \pi/4$, transmitted electrical sign of 1 V amplitude, medium attenuation in 10% of the maximum amplitude signal. Furthermore, it was considered the presence of an additive noise with different probability distributions. The detection of the received signal utilizing the Threshold Detection method was carried out considering a threshold of 0.8 V.

Thus, simulations were performed using the Monte Carlo Simulation procedure (MCS) with $M = 10000$ (number of samples) to meet the shape of the distribution ToF when considering the diverse types of noises for the model.

Figure 5 illustrates the signal-to-noise ratio (SNR) in function of standard deviation of the medium noise. Three kinds of noise probability density functions were considered in the simulations: Gaussian, Uniform and Triangular (according to figure 3). It was observed the Gaussian noises provide results with smaller SNR. On the other hand, a noise with uniform distribution provides results with bigger values for SNR. Meanwhile, triangular distribution produces values for SNR between the maximum and minimum of it.

Figures 6, 7 and 8 illustrate results for the ToF's PDF acquired using MCS considering 3 noise types of PDF with standard deviation, $\sigma = 0.3$ mV. For each scenario of simulation, normalization tests were utilized for the respective ToF's PDF and in all the cases the results for the tests were positive to normal distribution or Gaussian distribution

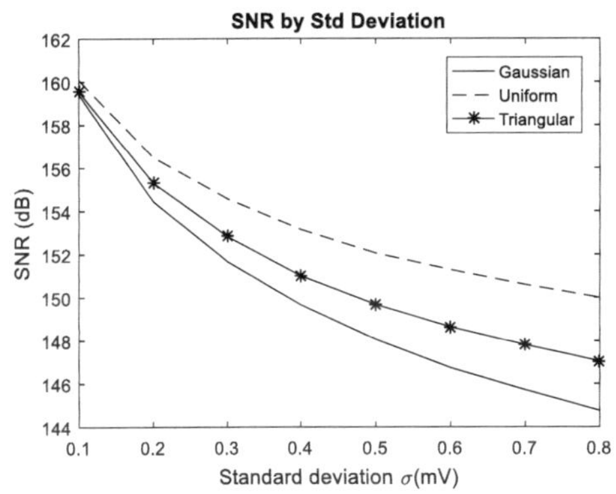

Figure 5. SNR by standard deviation.

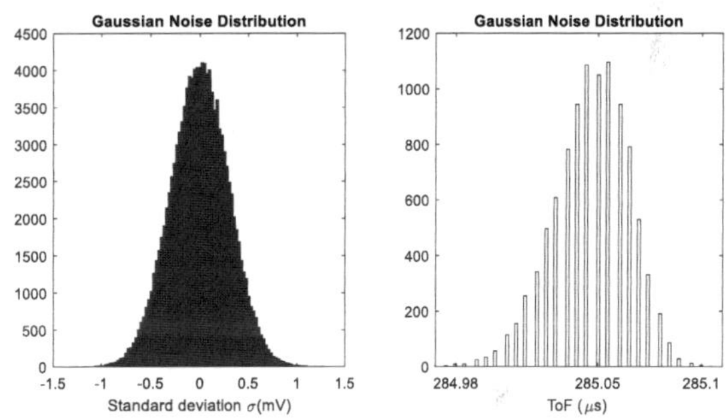

Figure 6. PDF of the ToF using Gaussian noise with σ=0.3 mV.

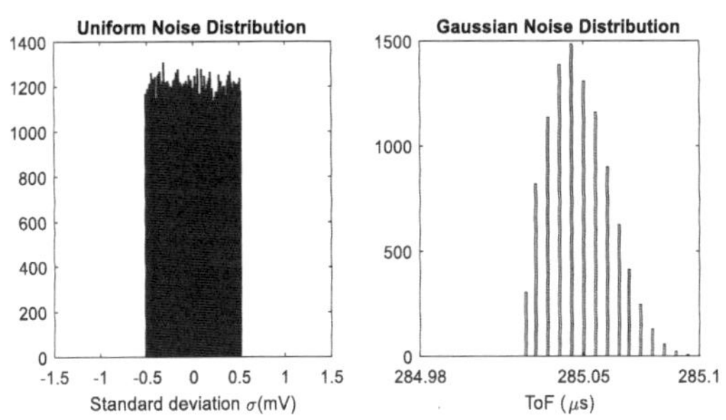

Figure 7. PDF of the ToF using uniform noise with σ=0.3 mV.

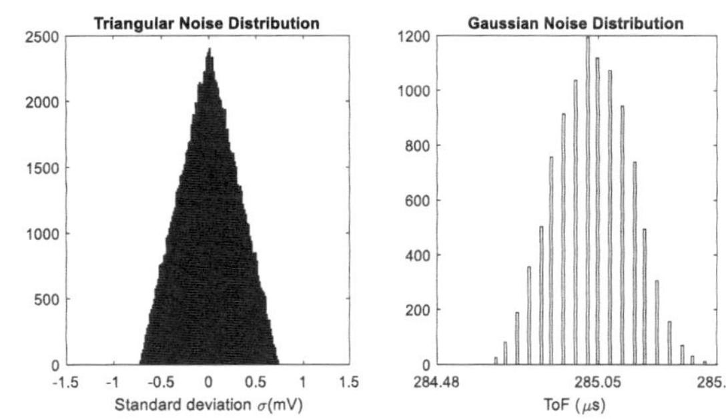

Figure 8. PDF of the ToF using triangular noise with σ=0.3 mV.

5. Conclusions

In this work, the implementation of the Monte Carlo simulation method has been shown to determine the probability distributions of transit of flight (ToF) for applications to wind speed measurement using ultrasonic transducers, when considering various forms of additive noise in the propagation medium (Gaussian, Triangular and Uniform). The proposed method considers the procedures of Supplement 1 to the Guide for the Evaluation and Expression of Uncertainty in Measurement (GUM). It was observed that the method of Monte Carlo simulation provided the ToF's PDFs which could be related to the shapes of the additive noise PDFs. Therefore, it is concluded that the shape of the noise PDF distribution influences the shape of the ToF's PDF. This result can be used to enhance the model of the system for measuring wind speed when used ultrasonic transducers whose measuring principle is the ultrasonic transit time (ToF).

Acknowledgments

The authors are thankful to the Pró-Reitoria de Pesquisa da Universidade Federal da Paraíba (PROPESQ) for the financial support to develop this research.

References

[1] JCGM 100:2008 *Evaluation of measurement data – Guide to the expression of uncertainty in measurement* Technical Report JCGM 2008

[2] Taylor B and Kuyatt C 1994 *Guidelines for Evaluating and Expressing the Uncertainty of NIST Measurement Results* NIST Technical Note 1297

[3] JCGM 101:2008 *Evaluation of Measurement Data – Supplement 1 to the Guide to the expression of uncertainty in measurement – Propagation of distributions using a Monte Carlo method* Technical Report JCGM 2008

[4] Billinton, R, and Allan R 1992 *Reliability Evaluation of Engineering Systems*. Second Edition pp 124 169

[5] Villanueva J, Catunda S and Tanscheit R 2009 Maximum-Likelihood Data Fusion of Phase-Difference and Threshold-Detection Techniques for Wind Speed Measurement *IEEE Transactions on Instrumentation and Measurement*. **58** 2189-7

[6] Duarte R, Villanueva J, Costa M, Freire R, Catunda S and Costa W 2013 Ultrasonic time-of-flight estimation for wind speed measurement based on time-frequency domain using STFT *19th IMEKO TC 4 Symposium* (*Barcelona, Spain, 18-19 July 2013*) pp 72-6

[7] JCGM 200:2012 *International vocabulary of metrology – Basic and general concepts and associated terms* (*VIM*).

Metrological concerns in multiphase flow measurement

E Silva[1], I T Chacon[1], D A Garcia[1], M Rosendahl [1,2,3], R S Teixeira[1,4] and M H Farias[1]

[1]Fluid Dynamics Metrology Division - Dinam, Directory of Scientific Metrology and Technology - Dimci, National Institute of Metrology, Quality and Technology - Inmetro, Av. Nossa Senhora das Graças, 50 - Xerém, 25250-020, Duque de Caxias - RJ, Brazil
[2] Rio de Janeiro State University – UERJ, Mechanical Engineering Department, R. Fonseca Téles, 121 - Rio de Janeiro - RJ, 20940-200 mila@uerj.br and [3] Estácio de Sá University – UNESA, R. Eduardo Luiz Gomes, 134 - RJ, 24020-340. mila.rosendahl@estacio.br . [4]Federal Rural University of Rio de Janeiro-UFRRJ, rsteixeira@ufrrj.br, Rodovia BR 465, km 07, s/n - Zona Rural, Seropédica - RJ, 23890-000

E-mail: mhfarias@inmetro.gov.br

Abstract. This work presents partial result of an introductory investigation on parameters of influence for the multiphase flow measurement reliability, which is going on in the Brazilian National Metrology Institute - Inmetro. Studies were performed with air-water mixture flowing inside a horizontal pipe. The flow pattern map of the experimental bench is identified, and parameters that influence for the flow pattern formation are analysed. Paper aims to disseminate knowledge on metrological bias on multiphase flow measurement. A brief overview on this theme is presented, along with the latest results of this research line at Inmetro.

1. Introduction

Static and dynamic fluid measurements play an important role in industrial, environmental, safety, medical, aerospace processes, among others. Moreover, metrological requirements are in increasing need for modern industrial solutions that assure exact measurements involving a wide range in volume, and the need for continuous or stepped sampling collection. Examples are the quantities of fluid that must be controlled in pharmaceutical and biological processing, or that one dealt in the oil and gas industry, as well as those related to environment pollutant monitoring, [1-2]. Nevertheless, regardless of the application, reliability of the measurement results is a must demanded.

The reliability in multiphase flow measurement has been receiving increasing attention in the last decades, and it has been driven by industrial demands and also by the need for sustainability. Important progress has been observed in this field, however, as the dynamics of the physical processes present in multiphase flow are very intricate, the metrological community has been facing significant challenges while pursuing to provide confidence in the measurement of this type of flow.

The metrological traceability chain for evaluating flowmeters to be employed in the measurement of so complex flow and also for evaluating the systems where these meters could be calibrated, claim for substantial advances on better understanding of the phenomena. Thus, aiming to contribute to this field, the Brazilian National Institute of Metrology, Quality and Technology - Inmetro has introduced in its activities the research line on multiphase flow measurement.

Until the authors of the present article could search, scarce works in the literature are focused in discussing metrological concerns on multiphase flow measurement. So, the goal of this work is contribute for knowledge dissemination about metrological bias related to multiphase flow measurement, by presenting a brief overview of the main consensuses on this theme, and also showing the lessons learned by the authors during the process of studying mixtures of liquid-gas flowing inside horizontal pipes.

2. General parameters which influence the uncertainty level and reliability of a multiphase flow measurement

Although advances of technology has been also motivated by interests on better understanding of the multiphase flow behavior, it is known that the complexity inherent to the physics of this flow and fluid characteristics influence the limits for using such technologies. Succinctly, the measurement result of a multiphase flow will depend on parameters as: i) the flow behavior (flow pattern, phase regime, instabilities, flow direction etc.); ii) phase properties (surface tension, viscosity, density, electrical and thermal conductivity, salinity etc.); iii) flow condition (emulsion, water cut, sediment, homogeneity, temperature, pressure, time dependency, stability etc.); iv) flow region (ex. geometry, scale, pipe material and roughness etc.); v) facility characteristics and associated boundary conditions (vibration level, connections, straight length, rigidity, temperature of the solid surfaces in contact with the fluid etc.); vi) environmental conditions; vii) instrument, technology and technique of measurement (dimensions, operation principle, equipment material, data rate, nominal range etc.), among others. Some of these parameters are usually correlated in the solution of equations involving multiphase flow.

A reliable quantification of a multiphase flow (i.e., totalized volume during a time interval, flowrate and velocity of each phase) can be strongly affected by several of those cited parameters, thus, the establishment of standardization for providing metrological traceability will request a compromise between technologies, metrological traceability of all involved quantities and the possibility of controlling and monitoring the flow.

In the specific case of multiphase meters, there are currently several commercially available models based on different measurement principles, and for each one a mathematical model is necessary to convert readings to flow patterns [3]. So, if the flow pattern is misinterpreted, all related quantities will not be reliable. This fact makes clear the need for better investigation of the effects on the propagation of measurement errors.

3. Aspects considered in the study of the two-phase flow in horizontal pipe

Aiming to understand how to get accuracy and reliability when using multiphase flowmeters, experimental investigations on two-phase flow were chosen as starting point in this direction. Two central questions have oriented the investigations:

- How to ensure the metrological traceability?
- What parameters must be reproduced in a laboratory bench, in order to create calibration conditions which would represent the field operating conditions of a multiphase meter?

To reach the goals, the first step was the construction and characterization of a horizontal bench. The two-phase flow patterns and other parameters for the available operating conditions were also analyzed as a way of creating repeatability conditions necessary for the execution of future tests.

The optical technique Shadow Sizer [4] was initially used to identify the flow patterns. This technique consists of a high resolution camera and a source of light, positioned in opposite one to other. The shadow of the particles (bubbles, for example) present in the flow is captured, and the diameter, area, perimeter and velocity of the bubbles can be identified after processing the images captured by the camera. Table 1 summarizes the characteristics of the bench and the main specifications.

Table 1. Test bench characteristics.

Description	Specification
Total length of the flow visualization field (acrylic piping)	10.3 m
Internal nominal diameters of acrylic pipes (D)	19 mm e 44 mm
Maximum liquid and air flowrates in 25 mm piping	20 L/min and 40 L/min, respectively
Maximum liquid and air flowrates in 50 mm piping	250 L/min and 60 L/min, respectively
Inner diameter of each air distributor valve injector	5 mm

In the bench, filtered water flows from the reservoir to the piping. Water flowrate is monitored by calibrated electromagnetic flowmeters, and the range is established by a valve and a hydraulic pump (controlled by frequency inverter). Compressed air is injected in the line through diametrically opposed four-way distributor valve. Calibrated thermal mass flowmeters monitor the air flowrate in each valve. Figure 1 shows a view of the bench. In the present study, the discussions were based on the flow inside the piping of 44 mm internal diameter.

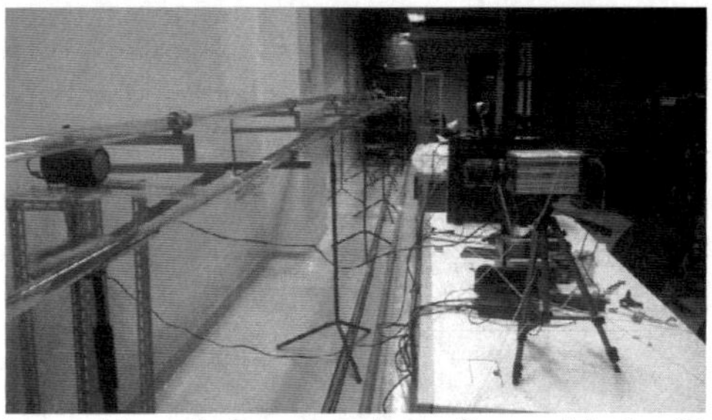

Figure 1. View of the horizontal bench for two-phase flow with the Shadow Sizer system.

4. Experimental procedure and results

Firstly, the mapping [5] of occurrence of two-phase flow patterns in the 50 mm piping was detailed, on which is indicated the water and air supply limits of the bench. Figure 2 shows such map.

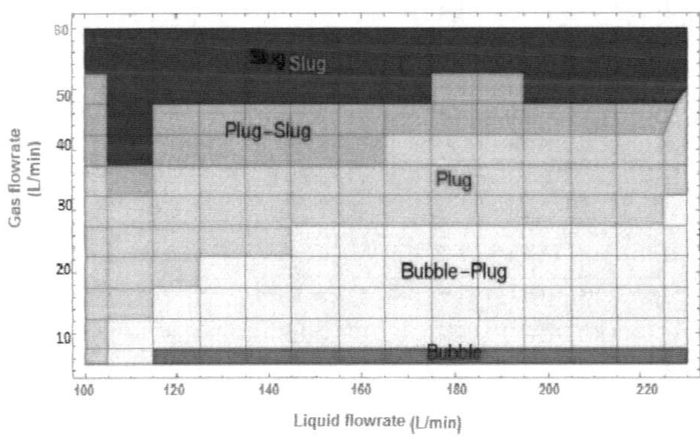

Conditions of the two-phase flow characterization:

Ambient temperature: 23 °C to 26 °C;
Humidity: 62 % to 80 % ;
Ambient pressure: 1007 hPa to 1017 hPa;
Liquid temperature: 25 °C to 31 °C ;
Gas temperature: 20 °C to 22 °C;
Range of air pressure at the pipe inlet: from 39.2 kPa to 147 kPa.

Figure 2. Map of flow patterns in the piping of diameter 50 mm.

Based on this map, after running some tests in the bench, for analizes were selected experiments considering the cases listed on Table 2, with some gas and liquid flowrates not indicated on the map. In such experiments, while maintaining the mean water flowrate value around 56 L/min, and then varying both the mean air flowrate and the air injection distribution at the tube entrance, the flow behavior at the tube entrance region and the established two-phase flow pattern far from the tube entry were observed. Air was injected at the tube entrance through two valves which were vertically aligned with each other (numbered 1 and 2, for north and south positions, respectively) and crossing the symmetry axis of the piping. The frequency level of the hydraulic pump motor was maintained, but the resulting water flowrate showed fluctuations in each case due to changes in the internal piping pressure promoted by changes in the amount of injected air flowrate. The acrylic tube of the bench can resist to internal pressure up to 0,26 MPa.

Table 2. Flow conditions.

Case	Liquid flowrate (L/min)	Gas flowrate (L/min)	
		Injector 1 (north)	Injector 2 (south)
1	56.4	10.9	0.31
2	56.6	10.7	1.5
3	56.5	10.8	3.0
4	57.4	10.9	6.28
5	58.5	10.55	12.18
6	57.4	6.28	10.9
7	56.4	1.52	10.8

Influences due to different air injection mode, considering a same air-water flowrate ratio, could be discussed by comparing case 2 with case 7, and case 4 with case 6. The following figures show the captured images through the Shadow Sizer system.

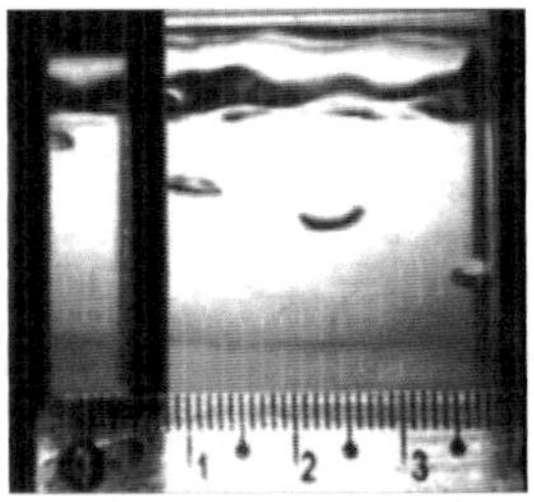

(a)

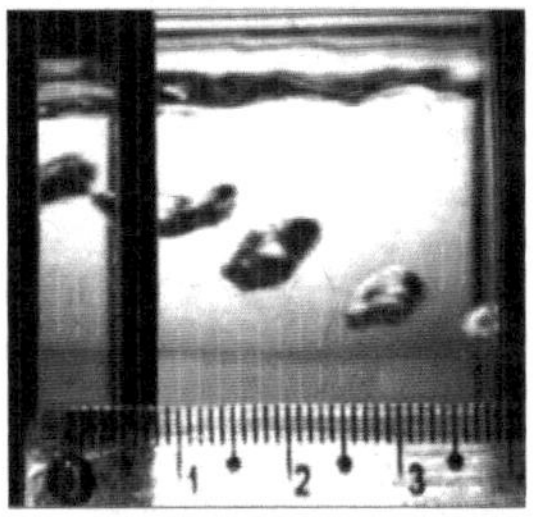

(b)

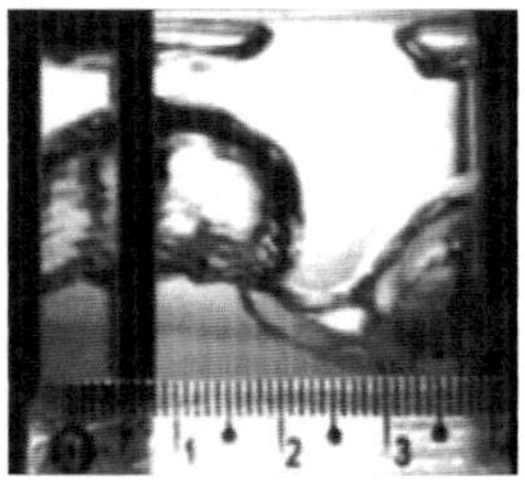

(c)

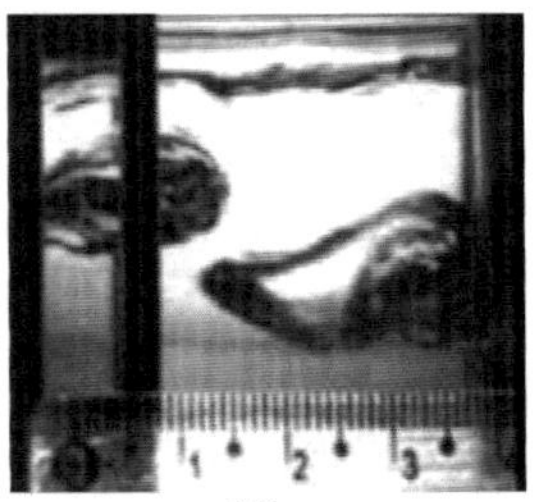

(d)

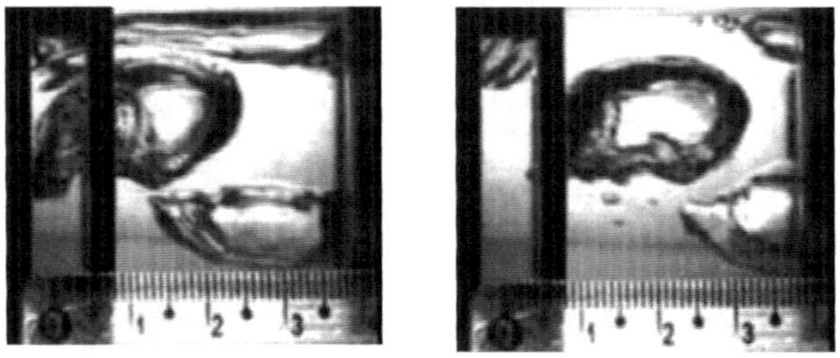

(e)

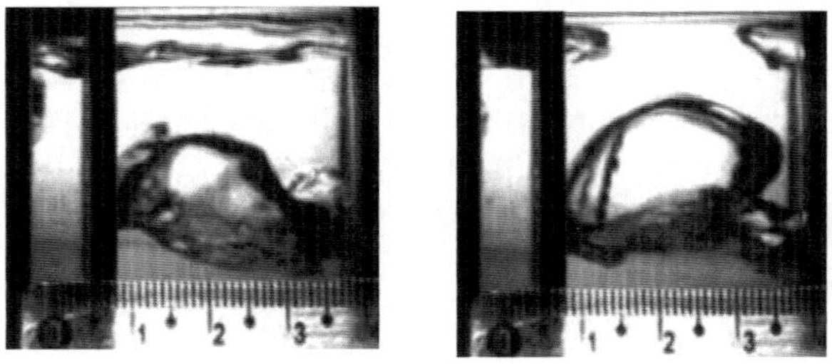

(f)

Figure 3. (a) Case 1 **(b)** Case 2 **(c)** Case 7 **(d)** Case 4 **(e)** Case 6 (two frames) **(f)** Case 5 (two frames).

Case 5 is depicted on Figure 8, on which can be identified the effect, to the flow behavior at the entrance region, of the increment in air flowrate when the amount of air injected through south is greater than that one injected through north.

Table 3 indicates the two-phase flow pattern established far from the tube entrance for each case, and associated comments.

Table 3. Established two phase flow (approximately 5 meters away from pipe entrance).

Case	Flow pattern	Comments
1	semi-slug	plug[a]
2	plug	well defined plug[b]
3	plug	well defined plug[b]
4	plug	well defined plug[b]
5	plug	plug (predominant)[b]
6	plug	almost slug[b]
7	slug	almost semi-slug[b]

[a] near air injection point.
[b] 5m far from pipe entrance.

5. Discussion

Among the several cases listed on Table 2, the most representative are those depicted on Figures 3 to 8. From Figures 3 and 4 it can be noticed the similarity of the flow appearance of cases 1 and 2 near the air injection point. However, although the mean air flowrate injection by south in case 1 is about 5 times less than that in case 2, in case 1 the flow pattern built up near the air injection point was plug, but the established flow pattern at 5 meters far from the entrance was semi-slug. The semi-slug pattern which was identified was very instable, marking a transitional condition of the flow behavior in that pipe position. In case 2, very close to the air injection point the plug pattern was already defined. Otherwise, just after the air injection point, in both cases the characteristics of the gas phase due to north air injection were similar, i.e., the gas bubbles generated by the air inserted through south injector did not separate the upper gas layer (it was observed a wavy flow kind at this region of the pipe, as shown on Figure 4, but the wavy flow characteristic was changed at the point where the bubbles touched the upper air layer, then developing the plug pattern).

In case 7, the air flowrates inserted through south and north were inverted in relation to case 2. Figure 5 shows the resulting effect on the flow characteristics just after the air injection point. As indicated on Table 3, the flow pattern built up close to air injection position was almost semi-slug, while 5 meters away this point the flow pattern was slug.

An equivalent comparison was made between cases 4 and 6 (Figures 6 and 7), but in these cases the lower air injection flowrate was four times higher than that one of case 2. In these cases, as seen on

Table 3, the flow pattern far from the air injection point were similar (plug), but in case 6 the plug flow was weakly outlined (trending to slug), showing that in such tube position the flow pattern probably was going to transition. At the air injection region, while in case 4 the liquid surface was wavy (Figure 6), in case 6 large air bubbles reach the water surface and disturbed the wavy profile in that point (see the two frames on Figure 7), then provoking the plug flow formation since this piping position. Figure 8 (case 5) also shows the changes of the upper air layer at the air injection region.

So, based on the cases observed up to here, an interesting finding by comparing mainly the cases 2 and 7, 4 and 6, is that, for some air and water flowrate ratios, the mode of air injection could influence the two-phase flow pattern formed inside the tube, even up to large distance after the air injection point (in this study, the position of visualization was 5 m, i. e., about 100D).

6. Conclusion

With the aim of getting elucidations on how to deal with some parameters which impacts on multiphase flow measurement reliability, experiments involving two-phase flow were conducted in a horizontal pipe. The flow mapping for the two-phase flow was pursued, and five flow patterns were developed inside the pipe of diameter 50 mm. It was observed that depending on the air-water flowrate ratio it can occur repeatability of the flow pattern at a same position far from the pipe entrance. In addition, when considering a same air-water flowrate ratio, within the range of air flowrate tested, the results indicated a tendency of the mode of air injection to influence the flow pattern characteristics at a fixed position far from the tube entrance. Besides, it was possible to verify that the wavy pattern generated upstream the air injection point was influenced by size, rotation and buoyancy of the air bubble injected at the lower part of the pipe. However, more investigation about such influences must be done in order to clarify several aspects of the flow behavior. In the continuity of this research other parameters for discussion should be included, such as variables of influence which must be monitored in a better way, as is the case of the liquid temperature increasing due to pumping. In fact, all these findings indicate the need to carry out multiphase flow measurements with a metrological approach, and those aspects will be investigated in the near future.

Acknowledgement

Authors are grateful to CNPq, Faperj, Finep, Petrobras, Pronametro, Prometro, Inmetro and Productivity Research Award - UNESA.

References

[1] Falcone G, Hewit G F and Alimonti C 2010 *Multiphase Flow Metering – Principles and Applications* Volume 54 (Great Britain: Elsevier) 329 p
[2] Brennen C E 2005 *Fundamentals of Multiphase Flow* (Cambridge: University Press) 345 p
[3] Primrose K M, Qiu C, Talmon A M, Glen N, Toss A and Learmouth D 2010 Visualisation and measurement of 3 phase (air water oil) flow and 2 phase flow (sand water) with electrical tomography *6th World Congress on Industrial Process Tomography*
[4] Settles G S 2001 *Schlieren and Shadowgraph Techniques* (Berlin: Springer) 376 p
[5] NOFGM and Tekna 2005 *Handbook of Multiphase Flow Metering* revision 2 (Norway: NOFGM and Tekna) 115 p

IOP Conf. Series: Journal of Physics: Conf. Series **1044** (2018) 012047 doi:10.1088/1742-6596/1044/1/012047

Modeling, characterization and test of a pediatric ventricular assist device

D.L. Sousa[1], T.D. Cordeiro[1], T.R. Melo[1], J.S. da Rocha Neto[1], I.A. Cestari[2], and A.M.N. Lima[1]

[1]Graduate Program in Electrical Engineering - PPgEE, Department of Electrical Engineering - DEE, Federal University of Campina Grande - UFCG, Campina Grande - PB, Brazil
[2]Bioengineering Division, Heart Institute, InCor, São Paulo - SP, Brazil

E-mail: {daniel.sousa, thiago.cordeiro, thamiles.melo}@ee.ufcg.edu.br, {zesergio, amnlima}@dee.ufcg.edu.br, idagene.cestari@incor.usp.br

Abstract. This paper presents a 0D model for the pulsatile pediatric Ventricular Assist Device (pVAD), as well as its characterization and test results obtained by using an experimental hydraulic system that provides operational conditions close to the physiological conditions observed in pediatric patients. The measurements of the physical quantities used in characterization of the pVAD have been obtained by using the LabVIEW platform. The air chamber pressure signal is used to model the pneumatic actuation system behaviour, whereas the pressure-volume relationship inside the blood chamber is applied to model the pVAD compliance. The set of test and validation results demonstrated the correctness of the proposed model.

1. Introduction

Heart transplantation is still the best alternative for the treatment of terminal heart disease. However, the number of patients on the heart transplant waiting list is large due to the limited availability of donor heart, what results in a high rate of death among these patients. Thus, mechanical circulatory support devices that assist in the pumping of the heart, also denominated as Ventricular Assist Devices (VADs), can be applied in these patients to maintain the hemodynamic variables at physiologically acceptable levels and to prolong the life expectancy [1].

Before the implantation of VADs, lumped parameter models, usually referred as 0D model, can be used to evaluate the interaction of these devices with the human cardiovascular system. A 0D model uses the concept of analog electrical circuit, in which the electric current represents the flow in blood vessels and the electrical voltage represents the pressures in the circulatory system [2].

The modeling of these devices is, in general, realized by physical tests in hydraulic system, to avoid *in vivo* experiments. These hydraulic systems are of great importance to correct problems and to improve the performance of VADs, as well as to assist in the development of control techniques [3].

In Brazil, a widely used pneumatic type of VAD is developed by Bioengineering Division of the Heart Institute, Hospital das Clínicas, Faculty of Medicine on the University of São Paulo (InCor - HCFMUSP). This device is used as a bridge to the transplantation procedure and is implanted in adult patients with chronic heart failure, which are on the heart transplant waiting

IOP Conf. Series: Journal of Physics: Conf. Series **1044** (2018) 012047 doi:10.1088/1742-6596/1044/1/012047

list. In the case of pediatric patients, a new pediatric device called pVAD is being developed at the same Institute.

To understand the influence of the pVAD in the circulatory system of pediatric patients, a 0D model of this device will be developed in this work. For that, the pVAD is coupled to a hydraulic system so that physical tests can be performed and measurements of pressure and flow of this VAD can be recorded and used in the estimation procedure. The organization of the paper is as follows. The hydraulic system and the methods used to model the pVAD are presented in Section 2. In section 3, physical measurements of the hydraulic system and the pVAD are compared with the estimated 0D model. Concluding remarks are presented in Section 4, and some directions for further work are suggested.

2. Materials and Methods

This section describes the hydraulic system and presents the methods to model the pVAD.

2.1. The hydraulic system

The hydraulic system that was used to evaluate the pulsatile pVAD is shown in figure 1. It is basically composed by: a pVAD, a pneumatic actuation driver, two acrylic reservoirs (aortic and atrial) and cannulas. A simplified schematic of this system can be seen in figure 2.

The pulsatile pVAD (15 mL ejection volume) has an air chamber, a blood chamber with two opening and a membrane between them. There are one-way biological valves on both the inlet and outlet that only allow the inflow or outflow liquid of the pVAD. The liquid used in the experiments is a solution composed by $\frac{2}{3}$ of saline and $\frac{1}{3}$ glycerin to reproduce the blood viscosity. To avoid fungal proliferation, 1% of benzyl alcohol was added in the solution.

The pneumatic actuation driver consists of one solenoid valve (SV), used for the control the filling and the ejection times of the pVAD, and of two pressure regulators that are used for the adjustment of both ejection pressure (P_e) and filling pressure (P_f). This driver generates periodically a membrane displacement which reproduces the pulsation and pumping of the liquid. The acrylic reservoirs are called aortic and atrial because they provide pressure levels close to pressure levels in a pediatric patient.

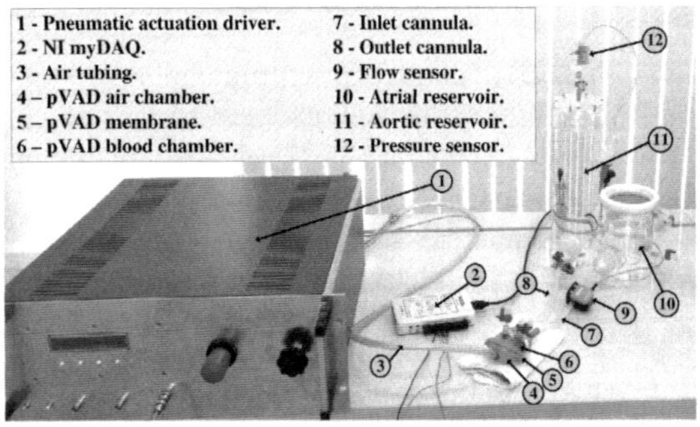

Figure 1. The hydraulic system coupled with the pVAD and pneumatic actuator driver.

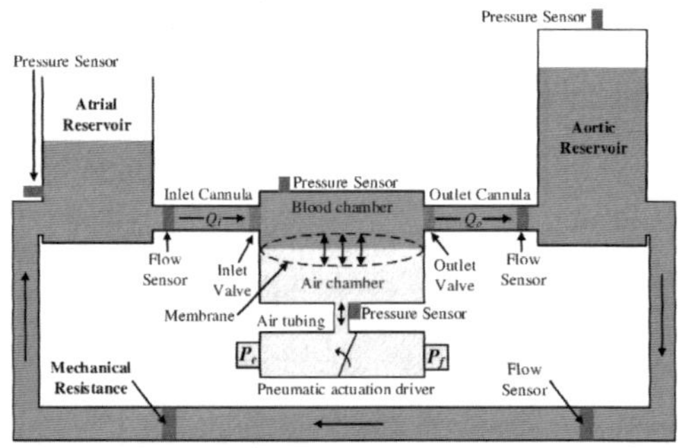

Figure 2. Simplified schematic of the hydraulic system coupled with the pVAD.

The available signals in this system are: aortic pressure (P_{ao}) at the aortic reservoir, atrial pressure (P_a) at the atrial reservoir, blood chamber pressure (P_{bc}) at the blood chamber of the pVAD, air chamber pressure (P_{ac}) at the air chamber of the pVAD, inflow (Q_i) at the inlet cannula and outflow (Q_o) at the outlet cannula.

To read and save these signals from sensors, a 16 bit National Instruments (NI) myDAQ module was used, which is connected to the computer by means of USB cable. The signals can be monitored by an Human Machine Interface (HMI) implemented on LabVIEW platform. The flow signals, Q_i and Q_o, are acquired using one non-invasive ultrasound flow sensor [4], which gives the mean flow (L/min).

There is one pressure sensor based on a strain gauge that is used to measure the pressure signals P_{ao}, P_{bc} and P_a (mmHg), one at a time in the system [5]. Inside of the pneumatic actuation driver, there is another dedicated pressure sensor, also based on a strain gauge, used to measure the signal P_{ac} [6], which is the signal used to synchronize all measurements, i.e., all signal are recorded along with the P_{ac}. This synchronization method is valid because the entire system operates in a periodic cycle, as shown in Torres *et al.* [7].

2.2. Model of the pneumatic actuation system

For the complete modeling of the pVAD behaviour, it is necessary to model the pneumatic actuation system. This system is similar to that described by Ferrari *et al.* [8], as highlighted in a simplified schematic shown in figure 3. The SV output, with pressure levels P_e or P_f, is connected to the air chamber by an air tubing.

Based on this description and using the analogy between hydraulic and electric systems, the 0D model of the pneumatic actuation system is modeled as a RC circuit where: the resistor R_{ac} represents the air tubing resistance; the capacitor C_{ac} represents the compliance of the air chamber; and the pressure P_{ac}, equivalent to the voltage across the capacitor C_{ac}, is the air chamber pressure in according to input pressure P_e or P_f. Since there is no fluid exchange through the membrane, an interface between the pneumatic actuation system and the blood chamber is necessary to separate them. The electrical element chosen for that purpose was a voltage controlled voltage source, αP_{ac}, where α represents the pressure drop across the membrane. In this work, this pressure drop was negligible and, therefore, $\alpha = 1$. Figure 4 shows the 0D model of the pneumatic actuation system.

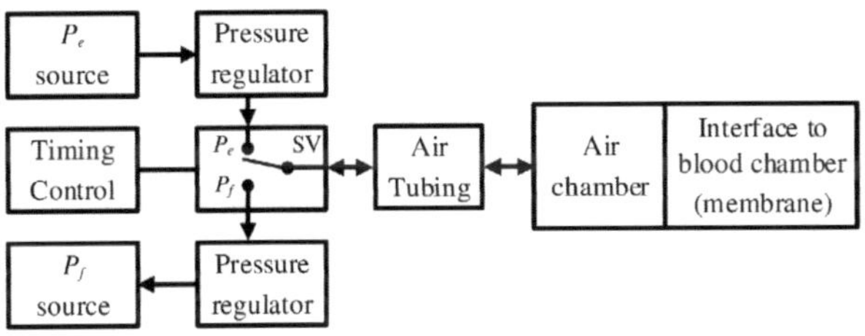

Figure 3. Simplified schematic of the pneumatic actuator system.

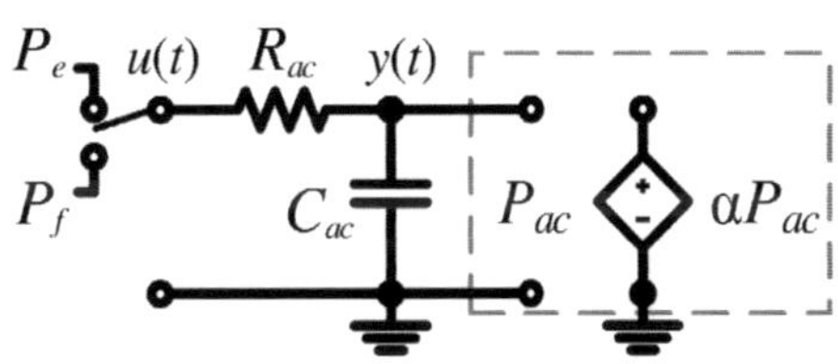

Figure 4. 0D model of the pneumatic actuation system.

The resistance R_{ac} can be calculated according to the air tubing resistance using the Poiseuille law as follow [9]:

$$R_{ac} = \frac{8\eta l_{ac}}{\pi r_{ac}^4} \tag{1}$$

where η is the air viscosity, l_{ac} is the length of the tube and r_{ac} is the radius of the tube.

The z-domain transfer function of the first order RC system, using the Euler discretization method, are the following:

$$\frac{Y(z)}{U(z)} = \frac{b_1}{1 - a_1 z^{-1}} \tag{2}$$

where $a_1 = R_{ac}C_{ac}/(R_{ac}C_{ac} + T_s)$, $b_1 = T_s/(R_{ac}C_{ac} + T_s)$ and T_s is the sampling time of the system. The least squares method (LSM) was used to determine the parameter vector $\theta = [a_1 \quad b_1]^T$. In turn, the time constant $\tau_{ac} = R_{ac}C_{ac}$ can be calculated using these parameters. With the value of R_{ac}, calculated by equation 1, the value of the parameter C_{ac} is calculated directly.

2.3. Model of the pVAD

The modeling of the pVAD is based on the study done by Hunsberger [10]. Firstly, the device was decoupled from the hydraulic system, the two valves were removed and the output of the pVAD was blocked. From these preliminary steps, a static test was performed to obtain the compliance of the blood chamber (C_{bc}) based on the pressure-volume (PV) relationship inside this chamber. The procedures of this test are:

(i) a syringe was coupled to the pVAD input to fill this device injecting 35 mL of solution;
(ii) quantities of 1 mL were withdrawn from the inside of pVAD using the same syringe;
(iii) the pressure (P_{bc}) and the blood chamber volume (V_{bc}) were recorded each 1 mL.

At the end of the static test, a PV curve is generated with the recorded values. The compliance C_{bc} is calculated according to the following expression:

$$C_{bc} = \frac{\Delta V_{bc}}{\Delta P_{bc}} \tag{3}$$

where ΔV_{bc} is the change in the blood chamber volume and ΔP_{bc} is the change in the blood chamber pressure. The pVAD is designed to work in a safe operating region, defined from 24 mL to 33 mL. Inside this operating region, the compliance C_{bc} can be approximated by a linear function and its value is given by the inverse of the PV curve slope.

In this work, a 0D model was developed for the cannulas, pVAD and pneumatic actuation driver. The hydraulic system was used to produce reasonable pressure levels at the input and output of the pVAD cannulas. The model structure and a simplified schematic of the pVAD, cannulas and the pneumatic actuation driver are shown in figure 5. The biological valves were considered ideal and the opening and closing state was modeled by the diodes D_i and D_o [10]. The inductances L_i and L_o are both calculated as follows [9]:

$$L_{i,o} = \frac{\rho l_{i,o}}{\pi r_{i,o}^2} \tag{4}$$

where ρ is the density of the fluid, $l_{i,o}$ is the length of the cannulas and $r_{i,o}$ is the radius of the cannulas. The resistance of the inlet and outlet cannulas, R_i and R_o respectively, were calculated by the following pressure relationship: $P_a - P_{bc} = R_i Q_i + L_i \dot{Q}_i$. At the instant of the maximum flow, the flow derivative is null and $\hat{R}_i = (P_a - P_{bc})/Q_i$ and $\hat{R}_o = (P_{bc} - P_{ao})/Q_o$. The value of the maximum instantaneous flow was obtained by the em-tec flow sensor using a sampling rate of 10 Hz. Thus, it is possible to obtain the values of theses resistance in a more accurate way then using equation 1.

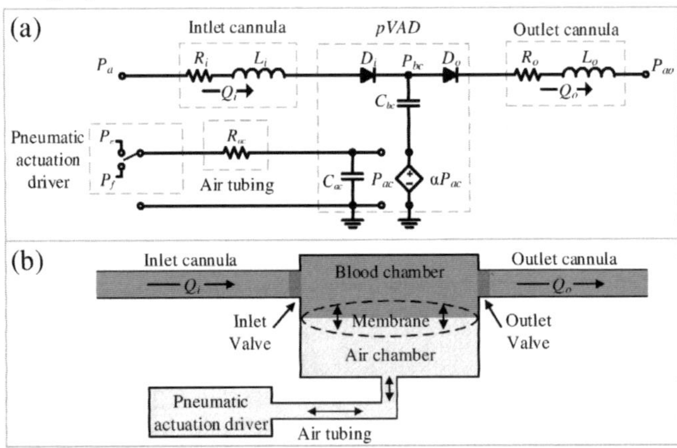

Figure 5. (a) 0D model and (b) simplified schematic of the pVAD, cannulas and pneumatic actuation driver.

3. Results and Discussion

Figure 6 shows the input pressure signal with values $P_e = 180$ mmHg and $P_f = 0$ mmHg. As can be seen in this figure, the measured pressure P_{ac} has a behavior very similar to a RC circuit response. That way, equation 1 leads to $R_{ac} = 0.25$ mmHg.s/mL. The value of the time constant, calculated using the LSM and the estimation dataset ($N_e = 1600$ samples), is $\tau_{ac} = 0.0343$ s. So, the compliance of the air chamber is $C_{ac} = 0.1370$ mL/mmHg. All measured signals were recorded with a sampling time $T_s = 0.001$ seconds.

Using the proposed 0D model depicted in figure 4, the estimated air chamber pressure $\hat{P}_{ac}$ was very close to the measured signal. The dataset used to validate this model was different from the estimation dataset, but with the same size ($N_v = 1600$ samples). The normalized mean square error (NMSE) between the measured and estimated curve is defined as $NMSE = \frac{1}{N_v}\sum_{k=1}^{N_v}\left(\frac{P_{ac}}{\max(P_{ac})} - \frac{\hat{P}_{ac}}{\max(\hat{P}_{ac})}\right)^2 = 6.28 \times 10^{-4}$.

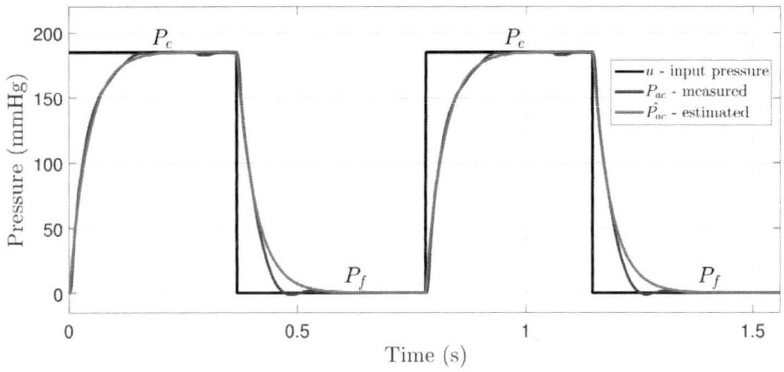

Figure 6. Input pressure (P_e and P_f) and pressure curves in the air chamber (P_{ac} and $\hat{P}_{ac}$).

Figure 7 shows the PV curve obtained by the static test. The value of the blood chamber compliance is $C_{bc} = 0.2521$ mL/mmHg, with $P_{ac} = 0$ mmHg, in according to equation 3.

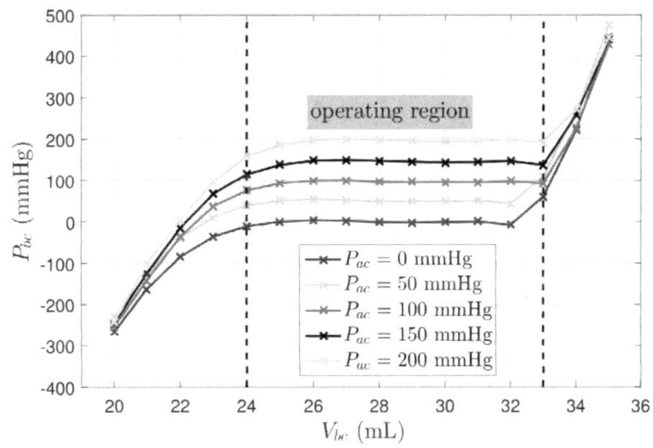

Figure 7. PV curve obtained by the static test and used to obtain the value of compliance C_{bc}.

Figure 8 shows the curves of the measured and estimated blood chamber pressures (P_{bc} and $\hat{P}_{bc}$), with $R_i = 0.5$ mmHg.s/mL, $R_O = 1.3$ mmHg.s/mL and $L_i = L_o = 0.04$ mmHg.s^2/mL. The P_{bc} curve was obtained with the pVAD coupled to the hydraulic system and the $\hat{P}_{bc}$ curve was obtained by numerical simulation of the 0D model proposed here. At the time interval between 0.2 and 0.4 seconds, or between 1.0 and 1.2 seconds, it can be noticed that the estimated curve does not reach the measured curve. This time interval refers to the start of filling phase, when the outlet valve is closing and the inlet valve is opening. Before closure of the outlet valve, it can occur a backflow that is not modeled in this work, once that the pVAD valves were modeled by ideal diodes. Moreover, the membrane displacement can influence the C_{bc}. These two phenomena must be investigated to improve the model response. Figure 9 shows the HMI implemented on LabVIEW to monitor the available signals of the hydraulic system.

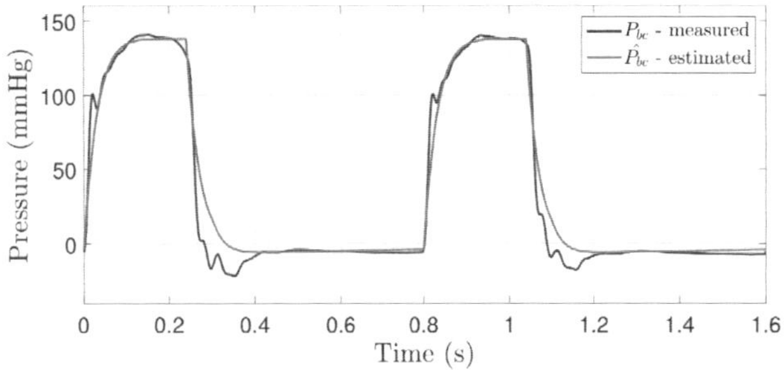

Figure 8. Validation of the 0D model using the pressure signal.

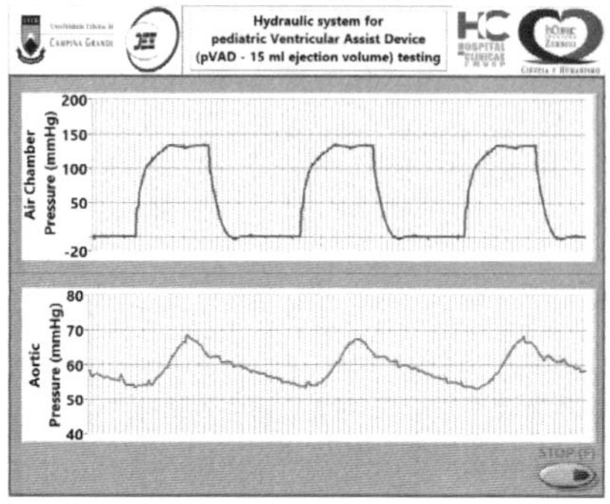

Figure 9. Snapshot of the HMI showing the signals P_{ac} and P_{ao}.

4. Conclusions

A 0D model for the pVAD, as well as for the cannulas and pneumatic actuation driver, was proposed in this work. The available signals in the experimental hydraulic system were defined somehow similar to a pediatric patient's health condition and these signals were monitored by using the HMI implemented on LabVIEW platform. The values of the parameters R_{ac} and C_{ac} were estimated using the least square method and based on the measurements of the air chamber pressure. The compliance of the blood chamber, C_{bc}, was obtained by the pressure-volume relationship inside the blood chamber (static test). The validation results using these parameters demonstrate the correctness of the proposed 0D model. This model will be very useful to design and test feedback control laws used to impose proper operating conditions for the pVAD. A more accurate model for the pVAD is currently being developed taking in account the backflow through the valves and the influence of the membrane displacement.

Acknowledgments

The authors thank PROEX-PPgEE and FAPESP for financial support to the development of this project.

References

[1] Benício A. *et al.*, Avaliação do desempenho hemodinâmico do dispositivo de assistência ventricular Incor como substituto do coração esquerdo, *Rev. Bras. Cir. Card.*, **14**, nº 3, pp. 237–46, 1999 (in portuguese).

[2] Shi Y., Lawford P., and Hose R., Review of zero-d and 1-d models of blood flow in the cardiovascular system, *BioMedical Engineering OnLine*, **1**, nº 33, pp. 1–38, 2011.

[3] Pantalos G.M. *et al.*, Characterization of an adult mock circulation for testing cardiac support devices, *ASAIO Journal*, **50**, nº 1, pp. 37–46, 2004.

[4] em-tec GmbH. Experts In Non-Invasive Flow Measurement.

[5] TruWave Disposable Pressure Transducer, Edwards Lifesciences LLC - Irvine, California, USA.

[6] Honeywell International Inc. Sensing and Control. Pressure Sensor SCX Series.

[7] Torres D.S. and Cestari I.A., Simulador Hidráulico da Circulação Sistêmica Pediátrica com Parâmetros Controlados Computacionalmente, XXV CBEB, 2016 (in portuguese).

[8] Ferrari, G. *et al.*, A modular numerical model of the cardiovascular system for studying and training in the field of cardiovascular physiopathology, *Journal of biomedical engineering*, **14**, nº 2, pp. 91–107, 1992.

[9] Olufsen, M.S. and Nadim, A. *et al.*, On deriving lumped models for blood flow and pressure in the systemic arteries, *Math Biosci Eng*, **1**, pp. 61–80, 2004.

[10] Hunsberger A.Z., Modeling and analysis of interactions between a pulsatile pneumatic ventricular assist device and the left ventricle, PhD Thesis, University of Pittsburgh, 2005.

IOP Conf. Series: Journal of Physics: Conf. Series **1044** (2018) 012048 doi:10.1088/1742-6596/1044/1/012048

Monitoring and evaluation of food allergen labeling regulation: a selection and ranking model for compliance indicators and respective metrics

L F R Criollo[1], M F L Almeida[1] and R F Calili[1]
[1]Pontifical Catholic University of Rio de Janeiro - PUC-Rio, Postgraduate Programme in Metrology, Rua Marquês de São Vicente, 225 - 22451-900, Rio de Janeiro - RJ, Brazil

E-mail: fatima.ludovico@puc-rio.br

Abstract - The aim of this article is to propose a selection and ranking model for compliance indicators and metrics concerning food allergen labeling regulations, after their implementation at national or regional levels. Within the institutional contexts in which the research is situated, it is assumed that this effort will provide the regulatory agencies as well as other actors involved in the implementation of food allergen labeling regulations, the proper tools for monitoring and evaluating their degree of compliance. During the applied phase, an empirical study on the food allergen labeling regulation in Brazil was developed, for validating the proposed model on a national basis. A logical framework for defining an initial list of indicators was combined with two multi-criteria decision making (MCDM) methods aiming at the selection and final ranking of indicators and metrics by category of legal requirements. The main results are a selection and ranking model for indicators and metrics for measuring the degree of compliance of a given food allergen labeling regulation, after its implementation; and a consistent set of indicators and metrics for monitoring and evaluating the food allergen labeling regulation in Brazil, ranked by each category of legal requirements.

1. Introduction

Nowadays, food labeling provides an important interface between manufacturing and distribution companies and consumers. Food labels bring information about the food and its ingredients, their nutrition value; origin; shelf life; and how to use and store the food.

To define the categories of information that must always be declared on a food label and to provide an appropriate framework for the control of voluntary labeling are the roles of food security regulators. In general, the information categories fall into two main classes: (i) information concerning public health and safety reasons; and (ii) information required to facilitate consumers' healthy choice or to prevent misleading conducts of food manufacturing and distribution companies [1].

In this context, the aim of this article is to propose a model for selecting indicators and respective metrics for monitoring and evaluating (ME) the degree of compliance with food allergen labeling regulations, after their implementation at national or regional levels.

2. Theoretical background

A first systematic literature search focused on the theme "monitoring and evaluation"(ME) was performed on peer-reviewed articles indexed in Scopus; Web of Science; and Science Direct databases, covering the period between January 1980 and December 2016. The search strategy focused on the most highly cited publications about the referred theme [2; 3; 4; 5; 6]. Some attempts concerning these works will be detailed in the final version of this paper.

A second search focused on multi-criteria decision making (MCDM) methods, by using the keywords "multiple criteria decision-making", "MCDM", "multi-criteria decision-making", with Boolean operator OR. This search strategy yielded a total of 5,011 publications and revealed that several researchers have attempted to combine MCDM methods for different applications.

Finally, a third literature search focused on food labeling regulation, in general, covering the period between January 2000 and December 2016, yielded 50 publications. However, when this result was combined with the previous results associated with MCDM methods, we conclude that these methods have not been used by academicians and practitioners to select indicators and metrics for monitoring and evaluating the degree of compliance with food labeling regulations in general.

Considering the characteristics of food labeling regulations and the review of MCDM methods, the following methods were combined to the logical framework designed according to the W. K. Kellogg Foundation guidelines [7]. They are (i) Analytical Hierarchy Process (AHP) and Technique for Order Performance by Similarity to Ideal Solution (TOPSIS).

AHP was proposed by Saaty [8], and the basic idea of this method is leaning on a pairwise comparison based on the eigenvector. Widely used for subjective assessments by practitioners, academics, and policymakers, the AHP method is a pairwise comparison in a small part of the hierarchical structure and then between the higher level of hierarchical structure. However, this method presents some disadvantages, as follows: weighting each criterion has a significant influence on the final alternative score, as weighting criteria in this method are judgmental and based on decision maker preference. .

The second selected method refers to the Technique for Order Preference by Similarity to Ideal Solution or TOPSIS, introduced by Hwang and Yoon [9]. It is based on the concept that the chosen alternative should have the shortest distance from the positive ideal solution (PIS) and the farthest from the negative ideal solution (NIS).

3. Methodology

The research methodology followed a procedural framework of analysis based on a model developed by Dibbern *et al.* [10] to provide an underlying structure and an approved course of action. Firstly, a literature review focused on the central research themes: (i) monitoring and evaluation, including ME logical framework design; (ii) multiple-criteria decision making (MCDM) methods, with an attempt to select the best methods to be considered in the context of food labeling regulations, in general.

The literature review covered the main sources of peer-reviewed scientific publications about these themes, such as Scopus; Web of Science; and Science Direct. Additionally, Google Scholar was accessed to complement the search results. The analysis of current state of research on these central themes led to the identification of research gaps and unsolved problems in the field of monitoring and evaluation of food labeling regulations.

Secondly, formal modeling was used to develop a selection and ranking model for compliance indicators and metrics concerning food allergen labeling regulations, after their implementation at national or regional levels.

Third, an empirical study focusing on the food allergen labeling regulation in Brazil involved managers and specialists on Food Quality and Metrology to validate the proposed model from the perspective of its application in another national or regional contexts.

The model encompasses seven main steps: (i) analysis of objectives and expected results of a given food allergen labeling regulation; (ii) identification of key stakeholders interested in the implementation of regulation; (iii) definition of types of indicators (key indicators; complementary indicators; and specific indicators); (iv) definition of criteria for selection and classification of indicators, considering eliminatory and classificatory criteria; (v) proposition of 'candidate' indicators; (vi) application of AHP method for weighting criteria concerning the selection and ranking of proposed ME indicators; and (vii) use of TOPSIS method for the final ranking of the proposed indicators and respective metrics for monitoring and evaluation a given food allergen labeling regulation.

4. Illustrative case: monitoring and evaluation of food allergenic labeling regulation in Brazil

The main results are highlighted as follows: (i) a selection and ranking model for compliance indicators and metrics concerning food allergen labeling regulations, after their implementation at national or regional levels; and (ii) a consistent set of indicators and metrics for monitoring and evaluating the compliance of the Brazilian food allergen labeling regulation, generated during the empirical study developed in this country.

The matrix creation of quantitative monitoring and evaluation indicators has to start characterizing the categories of legal labeling requirements in Brazil, identifying the relevant issues to be evaluated in each category, as can be seen in table 1, which is the first step of the proposed methodology.

Table 1. Basis to create the matrix of indicators

Category of legal requirements	Information to be presented on the allergenic food label
Attendance of mandatory information on packaged food labels, including allergens and derivatives, following Anvisa Resolution RDC nº 259, of September 20, 2002 [11].	Name of the sale food List of ingredients. Net contents. Origin identification. Name or business name and address of the importer in the case of imported food. Batch identification. Expiration date. Instructions on preparation and food use, when necessary.
The intentional presence of allergenic foods and their derivatives, according to Anvisa Resolution RDC nº 26, of July 2, 2015 [12].	The statement "Allergic: It contains [common names of foods that cause food allergies]". The statement "Allergic: It contains derivatives of [common names of foods that cause food allergies]." The statement "Allergic: It contains [common names of foods that cause food allergies] and derivatives," as appropriate.
The possibility of cross contamination with allergenic foods and their derivatives, according to Anvisa Resolution RDC nº 26, of July 2, 2015 [11].	The statement "Allergic: May contains [common names of foods that cause food allergies]."
Attendance of mandatory declaration on packaged food labels - "It contains gluten" or "It does not contain gluten" - according to Brazilian Law No. 10,674 of May 16, 2003 [13].	The statement "It contains gluten" or "It does not contain gluten," as appropriate.
Mandatory information adequacy on allergenic labels and derived packaged foods about applicable regulations' parameters.	Mandatory information adequacy - parameter "expression" Mandatory information adequacy - parameter "clarity." Mandatory information adequacy - parameter "visibility." Mandatory information adequacy - parameter "font size." Mandatory information adequacy - parameter "color contrast." Mandatory information adequacy - parameter "highlight."

IOP Conf. Series: Journal of Physics: Conf. Series **1044** (2018) 012048 doi:10.1088/1742-6596/1044/1/012048

After that, as a second step, it is necessary to identify the key stakeholders of the regulation to be monitored and evaluated, which in this case are: (i) National Sanitary Surveillance Agency (Anvisa); (ii) State and Municipal Sanitary Surveillance Agencies (Visa); (iii) food manufacturers and importers; (iv) allergic consumers, including associations.

Step 3 consisted of defining the types of indicators that will be required to meet different interests identified in step 2, namely: key indicators; complementary indicators; and specific indicators.

Step 4 comprised to determine the eliminatory criteria (must be attended) and classificatory criteria to select and rank the indicators, using the hybrid AHP-TOPSIS method. The criteria chosen for the addressed problem were:

- Eliminatory criteria: (i) stakeholder information needs' attendance; (ii) traceability over time; (iii) representativeness (in relation to regulation objectives and requirements);
- Classificatory criteria: (i) measurability; (ii) information availability; (iii) objectivity; (iv) simplicity.

In Step 5, it was to propose the initial matrix of the candidate indicators, considering the legal requirements described in table 1. In this step, the authors decided to deploy the ME matrix in five matrices associated to categories of legal requirements of the applicable regulation.

In Step 6 the AHP method was used to define the weights of the four classification criteria, namely; (i) measurability; (ii) availability, when necessary; (iii) objectivity; and (iv) simplicity. It was evaluated in Step 7 the existence of conflicts of choice between two or more options of indicators, through the quantitative analysis, with the support of the TOPSIS method.

The indicators chosen in Step 5, by using the hybrid AHP-TOPSIS method, are summarized in table 2.

Table 2. A general list of ME indicators for regulation in focus, ranked by legal requirements' category.

Legal requirements' category	Indicators (I)	Position in the ranking
The presence of general mandatory information on allergen packaged food labels and derivatives.	I2 - Label inadequacy due to the absence of ingredient list.	1st
	I3 - Label inadequacy due to information lack of net contents.	1st
	I4 - Label inadequacy due to the absence of origin identification.	2nd
	I6 - Label inadequacy due to the absence of batch identification.	3rd
	I7 - Label inadequacy due to information lack of expiration date.	4th
	I1 - Label inadequacy due to the absence of the name of seller.	5th
	I8 - Label inadequacy due to the absence of preparation instructions and use of packaged food, when necessary.	6th
	I5 - Label inadequacy due to the absence of the firm name or business name and importer address, in the case of imported food.	7th
The intentional presence of allergenic foods and their derivatives.	I9 - Label inadequacy due to the absence of the statement "Allergic: It contains [common food names that cause food allergies]".	1st
	I10 - Label inadequacy due to the absence of the statement "Allergic: It contains derivates of [common food names that cause food allergies]".	1st
	I11 - Label inadequacy due to the absence of the statement "Allergic: It contains [common food names that cause food allergies] and derivates, as appropriate".	1st
The possibility of cross-contamination with allergenic foods and their derivatives.	I12 - Label inadequacy due to the absence of the statement "Allergic: It may contain [common food names that cause food allergies]".	1st

The presence of the mandatory statement "It contains gluten" or "It does not contain gluten" on packaged food labels.	I13 - Label inadequacy due to the absence of the statement "It contains gluten" or "It does not contain gluten," as appropriate.	1st
Mandatory adequacy information on labels of allergenic and derivatives' packaged foods about the applicable regulations' parameters.	I17 – Label inadequacy in one or more mandatory information concerning parameter "font size."	1st
	I16 - Label inadequacy in one or more mandatory information concerning parameter "visibility."	2nd
	I14 - Label inadequacy in one or more mandatory information concerning parameter "expression."	3rd
	I18 - Label inadequacy in one or more mandatory information concerning parameter "color contrast."	4th
	I15 - Label inadequacy in one or more mandatory information concerning parameter "clarity."	5th
	I19 - Label inadequacy in one or more mandatory information concerning parameter "highlight."	6th

By way of illustration, in table 3 we present the metrics associated to the eight ME indicators ranked in the first position in each legal requirements' category.

Table 3. Metrics associated to the eight ME indicators ranked in the first position in each legal requirements' category.

Indicators (I)	Metrics (M)
I2 - Label inadequacy due to the absence of ingredient list.	M2.1 – Number of non-conforming labels of allergenic and derivatives' packaged foods due to absence of ingredient list (by food category or by health surveillance area) [without unit].
	M2.2 – Number of non-conforming labels of allergenic and derivatives' packaged foods due to absence of ingredient list / total of labels of allergenic and derivatives' packaged foods (by food category or by health surveillance area) [%].
I3 - Label inadequacy due to information lack of net contents.	M3.1 – Number of non-conforming labels of allergenic and derivatives' packaged foods due to information lack of net contents (by food category or by health surveillance area) [without unit].
	M3.2 – Number of non-conforming labels of allergenic and derivatives' packaged foods due to information lack of net contents / total of labels of allergenic and derivatives' packaged foods (by food category or by health surveillance area) [%].
I9 - Label inadequacy due to the absence of the statement "Allergic: It contains [common food names that cause food allergies]".	M9.1 – Number of non-conforming labels of allergenic and derivatives' packaged foods due to the absence of the statement "Allergic: It contains [common food names that cause food allergies]" (by food category or by health surveillance area) [without unit].
	M9.2 – Number of non-conforming labels of allergenic and derivatives' packaged foods due to the absence of the statement "Allergic: It contains [common food names that cause food allergies]" / total of labels of allergenic and derivatives' packaged foods (by food category or by health surveillance area) [%].
I10 - Label inadequacy due to the absence of the statement "Allergic: It contains derivates of [common food names that cause food allergies]".	M10.1 – Number of non-conforming labels of allergenic and derivatives' packaged foods due to the absence of the statement "Allergic: It contains derivates of [common food names that cause food allergies]" (by food category or by health surveillance area) [without unit].
	M10.2 – Number of non-conforming labels of allergenic and derivatives' packaged foods due to the absence of the statement "Allergic: It contains derivates of [common food names that cause food allergies]"/ total of labels of allergenic and derivatives' packaged foods (by food category or by health surveillance area) [%].

I11 - Label inadequacy due to the absence of the statement "Allergic: It contains [common food names that cause food allergies] and derivates, as appropriate".	M11.1 – Number of non-conforming labels of allergenic and derivatives' packaged foods due to the absence of the statement "Allergic: It contains [common food names that cause food allergies] and derivates, as appropriate" (by food category or by health surveillance area) [without unit].
	M11.2 – Number of non-conforming labels of allergenic and derivatives' packaged foods due to the absence of the statement "Allergic: It contains [common food names that cause food allergies] and derivates, as appropriate" / total of labels of allergenic and derivatives' packaged foods (by food category or by health surveillance area) [%].
I12 - Label inadequacy due to the absence of the statement "Allergic: It may contain [common food names that cause food allergies]".	M12.1 – Number of non-conforming labels of allergenic and derivatives' packaged foods due to the absence of the statement "Allergic: It may contain [common food names that cause food allergies]" (by food category or by health surveillance area) [without unit].
	M12.2 – Number of non-conforming labels of allergenic and derivatives' packaged foods due to the absence of the statement "Allergic: It may contain [common food names that cause food allergies]"/ total of labels of allergenic and derivatives' packaged foods (by food category or by health surveillance area) [%].
I13 - Label inadequacy due to the absence of the statement "It contains gluten" or "It does not contain gluten," as appropriate.	M13.1 – Number of non-conforming labels of allergenic and derivatives' packaged foods due to the absence of the statement "It contains gluten" or "It does not contain gluten", as appropriate (by food category or by health surveillance area) [without unit].
	M13.2 – Number of non-conforming labels of allergenic and derivatives' packaged foods due to the absence of the statement "It contains gluten" or "It does not contain gluten", as appropriate / total of labels of allergenic and derivatives' packaged foods (by food category or by health surveillance area) [%].
I17 - Label inadequacy in one or more mandatory information concerning parameter "font size."	M17.1 – Number of non-conforming labels of allergenic and derivatives' packaged foods due to inadequacy in one or more mandatory information concerning the parameter "font size" (by food category or by health surveillance area) [without unit].
	M17.2 – Number of non-conforming labels of allergenic and derivatives' packaged foods due to inadequacy in one or more mandatory information concerning the parameter "font size"/ total of labels of allergenic and derivatives' packaged foods (by food category or by health surveillance area) [%].

From the results presented in this section, it was possible to demonstrate the applicability of the proposed model. The definition of indicators and respectively metrics were developed and applied in the institutional context of the regulation on labeling of allergenic foods in Brazil [14]. The main result of this research consists of a list of 19 indicators ranked by degree of compliance with the classification criteria and legal category requirements. Due to space limitations, we presented only the metrics associated to the ME indicators ranked in first position in each legal requirements' category. Nevertheless, all the metrics can be found in detail in Criollo [15].

5. Conclusion

Evaluating the degree of compliance with food labeling regulations is still an open problem for academicians, practitioners, and policymakers. In this paper, an attempt was made to present a conceptual model for monitoring and evaluating food allergen labeling regulations, with the support of two multi-criteria decision methods.

The applicability of an integrated MCDM approach for selecting and ranking indicators and metrics for monitoring and evaluating the degree of compliance with food allergen labeling regulations at national or regional levels could be demonstrated through an empirical study on the Brazilian regulation on this issue. In general, the results of this research can contribute for the continuous improvement of the whole food

IOP Conf. Series: Journal of Physics: Conf. Series **1044** (2018) 012048 doi:10.1088/1742-6596/1044/1/012048

allergen labeling regulation process after its implementation and an effective communication of agencies involved in food safety research, regulation and education.

6. References

[1] Barnett J, McConnon A, Kennedy J, Raats, M, Shepherd, R, Verbeke, W, Fletcher, J, Kuttschreuter, M, Lima, L, Wills, J and Wall P 2011 *BMC Public Health* **11** 734
[2] Weiss C H 1998 *Evaluation* 1998 2.ed. Upper Saddle River, NJ: Prentice-Hall
[3] Patton M Q 1982 *Practical evaluation*. Beverly Hills, CA: Sage
[4] Wholey J S 1995 Assessing the feasibility and likely usefulness of evaluation. In: Wholey J S., Hatry H P and Newcomer K E (Eds.) *Handbook of practical program evaluation,* 15–39. San Francisco: Jossey-Bass
[5] Stake R E 1981 *Evaluation News* **2** 148
[6] Owen J M and Rogers P G 1999 *Program evaluation: forms and approaches*. London: Sage
[7] W. K. Kellogg Foundation 2004 *Logic model development guide: using logic models to bring together planning, evaluation, and action*. W. K. Kellogg Foundation, Michigan
[8] Saaty T L 2000 *Decision making for leaders*. Pittsburg: RWS Publications
[9] Hwang C L and Yoon K 1981 *Multiple attributes decision making methods and applications*. Berlin: Springer
[10] Dibbern J, Goles T, Hirschheim R and Jayatilaka B 2004 *The DATA BASE for Advances in Information Systems* **35** 6
[11] Brazil National Sanitary Surveillance Agency *Anvisa RDC Resolution 259* of September 20, 2002
[12] Brazil National Sanitary Surveillance Agency *Anvisa Resolution RDC nº 26* of July 2, 2015
[13] Brazil *Law nº 10,674* of May 16, 2003
[14] Brazil National Sanitary Surveillance Agency *Anvisa Resolution RDC nº 26* of July 2, 2015
[15] Criollo, L F R 2016 *Monitoring and evaluation of food allergenic labeling regulation in Brazil: a proposal of indicators and metrics*. 116 p. MSc. Dissertation – Programa de Pós-Graduação em Metrologia. Pontifícia Universidade Católica do Rio de Janeiro.

IOP Conf. Series: Journal of Physics: Conf. Series **1044** (2018) 012049 doi:10.1088/1742-6596/1044/1/012049

Monte Carlo simulation for the treatment of male breast cancer

N.A. Guimarães[1], A. X. da Silva [2], J. P.R. Junior[2], J. M. Lopes[2], W. J.L. Garção[1]
[1] Teacher, *Instituto Federal do Rio de Janeiro - IFRJ Campus Volta Redonda,*
Rua Antônio Barreiros, 212, Volta Redonda, Rio de Janeiro, BRA
[2] *Universidade Federal do Rio de Janeiro* - COPPE/ Programa de Engenharia Nuclear
Av. Horácio Macedo, 2030, Bloco G, Cidade Universitária, Rio de Janeiro, BRA

E-mail: nilmara.guimaraes@ifrj.edu.br

Abstract. The rate of male breast cancer occurrences has been increasing over the last years and, in the absence of proper protocols, the treatment proposed for male breast cancer follows the one established for the female breast neoplasia. The objective of the present study is to simulate the treatment of male breast cancer through the *Monte Carlo N-Particle* (MCNP) code, using the voxelized *Alderson Rando* phantom, following the protocol used on female breast cancer and compare the simulated results with the experimental values obtained with thermoluminescence dosimeters (TLD). A treatment configuration using the Alderson Rando phantom with placement of TLDs was performed, which had reproducibility on the simulation. The results obtained in the simulation were up to 12% greater than the doses planned, and up to 15% lesser than the measurements with TLDs. The results suggest that the treatment of male breast cancer using the treatment protocol for female breast neoplasia must be reviewed, once higher dose values are observed in comparison to those obtained by the planning system.

1. Introduction

The second most deadly type of cancer in the world is breast cancer, which may also affect men [1], [2] but, due to its smaller occurrence, it is little studied. It represents about 1% of every breast cancer, being responsible for less than 0.1% of deaths in men [3], [4].

The cancer may be treated by several means: surgery, chemotherapy, radiotherapy or a combination of those. In the lack of proper protocols, the proposed treatment for male breast cancer follows the one established for female breast neoplasia, based on the surgical treatment, followed or not by radiotherapy, chemotherapy and, mainly, hormone therapy [5].

The objective of the present study is to assess whether the proposed treatment for female breast cancer is really able to be used in male breast, besides performing the validation of the Monte Carlo simulation in order to aid future treatments. The concern with the fact that the structure and volume of the male breast are different from the female's, resulting in radiotherapy treatments with dose distributions that should be differentiated motivated this study.

2. Radiotherapy by Teletherapy and Dose Distribution

Radiotherapy is a method of application of ionizing radiation in a way to overcome or stop the reproduction of tumor cells. The treatment consists basically in the application of a (previously calculated) dose of radiation over a tissue volume which comprises the tumor, for a certain period of

time. The relation between the damage and the benefits of the treatment must be considered to cause the least possible damage to normal cells surrounding the tumor.
In teletherapy, the radiation emitted by a device towards the body region to be treated, with the patient lying down, is applied in sessions. The equipment used comprises cobalt (^{60}Co) therapy units and linear accelerators.

2.1. Dose distribution
Dose distribution refers to the dose values in points surrounding and within the irradiated volume. In radiotherapy, a detailed analysis of dose distribution allows to take conclusions on the preservation of healthy tissues bordering the target volume. Thus, it is important to make the dose prescribed by the doctor concentrate on the target volume, in such a way that the healthy tissue will not receive a dose above the tolerance limits proposed in the literature [6].

2.1.1. Ionizing radiation dosimeters
Ionizing radiation dosimeters are devices in which a physical phenomenon happens when exposed to that type of radiation. If there is a well-established relation between the intensity of the physical phenomenon and the amount of radiation energy absorbed per mass of the device, it can be used to dose the energy, as long as it is properly calibrated. Dosimetric systems are the set of dosimeters and equipment capable of quantifying the physical phenomenon induced by ionizing radiation. The result of a measurement is the value of a dosimetric magnitude expressed by a numerical value associated to its unit. There are several types of radiation dosimeters, but not all of them satisfy the characteristics necessary for the intended use. Thus, it is necessary to choose the most appropriate one, considering the demands of the measurement situation. Some types of dosimeters available in the market are: ionizing chambers, radiographic films, radiochromic films, luminescence dosimeters (TLD) and semiconductors (diode) [7].

2.1.2. Thermoluminescence dosimeter
The thermoluminescence dosimeter is a dose gauge resulting from the thermoluminescence phenomenon, which occurs when there is the emission of light after excitation of a material mean by thermal energy. The emission of light may occur by fluorescence, when it is spontaneous, and by phosphorescence, when it is necessary to apply energy in order to observe the light emission. TLDs may be found in several forms and compositions, as shown in Figure 1.

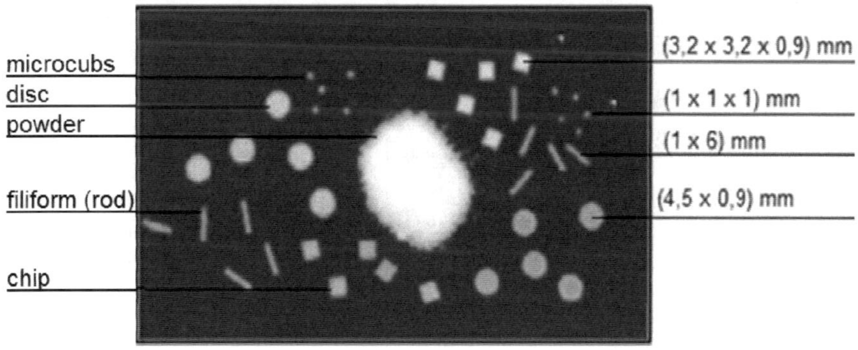

Figure 1. Types of thermoluminescence dosimeters [8].

Some ofthe main TLDs are: LiF:Mg,Ti; CaF_2:Dy; $CaSO_4$:Dy; CaF_2:Mn; LiF:Mg,P,Cu; and $Li_2B_4O_7$:Mn. The dosimeter-on-a-chip, composed by lithium fluoride, is generally the most used in dosimetry, for it is easy to be handled, besides presenting good repeatability and reproducibility conditions.

3. Materials and Methods

The thermoluminescence dosimeters were characterized and calibratedwith irradiations in a known dose rate in the SIEMENS teletherapy equipment ONCOR Expression, installed at *Clínicas Oncológicas Integradas* (COI). The accelerator used in the characterization was the same used for irradiation, that is, a photon beam of 6.0 MV and, therefore, it was not necessary to determine the energy dependence factor [6].

The thermoluminescence dosimeters used were the TLD-100 chip type (LiF:Mg,Ti), with dimensions of 3.2 mm x 3.2 mm x 0.9 mm, manufactured by Thermo Fischer Scientific Inc., commercially, Harshaw TLD.

3.1. Anthropomorphic Simulators or Phantoms

The anthropomorphic simulators are used to reproduce the characteristics of absorption and scattering of the human body, or parts of it, when subject to a radiation field. The main simulators are: mathematical, physical and voxel phantoms [9]. The physical phantoms are structures, normally manufactured in acrylic, filled with water to simulate the human body and they are used for dosimetry and imaging tests. The most popular is the Alderson Rando anthropomorphic simulator (Figure 2), which consists of a skeleton surrounded by rubbers which are equivalent to the soft and pulmonary tissues, besides following the standards of ICRU-44 [10]. This phantom is 175 cm high and weighs 73.5 kgand is horizontally sectioned in slices 2.5 cm thick. Each slice contains holes that are connected to pins of equivalent tissues, which may be: soft, pulmonary or bone tissues. Such pins may be replaced by powdered TLD pins or by pins with spaces for the insertion of the TLDs chip-type. The TLDs were distributed on the left side of the Alderson Rando phantom, between slices 13 and 20, a region of interest of the torso. Figure 3 shows a matrix of 5 mm diameter holes, 3 cm apart from each other, where the TLDs are placed.

Figure 2. *Alderson Rando* Phantom.

Figure 3. Slice with TLDs and holes [11].

3.2. Planning System of Male Breast Cancer

The planning system used to simulate the treatment of male breast cancer, following protocols of female breast cancer, was *Prowess* Version 5.11.

Prowess has an algorithm which can manage information on dose, monitor units, acquisition and image merging, besides transporting the planning directly to the linear accelerator where the treatment shall take place.

The planning was performed with the imaging obtained in the Alderson Rando phantom by the *Siemens Emotion Duo* scanner. The treatment consisted on the irradiation of two oblique fields with

IOP Conf. Series: Journal of Physics: Conf. Series **1044** (2018) 012049 doi:10.1088/1742-6596/1044/1/012049

prescription of 200 cGy in 100% of the isodose line. The first field had the following irradiation configuration: 303° angulation, source to surface distance (SSD) of 88.1 cm, 10 cm x 12 cm field, 30° filter, which resulted in a monitor unit with a value of 241 MUs. On the second stage, the gantry was positioned at 123°, with SSD of 88.5 cm, a 10 cm x 11 cm field and monitor unit of 241.8 MUs, remaining with the same filter.

For comparative purposes of the planned results with the results obtained by the TLDs, it was necessary to insert in the planning system some points of interest, that is, specific locations for the TLD, as presented in Figure 4. Thus, it was possible to generate a report in which the planned doses were determined, in the locations of the points of interest.

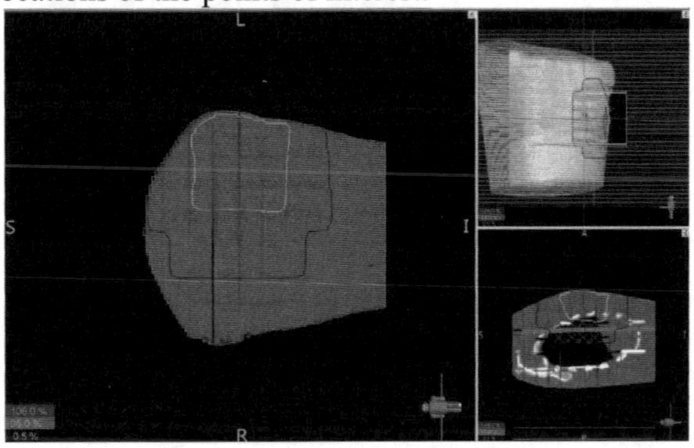

Figure 4. Prowess graphic display with the inserted points of interest [6].

3.3. Monte Carlo Simulation in MCNP

For validation purposes, the results obtained by the planning system and by the use of TLDs were compared to the results simulated in MCNP.

In MatLab, the tomographic images of the Alderson Rando phantom were converted in a *.txt input file, which MCNP was able to read and simulate. The phase space file was used for the reuse of the photon spectrum, reducing the computational time.

4. Results and Conclusions

4.1. Results Obtained with the Alderson Rando Voxel Phantom

The male voxel phantom used in the present study was implemented in the MCNP [12] code for the modeling and simulation of male breast cancer treatment, following the same experimental configuration used for the measurements with thermoluminescence dosimeters.

Figure 5 presents the image obtained by the *software* Moritz, of the male voxel phantom simulated in MCNP. Figure 5 represents the experimental scenario of slice 16 of the Alderson Rando phantom, where the tumor location was simulated, which is represented along with the wedge filter of the linear accelerator.

IOP Conf. Series: Journal of Physics: Conf. Series **1044** (2018) 012049 doi:10.1088/1742-6596/1044/1/012049

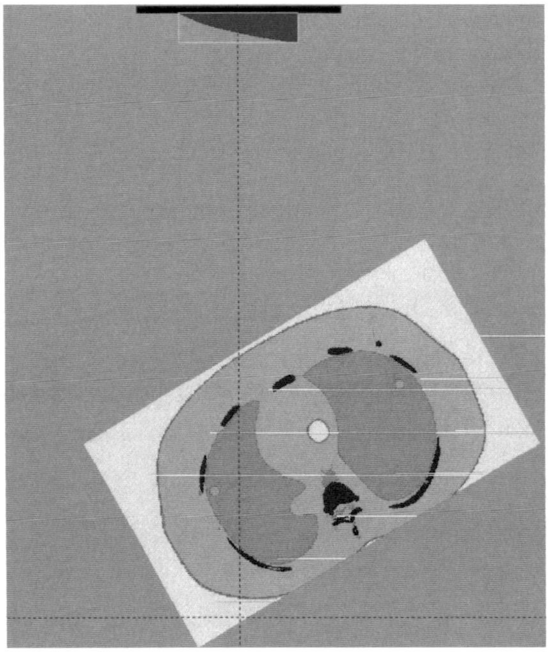

Figure 5. Visualization of the simulation of breast cancer treatment on MCNP obtained by the *software* Moritz [13].

Table 1. Comparison among the results obtained through the simulation of male breast cancer treatment and the results presented by the planning system and by the TLD dosimetry.

TLD	Planned Dose (cGy)	Simulated Dose (cGy)	Reference Dose of TLD (cGy)	Difference between planned and simulated doses (cGy)	Difference between planned and simulated doses (%)	Difference between reference and simulated doses (cGy)	Difference between reference and simulated doses (%)
e02	187,78	177,18	200,34	-10,60	-6%	-23,16	-13%
b01	181,31	194,69	199,77	13,38	7%	-5,08	-3%
n03	189,25	209,37	195,76	20,12	10%	13,61	6%
c04	197,45	225,15	202,26	27,70	**12%**	22,89	10%
d20	202,47	212,08	208,41	9,61	5%	3,67	2%
l01	199,45	199,45	198,92	0,00	0%	0,53	0%
h09	205,29	204,46	208,81	-0,83	0%	-4,35	-2%
e03	199,73	212,78	205,02	13,05	6%	7,76	4%
i11	207,10	210,44	221,54	3,34	2%	-11,10	-5%
b15	190,35	175,67	202,35	-14,68	-8%	-26,68	**-15%**
n20	203,82	197,37	208,63	-6,45	-3%	-11,26	-6%

Table 1 presents a maximum difference between planned and simulated doses of 12%; in relation to the reference (TLD) the maximum difference was of -15%. All simulated doses presented a combined uncertainty of ± 5%, given that such results were determined by two simulations, the first for a 123°

angle and the second for a 303° angle. In the simulations, only the places where the TLDs were positioned with results above 50 cGy were considered as points of interest.

5. Conclusion

The Monte Carlo simulation presented results up to 12% greater than the planned doses, and up to 15% lesser than the measurements using TLDs. Such a result is satisfactory and validates the use of the Alderson Rando phantom simulation by MCNP for further studies with male breast and other organs, aiding future implementations of a new specific treatment for the male breast.

The results suggest that the treatment of male breast cancer using the treatment protocol for female breast neoplasia must be reviewed, given the observation of greater dose values than those obtained by the planning system.

References

[1] PAIVA S. M M. *Avaliação da qualidade de vida de pacientes oncológicos em tratamento quimioterápico adjuvante*. Ribeirão Preto, 2006. 111 p. Dissertation.

[2] Hill T. D.; Khamis H. J.; Tyczynski J. E. and Berkel H. J *Comparison of male and female breast cancer incidence trends, tumor characteristics, and survival. Ann Epidemiol.* 2005; 15(10): 773-80.

[3] Giordano S. H. *A review of the diagnosis and management of male breast cancer. Oncologist. 2005*; 7(10): 471-9.

[4] Carmalt, H. L.; Mann, L. J.; Kennedy, C. W.; Fletcher, J. M. and Gillet, D. J. *Carcinoma of the male breast: a review and recommendations for management.* Aust and N Z J Surg. 1998; 68(10): 712-5.

[5] Leme L. H. S. and Souza G. A. *Câncer de mama em homens: Aspectos epidemiológicos, clínicos e terapêuticos*. Rev. Ciênc. Méd., Campinas, 15(5): 391-398, set./out., 2006.

[6] Guimarães N. A. and Crispim, V. R. (2014) M*ale Breast Cancer-Dosimetry Evaluation in Teletherapy. International Journal of Medical Physics, Clinical Engineering and Radiation Oncology*, 3.

[7] Izewska J. and Rajan G., *"Radiation Dosimeters" In: Podgorsak, E.B. (ed), Radiation Oncology Physics: A Handbook for Teachers and Students*, chapter 3, Vienna, Austria, International Atomic Energy Agency-IAEA, 2005.

[8] Da Rosa L.A.R., *Dosimetria Termoluminescente Aplicada à Física Médica, Instituto de Radioproteção e Dosimetria* - CNEN, Rio de Janeiro, 2001.

[9] Boia L..S..*Metodologia Computacional para Geração de Modelos Antropomórficos de Voxels para Simulação por Monte.Carlo de Tratamentos Radioterápicos/ Leonardo da Silva Boia. –* Rio de Janeiro: UFRJ/COPPE, 2010. (Thesis)

[10] ICRU, *International Commission on Radiation Units and Measurements. Tissue substitutes in radiation dosimetry and measurement*, Report 44, Bethesda, MD, USA (1989).

[11] Guimarães N. A. and CRISPIM V. R. *Dosimetric Evaluation in Teletherapy to Treatment of Male Breast*. INAC – 2013, Pernambuco. (Article)

[12] Pelowitz D. B. *MCNPX User's Manual Version 2.6.0*. Los Alamos, NM: Los Alamos National Laboratory; 2008.

[13] Guimarães, N. A. *Avaliação Dosimétrica em Teleterapia para o Tratamento de Câncer de Mama Masculina*, 2015 (Thesis) – Rio de Janeiro: UFRJ/COPPE, 2015

IOP Conf. Series: Journal of Physics: Conf. Series **1044** (2018) 012050 doi:10.1088/1742-6596/1044/1/012050

Multi-Application Platform for Education & Training Purposes in Photonical Measurement Engineering & Quality Assurance with Image Processing

P-G Dittrich[1,2], B Buch[1], A Golomoz[1], R Celestre[1], R Fütterer[1], P Votyakov[1], M Rosenberger[1], G Notni[1]

[1]Technische Universität Ilmenau,
Faculty of Mechanical Engineering,
Department of Quality Assurance and Industrial Image Processing
Gustav-Kirchoff-Platz 2, 98684 Ilmenau Germany

[2]SpectroNet c/o Technologie- und Innovationspark Jena GmbH
Wildenbruchstr. 15, 07745 Jena Germany

E-mail: pg.dittrich@spectronet.de

Abstract. Involving children as well as students in photonical measurement engineering and quality assurance with image processing is becoming increasingly important due to the fact that children and students nowadays are well equipped with powerful mobile cameras and computers, for example smartphones. The Department of Quality Assurance and Industrial Image Processing at the Technische Universität Ilmenau noticed that standard measurement equipment within the labs does no longer encourage children as well as students to develop and apply conventional laboratory equipment for image processing. One reason could be that children and students are missing the playful approach to image processing technology and their applications which appears familiar to them from everyday life. To tackle this issue, within the above-mentioned Department a multi-application platform for education and training purposes on photonical measurement engineering and quality assurance with image processing has been developed. To adapt to the accepted player experience of young users a LEGO Train within a LEGO City environment has been introduced. Integrated within the multi-application platform, there are several image processing technologies like color webcam setup for 2D QR code readings, laser line and monochrome line-scan camera setup for 3D measurements, spectral filter equipped cameras for spectral image acquisitions and an infrared camera for thermal image acquisitions. Thereby children as well as students can experience image processing technologies in a playful environment. The paper shows HOW the multi-application platform has been constructed and HOW the applied image processing technologies have been implemented. The necessary hardware- and software setups can be practiced.

1. Introducing the multi-application platform

The development and application of photonical measurement engineering and quality assurance with image processing requires an interdisciplinary understanding of illumination, optics, sensors, interfaces and image processing algorithms. The application of image processing is manifold and is practiced in industry, biology and medicine. To show a part of the broad spectra of applications to children and students the multi-application platform was engineered. The multi-application platform shows a LEGO Train within a LEGO City environment (Figure 1). The LEGO Train was mechanically and electronically improved with a Bluetooth wireless communication module and a stepper motor as well as a 3D printed housing for the implementation of the stepper motor. The rail bed of the LEGO "Railway Station" was equipped with two inductive sensors which interact when the LEGO Train passes. Furthermore, there are different LED illuminations and a designed "landscape" to improve visually the LEGO City environment. Within the application-platform a micro-controller board has been implemented for communication and control of the LEGO Train and the LEGO City environment. The firmware of the microcontroller board was written in standard C language and enables the communication and control of the LEGO Train and the LEGO City environment by a standard Windows based computer and the MATLAB programming software via standard COM-Port commands. With MATLAB, an intuitive touchscreen operable software was written. With this software, the LEGO Train and the LEGO City environment can be controlled. The implemented image processing can be accessed via USB and GigE interfaces. The MATLAB software processes and visualizes the acquired signals. Further information of the realized applications will be discussed in detail in the following sections.

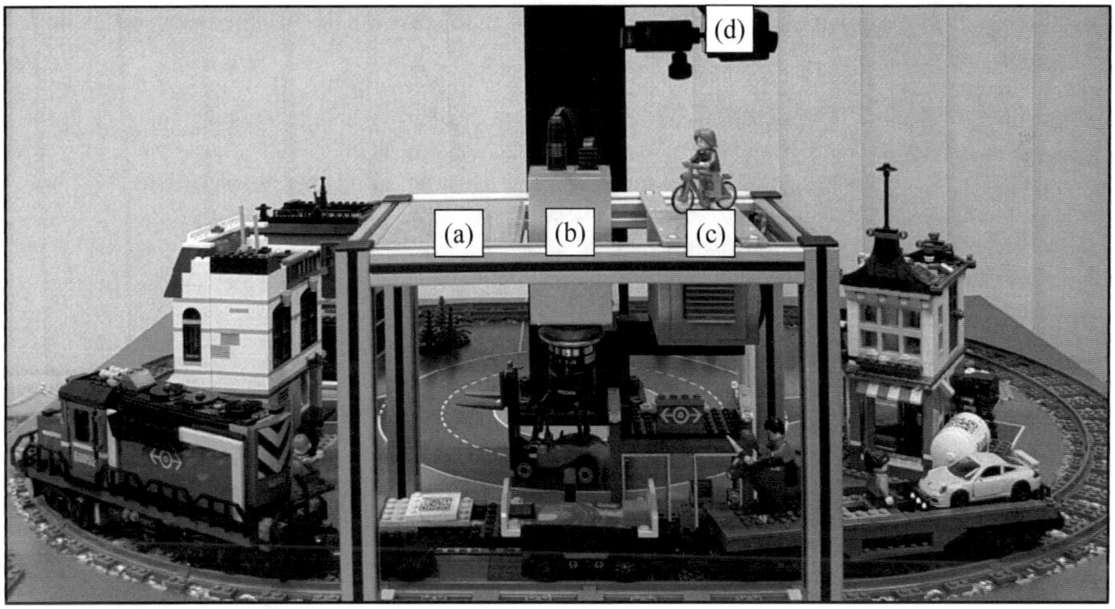

Figure 1. Multi-application platform with image processing
(a) color webcam setup for 2D QR code readings,
(b) laser line and monochrome line-scan camera setup for 3D profile measurements,
(c) spectral filter equipped camera for spectral image acquisitions and an
(d) infrared camera for thermal image acquisitions.

2. Developed and applied color webcam setup for 2D QR code readings

To tackle a standard 2D image processing task which is known by young children as well as students from real life and daily applications a module for 2D QR code readings was realized and applied. For the image acquisition a standard color webcam was equipped with a diffuse LED ring light (Figure 2, left). When the LEGO Train stops at the LEGO Railway Station the LED ring light at the first LEGO Train Wagon illuminates the measurement scene and the color webcam is acquiring an image. The acquired image is processed within the written MATLAB software in a try, catch - execute statement and catch resulting. The function for finding and decoding the QR code (Figure 2, center) within the acquired image locates the QR code embedded in the image and extract the string message embedded within the QR code (Figure 2, right). The used function to locate and decode the QR code works with added entries to dynamic Java class path in the open available Google zxing library [1]. For deeper insights, the MATLAB source code of the multi-application platform is open available on GitHub [2].

Figure 2. Color webcam with surrounded LED illumination (left), QR code example (center) and QR code decoding result (right)

3. Developed and applied laser line illumination and monochrome line-scan camera setup for 3D measurements

Laser triangulation, with its general low cost and simple implementation, is nowadays a widespread technology. The possible applications span from biomedical to industrial applications [3]. From a basic point of view, a laser-triangulation system consists a laser source with a lens that focalizes or collimates the laser beam on the target. A CMOS sensor camera and its optical system are used to acquire the laser line image on the target. The laser line is directed to the surface of the object at an oblique angle (Figure 3) [4]. It creates a visible line on the object. In areas where the object is lower, the beam is slightly shifted. By means of a camera this displacement of the bright laser lines can be determined in the image [4].

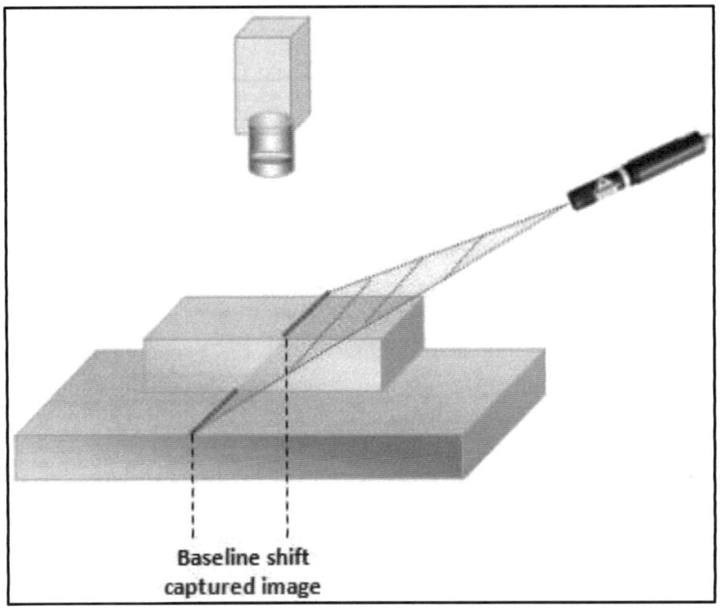

Figure 3. Basic principle of applied laser triangulation [4]

If the angle between camera and laser and the displacement of the line is known, the height of the object can be calculated by simple trigonometric formulae (right-angled triangle) (Figure 4) [4].

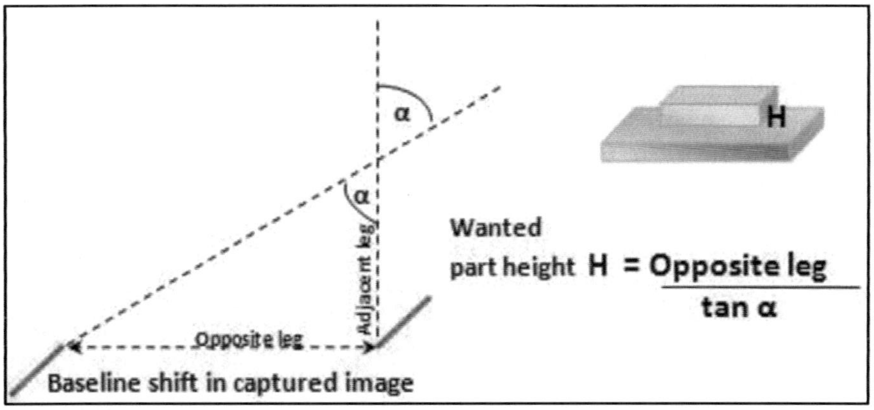

Figure 4. Calculation of the height of the object by using a simple trigonometric formula [4]

If the laser line runs on the object at a shallow angle, even smallest differences in height cause a strong displacement of the lines, however, the detected height range is low. Even differences of tenths of millimeters can be detected easily in practice without special effort. In case of obtuse angles, the measuring accuracy is low, but large differences in height can be detected. The principle of laser triangulation is brought to perfection when using 3D triangulation cameras and lasers. For this purpose, the test object is continuously moved along a camera-laser system and many triangulation profiles are generated. From the deflection of the single line profiles, a three-dimensional image of the object can be calculated using complex software algorithms [4].

For the realized 3D measurement setup, a self-developed laser line module and a camera both with 3D printed housings (Figure 5, left) have been used. The measurement object is a miniaturized toy car (Figure 5, center). During a special round of the LEGO Train - especially for the 3D laser line profile acquisition - the LEGO Train strongly decreases its driving speed. That allows the acquisition of more than 350 images within a measurement range of about 10 centimeters with the line based monochrome camera . That allows the 3D reconstruction of the 3D surface of the measurement object (Figure 5, right).

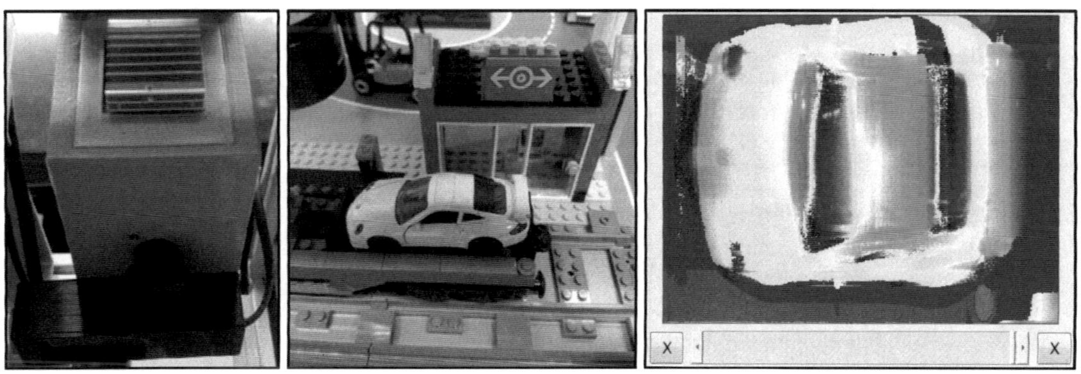

Figure 5. Laser and camera for applied 3D measurements (left), the measurement object (center) and the calculated three-dimensional image of the object (right)

4. Developed and applied spectral filter equipped camera for spectral image acquisitions

For the demonstration to children and students a spectral camera system was implemented into the application-platform to see things which are invisible for the unarmed human eyes. A monochrome CMOS camera equipped with a spectral bandpass filter at a center wavelength of 650 nm and a 50 nm Full-Width-Half-Maximum (FWHM) between the camera objective and the sensor (Figure 6, left). The application example is a 10 EUR bank note (Figure 6, center). When the LEGO Train stops at the LEGO Railway Station a broadband halogen illumination turns on and within the image of the spectral filter equipped camera one special security property of the bank note becomes visible and is shown on the screen by the MATLAB software (Figure 6, right).

Figure 6. Spectral filter camera with halogen illumination (left), bank note application example (center) and bandpass filtered image of the bank note (right)

5. Developed and applied infrared camera for thermal image acquisitions

Another technology of image processing is the so-called thermal imaging with an uncooled longwave-infrared camera. It uses a sensor that works by the changes of voltages when heated by infrared radiation. These changes of voltages are then measured and compared to the values at the operating temperatures of the sensor. The higher an object's temperature, the more infrared radiation is emitted as black-body radiation. A special camera can detect this radiation in a way like an ordinary camera which detects visible light. It works even in total darkness because ambient light level does not matter. This makes it useful for rescue operations in smoke-filled buildings and underground. Furthermore, some physiological activities, particularly responses such as fever, in human beings and other warm-blooded animals can also be monitored. The thermal imaging setup is used to show and explain children as well as students the huge variety of spectral intensity distributions in wavelength ranges from 5.000 nm up to 15.000 nm. The thermal camera can also be applied for relative temperature measurements of people who stand in front of the multi-application platform. It can be demonstrated that with image processing acquired quantities can be transferred into usable and comparable SI unit's equivalents.

6. Summary and Conclusions

It has been shown that photonical measurement engineering and quality assurance with image processing for education and training purposes can be demonstrated by applications with a LEGO Train, a LEGO City environment, LED illuminations, cameras for visible and infrared light, appropriate filters and specialized software for micro-computers with WINDOWS operating systems. In the Department of Quality Assurance and Industrial Image Processing at the Technische Universität Ilmenau the Multi-Application Platform (MAP) is used for education and training purposes in the before mentioned image processing application fields.

IOP Conf. Series: Journal of Physics: Conf. Series **1044** (2018) 012050 doi:10.1088/1742-6596/1044/1/012050

References

[1] GitHub - zxing/zxing: Official ZXing ("Zebra Crossing") project home,“ 13 March 2017. [Online]. Available: https://github.com/zxing/zxing.

[2] GitHub - TU-Ilmenau-MB-QBV/LegoDemo: Demonstration system with different kinds of cameras and image processings,“ 13 March 2017. [Online]. Available: https://github.com/TU-Ilmenau-MB-QBV/LEGODemo.

[3] Laser triangulation - Julight S.r.l.,“ 13 March 2017. [Online]. Available: http://www.julight.it/laser-triangulation/.

[4] L. Fermum, Principle of triangulation,“ 13 March 2017. [Online]. Available: http://www.vision-doctor.com/en/laser-illumination/principle-of-triangulation.html.

Multi-parameter fuzzy design space for QbD approach applied in the development of biomedical devices

D R T Velázquez[1], D R Louzada[1], E Costa Monteiro[1], L G S Fortaleza[1], C R H Barbosa[1], E C Silva[2], L A P Gusmão[2]

[1] Postgraduate Programme in Metrology, Pontifical Catholic University of Rio de Janeiro, PUC-Rio, Brazil.

[2] Department of Electrical Engineering, Pontifical Catholic University of Rio de Janeiro, PUC-Rio, Brazil.

dtasev88@gmail.com

Abstract. Since 2004, on the recommendation of the Food and Drug Administration (FDA), the concept of Quality by Design (QbD) has been widely applied by the pharmaceutical industry in drug development. The paradigm for quality assurance has been changed from inspections tests to focusing on increasing control within manufacturing processes. Recently, the methodology of QbD has gained space in the area of development of transducers for biomedical application. In order to advance on the adoption of this approach to the characteristics and needs of this new field, a methodology of representation of a multi-parameter Design Space with fuzzy inference was developed.

1. Introduction

The development of new sensors/transducers applied to the biomedical area must follow high standards of reliability, searching for high sensitivity, accuracy and precision; low-cost of fabrication and operation; low complexity; portability; as well as safety/innocuousness; in order to present efficacy in the treatment of patients, without generating adverse effects [1-2]. Based on these precepts, PUC-Rio's Biometrology Laboratory (LaBioMet) carries out research for the development of sensors and transducers based on properties such as the Giant Magnetoimpedance (GMI) effect. In this line of research, a system for localization of non-ferromagnetic metallic objects inside the human body, using GMI magnetometers, is currently in its advanced stages of development. This device is designed to help medical staff by providing an innocuous and more accurate foreign body location technique for effective surgical removal since all the currently available alternatives (radiography, computed tomography, and radioscopy) use ionizing radiation and lead to long and often unsuccessful procedures. [3-10]. The principle of operation of the developed system consists of using a solenoid as a time-varying primary magnetic field generator, which causes the occurrence of eddy currents in metallic bodies. These currents consequently produce a time-varying secondary magnetic flux density that can be measured by high sensitivity magnetometers in gradiometric (differential) configuration.

Since 2004, a new approach to quality assurance of products and processes has been used by the pharmaceutical industry. This new approach known as Quality by Design (QbD) is formed by a series of steps aimed at ensuring the quality of products manufactured more efficiently, both in operating costs and in their final quality [12-13]. According to the International Conference on Harmonization

(ICH), QbD is defined by the identification of product-related factors, such as the Target Product Profile (TPP), Critical Quality Attribute (CQA) and Critical Process Parameters (CPP) and the relationship between them (Design Space and Control Strategy).

In a simplified way, QbD seeks to establish the dependence that the critical attributes of the product (CQA) present with the critical process parameters (CPP) of the product. The understanding of these relationships is obtained by means of the Design Space (DS) that establishes the way in which two or more parameters interact to represent a CQA specification better. In this way, operation bands are defined optimizing the operation of the CPP_s and guaranteeing the final quality, safety, and efficacy of the product (TPP) [14-18].

Aiming at guaranteeing the quality of prototypes under development in LaBioMet, studies for the adaptation and implementation of the methodology proposed in QbD in the process of developing biomedical technologies were performed [19-21]. However, in this case, emerges the challenge of establishing a method that would allow the development of a Design Space formed by the significant number of parameters comprised in this new application, with multiple CQA_s and CPPs (represented by different units and ranges of operation).

This work describes the adaptations performed on the methodology based on fuzzy inference system [20, 21], with the objective of implementing a multi-parameter QbD approach to outline the Design Space for the development of the biomedical device for non-ferromagnetic metallic foreign bodies localization using low-cost GMI sensors.

2. Methods

The implementation of the QbD methodology required a series of discussions, conducted by the involved researchers, to identify all possible Critical Quality Attributes and Process Parameters (CPP_s and CQA_s) directly associated with the Target Product Profile (TPP_s) of the developing system for metallic foreign bodies localization. As a result, 21 CPP_s and 8 CQA_s representing the critical aspects of quality assurance for the localization system were identified.

The risk analysis between CPP_s and CQA_s, as characterized, as a preliminary step, for outlining the Design Space (DS) image representative of the relation between all critical indexes. Basically, it was established the degrees of influence that each CPP had on the CQA_s (high, medium or low); and the parameters classified with high impact were selected for consideration in the Design Space configuration.

After this first screening, all possible cases of CQA_s related to a given CPP were identified. The procedure was repeated for the other CPP_s, resulting in the complete group of parameters.

For a particular CPP (e.g., CPP_i) operating band`s determination, a fuzzy inference (FI) were performed between this CPP in respect to and each CQA with a high degree of influence with it. Each FI generates a two-dimensional graph representing the bands that optimize the given CPP under analysis.

Concerning CPP_1, 5 CQA_s had a high degree of influence, resulting in a fuzzy inference with five different CPP_1 bands (each one according to all the previously CQA_s identified as high related). In this particular case, the CPP_1 according to the CQA_4, CQA_6, CQA_7 and CQA_8 were the same, and the only case different appear for the CPP_1 according to the CQA_5. As a result, it was possible to obtain a two-dimensional graph (of the best bands of operation) that represent the CPP_1 (for CQA_4, CQA_6, CQA_7 and CQA_8) in one axis and the CPP_1 with respect to the CQA_5 in the other. The intersection between the result of CPP_1 in each axis of this two-dimensional graph enabled the description of the band of operation in which the CPP_1 guarantees the appropriate quality of the device.

The same methodology was applied to calculate the band of operation of the CPP_2 in this work.

When there are more than two operating bands of a given CPP (when analyzed in respect to all significant CQA_s), the visualization of the graph becomes restrict to a one-dimensional, where one axis (horizontal axis) represents the best's operation's band to the CPP in question, and the other axis (vertical axis) represents the degree of conformity the operating values.

3. Results

In this work, a full implementation of QbD approach was conducted for calculating the best bands of operation of the CPP_1 and CPP_2.

The Critical Quality Attributes, Process Parameters (CPP_S and CQA_S) directly associated with the Target Product Profile (TPP_S) identified by the researchers of the present project (developing system for metallic foreign bodies localization) were obtained through a brainstorm, and the result is presented in Table 1. As a consequence of this effort, it was possible to conduct a study of a risk analysis between all CPP_S and CQA_S, as shown in Table 2.

Table 1 – The relevant TPP, CQA and CPP identified for the metallic foreign bodies localization`s system.

Code	Target Product Profile (TPP)	Code	Critical Quality Attributes (CQA)	Code	Critcal Process Parameter (CPP)
TPP 1	Location of metallic foreign body (non-magnetic)	CQA 1	Spatial resolution	CPP 1	Frequency of the primary field
				CPP 2	Current intensity
				CPP 3	Current intensity (circuit)
TPP 2	High accuracy	CQA 2	Operation at ambient temperature	CPP 4	Circuit's supply voltage
TPP 3	Low measurement uncertainty			CPP 5	Sensor sensivity
				CPP 6	Solenoid's dimensions
TPP 4	Environmental sustainability			CPP 7	GMI supply voltage
		CQA 3	Transducer sensitivity	CPP 8	Oscillator supply voltage
TPP 5	Safety (harmlessness)			CPP 9	Sensor characteristics
		CQA 4	Signal to noise ratio	CPP 10	AC in GMI sensor
TPP 6	Low complexity manufacturing			CPP 11	AC frequency of the current in the sensor
TPP 7	Low manufacturing cost	CQA 5	Secondary magnetic flux	CPP 12	Sensor size
				CPP 13	Homogeneity of the sensor
TPP 8	Low operating cost			CPP 14	Gradiometric configuration
TPP 9	Reduced complexity of operation			CPP 15	Voltage Stability
		CQA 6	Resolution time	CPP 16	Signal amplification
TPP 10	Portability			CPP 17	Relative position between the sensors and solenoid
TPP 11	Low complexity (calibration included)	CQA 7	Circuit characteristics	CPP 18	Circuit noise
				CPP 19	Filter
TPP 12	Low cost of maintenance (calibration included)	CQA 8	Set up	CPP 20	Processing Software
				CPP 21	Acquisition software

Table 2 – Risk Analysis of all the parameters CPP and CAQ.

Code (CQA)	Code (CPP)																				
	CPP 1	CPP 2	CPP 3	CPP 4	CPP 5	CPP 6	CPP 7	CPP 8	CPP 9	CPP 10	CPP 11	CPP 12	CPP 13	CPP 14	CPP 15	CPP 16	CPP 17	CPP 18	CPP 19	CPP 20	CPP 21
CQA 1	L	L	L	L	H	M	L	L	L	L	L	H	L	H	L	L	L	L	L	L	L
CQA 2	L	L	L	L	L	L	L	L	H	L	L	L	L	L	L	L	L	L	L	L	L
CQA 3	L	L	H	H	L	H	H	H	H	H	H	H	H	H	H	H	H	H	H	H	H
CQA 4	H	H	L	H	H	H	H	H	H	H	H	H	H	H	H	H	H	H	H	H	H
CQA 5	H	H	L	L	L	H	L	L	L	L	L	L	L	L	L	L	L	L	L	L	L
CQA 6	H	L	L	L	H	L	L	L	L	L	H	M	L	M	L	L	L	L	H	L	H
CQA 7	H	L	H	H	H	L	H	H	L	H	H	H	H	H	H	H	H	H	H	L	L
CQA 8	H	H	H	H	H	H	H	H	H	H	H	H	H	H	H	H	H	H	H	H	H

The Design Space maps for all the associations between CPP_1 according to all high influenced CQA_S was built. The same was performed to the CPP_2.

Figure 1a presents DS relative to CPP_1, which is the critical process parameter that corresponds to the frequency of a primary magnetic flux density, used to induce eddy currents in the foreign object. The induced currents generate a secondary magnetic flux density that can be detected by the sensing component of the system.

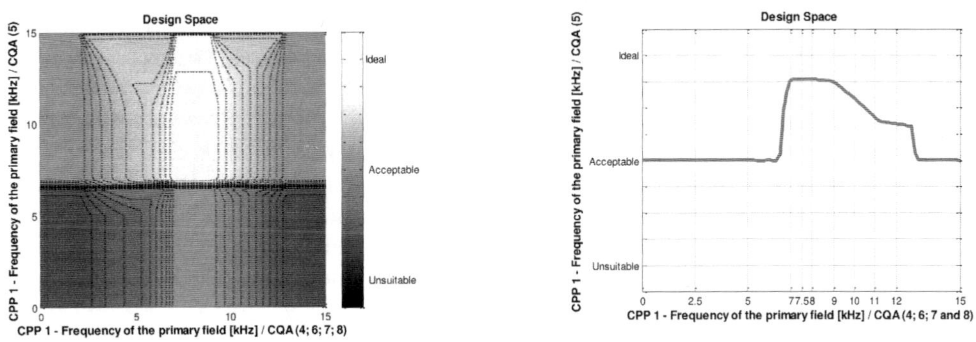

Figure 1 – Design Space maps obtained for the whole set of associations of the CPP_1 (frequency of the primary magnetic flux density) in respect to all CQA_S with high influence. In (a), the map of the Design Space obtained for the CPP_1 and, in (b), the results of its frequency range that optimizes the device's operation.

The map of Design Space for CPP_1 (figure 1a) allows identifying the range of the frequency of the solenoid's excitation current that ensures the appropriate quality of the localization system (figure 1b).

In this case, the operating band between 7 kHz and 11 kHz was well evaluated but the best operating band obtained was between 7 kHz and 9 kHz.

Figure 2a presents DS relative to CPP_2, which is the critical process parameter that corresponds to the current intensity of a primary magnetic generator, used to induce eddy currents in the foreign object. The bands of operating of this Process Parameter are best observed in the figure 2b.

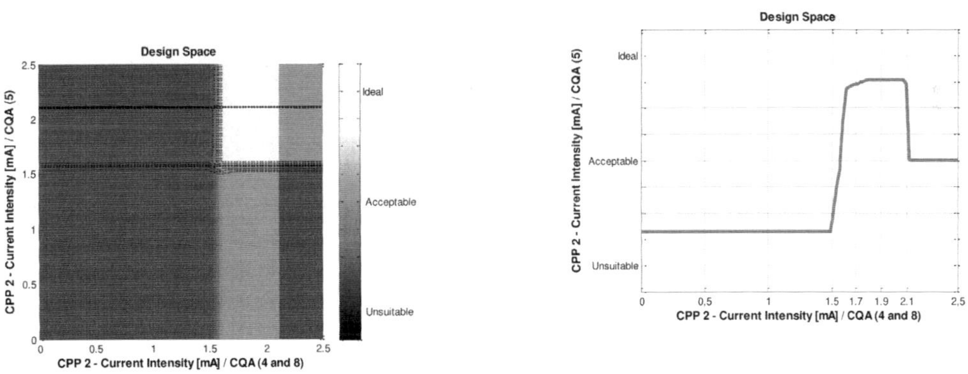

Figure 2 –Design Space maps obtained for the whole set of associations of the CPP_2 (Current Intensity) in respect to all CQA_S with high influence. In (a), the map of the Design Space obtained for the CPP_2 and, in (b), the results of its current range that optimizes the device's operation.

For the CPP_2, the map of Design Space (figure 2a) allows identifying the range of intensity of the solenoid's excitation current that ensures the appropriate quality of the localization system (figure 2b). It's possible to observe that the best operation band is approximately from 1.6 mA to 2.0 mA.

4. Conclusion

This work presents an evolved adaptation of the Quality by Design (QbD), already used in the pharmaceutical industry, for the quality assurance of biomedical technologies. The existence of a large number of parameters characterizing these applications has been considered in the fuzzy inference to configure a resulting representation of the multi-parameter Design Space.

Unlike what occurs in the pharmaceutical industry, the CPP_s for the QbD applied to the development of sensors/transducers has no dependence whatsoever. In this scenery, the analysis of the Design Space becomes restricted to the inference fuzzy of all the bands of operation of a given CPP regarding all the CQA_s with high relevance.

The developed approach was applied to the complete critical process parameter corresponding to the frequency of the primary magnetic flux density generated by the locating device, and the intensity of the current applied to the primary magnetic field generator (solenoid). The analysis allowed identifying the appropriate range of frequency between 7 kHz and 11 kHz (figure 1b), and the range of current intensity between 1.6 mA to 2.0 mA (figure 2b), for the performance of the transducer (considering all the other relevant requirements of the non-magnetic metallic foreign body localization system). Nevertheless, it is possible to restrict the operating bands of the CPP_1 to a better zone (7 kHz to 9 kHz). Since the present study performed a full analysis of $CPP_1$and CPP_2 associations with the whole set of CQA_s, and our previous study, in [21], evaluated the same critical parameters considering their interaction with just two quality attributes, a slight difference was observed in the best-operating bands obtained. This aspect stresses the importance of the complete analysis of the QbD application.

The presented advancements on the implementation of the QbD approach in the development of health technologies can contribute as a step forward in ensuring reliable diagnosis and treatments.

References

[1] Monteiro E C 2007. Biometrologia: confiabilidade nas biomedições e repercussões éticas. *Metrologia e Instrumentação*, **6**, 6-12.

[2] Monteiro E C and Leon L F 2015. Metrological Reliability of Medical Devices. *Journal of Phys. Conf. Series*, **588**, 012032.

[3] Monteiro E C, Barbosa C H, Lima E A, Ribeiro, P C and Boechat P 2000. Locating steel needles in the human body using a SQUID magnetometer. *Phys. Med. Biol*, **45(8)**, 2389.

[4] Barbosa C H, Monteiro E C, Lima E A, Santos S F, Cavalcanti E G and Ribeiro P C 2001. Improvement of a technique for localization of steel needles in humans using a SQUID magnetometer. Applied Superconductivity, *IEEE Transactions on*, **11**, 677-680.

[5] Barbosa C H 2004. Localization of firearm projectiles in the human body using a superconducting quantum interference device magnetometer: A theoretical study. *Rev. Sci. Instrum.*, **75(6)**, 2098-2106.

[6] Fortaleza L G S, Barbosa C R H, Monteiro E C, Siva E C and Gusmão L A P 2015. Detecting Non-Magnetic Metallic Foreign Bodies by GMR Sensors through the Use of Eddy Currents. *In: 8° Congresso Brasileiro de Metrologia*, 1-4.

[7] Fortaleza L G S, Silva E C, Barbosa C R H, Gusmao L A P and Monteiro E C 2015 Electronic circuit for excitation of inductive loads with high currents *Electron. Lett.* **55** 21 1808–1809

[8] Fortaleza L G S 2016 Nonferromagnetic metallic foreign body detection by eddy currents *MSc Dissertation Pontifical Catholic University of Rio de Janeiro doi:10.17771/PUCRio.acad.28358*

[9] Louzada D R, Monteiro E C, Gusmão L A P and Barbosa C H 2007. Medição não-invasiva de ondas de pulso arterial utilizando transdutor de pressão MIG. *In IFMBE proceedings Springer Berlin Heidelberg*, 436-439.

[10] Pompéia F, Gusmão L A P, Barbosa C R H, Monteiro E C, Gonçalves L and Machado F L A

2008. Ring shaped magnetic field transducer based on the GMI effect. *Meas Sci.Tech.*, **19(2)** 025801.

[11] Silva E C, Gusmão L A P, Barbosa C H and Monteiro E C 2013. Progress toward a hundredfold enhancement in the impedance phase sensitivity of gmi magnetic sensors aiming at biomagnetic measurements. *In V IFMBE proceedings, Springer Berlin Heidelberg*, 742-745.

[12] Us food and drug adminsration et al, 2015. US Department of Health and Human Services.

[13] ICH Harmonised Tripartite. Pharmaceutical development. Q8 (2R). 2009. *Int. conf. on harmonisation of tech. Requiremens for registration of pharmaceutical for human use*.

[14] Veljanovki I 2011. Integrating quality by design (QbD) in medical device manufacturing–concept, benefits, and challenges. White Paper. *Stevens Inst. of Tech.* 1-9.

[15] Lionberger R A et al 2008. Quality by design: concepts for ANDAs. *The AAPS journal*, **10(2)**, 268-276.

[16] Nadpara N P et al. 2012. Quality by design (QbD): A complete review. *Int. journal of pharmaceutical. Scie. Rev. and research.*

[17] Rathore A S 2009. Roadmap for implementation of quality by design (QbD) for biotechnology products. *Trends in biotechnology*, **27(9)**, 546-553.

[18] Lawrence X Y 2008. Pharmaceutical quality by design: product and process development, understanding, and control. *Pharmaceutical research*, **25(4)** 781-791

[19] Rivero A M, Monteiro E C, Leite D S, Tannenbaum F S, Pinto M A G and Leite K T F 2015. Implementing First Stages of Quality by Design Approach in The Development of an Assistive Robotics Technology. *In: XXI IMEKO World Congress*, 1295-1299.

[20] Rivero A M 2015. New Quality by Design – Fuzzy Method Applied in the Development of Biomedical Techonology. Masters dissertation, PUC-Rio.

[21] Louzada D R, Monteiro E C, Rivero A M, Fortaleza L G S, Barbosa C R H, Velázquez D R T, E C Silva and Gusmão L A P. (2016). Quality by Design approach in the development of a magnetic transducer for biomedical measurements: preliminary results on Design Space configuration. In Journal of Physics: Conference Series **772 (1)** 012016.

IOP Conf. Series: Journal of Physics: Conf. Series **1044** (2018) 012052 doi:10.1088/1742-6596/1044/1/012052

Network model for service providers at health sector in Brazil – SIBRATEC PRODSAUDE case

A L M de Oliveira[1] and E Margotti[2]
CERTI Foundation, Campus Universitário UFSC - Setor C, Florianópolis, SC, Brazil

E-mail: [1]aeo@certi.org.br, [2]elm@certi.org.br

Abstract. The term "network" is been used as a logic solution for complex systems with low resources and many actors. Although, only say that a system will work as a network is not enough. A network must have a strategic planning - as a business - with its own goals and organizational model, low hierarchy and different levels of connection and redundancy. The PRODSAUDE is a metrology laboratories network created in 2010 to share resources and connect technologic service providers on health sector in Brazil. Its progress is linked with the State economy and politics situation that is continuously changing. The network must be flexible to change with it, on its model and the positioning of the netweaving team. Strong ties are kinds of connections that should be motivated because those may bring companies and regulators closer a high level cooperative group of laboratories, creating singular impacts for the system.

1. Introduction – About networks

The creation of a sustainable network is a challenge in nowadays. Considering the evolution of information technology and the rise of business based on social networks, the term "network" has been understood as an adequate solution for complex projects with low resources, complex environments and many actors (public and private, central, local) [1]. Therefore, network concepts are being misapplied, leading to discredit the solution. According to Nohria [2], "this indiscriminate proliferation of the network concept threatens to relegate it to the status of an evocative metaphor, applied so loosely that it ceases to mean anything".

In this context, to adopt the networks model won't guarantee that the planned objectives for an organization will be consolidated. A "network organization" must be planned and treated like a business, with its own characteristics, strategic planning, goals, marketing, operation plan and quality assurance. It must consider a network as a complex organization focused on sustainability and continuous improvement.

In Brazil different kinds of laboratory's networks can be found. In a first approach, the national networks could be classified on:

- National networks for traceability and quality assurance, as the Brazilian Network of Calibration (RBC[1]) and the Brazilian Network of Analytical Laboratories (REBLAS[2]);

---

[1] *Rede Brasileira de Calibração*
[2] *Rede Brasileira de Laboratórios Analíticos em Saúde*

IOP Conf. Series: Journal of Physics: Conf. Series **1044** (2018) 012052 doi:10.1088/1742-6596/1044/1/012052

- Regional networks, as the Metrologic network of São Paulo (REMESP[3]) and the Metrologic Network of Rio Grande do Sul (RMRS[4]);
- Similar services networks, as the Brazilian Association for Conformity Assessment (ABRAC);
- Specific sectors service providers' networks [3] , as METRONET, for Oil and Gas, or PRODSAUDE, these one the focus of this report.

Considering the complexity of objectives and environments, the application of network concepts must be fully understood and implemented according to the system characteristics. The network's configuration influences directly the nodes controls level and cost. For example, a national network, generally in an "extended star" model, has reasonable control but also high costs of the central node, compared to other model.

1.1. Network Models

The application of a network concept in an environment or challenge should take into account as characteristics more adequate for its full achievement of the goals, varying in structure, connection types, positioning of members, size, density, diversity and quality [4].

Since the works of Paul Baran[5] in 1964 [8] , the concepts have been transferred between centralized and distributed networks, as extreme points of a normally network, which is decentralized. However, even this network can assume several different configurations that reduce or increase its redundancy and ability to reach established goals.

Depending on the model, a network can have a high cost of maintaining to keep the central node (such as metrology traceability networks) or a low management cost based on the expectation of a self-operation. Figure 1 presents the networks models, their strengths and challenges.

Topology	Location of knowledge	Stronger points	Challenges
Star	Node	Control over content and peripheral nodes	Exclusive dependency of the central node
Extended Star	Core e Node	Accompaniment and direct influence in the chain	Need for significant maintenance features
Line	Node	Strong bilateral relations	Low flexibility and speed
Mesh	Node	Low investment	Generate communication without overloading the system
Fully connected	Node	Dynamic, flexible and very redundant.	To create it and to Keep it

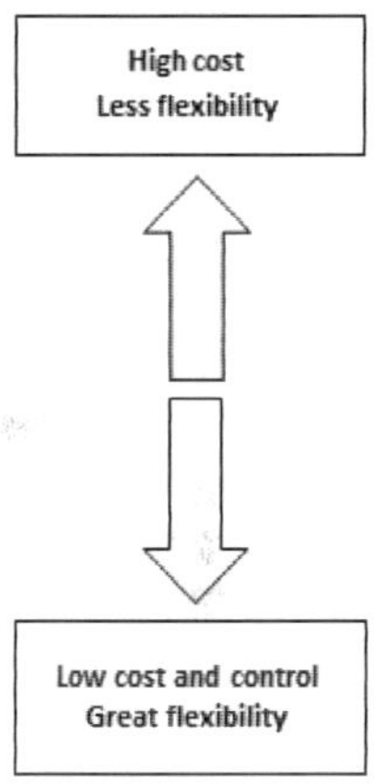

Figure 1 - Models, stronger points and challenges of networks [7] [8]

In the case of an extremely complex environment, with agents not governed by hierarchical structures, the network format must be implemented from stimuli for self-organization, but with netweaving directives, or network animation, that allow the development of the relationships within networks.

In a simple way, according to Granoveter [6], the relations within a network can be in strong ties or weak ties. Strong ties are those relations of neighbors or kinship, where individuals have similar goals

---

[3] *Rede Metrológica do Estado de São Paulo*

[4] *Rede Metrológica do Rio Grande do Sul*

[5] Paul Baran is considered one of the main actors of the creation of the Internet. He had an idea, in 1964, of creating a network with a form of a great web.

and perceptions. These strong relationships tend to accentuate over time, creating loops in networks and increasing the ability to exchange, integrate and share knowledge, as well as increasing trust.

However, it is in the weak ties that radical innovation is facilitated, as they promote the expansion of horizons and individual growth.

As mentioned by Granoveter [6], "... if two individuals have strong relationships, there is likely to be a superposition in their relationships, and the network as a whole is relatively limited. Thus, it is weak relationships that extend the boundaries of networks, connecting groups that have no links to each other."

The concepts of network format and connection types are central to this article, directly supporting the planning of the PRODSAUDE network.

2. About SIBRATEC and PRODSAUDE Network

The SIBRATEC – Brazilian System of Technology was created by initiative of MCTIC – Science, Technology, Innovation and Communication Department, through Decree No 6.259, 21 November 2007, with the objective of supporting the technology development of Brazilian companies, promoting research and development projects, process and product innovation through technological services and consulting.

SIBRATEC supports the creation of networks of institutions, and was structured in three distinct components [1] according to the resolution of its Steering Committee:

1. Innovation Centers - thematic networks formed by P & D institutions with the objective of generating knowledge until the prototype phase, which are intended to promote radical or incremental innovations.
2. Technological Services Networks - formed by laboratories with accredited services, thematically organized, with the objective of supporting the infrastructure and extension of accreditation scopes, support to regulatory structures and conformity assessment systems.
3. Technological Extension Networks - organized by state of the federation, focuses on the solution of small bottlenecks in technological management, product adaptation and improvement of the production of micro, small and medium enterprises.

The PRODSAUDE, a service providers network on health sector in Brazil, was the second network created from SIBRATEC, in 2010, integrating 17 institutions connected from their measurements and tests laboratories, in four different subjects on the sector: Implants, prostheses and surgical instruments; Electro-medical equipment; Dialysis supplies; and Medical supplies.

It is important to consider that other health-related sectors were already being addressed in other networks, such as blood and hemoderivative laboratories, for example. In addition, the system provided a cross-network capability from its ties, when a member participates in several other related networks, such as the Network of Transformed Plastics, Mechanical Manufacturing or Electronic Products and Devices, for example.

Due to diversity of subjects, at the first moment the PRODSAUDE was created using the extended star model, with a coordinator node on a central point, connected to the fund raiser, and sub coordinator for each thematic, connecting the labs on the tips (Figure 2).

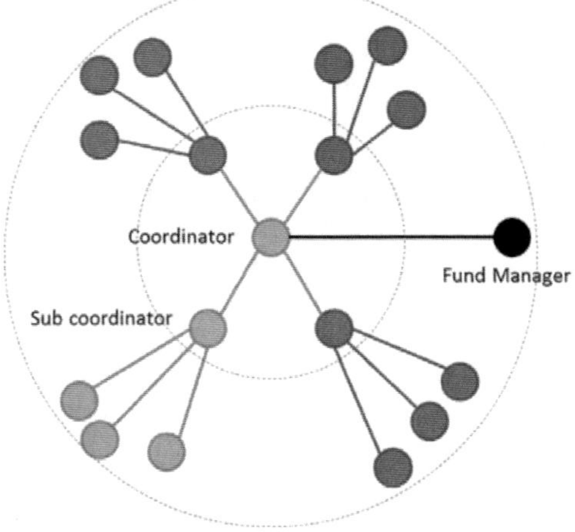

Figure 2 - Original model of PRODSAUDE configuration.

However, the dependence of the central node, in addition to not having enough human resources and funds to provide the animation at all levels, drove the network to an overloaded situation, showing that the extended star model was inefficient for that moment.

3. The reorganization of PRODSAUDE network's model.

In 2015, Brazil entered into a new economic and political crisis scenario. The funds have become scarce and the revision of the models was critical. Based on this, there is a need to revise PRODSAUDE's proposed coordination and operating model.

The PRODSAUDE model was revised based on network theories. Then three new recommendations was created:

- 1st – Focus on the animation of strong ties, maintaining or increasing the sub divisions.
- 2nd – Decentralization of operation activities and the connection with the fund management.
- 3rd – Implementation of quality management system to motivate weak ties focused on innovation and improvement of the health sector.

It means, the sub networks were kept and, initially, the strong ties were stimulated between similar institutions on the network, reducing the sub coordinator control level, which becomes an organizer. The information starts to come from each node, through an internet based system, for everyone on the network, or closed on sub networks, creating new groups based on their interests and work motivations. The fund bureaucratic activities were simplified and the coordinator could focus on the "netweaving" activities. The network redundancy increased expressively, and the central node cannot monitor all the activities anymore, but collect the results at the end.

The resources began to be addressed to those who submitted proposals aligned with the strategic needs of the network and the country's development, according to SIBRATEC's precepts. This meritocracy of the groupings, besides the economic crisis of the country, stimulated a race for the remaining resources and a new animation of the network, creating internal competition and demanding the last directive: implementation of the system of quality management of the network.

With the quality management system, the activities are executed according to procedures, bringing transparency and smoothness to the processes of resources utilization, to the network purposes and its contracting processes. The members' trust in management - now decentralized - depends directly on this tool.

As well as all the action in a complex system, the stimulus does not act equally in all the members, when it was perceived an evasion of some nodes of the network. The way of a new stimulus to the system, passes through the value analysis of the network. The integration of stakeholders at different levels: other private and public laboratories, companies, government agents and associations, for example, become accepted in the network - each in its own role.

The interaction of integrated institutions and laboratories, disputing internal resources, is now turning to the approximation of competencies with the demands in the market. The characteristic of the health sector was translated into a strategic positioning of market focus for the network, since many of the services offered were simply not recognized as useful by the current Brazilian market. Resources once destined to the acquisition of infrastructure or services to increase laboratory quality were redirected to activities - for example - training of professionals in industries and other laboratories in Brazil with the creation of the "PRODSAUDE Academy".

The strong ties between laboratories were then stimulated by constant and regular on-line meetings (instead of sporadic face-to-face meetings), with solution goals and targeted issues. The problems are categorized for the actions in network directed to the precompetitive approach of the laboratories, in actions of support to the national normalization and qualification of the consumer market of services, for example. Figure 3 illustrated this evolution of PRODSAUDE configurations.

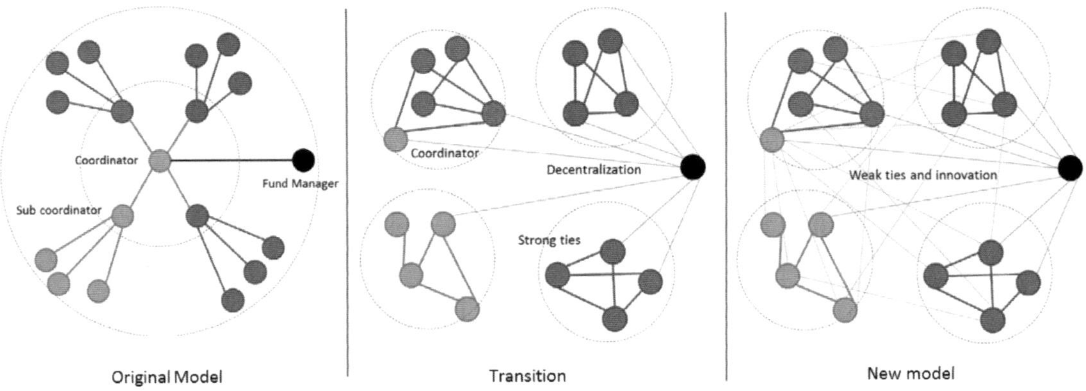

Figure 3 - Evolution of PRODSAUDE model.

4. The future of the network

The rise of network redundancy made PRODSAUDE starts to grow up with new laboratories and create new types of members: governmental agencies, associations and even companies. The combination of expectative brings new complexities as weak ties [6] , considering new resource fonts and the access to the market. The laboratory motivation is increasing as far as it grows the number of customers and regulators connected, as market pull.

It is expected, in this sense, that these disturbances in the original model spontaneously create new clusters, with specific focus, that can use multiple sources of development. In the same direction, previously obvious groupings can be dissolved without affecting the system. Redundancy grows from these groupings and creates new weak ties that, over time, can become strong bonds from new business models.

With the approach among the stakeholders of the sector, the network should seek thematic for the development of all. The increase in the perceived value of the network members will - certainly - be the development of the metrology consumer market and conformity assessment. In order to deal with the issues from the prior knowledge of the companies and to seek evolution, in the concept of the Market Pull, it is necessary to change the focus of laboratories working in the health sector, sharing their contributions in time and resource for the improvement of quality with training and the evolution of companies.

A proposed theme, for example, is the positioning of laboratories as inductors of advanced manufacturing. The discourse of the past (better measurement uncertainty, accredited quality,...) is directed towards the competitiveness of companies.

References

[1] Teixeira S M F 2002 O desafio da gestão das redes de políticas. (Rio de Janeiro, RJ). Available on: <http://unpan1.un.org/intradoc/groups/public/documents/CLAD/clad0043204.pdf>.

[2] Nohria N 1992 *Introduction: Is a Network Perspective a Useful Way of Studying Organizations? Networks and Organizations: Structure, Form and Organizations* (Harvard Business School Press: e.g.1 – 22).

[3] Oliveira A L M, Ferreira A, Donatelli G D Implementation of a R&D&I network in metrology for large corporations - Petrobras case study, *paper for XXI IMEKO World Congress "Measurement in Research and Industry"*.

[4] Julien, P A 2010 *Empreendedorismo regional e a economia do conhecimento* (São Paulo, SP: Saraiva).

[5] SIBRATEC 2008 Resolução Comitê Gestor SIBRATEC nº 001- Available on: <http://www.mct.gov.br/upd_blob/0023/23763.pdf>

[6] Granovetter M 1973 The strength of weak ties. (American Journal of Sociology v. 78 n. 6) e.g. 1360-80.

[7] Quinn, J. B., Anderson, P. E Finkelstein, S 2001 Novas Formas De Organização. In: Mintzberg, H. E Quinn, J.B. (Org.). *O Processo Da Estratégia*. 3ª Ed. (Porto Alegre: Bookman, 2001).

[8] P. Baran, *"On Distributed Communications: I. Introduction to Distributed Communications Networks",* The RAND Corporation, Santa Monica, CA, Memorandum RM-3420-PR, 1964.

IOP Conf. Series: Journal of Physics: Conf. Series **1044** (2018) 012053 doi:10.1088/1742-6596/1044/1/012053

New calibration procedure for differential pressure using twin pressure balances for flowrate measurement

M Rosendahl[1, 2,3], Nazareth R S[1], Magalhães M R[1], Silva W S[1], Ferreira P L S[1], Gouveia, J M[1], Farias, M H[1], Oliveira, J S[1], Couto, P R G[1]

1 National Institute of Metrology, Quality and Technology - Inmetro, Av. Nossa Senhora das Graças, 50 - Xerém, 25250-020, Duque de Caxias - RJ, Brazil

2 Rio de Janeiro State University–UERJ

3 Estácio de Sá University–UNESA

mrosendahl@inmetro.gov.br

Abstract. Reliable pressure measurements are essential in all activities of society and industry. Indeed, the concern with traceability has notably improved and has received considerable interest due to the intensification of international trade, the need for environmental issues, safety and health requirements, for example in medical instruments, as well as devices for measuring the flow of Fluids Based on the pressure change. Differential pressure sensors are widely used in fluid flow processes, such as for monitoring the flow conditions and to determine flowrate or fluid velocity (example are the Pitot tube, Venturi, orifice plate etc.).This study focused on the precise generation of differential pressures in the range of 500 kPa to 20 MPa using gas-operated twin pressure balances. The metrological characteristics of a differential pressure manometer were evaluated.

1. Introduction

Pressure measurements play an important role in industrial processes. There are several applications of pressure quantity; for example, in the petroleum, petrochemical, meteorological, aerospace, aviation, industries etc. The reliability of these measurements is associated with trade, quality, health, safety, etc.

Pressure measurement instruments can be classified into two main groups: fundamental and relative. The fundamental instruments measure pressure from the definition of quantity. In this group are included the liquid column manometer and the pressure balance.

The pressure balance that can measure absolute pressure, differential pressure, gauge pressure and vacuum. [8], is a fundamental group meter widely used as reference standard in the measurement of the pressure magnitude. It consists basically of a piston-cylinder assembly, a set of masses and a watertight system where the fluid is contained which will generate the pressure to be measured which acts at the base of the piston. Figure 1.

The magnitude of the pressure balance measured from the balance between the force resulting from the pressure of a fluid (gas or liquid) acting at the base of the piston with the force coming from the masses acting at the top of the piston. The equation of pressure measurement by a pressure balance is defined by equation 1

$$p = \frac{\left[m_p\left(1-\frac{\rho_a}{\rho_{mp}}\right)+\Sigma m\left(1-\frac{\rho_a}{\rho_m}\right)\right]\cdot g+\sigma C}{A_{0,20}\left[1+(\alpha_c+\alpha_p)\cdot(\theta-20)\right]\cdot(1+\lambda p_n)} \pm \rho_{fluid} g \Delta h \qquad (1)$$

where:
p is the measured pressure; m_p is the mass of the piston; ρ_a is the specific mass of the air; ρ_{mp} is the specific mass of the piston material; Σm is the sum of the remaining masses acting at the top of the piston; ρ_{m} is the specific mass of the material of the remaining masses; g is the acceleration due to local gravity; σ is the surface tension of the fluid; C is the length of the circumference of the piston; $A_{0,20}$ is the area of the cylinder piston set; $\alpha_c+\alpha_p$ are the coefficient of linear thermal expansion of the piston-cylinder assembly; θ is the piston temperature at the time of measurement; λ is the coefficient of deformation of the piston-cylinder assembly; p_n is the nominal pressure of the measurement; ρ_{fluid} is the density of the working fluid; Δh is the difference between the base of the piston and the point where the pressure is measured; In equation 1, the value of Δh is negative when the piston base of the pressure balance is below the pressure measurement point, and positive when the base is above the pressure measurement point.

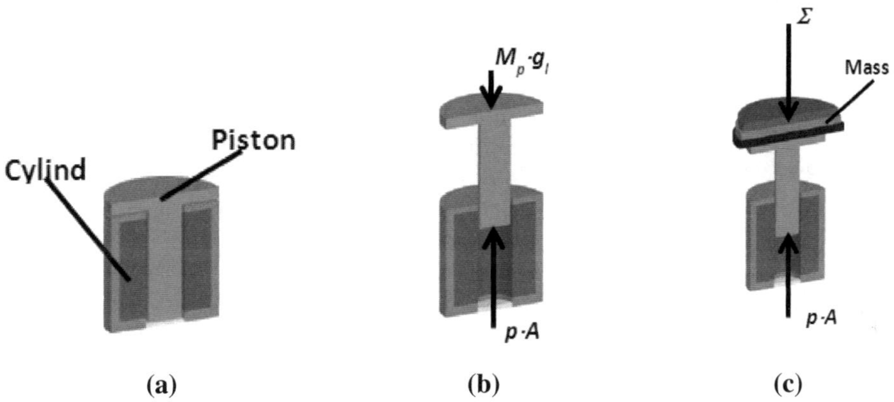

Figure 1. Principle of measuring the pressure balance [5].

The relative instruments measure the pressure in function of a physical property or a physical phenomenon. In this group are included manometers, vacuometers, manovacuometers, digital piston manometers, pressure transducers / transmitters, etc.

Differential pressure sensors are widely used in fluid flow processes, such as for monitoring the flow conditions and to determine flowrate or fluid velocity (example are the Pitot tube, Venturi, orifice plate etc.). The typical dependency of these meters on Δp is given by a square relationship, thus, the more large an expected measurement range of such meters, the more large a necessary range of a Δp sensor. Otherwise, while the flowrate range decreases, the respective uncertainty increases due to the difficulty of measuring accurately the low level of its related differential pressure. In order to minimize this impact, it could be used more than one instrument for differential pressure measurement, of for each range. However, independent on the range, the calibration of the differential pressure sensor is fundamental for the reliability of flow measurements.

2. Objectives

The objective is to propose a new methodology to be applied to differential pressure measurement using two reference pressure balances.

3. Methodology

3.1. Calibration Procedure

In calibration care must be taken regarding the instrument in calibration such as: cleaning; being it digital must be connected before to the electrical network for a period of 30 minutes and in which fluid it should be calibrated. The standards used must have at least one accuracy class four times better than the instrument to be calibrated. Before starting calibration with the instrument connected to both standards, its maximum pressure must be applied for 10 minutes to check for system leakage. The meter shall be calibrated in its working position (vertical or horizontal) at ten points at increasing and decreasing pressure and the measurements shall be recorded on the calibration worksheet.

3.2. Differential Reference Pressure Calculation

When two pressure balances are used in calibration the reference differential pressure is defined by equation 2.

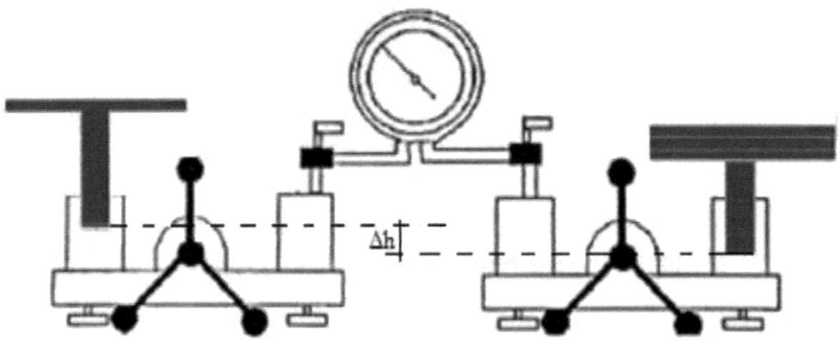

Figure 2. Principle of measuring the pressure balance.

$$\Delta Pref = P_{ref1} - P_{ref2} = \frac{\left[M_{p1}\cdot\left(1-\frac{\rho_a}{\rho_{Mp1}}\right)+\sum M_1\left(1-\frac{\rho_a}{\rho_{M1}}\right)\right]\cdot g+\sigma C_1}{A_{10,20}[1+(\alpha_{c1}+\alpha_{P1})\cdot(t_1-20)]\cdot(1+\lambda_1\rho_n)} - \frac{\left[M_{p2}\cdot\left(1-\frac{\rho_a}{\rho_{Mp2}}\right)+\sum M_2\left(1-\frac{\rho_a}{\rho_{M2}}\right)\right]\cdot g+\sigma C_2}{A_{20,20}[1+(\alpha_{c2}+\alpha_{P2})\cdot(t_2-20)]\cdot(1+\lambda_2\rho_n)} \pm \rho_{fluid}\cdot g\cdot\Delta h \quad (2)$$

where all the variables of the equation 2 have already been described in the previous item.

3.3. Measurement Uncertainty

The uncertainty of measurement in the calibration of a pressure balance is estimated according to "JCGM 100: 2008 - GUM 1995 with minor corrections - Evaluation of measurement data - Guide to the expression of uncertainty in measurement - First edition September 2008 ".

The measured in a calibration is the value of the quantity that defines the error of the instrument. Thus in the case of calibration of the differential pressure manometer the error is defined by the expression 3.

$$e_{(p)} = \Delta \mathrm{P}i - \Delta Pref \ (3)$$

At where: $e_{(p)}$ is the error of the instrument; ΔPi is the magnitude value indicated by the instrument and $\Delta Pref$ is the magnitude value measured by the standard.

Thus, considering equations 2 and 3, show the cause-effect diagram of the calibration performed by two twin pressure balance.

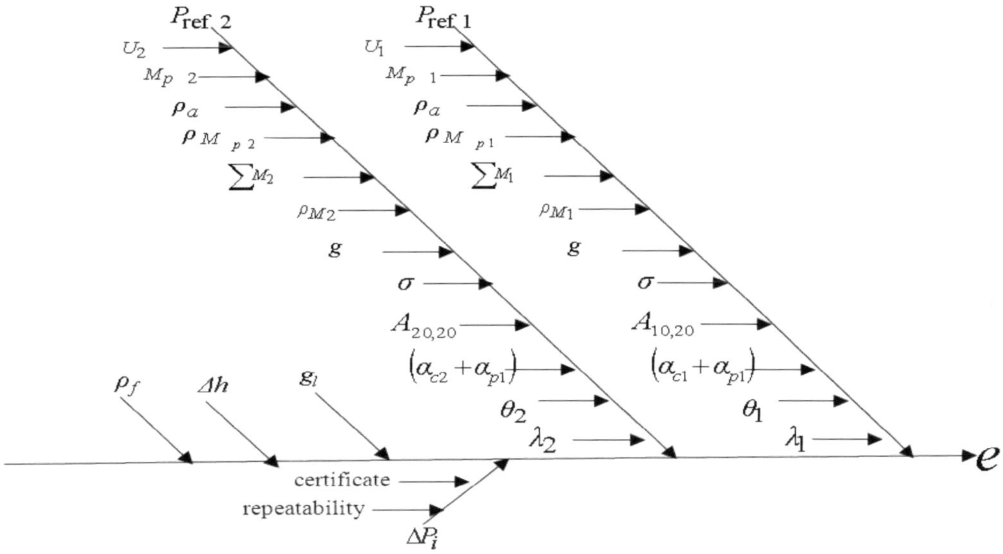

Figure 3. Diagram cause-effect of the calibration

4. Results and Discussion

4.1. Errors

Table 1 shows the instrument errors according to equations 3.

Table 1- Instrument errors.

Δpi	Error			
bar				
0,00	0,00	0,01	0,01	0,01
1,00	-0,01	-0,01	-0,04	-0,01
2,00	-0,05	-0,04	-0,04	-0,04
3,00	-0,06	-0,11	-0,11	-0,11
4,00	-0,09	-0,13	-0,14	-0,13
5,00	-0,12	-0,16	-0,13	-0,16
6,00	-0,15	-0,20	-0,15	-0,20
7,00	-0,28	-0,24	-0,18	-0,24
8,00	-0,28	-0,24	-0,16	-0,24
9,00	-0,33	-0,36	-0,29	-0,36
10,00	-0,37	-0,39	-0,27	-0,39

IOP Conf. Series: Journal of Physics: Conf. Series **1044** (2018) 012053 doi:10.1088/1742-6596/1044/1/012053

The graph of figure 4 shows that errors are contained in the intervals defined by their systematic errors plus or minus half resolution of the calibrated instrument (±0,1 bar).

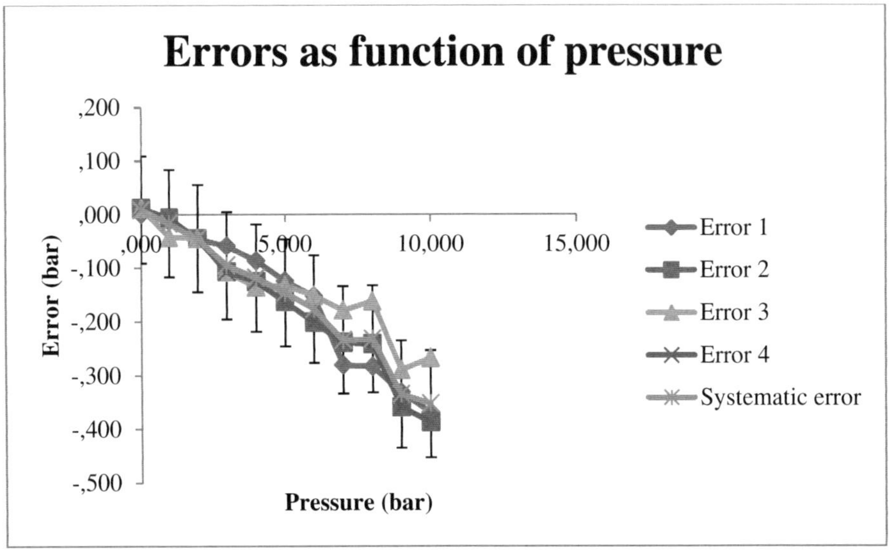

Figure 4 Instrument error curve -

This step is very important because it can be used as a tool for the evaluation of calibration quality control. A calibration is under control when all punctual instrument errors are contained within the range defined by punctual systematic errors plus or minus half the resolution of the calibrated instrument.

4.2. Hysteresis, repeatability and linearity

Table 2 presents the metrological characteristics of the manometer: maximum error, hysteresis, repeatability, linearity and the calibration curve of the instrument. The percentage values shown are in relation to the amplitude of the measuring range of the instrument.

Table 2- Maximum error, hysteresis, repeatability and linearity

Maximum error (%):	2,3
Hysteresis (%):	0,7
Repeatability (%):	0,7
Linearity (%):	0,6
Calibration curve	ΔPref = 1,0359ΔPi -0,0212

4.3. Measurement Uncertainty

The graphs of Figures 5A and 5B show, respectively, the budget of uncertainties in the lower and upper limits of the manometer measurement range. It is observed that in the two limits the uncertainty that most impacts is that coming from the resolution of the instrument.

IOP Conf. Series: Journal of Physics: Conf. Series **1044** (2018) 012053 doi:10.1088/1742-6596/1044/1/012053

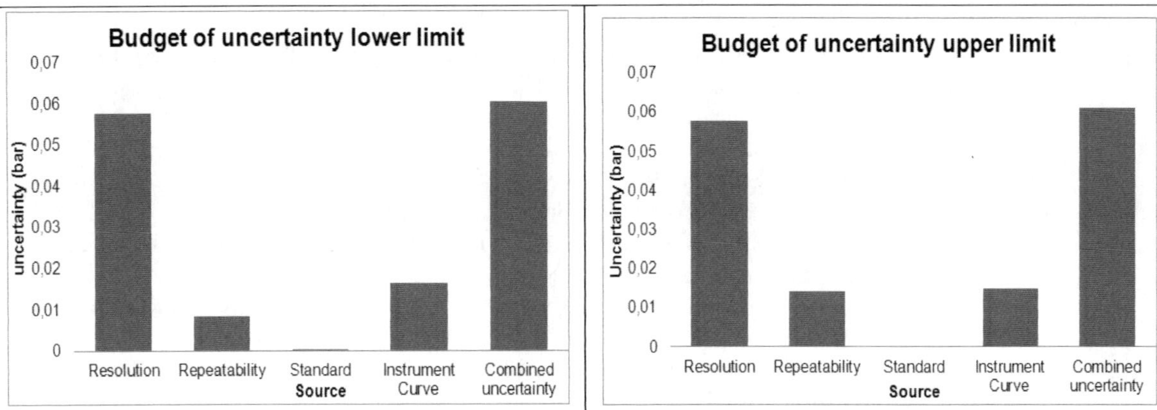

A	B

Figure 5- Budget of uncertainties in the lower and upper limits

Table 3 presents the expanded uncertainty (U) of the manometer with the respective coverage factors for the 95% confidence level.

Table 3 Expanded uncertainty

Pressure (psi)	Coverage factor (k)	U (bar)
0,00	2,00	0,12
1,00	2,00	0,12
2,00	2,00	0,12
3,00	2,00	0,12
4,00	2,00	0,12
5,00	2,00	0,12
6,00	2,00	0,12
7,00	2,01	0,13
8,00	2,02	0,13
9,00	2,00	0,12
10,00	2,03	0,14

5. Conclusion

A new methodology to be applied to differential pressure measurement using two reference pressure balances is proposed. Manometer errors are contained within the intervals defined by point systematic errors ±0,1 bar. The source of uncertainty that most impacts the calibration is that coming from the resolution of the instrument. The calibration uncertainty varied from 0.12 bar to 0.14 bar.

Acknowledgement
Authors are grateful to CNPq, Faperj, Finep, Petrobras, Pronametro, Prometro, Inmetro and Productivity Research Award - UNESA.

References

[1] C. R. Tilford and R. W. Hyland, Proc. XI IMEKO World Congress, Houston, Texas, 1988; C. R. Tilford, Proc. Workshop and Symposium of the National Conference of Standards Laboratories, 1988.

[2] Buonanno, et al, Ten Years of Experience in Modelling Pressure Balances in Liquid Media up to few GPa; ISBN: 978-88-8317-037-9

[3] Oliveira, J. S.; Couto, P. R. G. Metrologia da Grandeza Pressão – Volume 2

[4] ASSOCIAÇÃO BRASILEIRA DE NORMAS TÉCNICAS. Guia para a Expressão da Incerteza de Medição. ISO GUM 95. Terceira Edição Brasileira

[5] INMETRO, "Calibração de Balança de Pressão. Guia de Calibração". INMETRO-DIMCI-DIMEC-LAPRE INMETRO. Nota técnica, DIMEC/gc-03/v.01. 22p. (2010).

[6] Guide to the Expression of Uncertainty in Measurement. Rio de Janeiro :ABNT , INMETRO. Edição Revisada. Agosto de 2003.120p.

[7] JCGM 100:2008 GUM 1995 with minor corrections Evaluation of measurement data — Guide to the expression of uncertainty in measurement-First edition September 2008

[8] INSTITUTO NACIONAL DE METROLOGIA, NORMALIZAÇÃO E QUALIDADE INDUSTRIAL (INMETRO).Vocabulário Iinternacional de Metrologia- Conceitos Fundamnatis e Gerais e Termos Associados 1.ed. Rio de Janeiro, 2008.78p.

[9] Versão Brasileira do Documento de Referência EA-4/02 - Expressão da Incerteza de Medição na Calibração, INMETRO e ABNT e SBM, Rio de Janeiro, 1999.

[10] Versão Brasileira do Documento de Referência EA-4/02-S1, Suplemento 1 ao EA-4/02 - Expressão da Incerteza de Medição na Calibração - Exemplos, INMETRO e SBM, Rio de Janeiro, 1999.

IOP Conf. Series: Journal of Physics: Conf. Series **1044** (2018) 012054 doi:10.1088/1742-6596/1044/1/012054

Nonlinear anisotropic diffusion filtering for the characterization of stochastic structured surfaces

N Loftfield, M Kästner and E Reithmeier

Institute of Measurement and Automatic Control, Leibniz Universität Hannover, Nienburger Strasse 17, 30167 Hannover

E-mail: nina.loftfield@imr.uni-hannover.de

Abstract. Structured surfaces enhance the functionality of components. Well known is the influence of the surface structure on friction and wear behavior. Beyond this, structured surfaces are widely used for various purposes such as optical, biological or mechanical applications. Therefore, the characterization of structured surfaces and surface features becomes increasingly important. The functionality of a surface can either be tested directly or indirectly. Due to the correlation of geometric surface features and its functionality, an indirect and self-evident way is by measuring the surface topography. To obtain the geometric essentials of these features, they need to be separated from the raw surface data. The standard procedure of decomposing a surface topography is by the use of a Gaussian filter bank, gaining so called scale-limited surfaces. This procedure shows drawbacks when characterizing structured surfaces by introducing distortions to the feature boundaries. To overcome these limitations, this work proposes the use of an automatic nonlinear anisotropic diffusion filter as an initial step to separate the features from the residual surface topography because of its edge preserving properties. It is shown that the nonlinear anisotropic diffusion serves well the separation of the features and their geometrical characterization.

1. Introduction

It is widely known that structured surfaces serve the functionality of components. Well-known functions are of optical or mechanical nature like the Fresnel lens or the influence of surface features on friction and wear behavior. Therefore, structured surfaces become increasingly important, not only technologically but also economically [1]. Testing the functionality of a surface can either be done directly or indirectly. Due to the correlation of surface features and its physical functionality, an indirect yet obvious characterization is possible by geometric measurements of the surface. However, Evans et al. stated already in 1999 that the function of structured surfaces cannot be related to traditional surface finishing parameters [2]. Further, Mathia et al. showed more recently that the traditional concept of determined parameters for roughness and waviness does not satisfy the characterization of structured surfaces [1]. A surface is commonly considered as a superposition of structures with different scales. Scale-limited surface descriptions are most widely gained by spatial linear Gaussian filtering with determined cutoff frequencies. For structured surfaces this holds some drawbacks. For example, when filtering noise: noise is usually represented by high frequencies, however the converse is not necessarily true, not all high frequency portions of the signal are solely due to noise. Therefore, filtering the surface data with a linear Gaussian filter can diminish slopes at crucial

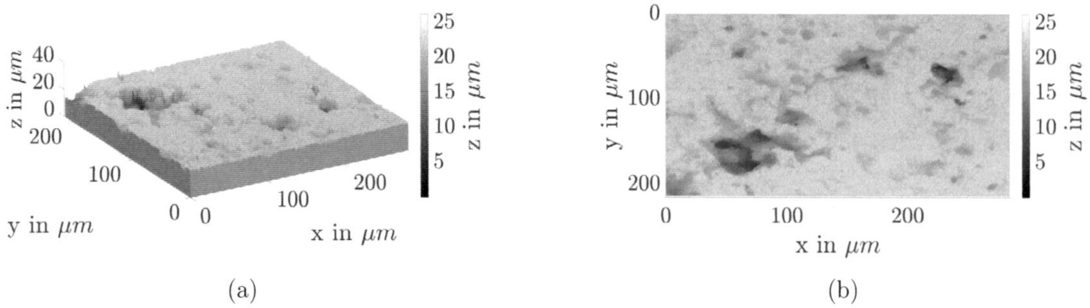

Figure 1: Raw measurement data of porous aluminum oxide layer: (a) three dimensional plot, (b) top view.

feature boundaries. In contrast, nonlinear anisotropic diffusion has very good edge preserving properties. In the research field of metrology different approaches have been pursued. Zeng et al. use the advantage of the nonlinear anisotropic diffusion to separate the geometrical features of MEMs surfaces [3]. Rather than investigating the roughness or waviness components of the surface data, Zeng et al. state that the focus is on geometrical features like the line width, step height, etc. Their work shows good results but no setting parameters are recommended and the problem of an automatic stopping criterion is not addressed. Also, Wang et al. apply the nonlinear anisotropic diffusion filter to discontinuous surfaces, thereby avoiding smoothing effects around the edges [4]. However, the used stopping criterion is based on the cutoff frequencies of the scale limited surfaces using the correlation of diffusion and wavelengths. As they state, this can only be used as an approximation. Nonlinear filtering cannot be assigned to one specific cutoff frequency. Based on the promising results of the work of Zeng et al., we propose a nonlinear anisotropic diffusion as a preliminary filtering step for structured surfaces. For the stopping criterion of the diffusion process a frequency based approach similar to the one proposed by Ilyevsky et al. is pursued [5].

2. Materials and Methods

Within this work, stochastic structured, porous ceramic layers are characterized. The measurement data of the ceramic layers are filtered to separate the structure from noise and roughness components. This is done by nonlinear anisotropic diffusion. Following, to identify surface features, the data needs to be segmented, e.g. by watershed segmentation. In this work the focus lies on the filtering process and the segmentation is not further discussed.

2.1. Ceramic Layers

The investigated surfaces are comprised of porous aluminum oxide layers with a nominal planar surface. The layers are manufactured by a thermal spraying process and are subsequently polished. We will refer to them as stochastic structured surfaces. The porosity is influenced by the process parameters and the material composition. The measurement data are acquired by a confocal laser scanning microscope VK-X210 from Keyence. All data shown in this work is acquired with a 50x magnification lens, respectively with a lateral pixel resolution of 0.277 μm.

2.2. Anisotropic Diffusion

Anisotropic diffusion was first published by Perona and Malik, suggesting a new definition of scale-space. The scale-space is modeled as a diffusion process [6]. The linear heat diffusion equation describes the dissipation of heat in a specific region. In image processing or signal processing this can be transferred to a smoothing process at a constant rate dependent on the diffusion time, similar to Gaussian filtering depending on the variance. With the topography height data $I(x, y, t)$, where $I(x, y, 0)$ is the initial surface topography, the linear diffusion is described by:

$$\frac{\partial I(x,y,t)}{\partial t} = \operatorname{div}(c \cdot \nabla I(x,y,t)), \tag{1}$$

where div is the divergence operator, ∇ is the gradient operator and c is a constant conductance coefficient indicating the diffusion speed. Though linear diffusion reduces noise, it also blurs the signal, which in turn results in smoothed structure boundaries.

The nonlinear anisotropic diffusion is a modified linear diffusion process with an adaptive conductance coefficient c which is updated in every iteration step as a function of the gradient:

$$c(x,y,t) = g(\nabla I(x,y,t)). \tag{2}$$

Perona and Malik suggest two conductance functions:

$$g_1(\nabla I(x,y,t)) = e^{(-(||\nabla I(x,y,t)||/K)^2)} \tag{3}$$

and

$$g_2(\nabla I(x,y,t)) = \frac{1}{1 + (\frac{||\nabla I(x,y,t)||}{K})^2}, \tag{4}$$

with the diffusion coefficient K controlling the rate of diffusion. The first conductance function $g_1(\nabla I(x,y,t))$ privileges high-contrast edges over low-contrast ones, the second one $g_2(\nabla I(x,y,t))$ privileges wide regions over small ones [6]. In the following $g_1(\nabla I(x,y,t))$ and $g_2(\nabla I(x,y,t))$ will be denoted as $g_1(\nabla I)$ and $g_2(\nabla I)$. In any case the conductance function is chosen to provide fast diffusion within regions and a slow diffusion at sharp edges with a high gradient to avoid smoothed structure boundaries.

With I_t the discrete sampled image at the iteration step t the discretized anisotropic diffusion is given by Perona and Malik by:

$$I_{t+1} = I_t + \lambda \cdot [c_N \triangledown_N I_t + c_S \triangledown_S I_t + c_E \triangledown_E I_t + c_W \triangledown_W I_t] \tag{5}$$

with $0 \leq \lambda \leq 1/4$ for the numerical scheme to be stable, c_i the conductance function and $\triangledown_i I_t$ the difference of the neighboring pixels. The indices $i \in \{N, E, S, W\}$ indicate the 4-neighbor pixel: North, East, South, West, e.g. in the North direction:

$$\triangledown_N I(x,y,t) = I(x,y-1,t) - I(x,y,t). \tag{6}$$

As Perona and Malik state this is not an exact discretization of the anisotropic diffusion but a good approximation and is chosen due to its simplicity.

2.3. Choice of Parameters

The choice of the conductance function, the diffusion coefficient K and the number of diffusion iterations n is very decisive for the result of the filter operation. Therefore, these have to be chosen carefully. Oftentimes, the parameters are determined by trial and error and appointed manually, however, an automated parameter selection is desirable. A detailed overview of different approaches on the choice of parameters for anisotropic diffusion is given by Tsiotsios and Petrou, further, they discuss the importance and influence of the different parameters [7].

The conductance function characterizes the diffusion process in terms of which parts of the signal are favored. In any case the conductance function is a monotonically decreasing function of the gradient. Other conductance functions have been proposed, e.g. by Black et al. or Weickert adjusting the diffusion speed or even stopping the diffusion at a given gradient, in respect to avoiding an over-smoothed filter result [8, 9]. However, in this work we focus on the proposed conductance functions of Perona and Malik. The input of the conductance function is the gradient of a smoothed image rather than of the original image to avoid the influence of outliers. The regularization process is done by Gaussian smoothing. The cutoff frequency of the Gaussian filter is picked depending on the resolution of the measurement data according to the DIN Norm 25178 [10].

The diffusion coefficient K controls the rate of diffusion. K is a soft threshold between the image gradients that are attributed to noise or edges. K can either be set to a constant related to the original data or be updated in every iteration step related to the diffused image. Most approaches are based on the gradient distribution, e.g. K is updated as the mean value or the maximum value of the gradients in every iteration [3, 4]. In this work, K is set adaptively in every iteration step according to the Canny noise estimator: the histogram of gradients is determined and the threshold is set to the 90% value of its integral [11]. Further Tsiotsios and Petrou suggest the use of two separate thresholds, evaluated in the horizontal (EW) and in the vertical (NS) direction to individually consider the gradients in the smoothing process oriented along the separate axes [7]. As we determine the 8-neighbors gradient, four different thresholds are used, differentiating also the diagonal directions, NW_SE and NE_SW. This is done with respect to filtering structured surfaces in general. Isotropic surfaces show a predominant direction, different gradients respectively. The thresholds K_j with $j \in \{NS, EW, NW\_SE, NE\_SW\}$ are, therefore, evaluated based on four different histograms of gradients.

As stated, anisotropic diffusion is an iterative process and the filtering result is highly dependent on the number of iterations. The used stopping criterion for t is based on a frequency approach similar to Ilyevsky et al. by estimating the energy of the signal's high frequency components and stopping at an ideal stopping time [5]. They define the determination based on a highly denoised smoothed signal and a ratio at which to stop. Commonly, high frequencies are labeled as frequencies in the range of N/4 and N/2 in the frequency band. The energy of the frequencies is the Euclidean norm of the frequency section, denoted by L_2^h. Noise is mainly composed of high frequencies but the opposite is not always true. In our approach we determine the number of iterations n in which the high frequency component energy falls in the first local minimum with a minimum distance of eight or, alternately, when the function runs approximately in a constant course. Instead of choosing the high frequency components in the range of N/4 and N/2, the cutoff frequency could also be defined according to the DIN 25178 [10]. Just as before, the stopping criterion for the diffusion time would be set to the local minimum of the frequencies' energy of the specified band width.

3. Results

The results show well preserved features of the stochastic structured surface together with high-level noise suppression or, if desired, roughness separation. In comparison to linear filtering, e.g. Gaussian filtering, the edges are well preserved and the filtered signal follows the features' structure even in very steep regions. The surfaces were filtered according to the proposed approach with the use of both conductance functions $g_1(\nabla I)$ and $g_2(\nabla I)$. Different results for the decreasing energy and diffusion coefficients were detected for the different conductance functions.

From the filtering results, it can be seen that the energy portion of the high frequencies

IOP Conf. Series: Journal of Physics: Conf. Series **1044** (2018) 012054 doi:10.1088/1742-6596/1044/1/012054

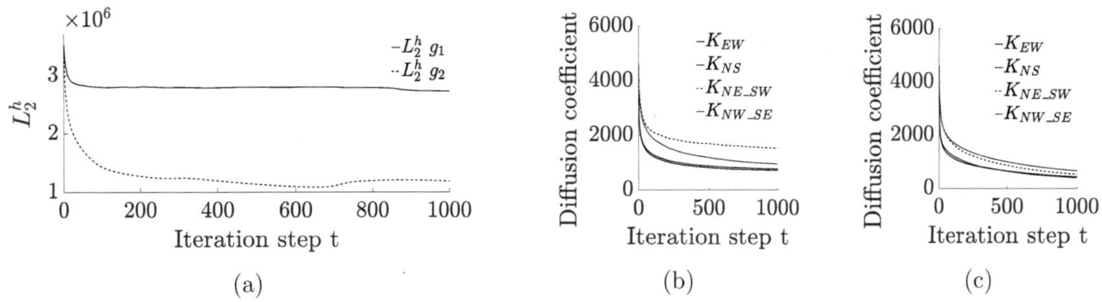

Figure 2: Course of energy L_2^h and diffusion coefficients K_j of the filtering process with $g_1(\nabla I)$ and $g_2(\nabla I)$ over 1000 iteration steps: (a) L_2^h, (b) K_j with $g_1(\nabla I)$, (c) K_j with $g_2(\nabla I)$.

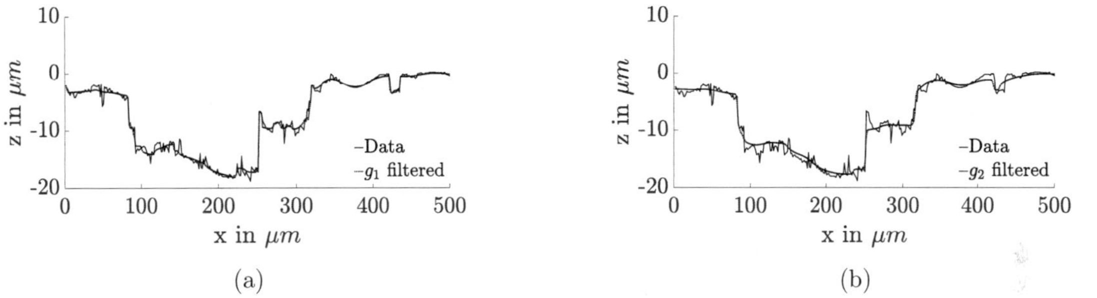

Figure 3: Profile sections at $y = 175\ \mu m$ of the aluminum oxide layer (figure 1) and the filter results: (a) with $g_1(\nabla I)$, (b) with $g_2(\nabla I)$.

L_2^h falls with increasing time. Due to the different characteristics of the conductance functions $g_1(\nabla I)$ and $g_2(\nabla I)$ the course of L_2^h also differs. The results of the filtering process with the conductance function $g_1(\nabla I)$, privileging high contrast edges over low-contrast ones, show higher L_2^h preserved in the data, see figure 2a. The sequence of the four different diffusion coefficients K_{NS} K_{EW}, K_{NW_SE} and K_{NE_SW} are shown in figures 2b and 2c. The diffusion coefficients show a similar decreasing course. The different directions vary just slightly due to the form of the micro-structure, there is no preferable structure direction, see figure 1.

The filtering results of the stochastic structured aluminum oxide surfaces of the proposed approach are shown in figure 3. With the conductance function $g_1(\nabla I)$ the micro-structure edges are well preserved and the high frequency portion is suppressed. With $g_1(\nabla I)$ the diffusion process is stopped at the iteration step $t_1 = 116$, the first local minimum in the L_2^h course. In contrast, when filtering with $g_2(\nabla I)$, also the edges of big structures are well preserved, however, small regions get smoothed and modified by the filtering process. The diffusion is stopped at the iteration step $t_2 = 273$.

Concluding, filtering results show the advantage of the proposed method with respect to Gaussian filtering. The surface micro-structure shows steep edges, see figure 4a. Filtered by the proposed nonlinear anisotropic diffusion filter, the edges are preserved very well and the roughness and noise components are filtered. Because the advantage to linear filtering is best seen at steep edges the filtered signal is compared to Gaussian filtering. With the same filter result for the roughness and noise with the Gaussian filter the micro-structure edges are smoothed, see section $x = 800\ \mu m$.

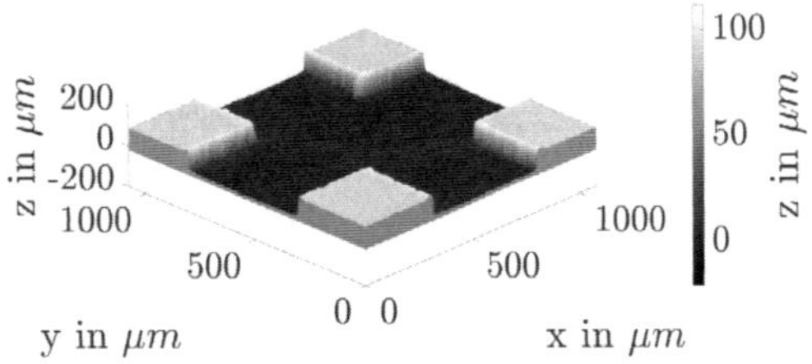

(a)

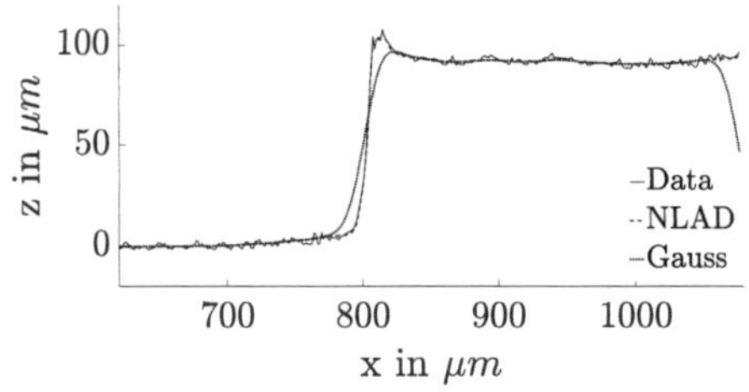

(b)

Figure 4: Filter results of a deterministic micro-structure: (a) deterministic structure, (b) profile section at $y = 140$ μm of raw data, filtered by nonlinear anisotropic diffusion and by Gaussian.

4. Discussion and Conclusion

Nonlinear anisotropic diffusion with a frequency based stopping condition was used for the scale-space representation of structured surface data with focus on the features. It is shown that the micro-structures can be well segmented from the other surface components with good edge preserving conditions. The smoothed edges impede the following segmentation of structures, e.g. by watershed segmentation.

However, as stated the filter result of the surfaces is dependent on the choice of the conductance function. Filtered with $g_1(\nabla I)$, the results show sharper edges and preserves small regions, whereas when filtered with $g_2(\nabla I)$ the filter result smooths small regions. For our proposed stochastic structured surfaces we recommend filtering with $g_1(\nabla I)$ rather than $g_2(\nabla I)$. Although, filtering large micro-structures with wide region, good results can also be achieved with $g_2(\nabla I)$, see figure 4b.

While mainly shown based on porous surfaces yet, we believe that for other structured surfaces the anisotropic diffusion is a powerful tool to separate the features from the surfaces.

Acknowledgments

Acknowledgments for the funding of this work go to the "Dr. Jürgen und Irmgard Ulderup Stiftung" and for his contribution to this work to Stefan Siemens.

References

[1] Mathia T G, Pawlus P and Wieczorowski M 2011 *Wear* **271** 494–508 ISSN 00431648
[2] Evans C J and Bryan J B 1999 *CIRP Annals - Manufacturing Technology* **48** 541–556 ISSN 00078506
[3] Zeng, W, Jiang, X, Scott, P J, and Blunt, L *Proc. of the 10th International Symposium of Measurement Technology and Intelligent Instruments, 2011*
[4] Wang M, Shao Y P, Du S C and Xi L f 2015 *International Journal of Precision Engineering and Manufacturing* **16** 2057–2062 ISSN 2234-7593
[5] Ilyevsky A and Turkel E 2010 *Journal of Scientific Computing* **45** 333–347 ISSN 0885-7474
[6] Perona P and Malik J 1990 *IEEE Transactions on Pattern Analysis and Machine Intelligence* **12** 629–639 ISSN 01628828
[7] Tsiotsios C and Petrou M 2013 *Pattern Recognition* **46** 1369–1381 ISSN 00313203
[8] Black M J, Sapiro G, Marimont D H and Heeger D 1998 *IEEE transactions on image processing : a publication of the IEEE Signal Processing Society* **7** 421–432 ISSN 1057-7149
[9] Weickert J 2001 *Pattern Recognition* **34** 1813–1824 ISSN 00313203
[10] DIN Deutsches Institut für Normung e V November 2012 DIN ISO 25178-3 Geometrische Produktspezifikation (GPS) - Oberflächenbeschaffenheit: Flächenhaft URL `www.din.de` ; `www.beuth.de`
[11] Canny J 1986 *IEEE Transactions on Pattern Analysis and Machine Intelligence* **PAMI-8** 679–698 ISSN 0162-8828

IOP Conf. Series: Journal of Physics: Conf. Series **1044** (2018) 012055 doi:10.1088/1742-6596/1044/1/012055

On a law of ordinal error

David Andrich and Pender Pedler

Graduate School of Education, The University of Western Australia, Crawley WA 6009, Australia

Email: david.andrich@uwa.edu.au

Abstract. When no systematic factor disturbs replicated measurements of the same entity with the same instrument, the observed or inferred distribution is expected to satisfy the Gaussian *law of measurement error*. A characteristic of this distribution, which ensures it is unimodal with a smooth transition between adjacent probabilities, is that it is strictly log-concave. Many assessments in the social sciences begin by analogy to measurements in that an idealised linear continuum is partitioned by successive thresholds into contiguous, ordered categories. However, the distances between successive thresholds, generally finite, are not equal and assessments remain ordinal. The paper establishes that if the thresholds in the probabilistic Rasch measurement model used to transform ordinal assessments into measurements are in their natural order, then the distribution is also strictly log-concave. Therefore it is proposed that the Rasch model with ordered thresholds be referred to as *the law of ordinal error*. Accordingly, by analogy to the expectation that the distribution of replicated measurements satisfy the law of measurement error, it is proposed that the observed or inferred distribution of replicated ordinal assessments be expected to satisfy the proposed law of ordinal error.

1. Introduction

The prototypic measurement, that of mapping the amount of a property on a real line divided by successive thresholds from a conventional or natural origin into equal intervals called a unit, is understood by elementary school children. However, the *function* of measurement in science, which is the basis of the many remarkable advances in the physical sciences, is the understanding of variables and their mathematical formalization into scientific laws [1]. Although measured variables are conceived of as continuous, every measurement is discrete in the unit of measurement. Moreover, these measurements have errors and the role of errors continue to exercise scientists at both the conceptual and instrumental or empirical levels [2].

The long history of the recognition, acceptance, formalization and exploitation of random errors of measurement has been studied in Eisenhart [3].

> "Laws of error," i.e., probability distributions assumed to describe the distribution of the errors arising in repeated measurement of a fixed quantity by the same procedure under constant conditions, ...culminating in the quadratic exponential law of Gauss, $f_X(x) = (h/\sqrt{\pi})\exp(-h^2x^2)$, upon which Gauss based his first formulation of the method of

least squares, which became almost universally regarded in the nineteenth century as "the law of error''. [3: p.1].

When no measuring instrument of the kind used in the physical sciences is available, measurement in the social sciences generally begins with assessments in a set of *ordered* categories. A hypothesized linear continuum, by analogy to a unit of measurement, is partitioned by successive thresholds into ordered categories. However, unlike in measurement, the number of categories is finite and small and the successive thresholds are not expected to be equidistant. In advanced analyses these ordinal assessments are modelled by probabilistic models. Although such modelling has a substantial history [4], there is *no law of ordinal error* analogous to the Gaussian law of measurement error. This paper proposes such a *law* based on the Rasch measurement model (RMM). The function of the proposed law for ordinal assessments is the same as the function of the law of error in measurement – to disclose factors that generate errors which are systematic rather than random.

2. The Rasch measurement model

There is not the space in this paper to detail the case that when data conform to the RMM, its transformation of the ordinal assessments provides measurements which in principle are of the kind obtained in the natural sciences. The case is made in Rasch [5], [6], and developed further in Andrich [7], Fisher & Stenner [8], Wright [9] and Stone & Stenner [10]. The model was derived from the theory that measurement arises from invariant comparisons – that within a specified empirical frame of reference, the comparisons of entities are independent of the instruments of measurement, and that the comparisons of instruments are independent of the entities [6]. Without loss of generality, the ordinal assessments can be assigned successive integers beginning with 0, giving ordinal counts $x = 0, 1, 2, ..., m$.

Then the RMM used to model such ordinal counts is expressed in the form

$$P_{nix} = \Pr\{X_{ni} = x; \beta_n, (\delta_i)\} = \exp\{(-\delta_{i0} - \delta_{i1} - \delta_{i2} - ... - \delta_{ix}) + x\beta_n\}\gamma_{ni}^{-1}, \quad (1)$$

where (i) X_{ni} is the random variable of the ordinal counts of assessed entity n with instrument i which take the values $x = 0, 1, 2, ..., m$ for the $m+1$ successive categories, (ii) β_n is the measure of person n, (iii) $\delta_{i1}, \delta_{i2}, ..., \delta_{ix}, ... \delta_{im}$ is a vector of m_i thresholds where δ_{ix} is the point on the continuum where the probabilities in the two adjacent categories $x-1$ and x, $x > 0$ are equal, with $\delta_{i0} \equiv 0$ for convenience, and (iv) where $\gamma_{ni} = \sum_{x=0}^{m_i} \exp(-\delta_{i0} - \delta_{i1} - \delta_{i2} - ..... - \delta_{ix} + x\beta_n)$ is a normalizing factor which ensures Equation (1) is a probability distribution. In the rest of the paper we focus on the responses to a single instrument and drop the subscript i from m_i. Subscripts n, i are retained to emphasise that the replications of concern are of the assessment of a single person with a single instrument. Incidentally, in the social sciences, a set of ordered categories is referred to as an *item*.

3. The functions of the laws of measurement and ordinal errors

The Gaussian *law of measurement error* (LME) is derived theoretically and not to fit any observed distribution of measurements. Like hypothesized variables of measurement, the Gaussian distribution is continuous. However, because all measurements are discrete in the unit of measurement, we immediately consider the discretised Gaussian distribution in the unit of measurement. It is discretised by treating the ordinate at each possible measurement as its probability and to ensure a probability distribution, the ordinates in the range of possible measurements are normalised by their sum. Panel A of Figure 1 shows such a discretised distribution. The *RU ratio* referred to in Figure 1 is considered later in the paper as is the distribution of Panel B.

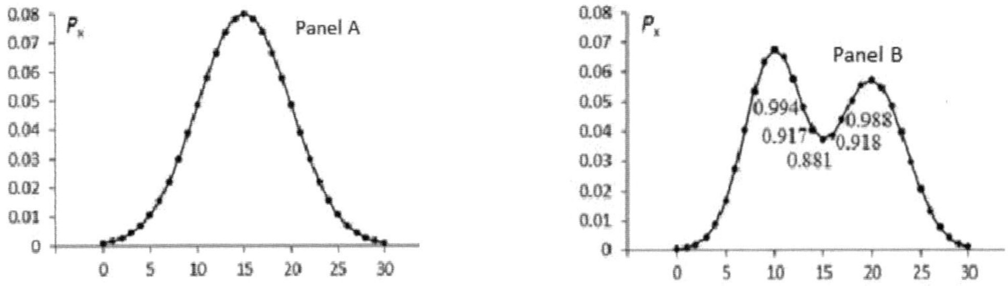

Figure 1. Panel A: $RU\ ratio = 1.041,\ \forall 0 < x < m$. Panel B: $RU\ ratio < 1$ for $x = 13,14,15,16,17$.

The LME is *unimodal* with a *smooth* transition between the adjacent probabilities. These properties are generally taken for granted. For the purpose of this paper in which these features are to be generalised to ordinal assessments, they need to be made explicit. To convey smoothness and unimodality, such a distribution will be referred to as *randomly unimodal*.

The functions of the LME arise from its comparison with a real or inferred empirical distribution of replicated measurements. If the empirical distribution is deemed to deviate significantly from the LME then two implications follow. First the mean of the measurements cannot be taken as an estimate of the measure of the entity. Second, no less importantly, an unknown factor(s) have interfered with the measurements. That interference can lead to an empirical and theoretical search which can further the understanding of the variable and its measurement.

Like the Gaussian distribution, the RMM was derived theoretically. Its present form was derived with no data analyses in three successive papers [5], [11], [12]. The present paper shows that if the thresholds in the RMM are ordered, then the distribution satisfies the definition of random unimodality and therefore can be referred to as the *law of ordinal error* (LOE). By analogy to the functions of the LME, if it is deemed that real or replicated ordinal assessments deviate significantly from the LOE, then a scalar cannot be taken as an estimate of the measure of an entity and some unknown factor(s) has interfered with the assessments. By analogy to the functions of the LME, a search for a factor(s) can lead to further understanding of the variable and its assessment.

4. The definition of random unimodality: Strict log-concavity

The random unimodality of the continuous Gaussian distribution arises from its property of *strict log-concavity* [13]. Keilson and Gerber [14] generalise strict log-concavity to discrete distributions showing, for example, that the Poisson and binomial distributions are strictly log-concave. A discrete distribution P_{ix} of counts $x = 0,1,2,\ldots m$ of instrument i is strictly log-concave if

$$P_{ix}^2 / (P_{i(x-1)} . P_{i(x+1)}) > 1, \quad 0 < x < m. \tag{2}$$

We refer to the ratio on the left of Equation (2) as the *RU ratio*. The term *log-concave* arises from Equation (2) where $\ln P_{ix}$ is greater than the mean of $\ln P_{i(x-1)}$ and $\ln P_{i(x+1)}$.

4.1. Strict log-concavity and the law of measurement error

Figure 1 shows two distributions. The distribution in Panel A, with mean 15.0, satisfies the LME with measurements in the range of 0 to 30. Its RU ratios are not only greater than 1 (1.041), $0 < x < 30$, but constant. It can be shown readily that for the LME, the

$$RU\ ratio\text{: } P_{ix}^2 / (P_{i(x-1)} . P_{i(x+1)}) = \exp[(/\sigma^2)^{-1}] \ \forall 0 < x < m. \tag{3}$$

The constant ratio of Equation (3) shows that the LME is uniformly smooth. The mean of 15.0 estimates the entity's measure and there is no evidence of a systematic factor(s) interfering with the measurements. In contrast the distribution in Panel B, also with mean 15.0, is not randomly unimodal with five measurements with RU ratios less than 1 – the distribution is also bimodal. Its mean of 15.0 does not estimate the entity's measure and some unknown factor(s) has interfered with the measurements. This extreme non-randomly unimodal distribution, unlikely in measurement, provides a basis for identifying non-randomly unimodal distributions of ordinal assessments.

4.2. Strict log-concavity and the law of ordinal error

It can be shown readily that the RU ratio in the RMM of Equation (1) is given by

$$RU\ ratio: (P_{nix}^2)(P_{ni(x-1)} . P_{ni(x+1)})^{-1} = \exp(\delta_{i(x+1)} - \delta_{ix}), 0 < x < m. \quad (4)$$

Therefore if $\delta_{i(x+1)} > \delta_{ix}$, in which case the thresholds are in their *natural* order, then independently of the measure β the distribution is strictly log-concave and randomly unimodal. It justifies the proposal that the RMM with ordered thresholds characterises the *law of ordinal error*.

5. Replications and analyses of ordinal assessments with the RMM

Many assessments in the social sciences, such as those of performance and attitude, involve multiple instruments (items) which assess the same variable for a well-defined class of persons. Multiple instruments provide a kind of replication and are used, by analogy to replicated measurements in the natural sciences, to improve the precision and validity of the assessment.

5.1. Estimating the parameters of an instrument

One distinctive property of the RMM is that the vector of each instrument's parameters can be estimated independently of the parameters β_n of any sample of $n = 1,2,...,N$ persons. The estimation exploits the property that the sum score $r_n = \sum_{i=1}^{I} x_{ni}$ of each person across the I instruments is sufficient in the sense that, if the profile of responses fits the model, then there is no further information about β_n in the vector of assessments across the instruments. For example, for two instruments (i, j), the conditional estimation equation takes the form

$$\Pr\{(x_i, y_j),(\delta_i),(\delta_j) \mid r_n\} = \exp\{(-\delta_{i0} - \delta_{i1} ... - \delta_{ix}) + (-\delta_{j0} - \delta_{j1} ... - \delta_{jy})\}\gamma_{nijr}^{-1} \quad (5)$$

where $\gamma_{nijr} = \sum_{r=1}^{2m-1} \exp\{(-\delta_{i0} - \delta_{i1} ... - \delta_{ix}) + (-\delta_{j0} - \delta_{j1} ... - \delta_{jy})\}$ is the sum of the exponent in Equation (5) over all meaningful scores of r_n. Because there is only one possible response pair for each of $r_n = 0$, $r_n = 2m$, these conditional probabilities are uninformative. The distinctive property of Equation (5) is that it contains only the vectors of thresholds $(\delta_i),(\delta_j)$ of i, j and does not contain β_n. Equation (5) shows the first distinctive property of the RMM. Even though each response pair (x_{ni}, y_{nj}) is governed respectively by a different parameter, $\lambda_{ni} = f((\delta_i), \beta_n)$, $\lambda_{nj} = f((\delta_j), \beta_n)$, the distribution of conditional responses is a function of the same set of instrument parameters $\{(\delta_i), (\delta_j)\}$; therefore across a sample of persons it reflects genuine replications. Second, when these replications are used to obtain the estimates $\{(\delta_i), (\delta_j)\}$, no assumptions need be made regarding the distribution β of the persons in the sample. Third, the instrument parameters characterize the assessment features of the instrument for any potential measure β. These distinctive properties are relevant when concerned with distributions of replications of assessments of a single person with a single instrument. Equation (5) readily generalises to the total score of many instruments and variable numbers of thresholds among them. Therefore, given estimates $(\hat{\delta}_i)$ of instrument i, and assessments with one or more instruments, each estimate $\hat{\beta}_n$ can be obtained individually and is taken as a measurement with an arbitrary conventional unit and an arbitrary conventional origin which can be transformed linearly for convenience.

5.2. Inference of replicated ordinal assessments

Substituting the *estimates* $(\hat{\delta}_i)$ into the RMM of Equation (1) gives

$$\hat{P}_{nix} = \Pr\{X_{ni} = x; \beta_n, (\hat{\delta}_i)\} = \exp\{(-\hat{\delta}_{i0} - \hat{\delta}_{i1} - \hat{\delta}_{i2} - ... - \hat{\delta}_{ix}) + x\beta_n\}\hat{\gamma}_{ni}^{-1}. \quad (6)$$

for any β. Equation (6) is the distribution of *inferred* replications of the assessment of person n with instrument i. Given the parameter values $(\hat{\delta}_i)$, it is the distribution of the assessment of any person with any β. It can be interpreted as if the person was assessed independently with the instrument on multiple occasions even though each person is assessed only once by each instrument. Clearly the response of a person to the instrument provides an observation in this distribution.

A second distinctive property of the RMM is that the threshold *estimates* may not be in their natural order even if the assessments fit the RMM statistically. That is, the inferred distribution of replicated assessments is not forced by the RMM to be randomly unimodal - random unimodality is an empirical property of the assessments as disclosed by the RMM, and not a property of the RMM. However, it is expected that the threshold estimates of any instrument will be in their natural order, that is, that the distribution for any measure will be deemed to satisfy the LOE. This is analogous to the expectation that replicated measurements, which are not forced to be Gaussian, will satisfy the LME.

5.3. *Examples of the analysis of ordinal assessments*

Two examples of ordinal assessments analysed with the RMM are shown below in Figure 2. Panel A shows a distribution from R. A. Fisher [15: p.290] which involved 12 actual replicated serological readings for five cells classified into five ordered levels of reaction. The threshold estimates are $(\hat{\delta}) = (-3.543, -2.095, 2.357, 3.282)$, though not equidistant, they are in their natural order - $\delta_{i(x+1)} > \delta_{ix}, x = 1, 2, 3$. The distribution shown in Panel A, which includes the RU ratios, is for a cell measurement of $\beta = 0.00$. Clearly for this measurement, and from Equation (5) for any other cell location, the RU Ratio is greater than 1 and the distribution is randomly unimodal. If the profile of responses across five readings of a cell fits the RMM, a calculated value of $\hat{\beta}$ can be taken to estimate β for the cell. In addition, there is no evidence of a systematic factor interfering with the assessments. Panel B shows a distribution from Embretson and Reise [16] concerning a 12-item (instrument) neuroticism questionnaire where each instrument involves ratings in five ordered categories. The distributions are from an analysis of responses of persons to their *Item 3*, highlighted by the authors. The vector of threshold estimates is $(\hat{\delta}) = (-1.800, 0.167, -1.168, 1.117)$. In this example the thresholds are not in their natural order; specifically, $\hat{\delta}_{i3} < \hat{\delta}_{i2}$. Therefore for $x = 2$, the $RU\ Ratio = 0.263 < 1$ and the distribution $\hat{P}_{nix}, x = 0, 1, 2, 3, 4$ for any β is not randomly unimodal. The distribution in Panel B is specifically for $\beta = -0.50$ and in this case the distribution is even bimodal. In addition to Item 3, all the instruments in the questionnaire exhibit similarly non-naturally ordered thresholds. Therefore, even if the profile of responses of a person across instruments of the questionnaire fits the RMM, a calculated value of $\hat{\beta}$ cannot be taken to estimate β. In addition, there is evidence that there is a systematic factor(s) interfering with the responses and that the empirical ordering of the categories, specifically the middle category, is not the intended ordering. This factor(s) needs to be studied theoretically and empirically. Understanding and correcting the interference which renders the middle category to malfunction would add to the understanding of the assessment of neuroticism.

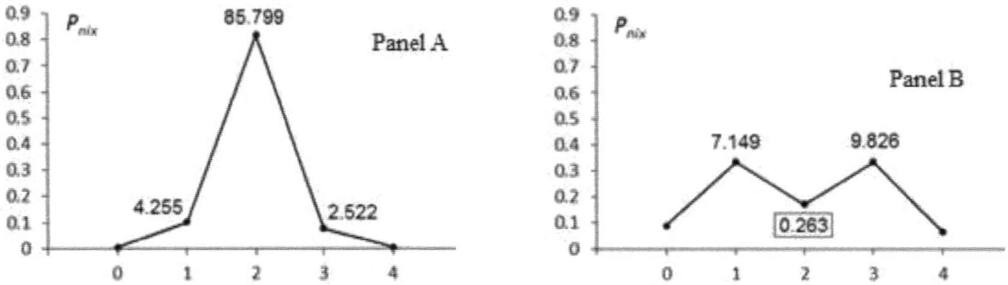

Figure 2. Panel A: R. A. Fisher distribution P_{nix} for $\beta = 0.00$. Panel B: Embretson and Reise distribution for $\beta = -0.50$. The *RU Ratio*, $x = 1,2,3$ is also shown.

6. Conclusion

Surprisingly, many ordinal assessments when analysed with the RMM exhibit distributions of the kind shown in Panel B of Figure 2. Perhaps this situation arises from the lack of formalisation, up to this point, of a LOE. With its formalisation, the properties of errors of ordinal assessments used in the social sciences to construct measurements can be studied from the same perspective as errors of measurements in the physical sciences.

References

[1] Kuhn T S 1961 The function of measurement in modern physical science *Isis* **52** 161-93
[2] JCGM 2008 *Evaluation of measurement data - guide to the uncertainty in measurement (GUM)*
[3] Eisenhart C 1983 Law of error I: development of the concept *Encyclopedia of statistical sciences* vol 4, ed S Kotz and N L Johnson (Toronto: Wiley) pp 530-47
[4] Clogg C C and Shihadeh E S 1994 *Statistical models for ordinal variables* (Thousand Oaks, California: Sage Publications)
[5] Rasch G 1961 On general laws and the meaning of measurement in psychology *Proc. Fourth Berkeley Symp. on Math. Statis. and Prob.* vol 4, ed J Neyman (Berkeley CA: University of California) pp 321-34
[6] Rasch G 1977 On specific objectivity: an attempt at formalising the request for generality and validity of scientific statements *Danish Yearbook of Philosophy* **14** 58-94
[7] Andrich D 2017 Advances in social measurement: a Rasch measurement theory *Perceived health and adaptation in chronic disease: stakes and future challenge* ed F Guillemin, A Leplège, S Briançon, E Spitz and J Coste (Taylor and Francis) *In Press*
[8] Fisher W P Jr and Stenner A J 2016 Theory-based metrological traceability in education: a reading measurement network *Measurement* **92** 489-96
[9] Wright B D 1977 Solving measurement problems with the Rasch model *J. Educ Meas.* **14(2)** 97-116
[10] Stone M H and Stenner A J 2014 Comparison is key *J. Appl. Meas.* **15(1)** 26-39
[11] Andersen E B 1977 Sufficient statistics and latent trait models *Psychometrika* **42** 69-81
[12] Andrich D 1978 A rating formulation for ordered response categories *Psychometrika* **43(4)** 561-74
[13] Ibragimov I A 1956 On the composition of unimodal distributions *Theor. Probab. Appl.* **1(2)** 255-60
[14] Keilson J and Gerber H 1971 Some results for discrete unimodality *J. Am. Stat. Assoc.* **66** 386-9
[15] Fisher R A 1958 *Statistical methods for research workers 13th edition* (New York: Hafner)
[16] Embretson S E and Reise P R 2000 *Item Response Theory for Psychologists* (Mahwah: Lawrence Erlbaum)

IOP Conf. Series: Journal of Physics: Conf. Series **1044** (2018) 012056 doi:10.1088/1742-6596/1044/1/012056

Participation of universities in the diffusion of metrology research: the case of the Metrology Symposium from CENAM

Omar-Jair Purata-Sifuentes[1], **Andrea-Guadalupe Fernández-Torres** and **Anahí-Yridian Belman-Hernández**

Departamento de Estudios Organizacionales, Universidad of Guanajuato, Fraccionamiento I, El Establo, Campus Guanajuato, C.P. 36250, Guanajuato, México

Email: opurata@ugto.mx

Abstract. An exploratory study on the participation of universities on the CENAM Metrology Symposium was done. Nine editions of the symposium have been carried out since 2001 until 2016, mostly in a biannual scheme. Although the symposium has a high reputation in Mexico because it is organized by the national metrology institute, and even when there are more than 2 000 universities all over the country, less than 100 universities have been doing metrology research diffusion in that forum. Exploration of possible causes was done. Among the results, geographic localization, presence of metrology related research groups, and metrology related subjects in the careers offered arise as some of the reasons for the participation of universities in the CENAM Metrology Symposium.

1. Introduction

In Mexico, the scientific authority in metrology is the National Metrology Center (CENAM, for the Spanish 'Centro Nacional de Metrología'), among whose functions are: "to promote and carry out research and technological development activities in the different fields of metrology, as well as to contribute to the formation of human resources for the same objective; and organize and participate, where appropriate, in congresses, seminars, conferences, courses or in any other type of events related to metrology" [1]. Even though metrology is a relatively young science in México (CENAM just celebrate its 20th anniversary in 2014), its diffusion has expanded due to its relevance and preponderant role in the scientific and technological development of the country.

By far, the most important metrology event in Mexico is the CENAM's Metrology Symposium (CMS), which CENAM has been organizing since 2001, and since 2002 every other year, so that until 2016 nine editions of the symposium have been carried out. The CMS, an open event, has been showing to the Mexican metrological community the technological advances and the state of the art of the science of measurements. Given the youth of the discipline in the country, the participation of the scientific community in the CMS has been growing gradually (see figure 1). Naturally, in the first editions of the CMS, the participation was practically covered by the staff of CENAM and some guests of other national metrology institutes (NMIs), specially for plenary conferences [2].

Therefore, most of the Mexican universities have not been present at the CMS as presenters at the oral or poster sessions. Only few universities have carried-out science diffusion activities in the CMS. This paper aims to make an exploratory study about which universities have been participating in CMS, its research areas, frequency of participation, relation with the CENAM and even the geographical proximity to the CENAM location, in search of the reasons for their participation in the CMS.

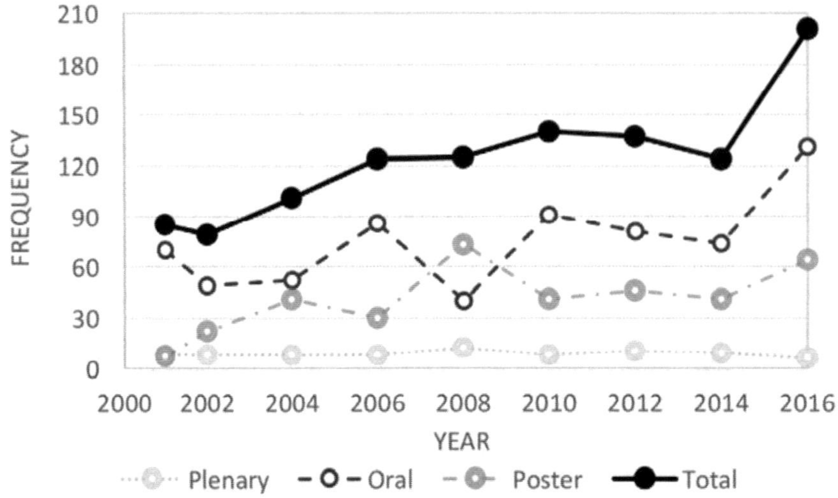

Figure 1. General participation in the CMS [2].

2. Participation of universities in the CMS

Even when in fully developed countries the role of the universities in metrology is a primary one (e.g., in Denmark, the Technical University of Denmark owns a private metrology company [3]), in México it is not the same. To reach such levels of participation from the universities in the metrology research and development of México, there is a need to start disseminating research in metrology carried out by the universities. It is important to note that the concept of university here includes all kind of corporation of people that deals with teaching at undergraduate level and makes research too [4].

One of the indicators that can be used to monitor the participation of universities in metrology research activities is their participation in the CMS. Figure 2 shows the trend of the universities participation in CMS oral presentations, while figure 3 shows the universities participation in CMS poster presentations. As can be seen, it has been a growing trend in both universities participations since the first edition of the CMS in 2001.

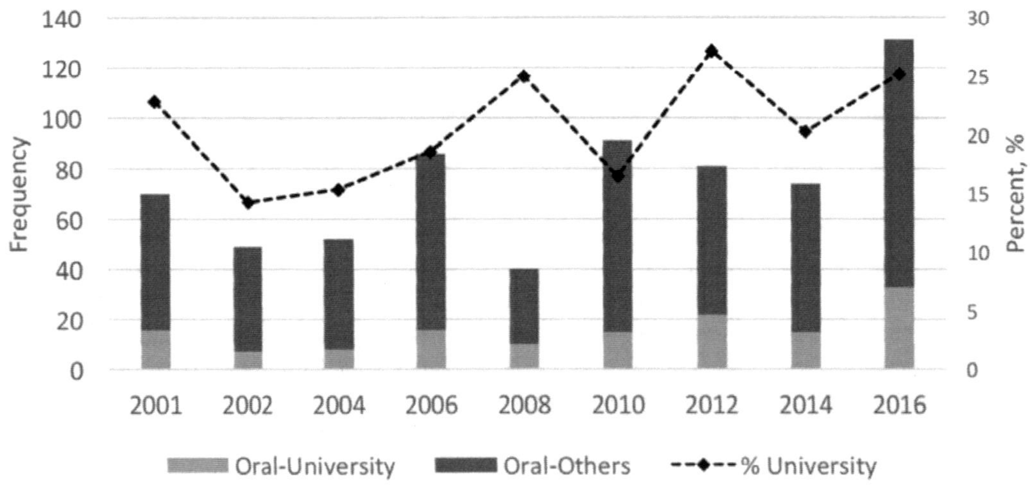

Figure 2. Participation of universities in oral presentations in the CMS [2].

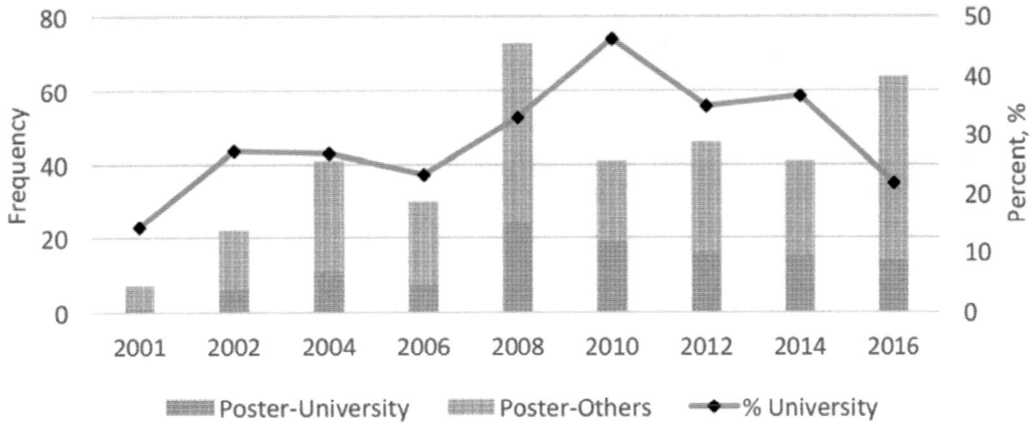

Figure 3. Participation of universities in poster presentations in the CMS [2].

However, it is important to note that average percent of poster presentations is higher by ten points than average percent of oral presentations. This could be natural for presentations coming from universities research groups, specifically in a metrology event hosted by the CENAM. Additionally, a kind of complementary behavior could be seen on the graphs, that is, more oral participation implies less poster participation and vice versa, this behavior follows the general trend in the CMS, as can be seen in figure 1. A graph of plenary sessions is not included because no speaker from any university has been invited to present a plenary conference in the CMS.

The next important question to answer could be: How many different universities originate the figures in figures 2 and 3? The answer is illustrated in figure 4, where can be seen that 50 % of the 306 works presented in the CMS with authors coming from universities, belong to only 12 of the 94 universities participants. From that 12 universities 3 of them are not Mexican ones [2]. It is important to note that there are almost 2 300 universities in Mexico, and more than 900 of them offer careers (e.g., Engineering, Physics) related with or that includes metrology or related subjects (e.g., instrumentation, physics laboratories, etc.) [5].

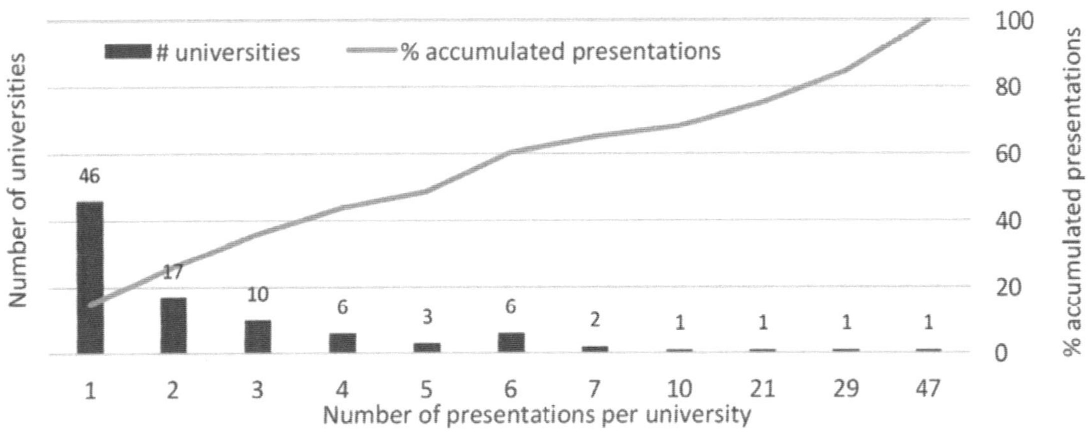

Figure 4. Distribution of the general participation of universities in the CMS [2].

3. Results and Discussion

Only twelve universities were responsible for the fifty percent of the works presented at CMS between 2001 and 2016. Figure 5 shows the relation between the participation in the CMS of the twelve universities and their general publications rate in Scopus® indexed journals (source: Scival®) [6]. Even when there seems to be a direct relationship between the number of articles published with a certain university affiliation, and the presence of that university in the CMS, it is not true for the Mexican universities UANL, UPSRJ or ITC, nor for the foreign university UCR (University of Costa Rica). Additionally, it must be considering the size of the university and its researchers population. Of course, there is the possibility that the researchers made most of its publications in journals not indexed in Scopus®.

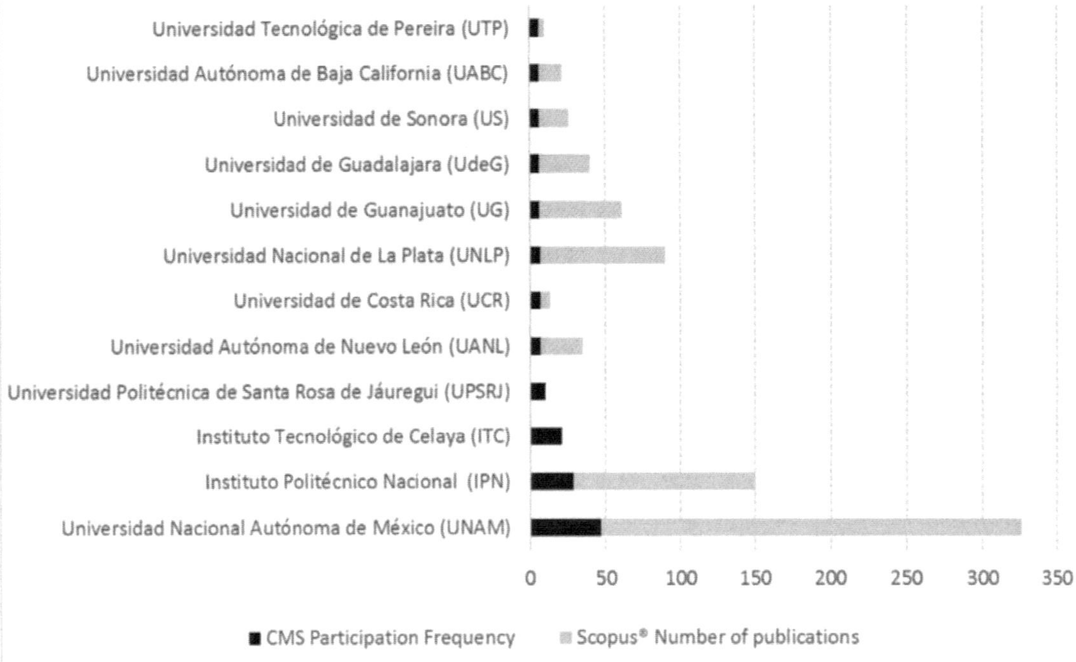

Figure 5. The twelve universities that own 50 % of universities' presentations in the CMS [2].

With respect of the localization of the universities that have been making diffusion in the CMS, figure 6 shows the number of presentations per localization in a Mexico map. It is clear that the nearness of CENAM has an important influence on participation, because the 3 most prolific states (Ciudad de México: 82 presentations, Guanajuato: 29 presentations, and Querétaro –where CENAM is–: 28 presentations). The exceptions, besides the foreign participations, are the next two more prolific states (Sonora: 16 presentations and Chiapas: 9 presentations), which are too far from the center of the country. It is worth to note that Mexico City houses UNAM and IPN, Guanajuato houses ITC and UG, and Querétaro houses UPSRJ.

The other prolific national universities: UANL, UdeG, US, and UABC, are housed in the states Nuevo León (8 presentations), Jalisco (in the Pacific Ocean middle side, 6 presentations), Sonora and Baja California (norther point, 6 presentations), respectively. UNLP is from Argentina (9 presentations), UCR is from Costa Rica (8 presentations), and UTP is from Colombia (19 presentations).

Another possible cause for the participation of universities in the CMS is the presence of research groups related with metrology. One indicator about research groups present in a Mexican university is the presence of recognized researchers by the National System of Researchers ('Sistema Nacional de Investigadores', SNI for the Spanish term) [7]. SNI is divided in 7 areas and everyone has between

10 % and 17 % of the total Mexican recognized researchers. Area I includes Physics (16 %) and Area VII includes Engineering (14 %). The table 1 shows the presence of SNI researchers in the states with the most prolific metrology research presented in the CMS. Even when there is no way to claim that a causation exists, in the cases of Ciudad de México, Guanajuato and Querétaro, a good level of SNI researchers favors the higher levels of participation in the CMS.

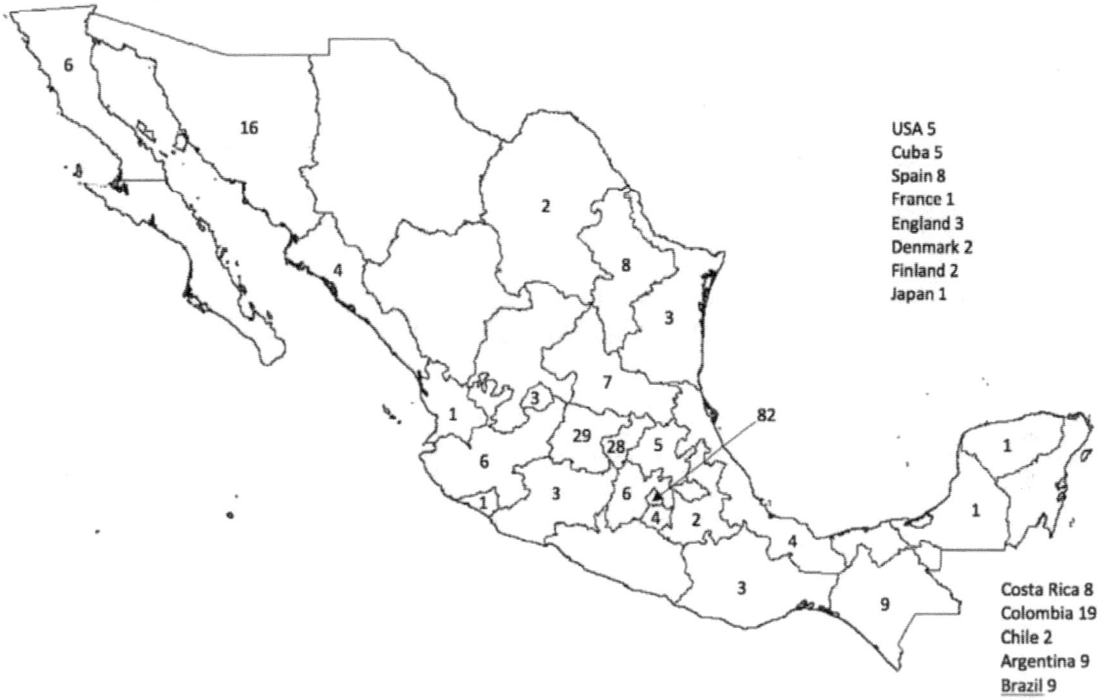

Figure 6. Geographical distribution of 307 presentations in the CMS between 2001 and 2016 [2].

Table 1. Geographical distribution of SNI Researchers and States participation in CMS [7].

State	SNI Researchers	Participation in CMS
Ciudad de Mexico	7 524	82
Guanajuato	719	29
Querétaro	548	28
Nuevo León	857	8
Jalisco	1 083	6
Sonora	454	16
Baja California	659	6

Finally, a web-based research was carried out on the undergraduate programs of the 94 universities that have been participating in the CMS. Subjects called 'Metrology', 'Instrumentation' or 'Laboratory of Physics' were found in 87 % of the 94 universities. For the remaining 13 % it was not possible to find out a metrology related subject in the specific engineering or sciences programs.

4. Conclusions

Participation of universities in the CMS has been growing. The expected behavior of more participation in poster than oral, has reached 10 % average. However, a better diffusion of metrology research need to be done through more Mexican universities, provided that only 10 % of Mexican universities with engineering or sciences programs have been at least one time in the CMS. Some possible causes about this lack of participation were studied. Geographical localization was found as an important factor that promotes the participation in the CMS: in general, universities located near to CENAM are more participative in the CMS. The presence of nationally recognized metrology researcher, with the subsequent existence of a metrology research group in the universities was also analyzed, however it was not found as a determinant factor. May be this is because many of the works presented by the universities in the CMS had CENAM's personnel as coauthors, so they provide the metrology expertise. Finally, on a web-based research a very high relation between the presence of a subject related with metrology in a university undergraduate program, and the participation in the CMS of that university was found, for the 94 universities that have been at least one time in the CMS.

Acknowledgments

Fernández-Torres acknowledes the financial support of EDUCAFIN and Presidencia Municipal de Silao, Guanajuato, México. Belman-Hernández acknowledes the financial support of EDUCAFIN and Presidencia Municipal de Romita, Guanajuato, México.

References

[1] LFMN 2005 Ley Federal sobre Metrología y Normalización {Federal Law on Metrology and Standardization} (México, D.F.: Diario Oficial de la Federación, Gobierno Federal)

[2] CENAM 2001-2016 Proceedings of the Metrology Symposium {Memorias del Simposio de Metrología, años 2001 a 2016} http://www.cenam.mx/memorias/

[3] Barker K E, Cox D and Sveinsdottir T 2012 *R&D Management* **42** 170–9

[4] Nangia V K 2011 *International Journal of Social, Behavioral, Educational, Economic, Business and Industrial Engineering* **5** 21–30

[5] ANUIES Anuario Estadístico de Educación Superior 2015-2016. Recovered from: http://www.anuies.mx/informacion-y-servicios/informacion-estadistica-de-educacion-superior/anuario-estadistico-de-educacion-superior , on September 6th, 2016.

[6] SciVal Database. Recovered from: https://www.elsevier.com/solutions/scival , on April 30th, 2017 .

[7] Conacyt. Recovered from https://www.conacyt.gob.mx/images/SNI/BENEFICIARIOS2017.xlsx on April 30th, 2017.

Patient-centred outcome metrology for healthcare decision-making

S J Cano[1,5], L R Pendrill[2], S P Barbic[3] and W P Fisher Jr[4]

[1] Modus Outcomes, Spirella Building, Letchworth Garden City, SG6 4ET, UK
[2] RI.SE, Eklandagatan 86, S-412 61, Göteborg, Sweden
[3] Faculty of Medicine, Department of Occupational Science and Occupational Therapy, the University of British Columbia, Vancouver, BC V6T 2A1, Canada
[4] BEAR Center, Graduate School of Education, University of California, Berkeley, CA 92720 USA

E-mail: stefan.cano@modusoutcomes.com

Abstract. Valid and precise quantification of clinical variables is essential for appropriate interpretation to inform healthcare decision making. The outcomes produced from different measurement procedures and instruments, purporting to quantify the same measurand, should be directly comparable. This ensures the appropriate application and widespread adoption of clinical research findings. Metrology provides a framework for the development of a *common language* of reference measurement systems, which have the potential to improve the accuracy and comparability of patients' results. However, the practices, procedures and instruments used in social measurement are currently excluded from any formal metrological framework. In this paper, we build on previous arguments, and propose a new international body to bring together metrology, psychometrics, philosophy, and clinical management to support the global comparability and equivalence of measurement results in patient centred outcome measurement to improve healthcare.

1. Introduction

Previously, we put social measurement 'on trial', and provided two perspectives arguing why measurement in the social and in the physical sciences are incompatible and countered with two perspectives supporting compatibility [1]. We concluded that measurement, whether physical or social, should have the same definition (i.e., the ratio of two magnitudes of the same *thing* in which the denominator is the unit, providing for invariant comparisons) and the same broad goal (i.e., quantification of meaningful variables). The role of social measurement methods in high-stakes decision making for education, psychology, and health more generally, highlights the need for the field to advance. We offered a way forward potentially applicable to both physical and social measurement (see section 3 below). In this paper, we proffer potential next steps, focussing on the area of healthcare. We

[5] To whom any correspondence should be addressed.

IOP Conf. Series: Journal of Physics: Conf. Series **1044** (2018) 012057 doi:10.1088/1742-6596/1044/1/012057

propose establishing a joint international effort aimed at delivering patient-centred outcome measures (PCOMs) in support of healthcare policy decision-making.

2. The quest for better healthcare

Patient experience affects patient safety, clinical effectiveness, allocation of funding [2], and is associated with: health outcomes; adherence to therapeutic regimens; preventative care; healthcare resources; quality of care delivery; and adverse events [3]. It is unsurprising, then, that patients who are more engaged in decision-making have better outcomes [4]. Therefore, decision makers within health systems increasingly rely on timely and relevant information about patient outcomes to improve policy and practice.

PCOMs are instruments (e.g., questionnaires, biometric equipment, wearables) that quantify patients' health, health-related quality of life and other latent health constructs, such as pain, mood, and function. They can also be used to quantify patient experience and satisfaction of healthcare, which play an increasingly central role in evidence-based medicine [5]. PCOMs have the potential to 'speak' to patients, carers and clinicians. Used alone, or in tandem with surrogate data (e.g., analysed in the laboratory), PCOMs offer the opportunity for more meaningful and interpretable measurement of patient outcomes. This helps to improve our understanding of the natural course of disease and guiding treatment decisions.

There is increasing interest in embedding PCOMs within health information systems [6]. It has been suggested that they have the potential to transform healthcare into a more patient-centred model [7]. In 2017, the Organisation for Economic Cooperation and Development (OECD) went so far as to claim that a specific type of PCOM (viz., patient reported outcome measures; PROMS) have a role in benchmarking the performance of whole healthcare systems [8]. The routine collection of PROMs has already been implemented internationally, including United Kingdom, Australia, United States of America, Sweden, Canada, and the Netherlands [8].

Thus, PCOMs have a potentially significant role in modern healthcare policy decision-making and evidence-informed practice. But are we able to deliver high-stakes measurement of this kind? Some suggest not. For example, it has recently been estimated that only 10% of existing PROMs have been developed based on patient-identified priorities [9]. In addition, reflecting upon the current OECD initiative, Angela Coulter suggests that using PCOMs for high stakes decision making may be premature [5]:

> *"Multi-purpose applications of [PCOMs] - using them in individual clinical care and aggregating the data for performance assessment - remains largely aspirational at present..."*

Elizabeth Teisberg provides a potential reason as to why [10]:

> *"The quest for better health care, driven by measuring safety and quality, is well intentioned and has notable achievements. But like the Biblical story about building a better city, the measurement effort has become a cacophonous muddle that is distracting clinicians, raising the cost-of-care delivery, and not helping consumers make better health care choices...The lack of measurement standards [has resulted]...in contradictory conclusions...The problem isn't requiring measurement; health care needs meaningful measures...But...despite their vast number, the [existing] metrics don't measure the things that define health for patients and success for clinicians."*

The absence of standardization of PCOMs, and lack of recourse to formal metrological frameworks, can be ascribed to historical, ideological, and political differences [11]. But the deleterious sequalae of this situation is profound [12-14]. PCOMs are currently in a position not dissimilar to the measurement of temperature in the 17th Century [15]. As in other areas of clinical sciences [16], recourse to a metrological framework can help to provide a resolution.

3. The promise of the synthesis of metrology, psychometrics and philosophy

To ensure decision makers within health systems have access to the necessary objective evidence, we need the same kind of quality-assurance of PCOMs as is established in physics and engineering [17]. This translates into assurance with objective metrological comparability ('traceability') and declared measurement uncertainty. A possible way forward involves the synthesis of metrology, psychometrics and philosophy [1]. Essentially, evidence-based measurement for evidence-based clinical management would involve:

- extending traditional metrological concepts via an operational model of a specific measurement system, in which the output of the instrument (which could be human) in response to probing the object ('entity') is a *performance* metric, i.e., how well the set-up performs an assessment [17];
- exploiting the unique properties of Rasch Measurement Theory [18], which enables key metrological components (e.g., references for traceability and means of evaluating measurement uncertainty) to be established;
- establishing metrological references founded on objective and clinically meaningful measurement against the backdrop of the specific challenges of social measurement methods.

4. International co-operation, innovation and communication

In laboratory medicine, the standardization of measurements is a high priority, with the goal of comparability of results obtained using routine procedures [16, 19]. Recommendations of the IFCC Task Force on the Impact of Laboratory Medicine on Clinical Management and Outcomes (TF-ICO) include: "Effective collaboration with clinicians, and a determination to accept patient outcome and patient experience as the primary measure of laboratory effectiveness." [14]. The benefits of this are clear. For example, the standardization of cholesterol testing in the evaluation of cardiovascular risk has led to savings in the region of 100 million USD/year in treatment costs with a parallel significant reduction of untreated "false-negative" individuals who are at increased risk [20].

But, international cooperation for developing reference measurement for PCOMs is in its infancy. Its aims could include traceability, comparability, and the inclusion (and possible reduction) of measurement uncertainty. This would require the consideration of five key components: 1) key stakeholder involvement (including patients, clinicians, caregivers, policy makers); 2) a clear definition of the measurand in regards to the intended clinical use [21]; 3) a clear definition of the clinically allowable error of measurements; 4) international cooperation and consensus to navigate the complexities of the development of metrologically sound reference measurement systems; and 5) continued clinical validation of newly calibrated instruments.

International cooperation would require the involvement of international metrology treaty organizations, professional societies and federations devoted to improving measurement quality in health sciences. One of its central functions would be an online database listing: reference materials; reference methods; and reference measurement services meeting appropriate international standards. This database would allow PCOM providers to select references for calibration traceability and provide support for suppliers of these services. In addition, it would be important to consider three key components to navigate the complexity of the scientific and political landscapes.

First, the development and evaluation of new PCOMs is time consuming and resource intensive. As such, we should move towards a non-competitive exercise where expertise and resources are pooled to avoid the development of multiple instruments for the same purpose, thus engendering wider reaching collaboration. Second, as our understanding of the etiology, sequalae, and natural history of many diseases expands, we should be open to innovative ways to capture impact and patient benefit. One size does not fit all, and the most appropriate fit-for-purpose instrumentation should be established early on (e.g., de novo instruments, adaptation of existing instruments, exploitation of information technology resources [e.g., computer adaptive testing], or mobile health technologies). Third, open communication is key to establish wider dissemination of important patient research, further avoiding the duplication of efforts, and involving a wider range of audiences. Ultimately, PCOMs will lead to patient-based evidence and the realization of patient-centered care. Patients should always be involved as key stakeholders in the process of identifying and interpreting meaningful clinical outcomes that resonate with their experience, preferences, expectations and values.

5. Towards a common language for patient-centred outcome metrology

The comparability of findings among measurement procedures and instruments will result in harmonization; a *common language* to enable interchangeability geographically and temporally. However, as described above, the situation is complex. For example, as Angela Coulter [5] flags, different contexts will require different information needs. It is unlikely, that these needs are amenable to perfect alignment. But with several individual and separate global initiatives currently happening (e.g., ICHOM, COMET [22-25]) this will require more coherent coordination and alignment than is currently the case. There are two components to this situation.

First, universality is not an *a priori* fact of objective existence for some measured constructs. This is typically assumed for physical variables, but not for others, such as psychological and social variables. However, contrary to common assumptions, universality is not an inherent property, even in the domains of physical variables, apart from language and culture. 'Things' in themselves do not retain constant and invariant forms over time, space, and levels of information and organizational complexity. Instead, universality is a social construct that emerges as a product of a particular way of organizing experience. It arises via complex processes in which signal-noise ratios are lower than would be expected if physical variables were objectively constituted in and of themselves [26]. The counter-intuitive conundrum is that:

> *"In order for a statistical logistics to enhance precise decision making, it has to incorporate imprecision; in order to be universal, it has to carefully select its locales." [26], p. 55*

This kind of psychosocial stochastic resonance (interdependency of noise and signal) has been posited in other studies of scientific standards and universals [27], pp. 843-844; [28], pp. 69-71.

Second, the incorporation of standards in information infrastructures must be sensitive to the discontinuous shifts in context that occur across applications:

> *"With the rise of decentralized technologies used across wide geographical distance, both the need for common standards and the need for situated, tailorable and flexible technologies grow stronger. A lowest common denominator will not solve the demand for customized possibilities; neither will rigid standards resolve the issue." [29], p.111*

In relation to PCOMs, the general assumption is that common standards require common content; the same questions asked of every patient in every health care context and no matter what the individual patient's needs and conditions might be. The unworkable rigidity of this approach has prevented the development of information infrastructure standards in health care that would be of great value in patient care, quality improvement initiatives, and accountability. With the increasing availability of computerized adaptive administrations of PCOM items from calibrated banks [30], this lowest common denominator approach is being supplanted by methods better able to meet the demand for customized information relevant to the needs of individual patients.

But this shift from *fixed forms* to dynamically configured PCOMs is insufficient for the need of a health care information infrastructure integrating local sensitivities with universal standards across applications. In the same way public health has adopted an ecological frame of reference distinguishing micro-, meso-, and macro-levels of complexity in its policy and practice implications [31], so too, must PCOMs be used to similarly attend to the relational properties obtained in applications requiring different forms of information and organization [29]. For example, spanning the gulf of complexity from individual interpretation of PCOMs (micro-), aggregating outcomes for quality improvement purposes (meso-), and producing theories of health status and the impact of care (macro-). Each level demands a learning and subsequent relearning about patients' health status.

In other words, denotative statements of fact at the local, individual level (e.g., "your PCOM indicates your health is good") are not automatically commensurable with metalinguistic statements (e.g., "your patients' PCOM survey results indicate your care is producing satisfactory outcomes relative to the outcomes obtained by your peers") and neither are automatically commensurable with metacommunicative statements (e.g., "if I adopt that new treatment modality perhaps my patients' PCOMs will be superior to my colleagues' PCOMs"). An individual's array of clinical indicators and responses to a health status survey may correspond well with expectations formed regarding the natural history of a chronic disease or the developmental effects of aging and growth. The information relevant and applicable at that level, however, is not the same as the information needed at the metalinguistic or metacommunicative levels.

The consequences of not attending to the specific demands that must be met to coordinate common standards for expressing the information needed across these levels is well stated by Star and Ruhleder:

> *"If we, in large-scale information systems implementation, design messaging systems blind to the discontinuous nature of the different levels of context, we end up with organizations which are split and confused, systems which are unused or circumvented, and a set of circumstances of our own creation which more deeply impress disparities on the organizational landscape." [29], p. 118*

Suggestions for coherently coordinating information across these levels are offered elsewhere [32, 33]. But, in the end, a common language for patient-centered outcome metrology would significantly contribute to improvements in patient-centered care by allowing results of clinical studies undertaken in different locations or times to be universally applied. The 'semantic interoperability' has to be good i.e. the "the ability to ensure that the precise meaning of exchanged information is unambiguously interpretable by any other system, service or user", where we borrow the term from the context of digital solutions that enable public administrations, businesses and citizens to benefit from interoperable cross-border and cross-sector public services [34]. It would also support an effective application of evidence-based medicine and use of guidelines (established by scientific or professional bodies) to define specific decision limits for diagnosis and therapeutic interventions.

Thus, a joint international effort aimed at delivering PCOMs in support of modern healthcare policy decision-making could lead to the appropriate embedding of PCOMs in health systems. This would allow us to evaluate the effectiveness and cost-effectiveness of major policy initiatives, to identify conditions or interventions where substantial variation in patient outcome exists across providers, and improve the choice of treatment plans for individual patients.

References

[1] Cano S, Vosk T, Pendrill L and Stenner A J 2016 *J. Phys. Conf. Series* **772** 012025

[2] Basch E 2017 *N. Engl. J. Med.* **376** 105-8

[3] Doyle C, Lennox L and Bell D 2013 *BMJ Open* **3** e001570

[4] Greenfield S, Kaplan S and Ware J 1985 *Ann. Int. Med.* **102** 520-8

[5] Coulter A 2017 *BMJ* **356** j816

[6] Devlin N and Appleby J 2010 *Getting the most out of PROMs: putting health outcomes at the heart of NHS decision-making* (London, UK: Kings Fund)

[7] Black N 2013 *BMJ* **346** f167

[8] OECD 2017 *Recommendations to OECD Ministers of Health from the high-level reflection group on the future of health statistics* (Paris: OECD)

[9] Wiering B, de Boer D and Delnoij D 2017 *Health Expectations* **20** 11-23

[10] Teisberg E 2015 *The Quality Tower Of Babel* http://healthaffairs.org/blog

IOP Conf. Series: Journal of Physics: Conf. Series **1044** (2018) 012057 doi:10.1088/1742-6596/1044/1/012057

[11] Cano S and Hobart J 2011 *Pat. Pref. Adherence* **5** 279-90

[12] Hunt S 1997 *Qual. Life Res.* **6** 205-12

[13] Boyce M and Browne J 2013 *Qual. Life Res.* **22** 2265-78

[14] Hobart J, Cano S, Zajicek J and Thompson A 2007 *Lancet Neurol.* **6** 1094-105

[15] Chang H 2007 *Inventing Temperature: Measurement and Scientific Progress* (Oxford: Oxford University Press)

[16] Jones G and Jackson C 2016 *Clin Chim Acta* **453** 86-94

[17] Pendrill L 2014 *NCSLi Measure: The Journal of Measurement Science* **9** 22-33; Pendrill L R and Petersson N 2016 *Measurement Science & Technology*, **27**, 094003

[18] Andrich D 2017 *J. Phys. Conf. Series* in press

[19] Hallworth M, et al 2015 *Med Clin Chem* **61** 589–99

[20] Panteghini M 2007 *Clin. Biochem. Rev.* **28** 97-104

[21] FDA 2013 *Roadmap to Patient-focused Outcome Measurement in Clinical Trials* http://www.fda.gov/downloads/Drugs/DevelopmentApprovalProcess/DrugDevelopmentToolsQualificationProgram/UCM370174.pdf

[22] Gargon E et al 2015 *Trials* **18** 54

[23] International Consortium for Health Outcomes Measurement (ICHOM) 2017. *ICHOM and OECD sign Letter of Intent to collaborate* http://www.ichom.org/news/ichom-and-oecd-sign-letter-of-intent-to-collaborate-on-the-collection-analysis-and-publishing-of-patient-reported-outcomes/.

[24] Williamson P R, et al 2012 *Trials* **13** 132

[25] Porter M E, Larsson S and Lee T H 2016 *N. Engl. J. Med.* **374** 504-6

[26] Berg M and Timmermans S 2000 *Configurations* **8** 31-61

[27] Galison P 1997 *Image and logic* (Chicago: University of Chicago Press)

[28] Fisher W P and Wilson M 2015 *Pensamiento Educativo: Revista de Investigacion Educacional Latinoamericana* **52** 55-78

[29] Star S and Ruhleder K 1996 *Info. Sys. Res.* **7** 111-34

[30] Cook K, O'Malley K and Roddey T 2005 *Health Serv. Res.* **40** 1694-711

[31] Susser M and Susser E 1996 *Am. J. Public Health* **86** 674-7

[32] Fisher W P, Oon E and Benson S 2017 *J. Phys. Conf. Series* in press

[33] Fisher W P and Stenner A J 2017 *J. Phys. Conf. Series* in press

[34] ISA 2017 *Interoperability Solution for European Public Administrations* (EU Commission: https://ec.europa.eu/isa2/isa2_en)

IOP Conf. Series: Journal of Physics: Conf. Series **1044** (2018) 012058 doi:10.1088/1742-6596/1044/1/012058

Principal investigation for the application of structured light scanning for the measurement of thickness using differential measurement method

R. Illmann , M. Rosenberger[1] , G Notni[1]

[1]Ilmenau University of Technology, Department Quality Assurance and Industrial Image Processing, Gustav Kirchhoff Platz 2, 98693 Ilmenau, GERMANY

maik.rosenberger@tu-ilmenau.de

Abstract. As an industrial approach, structured light scanning is an established technique to handle length measurement by using light as information carrier and a CCD sensor to acquire that data into the digital domain. By means of triangulation, contours can be scanned and initiating uniquely defined relative movements of the object add another dimension. So it is possible to scan surfaces as well. That technique is basically known by scientists and manufactures. However, the very high application diversity leads to problems that needs good engineering and interdisciplinary work. The aim of this paper is to explain how structured light scanning could deal with complex objects like elastic and quite flexible materials. To examine if that approach is principally possible and how uncertainties are scalable is the primary purpose. Furthermore, that fact and the measuring principle itself requires well-considered approaches in calibration, adjustment, data management until a needed construction to minimalize expansions and keep that system working stable within industrial environments.

1. Introduction, state of the art and short system description

Image processing is one of the key technologies for optimization of industrial processes, the option for 100% quality assurance and many other essential industrial and research topics. Not least, by regarding the central ideas and the spirit of internet of things, applications featured image processing are omnipresent. A major application is the acquisition of projected structured forms, such as lines and other shapes. The geometrical conditions to use the concept of triangulation are complied with a two-dimensional CCD camera and a laser line [1],[2]. The aim of this concept is to get three-dimensional information by projecting a line as a shape on an object. On condition that distance and triangulation-angle being known, a function to calculate height can be defined. By relative movement between probe and structured illumination, it is possible to acquire surfaces and other 3D-informations. Assuming the object is symmetrical (or only one surfaces is the point of interest), the concept of one triangulation system works fine. However, if the object is unsymmetrical to the plain that was defined after calibration, information would be lost while the object covers partially his shape himself. Due this work, the outline of a cutaway view is the point of interest, provided that prismatic or cylindrical shapes are also manageable. So, it is necessary to apply the concept of differential measurement.

IOP Conf. Series: Journal of Physics: Conf. Series **1044** (2018) 012058 doi:10.1088/1742-6596/1044/1/012058

2. Experimental setup and measurement principle

A schematic view of the experimental setup is given in Figure 2. In this case, two triangulation systems are required to practice differential measurement. The basically needed components are: (1)-upper CCD camera, (2)-lower CCD camera, (3)-flexible sample, (4)-upper laser and (5)-lower laser. To find the optimal extensions, dimensions and triangulation angles of the setup and its components depending on the sample geometry is also part of the investigation. Due to a specific industrial demand, the working distance between the laser source and the calculated reference plane should be scalable up to one meter as maximum.

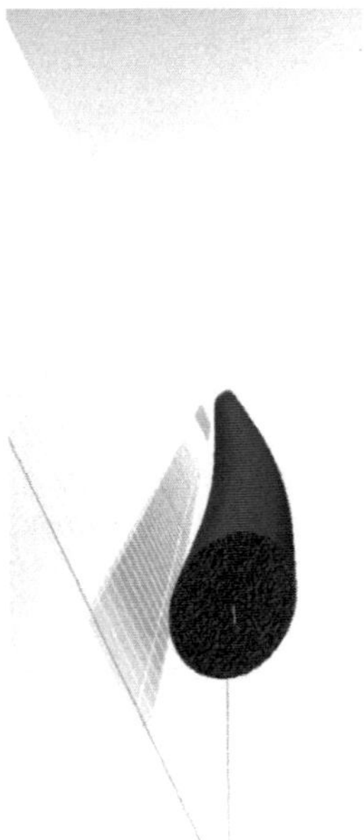

Figure 1. Possible sample geometry.

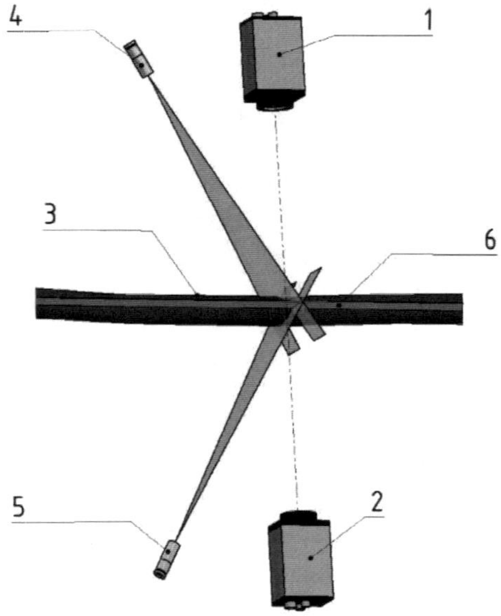

Figure 2. Experimental schematic view.

Figure 1 shows a possible sample geometry that represents properties of the interesting objects. In this example, the cross-section profile is elliptical, but longitudinally not constant, what means that it sweeps on a not linear path. This circumstance is a succession of flexibility in material and causes another relative movements. The advance against the triangulation system must be respected in the incoming signal processing as well. With respect to a working distance up to one meter, it is necessary to observe special laser light sources, which insure a much as is necessary accurate laser line with minimal diverging optical behaviour in the projected contour as well. Additionally, results of researches show that, regarding to adjustments in laser linewidth, off the shelf laser light sources and optics are insufficient. Therefore, the required linewidth is not realizable without additionally optical instruments. So, another research in this publication deals with modulations inside the optical path to make that linewidth adjustable. Figure 3 to Figure 5 shows some chosen approaches to design the optical path. To rank them, it is necessary to compare the individually advantages and disadvantages. One first simple solution is to convolve the optical path by integrating mirrors. The light source must produce a orthogonal linear laser beam by itself. Changing angles and distances in this setup leads to adjustments in line width. But respecting the experimental setup, approximately calculated geometrical extensions show that this solution is not practicable for that investigation. The necessary diameters of that mirrors need too much space, especially by the fact that to optical systems are needed.

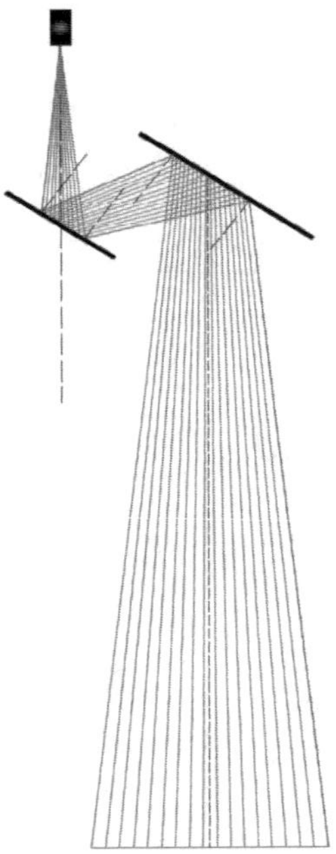

Figure 3. Convolved optical path.

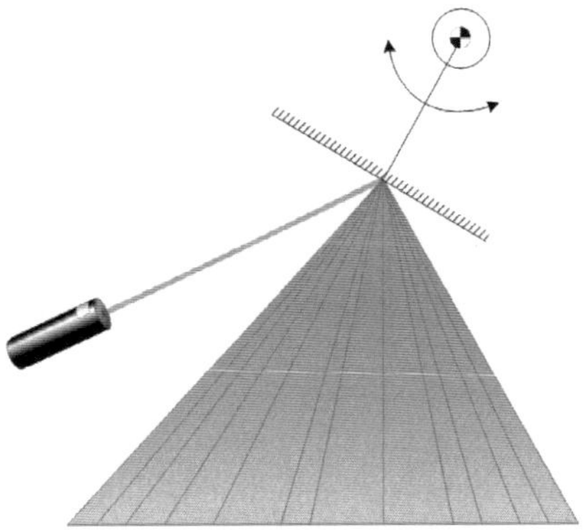

Figure 4. Modulated by oscillating plane mirror.

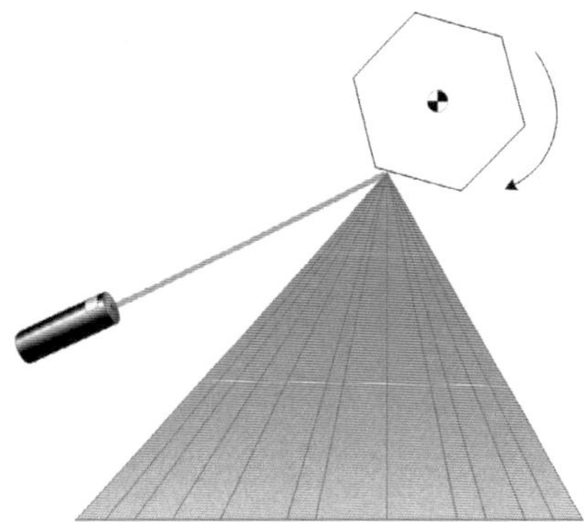

Figure 5. Modulated by rotating hexagonal polygon mirror.

The system shown in Figure 4 produces the laser-line by direct a point-like laser source at a one direction oscillating plane mirror. A main disadvantage of this system is that the actor must also be oscillated. Consequently, much attrition appears and pulsed signals must be produced and initiated too. Another disadvantage is the (mechanical) limited frequency that leads to limitations in variation of the integration time of the CCD-cameras. The third approach, shown in Figure 5, is to focus a point-like laser source at a continuously rotating hexagonal mirror. That system have nearly the best preconditions to be integrated in the explained experimental setup. By using a brushless DC motor, less attrition appears and repetition time could be manipulated by modifying rate of rotation. Standard laser diodes could be used as a low-cost light source. A simple electrical actuator is schematically shown in Figure 6. Only one voltage source is required to supply all components with electrical energy. A few different stages are needed to control all components too. A major advantage of this approach is that it is long-term stable, easy to adjust and a high laser line length could be realized by using few components.

3. Principle required calculations for the method of measurement

All needed trigonometrical basics are shown in Figure 7. Transferring that relationships into analytical formulas leads to an expression that contains the parameter of interesting z_p as a function of two different linear displacements x_1 and x_2 and the triangulation angles α_1 and α_2.

$$z_p(x_1, x_2, \alpha_1, \alpha_2) = \frac{x_1}{\tan \alpha_1} + \frac{x_2}{\tan \alpha_2} \tag{1}$$

Assuming that uncertainties in distances are included in the angles, the significant uncertainty is theoretical dependent by the pixel-scales dx_1 and dx_2 and the two triangulation-angles and can be approximated [3] by using:

$$dz_p = \frac{1}{\tan\alpha_1}dx_1 + \frac{1}{\tan\alpha_2}dx_2 + \frac{x_1}{(\tan\alpha_1)^2(\cos\alpha_1)^2}d\alpha_1 + \frac{x_2}{(\tan\alpha_2)^2(\cos\alpha_2)^2}d\alpha_2 \quad (2)$$

For a reliable contour representation within production process, it is necessary to have an intelligent and stable working signal processing concept. Based on reference values, which result from calibration with a calibration object, the system must be adjusted. Therefore, it is possible to compare all the real-time data with a reference model to compute the real shape of the cutaway contour. Finding an acceptable calibration model is also part of the investigation and annotated in chapter 4. An overview of the principle necessary computing steps is given in Figure 8. Based on the elliptical example that is given in Figure 1, Figure 8 (a) shows the separate acquired shapes.

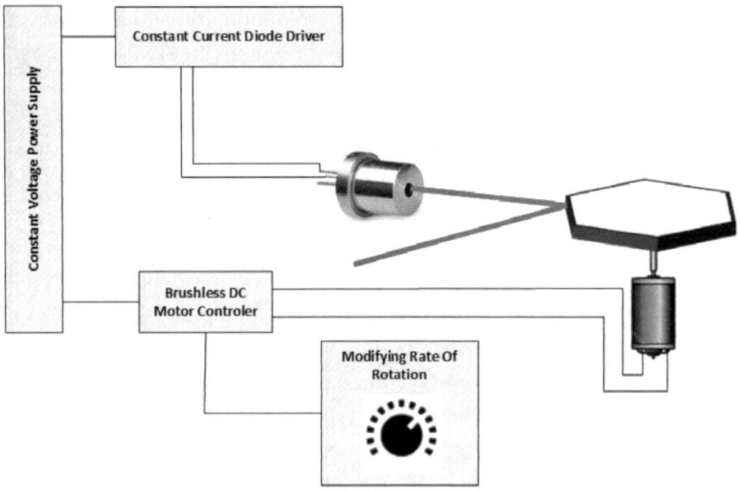

Figure 6. Electrical components.

Firstly, the difference must be computed, perhaps by applying operations and affine transforms like rotating or stretching relatively [4]. The combined result is shown in Figure 3 (middle). Secondly, static parameters from the triangulation system are used to reproduce the real contour by adapting the defined function. The final result is shown in Figure 3 (right). With respect to fast image processing, it is necessary to develop efficient algorithms - minimizing computing time is also an aim. So, the correction algorithms were programmed as efficiently as possible [5] . Furthermore, the basic concept of the affine transformation– showed in chapter 4 - will be explained. The main property of an affine transformation is that a point will be mapped to a point and a line will be mapped to a line. Respected to homogeny coordinates $(x, y, 1)^{\mathrm{T}}$, the following transformation matrices will be used [3] :Displacement in x by respecting parameter a, displacement in y by respecting parameter b:

$$M_{a,b} = \begin{pmatrix} 1 & 0 & a \\ 0 & 1 & b \\ 0 & 0 & 1 \end{pmatrix}. \quad (3)$$

Rotating against the origin by respecting angle α:

$$M_{\alpha} = \begin{pmatrix} \cos(\alpha) & -\sin(\alpha) & 0 \\ \sin(\alpha) & \cos(\alpha) & 0 \\ 0 & 0 & 1 \end{pmatrix}. \quad (4)$$

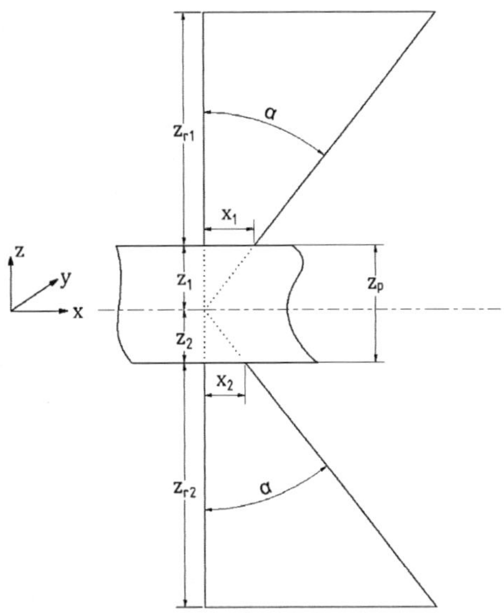

Figure 7. Trigonometrical relationships of differential measurement by triangulation.

The complete affine transformation matrix results by multiplying $M_{a,b}$ and M_α .

$$M_{affine} = \begin{pmatrix} 1 & 0 & a \\ 0 & 1 & b \\ 0 & 0 & 1 \end{pmatrix} \cdot \begin{pmatrix} \cos(\alpha) & -\sin(\alpha) & 0 \\ \sin(\alpha) & \cos(\alpha) & 0 \\ 0 & 0 & 1 \end{pmatrix}. \quad (5)$$

Finally, the needed affine transformation can be done by multiplying each (homogeny) pixel coordinate with the affine transformation matrix:

$$I_{pixel,transformed} = I_{pixel} * M_{affine} = \begin{pmatrix} x \\ y \\ 1 \end{pmatrix} \cdot \begin{pmatrix} 1 & 0 & a \\ 0 & 1 & b \\ 0 & 0 & 1 \end{pmatrix} \cdot \begin{pmatrix} \cos(\alpha) & -\sin(\alpha) & 0 \\ \sin(\alpha) & \cos(\alpha) & 0 \\ 0 & 0 & 1 \end{pmatrix}. \quad (6)$$

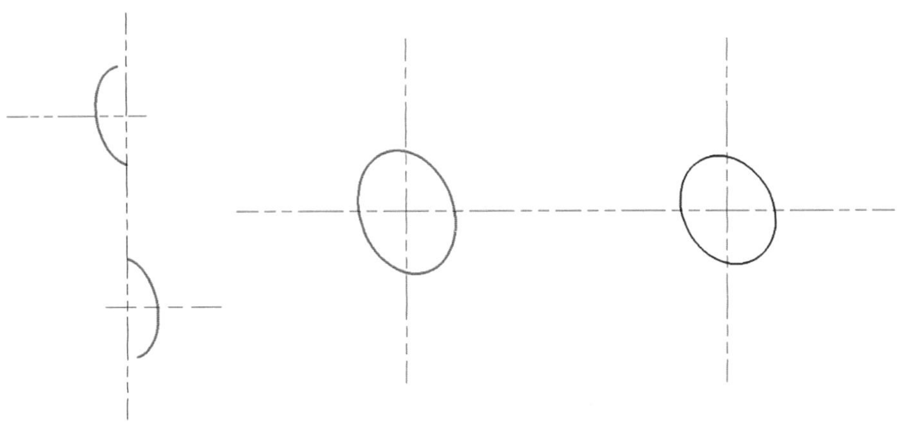

Figure 8. Separate shapes from upper and lower CCD (a), combined shapes after correction (b), finally calculated shape (c).

IOP Conf. Series: Journal of Physics: Conf. Series **1044** (2018) 012058 doi:10.1088/1742-6596/1044/1/012058

In addition, it is obvious that it is necessary to iterate that calculation with all pixels to get the complete transformed image I'.

$$\sum I'(x_i, y_i) = \sum\left[I(x_i, y_i) \cdot M_{affine}\right] = \sum\left[\begin{pmatrix} x_i \\ y_i \\ 1 \end{pmatrix} \cdot \begin{pmatrix} 1 & 0 & a \\ 0 & 1 & b \\ 0 & 0 & 1 \end{pmatrix} \cdot \begin{pmatrix} \cos(\alpha) & -\sin(\alpha) & 0 \\ \sin(\alpha) & \cos(\alpha) & 0 \\ 0 & 0 & 1 \end{pmatrix}\right]. \quad (7)$$

4. Image processing, adjustment and calibration concept

Assuming that both cameras are displaced, it is necessary to adjust each other. One camera (i.e. the upper one) is selected as a reference camera. In this investigation this task should be done by software adjustment. That procedure is detailed shown in Figure 9. Initially, static pictures was acquired by both cameras. Based on this pictures, images undergoes edge detection to subtract them later from next pictures including a first calibration object. The first step is to bring in a physical reticle inside the optical path. Secondly, the both pictures undergoes separately edge detection again. After adding those pictures and subtracting the initial taken static pictures, the result contains all information that was necessary to adjust. In the next step, gaps in the lines were filled by interpolation. In the calibration step, the two intersections of every reticle are computed and set to an initial defined reference point which represents the basic Cartesian point of origin and the rotation point for all following steps. Furthermore, all parameters (rotation angles, linear displacement values) that are necessary to apply affine transformations can be computed and combined to an image rectification matrix. That matrix contains all transformation information for each camera channel. Therefore, both pictures can be rectified separately. After the image rectification matrix is deposited, the first calibration procedure is completed. The third and last procedure is to find and correct the displacements of the two projected laser lines. This can be done by bringing in a rectangular calibration object with known thickness inside the optical path. The displacement occurs as a non-parallelism and could be detected by using the first calibration procedure again, but now with focusing on the laser lines. Therefore, both fitted line vectors have an intersection. Its vertical value must be compared by that of the real calibration object and the resulting difference is the coefficient to correct the vertical displacement. By correcting the point of origin this way, the Cartesian coordinate system is now scaled to calculate thickness information. Therefore, the abscissa is set as a zero order reference.

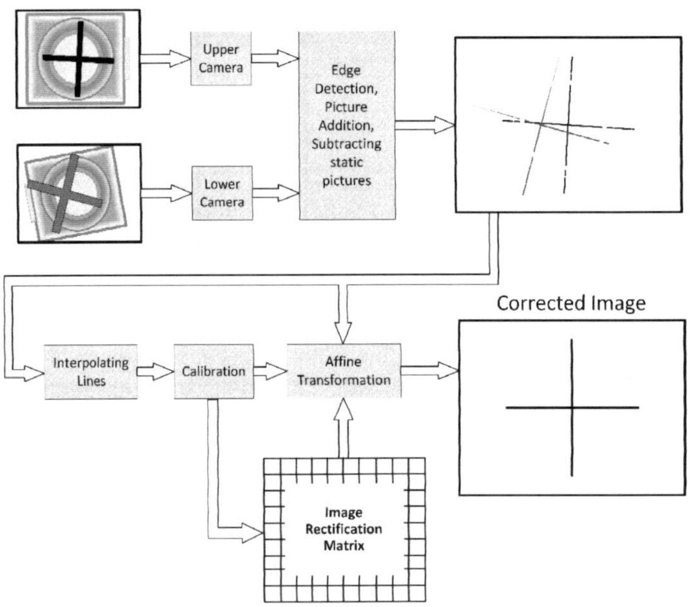

IOP Conf. Series: Journal of Physics: Conf. Series **1044** (2018) 012058 doi:10.1088/1742-6596/1044/1/012058

Figure 9. First calibration procedure to getting affine transformation parameters.

Additionally, the value must be deposit as a transformation parameter in the extended image rectification matrix as well. Now, the reference-plane is shown in Figure 2 (6). The last step is to equalize the parallelism of the two lines by finding the rotation angles against the corrected Cartesian coordinate system and deposit them in the extended image rectification matrix as well. Finally, the lines can be corrected by applying the affine transformation including the extended image rectification matrix. This procedure is exemplarily shown in Figure 10. Annotating, the maximal distance between that two straight lines inside the area of interest is a systematic error caused by warping profile and ignoring a physical area of material.

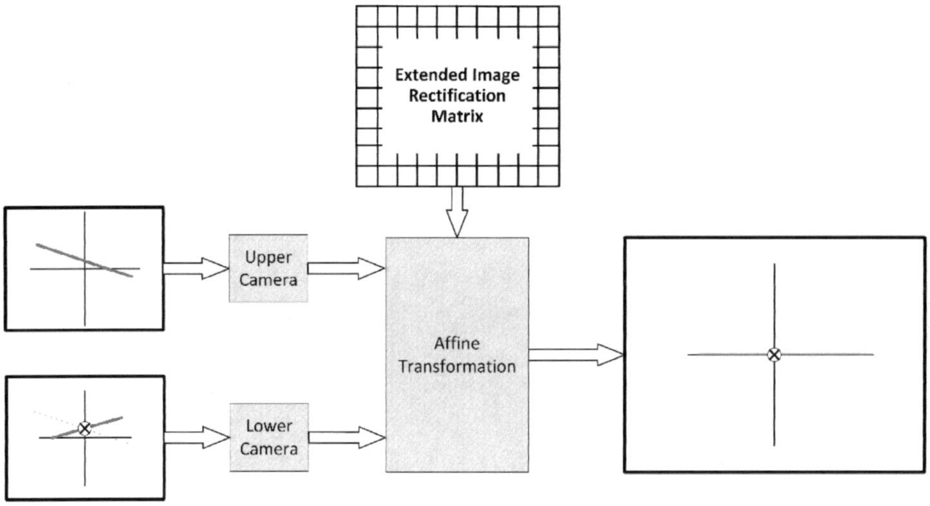

Figure 10. Equalizing the parallelism of the two laser lines.

5. Measurement data and discussing uncertainty

Uncertainty caused by pixel size could be evaluated by consulting the datasheets of the used optical equipment. Another method is to compute pixel-scaling by using a 2D chess-pattern, which was applied in that first prove of concept. By shifting that pattern inside the optical path (orthogonal to the optical axis) it is possible to develop a correcting-polynomial to scale optical parameters as well. That practice is useful if the objective is not telecentric. Beside the fact of systematic uncertainty caused of the pixel size, all the rest influences are coded in the angle. By respecting the complexity, it was not possible to estimate, all (geometrical) uncertainty in that investigation. The given data was calculated by trigonometrical values estimated with a calibration object. Therefore, the uncertainty of other trigonometrical parameters coded in that triangulation angles could not be discussed in that investigation.

Figure 11 shows the data which was taken in that prove of concept. All previously mentioned concepts, such as affine transformation and image rectification was applied after capturing that data. The object in that image has a rectangular shape with the dimension of 20 x 20 mm. That simple shape was used to get calibration-data as well. So it is obvious that both sides fit in the 10 mm lines relatively well. Additionally, there is a groove at both sides that decreasing thickness from 20 mm to 5 mm . Focusing that grove, the upper measurement deviation amounts to 1.21 mm average and the lower measurement deviation amounts to 1.76 mm average. Both values are not satisfying for precision measurement applications. Furthermore, experiments done with other materials offers that deviations increasing by enlarging thickness.

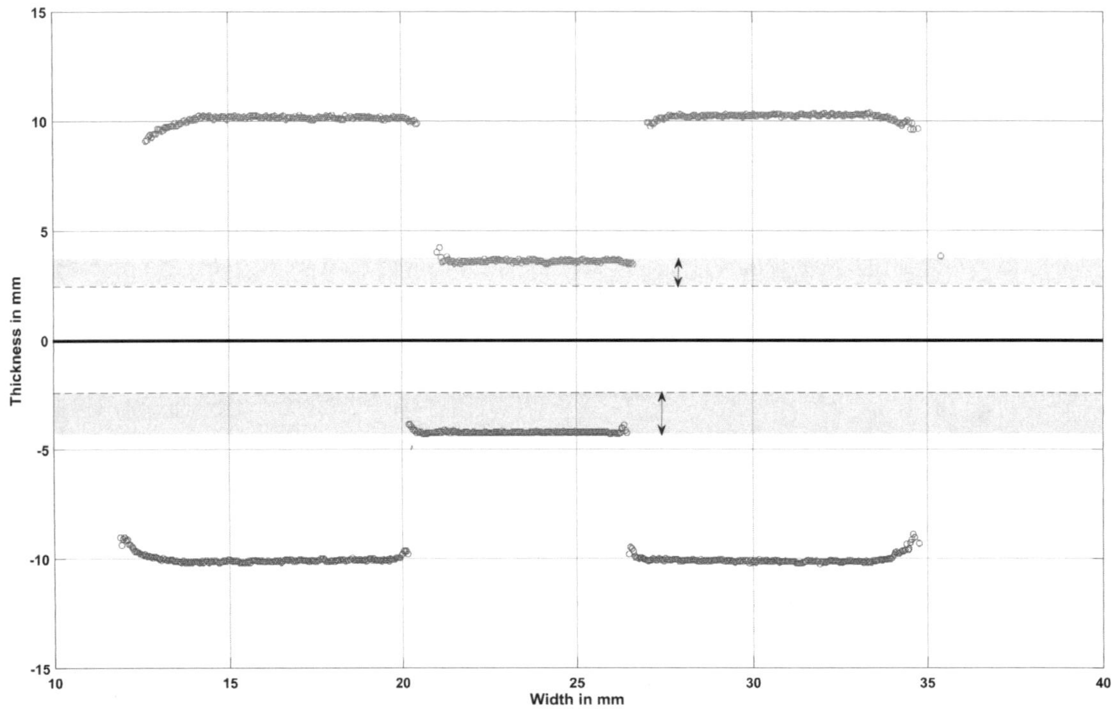

Figure 11. Captured measurement data of the discussed object.

6. Conclusion

This system investigation shows that online differential measurement for evaluating the thickness of elastic material is principal possible. Basically, almost all result adulterating influences are respected in developing a much as is necessary accurate testing measuring system. However, measurement data and estimated uncertainty are not satisfactory. On the basis of this knowledge, the next development steps can be identified. Upcoming researches should deal with possibilities in mechanical adjustments to minimize systematic errors of measurement. Furthermore, a more precise investigation to determine all parameters of uncertainty must be done. Therefore, a separate mechanical design process to ensure stability inside industrial environments must be done as well. Then, the upcoming result of the system investigation will be able to take measurements that are nearly independent by contour, deformation and position inside the scanning area.

References

[1] Pfeifer T and Schmitt R 2010 *Fertigungsmesstechnik* (De Gruyter)
[2] Christoph, R ; Neumann H J 2013 *Multisensor-Koordinatenmesstechnik* (München)
[3] Bronstein I N, Semendjajew K A, Musiol G and Mühlig H 2008 *Taschenbuch der Mathematik* vol 23
[4] Jähne B 2002 *Digital Image Processing, 5th revised and extended edition* vol 31
[5] Szeliski R 2010 Computer Vision : Algorithms and Applications *Computer (Long. Beach. Calif).* **5** 832

IOP Conf. Series: Journal of Physics: Conf. Series **1044** (2018) 012059 doi:10.1088/1742-6596/1044/1/012059

Recycled aluminium used in low cost shielding

A V Pantoja[1,2],L G Padilha fh[2], R S Machado[2], C H S Sousa[1,2], J G P Peixoto[1].

[1]Insituto de Radioproteção e Dosimetria
Av. Salvador Allende S/N, Barra da Tijuca,Rio de Janeiro
[2]Faculdade Tecnico-Educacional Souza Marques
Av. Ernani Cardoso, 335, Cascadura, Rio de Janeiro

guilherm@ird.gov.br

Abstract. The used high atomic weight and density materials, like lead and barite, contribute with the environmental problem, when they will be hand and disposal. Therefore we showed the recycled aluminiumis possible as shielding material for mammography X ray. In addition to the environmental gain, we will have the social gain, since the aluminium used came from recycled. Attenuation measurements could made with recycled and high pure aluminium (99.999%), the results shows better than 1%.

1.Introduction

The human action has an environment impact that could be greater or lesser gravity, which ecological imbalance generates, that could be assimilated or even generates catastrophic changes. Concern with sustainable development is becoming inevitable, since world population increase,therefore forced increasing production and then consequent residues increase that are environment dispensed, which could generate an imbalance making life impossible on Earth as we know [1].

The 3 R's policies (reduce, reuse and recycle) is the most accepted as the best way to achieve sustainable development. Priorities hierarchy is exactly the order which it presents: we must reduce environment exploitation, as a priority; reuse what has been produced as long as possible and recycle what cannot be reused [2].

In this way we are concerned with the environmental impacts caused by used materials in the shielding of ionizing and electromagnetic radiation, which leads us to seek new ways of accomplishing them with the reuse of materials. The use change of the classic shielding material, leaving heavy metals manipulate with, whose environment exemption can cause serious damages [3]. In additional to fail use high environmental impact material, we think to be created a new way for used aluminium packaging, avoiding the inappropriate disposal of these packages in the environment.

Shielding in X ray services has been chosen because its peculiar characteristics and different examination types, whose X rays generation is projected in order to produce an image through low energy photons [4].

These reused material proposed brings an addition environmental gain advantage. As proposed to take advantage of used aluminium cans there will be added value to this material, created more one way to support for the excluded population society.

The recycled material should be flat and the new way of collecting should be used, creating a new business, the social inclusion through work is another indirect gain from this research [5].

The objective is demonstrated technically feasible to use aluminium, from recycled beverage cans, for X ray room shielding.

The attenuation factors, in the mammography energy range, of recycled aluminium fois, with several thicknesses, were compared to ones obtained by high pure (99,999 %) aluminium, used in calibration laboratories.

2.Ionizing radiation avoidance techniques

There are three effective ways to avoid ionizing radiations: time, distance and barrier. It should be kept as far as possible from the radiation source, time exposed shorted as possible and barrier interposed between the source and individual. Among these three forms the most effective is the distance [6].

Certain materials and its thicknesses shown that reduced the radiation effects. Using this principle to be used as X rays shielding, beginning to notice how higher denser and atomic number materials are block them and more effectively [7].

The better result material which has ease obtain a controlled thickness and which has an abundance that guarantees a reasonable price for X ray shielding is lead [8]. The wall shield can also be made by incorporating barite, a barium compound, which ensures the sealing and reduces costs [9].

The conventional shields,made with lead and barite, no longer take into account that X ray radiation is attenuated by interaction with different other materials. The conventional screening materials used could have some draw backs, both materials are heavy metals that environment pollute when their wastes are not properly disposed off, including since mining to disposal.

Each radiology service will require specific calculations for each wall barriers, on each screen, taking into account the photons energy to be shielded, the workload to be executed on site, the occupancy type behind the barrier and the shield material type used [4]. These calculations are indispensable not only for the radioprotection but also for the environmental view, since a precise calculation reduces the shield material use.

2.1.Aluminium material

It is the third most abundant element and nonferrous metal in the earth's crust, found in bauxite. Bauxite processing, which includes grinding, digestion, filtration, precipitation and calcination phases, produces the alumina, that becomes the primary aluminium. Followin the secondary aluminium is when obtained by processing aluminium scrap. One of the most important characteristics of aluminium is its ability to be infinitely reprocessed without losing its properties [10].

Aluminium is a light metal whose density is 2.7 g / cm^3, whose melting point is 660 ° Celsius, non-toxic, with high thermal and electrical conductivity, impermeable and opaque, malleable and resistant to corrosion [11].

Due to its characteristics, it is widely used in the most diverse applications, from the aeronautical industry to the packaging of food, passing through the production of electricity conducting wires, combustion engine blocks, doors, windows, heat exchangers, vehicle chassis, thermal insulation and even in the treatment of water [10].

Aluminium is made by an electrolytic process that consumes very large amounts of electrical energy, about 15 MWh is consumed [10]. The aluminium recycled saves 95% of the energy needed for production, in comparison to bauxite processing. In addition to the large energy savings, five bauxite tons produce one ton of primary aluminium. The recycling process reduces the mining activity, that is environment harmful [12]. Brazil is the second largest bauxite producer and the sixth largest primary aluminium producer [10].

The aluminium recycle cans is considered as example, in 2008 it reached a Brazilian record, with 96.27% of the beverage cans returning to the production chain, as shown in Figure 1. In addition to the environmental implications, there are social implication as well, since about 150,000 workers live on aluminium cans recycling and their incomes are higher than the minimum wage [10].

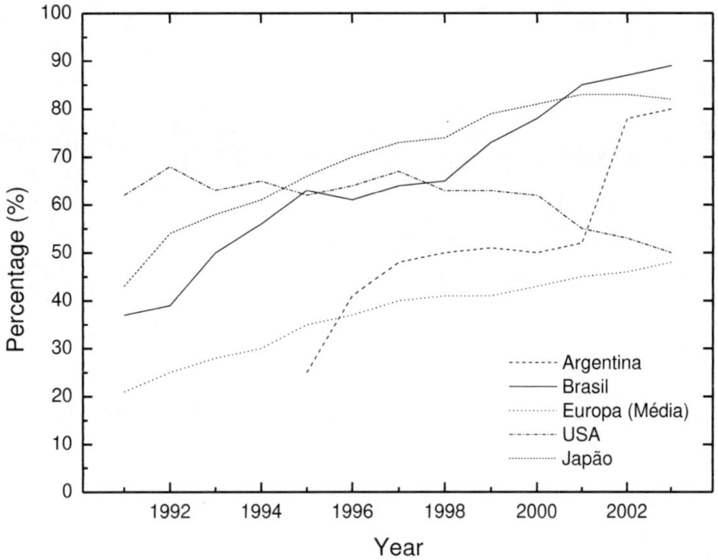

Figure 1. Recycle percentage of beverage aluminium cans.

3.Experimental set up

The used aluminium results from beverage cans collection by different brands [13]. The cans caps and bottoms were removed and their bodies opened to leave them in sheet form. The sheets were simply washed with soap and water to remove any beverage residues, the average thickness was measured with a micrometer, the mean value were found 0.1 mm per leaf. Measures were taken to ensure the uniformity for aluminium attenuation, with differences smaller than 4% between different leaf regions, Figure 2.

Figure 2. Transformations aluminium material from can to foil.

The high pure aluminium (99.999%) used for comparison was the same as used in the radiation beams implementation for calibration, where ten sheets of this aluminium with 0.1 mm thickness per sheet were used.

The ionization chamber reference was used, PTW model 23342, suitable for use in low energy X rays with 0.02 cm^3 sensitive volume, connected to the electrometer Keithley model 6517A. The X ray tube was the Panalytical with side window model PW 2185/00 with molybdenium anode, water cooling and beryllium window, without additional filtration.

3.1.How done

The ionization chamber was positioned and aligned to the tube window at 600 mm focus-chamber distance. The attenuators support was placed at 300mm from the tube, where large distance fader-chamber, which will serve to reduce the effects of measurement scattering. The central region of the aluminium sheets was aligned with the focus-chamber axis, according to experimental set up present in Figure 3.

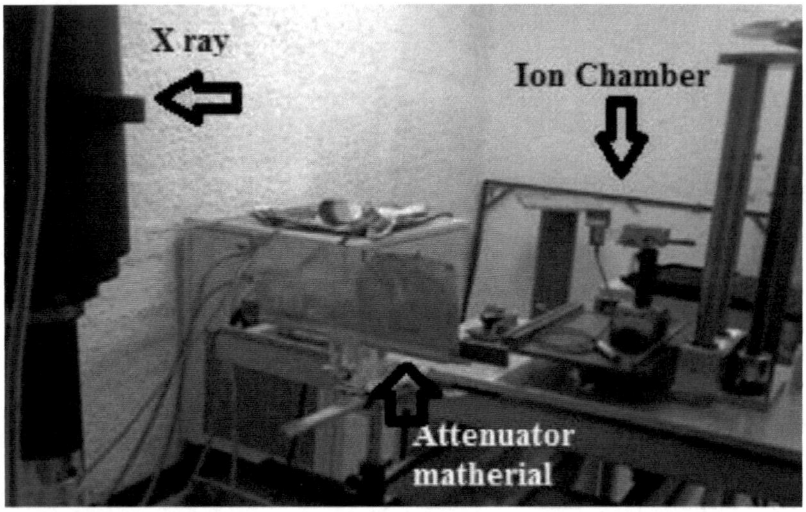

Figure 3. Set up used at de Brazilian Metrology Laboratory.

The tube current and high voltage was set at 50 mA and 28 kV, respectively. Ten measurements were performed for each aluminium included at the additional filtration, without attenuator and with the addition of one to ten aluminium foil, both from cans and hyper pure aluminium filtrations.

4.Results

The measurements results, shown in Figure 4, indicate that the comparison between the high pure (99.999%) and the recycled aluminium has a 1 % maximum relative difference.

Since ten measurements of the attenuation factor were performed for each thickness for high pure (99.999%) and the recycled aluminium are shown the mean values and standard deviations at table 1.

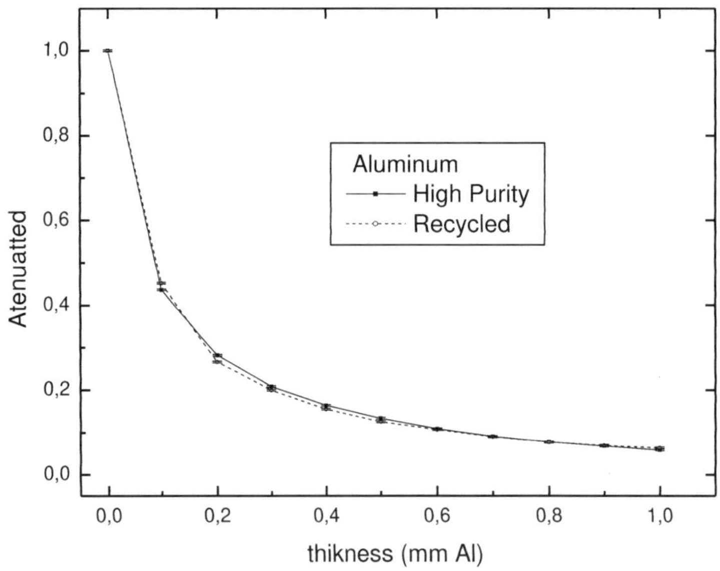

Figure 4: Comparison of the attenuationfactors achieved with high pure (99,999 %) and recycled aluminium[13].

Table 1: Mean values and standard deviations from ten measurements of the attenuation factor were performed for each thickness for high pure (99.999%) and the recycled aluminium.

thickness	0,1	0,2	0,3	0,4	0,5	0,6	0,7	0,8	0,9	1,0
high pure (99.999%) aluminium										
Mean (nC)	0,3279	0,1432	0,0925	0,0682	0,0538	0,0437	0,0357	0,0298	0,0254	0,0222
SD	0,0033	0,0004	0,0002	0,0006	0,0001	0,0002	0,0002	0,0001	0,0002	0,0002
recycled aluminium										
Mean (nC)	0,3451	0,1561	0,0918	0,0688	0,0535	0,0433	0,0368	0,0310	0,0266	0,0240
SD	0,0003	0,0002	0,0002	0,0003	0,0002	0,0002	0,0001	0,0001	0,0002	0,0001

5.Conclusion

The results confirm the possibility to use recycled aluminium material for X ray shielding.

The primary beam attenuation is already sufficient to confirm the research hypothesis and taking into account that the dispersed radiation energy is smaller than the primary radiation beam.

The focus-attenuater distance is the same as used by calibrations laboratory and shows the efficient as a attenuatiom material, that became as a shield wall.

The commercial alluminium used to produce the foils use at the can is not high purity as used at the calibrations laboratories, but the impurity into the foil shows the possibility material shield could be used.

The hypothesis confirmed in this research makes us seek deeper results that will be tested in different energies and field, producing a material prototype for shielding with the embedded finish.

References

[1] Sperandio SA, Gaspar MA. Gestão socioambiental em empresas industriais. 2006.
[2] Tocchetto MRL. Química Industrial – caderno didático. Santa Maria: 2005.
[3] Silva JPS. Impactos ambientais causados por mineração. Rev Espaço Da Sophia 2007.
[4] Bushong SC. Ciência Radiológica para Tecnólogos. São Paulo: 2010.
[5] Romansini SRM, Viana E. O catador de resíduos sólidos recicláveis no contexto da sociedade moderna. 2005.
[6] Scaff LAM. Bases físicas da radiologia – diagnóstico e terapia. Sarvier. São Paulo: 1979.
[7] INCA. Blindagem em radioterapia: técnica e normas. Rio de Janeiro, Brazil: 2000.
[8] Almeida Jr AT. Caracterização da barita e do vidro cristal como atenuadores na blindagem das radiações X e gama. Recife: 2005.
[9] INCA. Estimativa 2012: incidência de câncer no Brasil. Rio de Janeiro, Brazil: 2011.
[10] Costa TS. Experimentos com alumínio.Química na Nova Escola 2006.
[11] Oliveira AS, Gomes A das N. Logística reversa: geração de emprego e renda através de coleta de lixo residencial – Estudo de caso: Cooperativa Cruma. São Paulo: 2009.
[12] Prado MR. Análise do inventário do ciclo de vida de embalagens de vidro, alumínio e PET utilizadas em uma indústria de refrigerantes no Brasil. Universidade Federal do Paraná, 2009.
[13] Pantoja AV. Uso de alumínio reciclado em blindagem de baixo custo em serviços de mamografia. 2012.

IOP Conf. Series: Journal of Physics: Conf. Series **1044** (2018) 012060 doi:10.1088/1742-6596/1044/1/012060

Standardization of ^{59}Fe by efficiency extrapolation 4πβ-γ anticoincidence method

C J da Silva[1], P A L da Cruz[1], A Iwahara[1], J S Loureiro[1], A. Ruzzarin[2] and L Tauhata[1]

[1]Laboratório Nacional de Metrologia das Radiações Ionizantes/Instituto de Radioproteção e Dosimetria/Comissão Nacional de Energia Nuclear (LNMRI/IRD/CNEN), Av. Salvador Allende, s/no – Barra da Tijuca, Rio de Janeiro, Brazil, CEP 22783-127
[2]Programa de Engenharia Nuclear/Coordenação de Programas de Pós-Graduação em Engenharia/Universidade Federal do Rio de Janeiro (PEN/COPPE/UFRJ)-Rua Moniz de Aragão, 360, Bloco 1, Ilha do Fundão – Cidade Universitária, Rio de Janeiro, Brazil, CEP 21941-594

E-mail: iwahara@ird.gov.br

Abstract. A solution of ^{59}Fe was standardized in the frame of a Bureau International des Poids et Mesures (BIPM) international key-comparison piloted by Asian Pacific Metrology Program (APMP) regional metrology organization. The method used was the live-timed 4πβ-γ anticoincidence counting with extended dead-time with efficiency extrapolation method. The confirmation of these results was performed by conventional 4πβ-γ coincidence counting with efficiency extrapolation and CIEMAT/NIST liquid scintillation counting methods. The value of the evaluated anticoincidence measurement was used to calibrate the LNMRI second standard ionization chambers for providing ^{59}Fe solutions to nuclear medicine canters in the procedures of anaemia diagnosis. Besides this value was adopted as the comparison value of LNMRI and sent to APMP organization. The half-life of ^{59}Fe was determined following the decay of one ampoule sample with ^{59}Fe solution in an ionization chamber (Centronic model IG12).

1. Introduction

The radionuclide ^{59}Fe decays by beta emission mainly to the excited levels of 1099 and 1291 keV of ^{59}Co followed by the emission of two major gamma rays with energy of 1099.3 and 1291.6 keV, respectively. It is a radionuclide used in nuclear medicine for detection of problems of bone joints and allows the diagnosis of anaemia. The method used at LNMRI for the activity standardization of ^{59}Fe solution was the live-timed 4πβ-γ anticoincidence counting with extended dead-time with efficiency extrapolation technique. The confirmation of this result was checked by conventional 4πβ-γ coincidence counting with extrapolation technique and CIEMAT/NIST liquid scintillation counting methods whose results agreed with the anticoincidence result within the respective uncertainties (coverage factor k=2). The result of anticoincidence measurements was used to calibrate the LNMRI secondary standard ionization chambers for providing ^{59}Fe solutions to nuclear medicine canters in the procedures of anaemia diagnosis. The half-life of ^{59}Fe was determined following the decay of ampoule sample with ^{59}Fe solution in an ionization chamber during 20 days of measurements giving result consistent with those found in literature.

2. Methodology

2.1 4πβ-γ(NaI) anticoincidence counting method

The anticoincidence method was originally developed by Bryant [1] and improved by de Carlos and Granados [2] that incorporated the concept of shared dead time for beta and gamma channels. Due to dead-time correction difficulties, Baerg et al. [3] developed an anticoincidence system based on extendable dead-times and the lived-time technique. Using the same concept, the live-timed 4πβ(LS)-γ(NaI(Tl)) anticoincidence system and extendable dead-time operating at LNMRI, is based on specialized modules (MTR2, MI1, MI2 and MI3) developed for the dead-time and counting implementation in the different channels [4,5,6]. Basically the anticoincidence counting method consists in obtaining an experimental extrapolation curve, that links the activity of the source N_o to the beta (N_β), gamma (N_γ^W)and coincidence count rates N_c given by equation (1). Because coincidences are measured indirectly, N_c is calculated as $N_\gamma^w - {}^iN_\gamma^w$ in equation (1) where N_γ^W is the count rate for a given gamma window and the ${}^iN_\gamma^w$ is the gamma count rate for uncorrelated events. In this case it is not necessary the use of a coincidence unit by eliminating the uncertainties introduced by the inevitable accidental counts inherent to that unit. Using an extrapolation technique in equation (1), N_{Ap} tends to N_0, the activity, when beta efficiency given by ratio N_c/N_γ tends to 1. The variation of beta efficiency in this work was carried out by electronic discrimination using a single channel analyzer module in beta counting channel.

$$N_{Ap} = \frac{N_\beta N_\gamma^w}{N_\gamma^w - {}^iN_\gamma^w} \qquad (1)$$

The detector for beta events consists of a scintillation cell coupled to two photomultipliers placed at an angle of 180 degrees relative to each other. The counting cell of the liquid scintillation counter is made of polyethylene painted inside with TiO in order to reflect light. A scintillation crystal of NaI(Tl) for gamma counting is coupled to the scintillation cell forming the anticoincidence detection set-up.

2.2 4πβ(PC,LS)–γ(NaI) coincidence counting method

The coincidence method was precursor to the anticoincidence method, but basically the operation of them is quite similar. The fundamental difference is that the coincidence method is based on the actual beta-gamma coincidence count whereas in the anticoincidence method the coincidence count is obtained through uncorrelated beta-gamma events. In the same way it consists in obtaining an experimental extrapolation curve, that links the activity of the source N_o to the count rates of beta, gamma and coincidence N_β, N_γ and N_c, respectively, given by equation (2) where the ratio N_c/N_γ represents a measure of the detection efficiency in the beta-channel.

$$N_0 = \frac{N_\beta N_\gamma}{N_c} * \left[1 + C\left(\frac{1 - N_c/N_\gamma}{N_c/N_\gamma}\right)\right]^{-1} \qquad (2)$$

The extrapolation technique is applied by varying the detection efficiency preferably in the region of high values of N_c/N_γ The parameter C represents the slope of expression (2) and it depends on the radionuclide decay scheme. Using a linear fitting, the source activity N_o is obtained from the extrapolation of expression (2) when the ratio N_c/N_γ tends to unity. Two detectors configurations set-ups have been used in the measurement of ^{59}Fe. The first one consisted of a gas flow 4π proportional counter (PC) coupled to a 10.2 cm ×10.2 cm crystal of NaI(Tl) and to a coaxial HPGe. The counting gas of the 4π counter is a mixture of 90% argon and 10% methane operating at 0.1 MPa pressure. The second set-up consists of the same scintillation cell described in 2.1 and coupled to the same NaI(Tl) scintillation crystal of first set-up. Both set-ups are coupled to the same electronic chain comprising the coincidence system used in this work. Beta particles originating from ^{59}Fe are counted in the beta channel and the gamma rays of 1099 keV and 1291 keV are counted in the gamma channel for different gamma energy windows. The gamma window condition in the PC-NaI set-up counting was done around the 1099 and 1291 keV. For PC-HPGe the gamma window was set around 1291 keV photopeak and for LS-NaI counting was around the two photopeaks simultaneously. The range of beta efficiency variation in PC counting was 0.90 to 0.65 using inactive carriers of $FeCl_3$ at various concentrations in source preparation. For liquid scintillation counting the efficiency variation, from 0.91 to 0.76, was carried out using electronic discrimination. The experimental results of the coincidence measurements were fitted using nonlinear regression Levenberg-Marquardt algorithm [7]. The final result reported was adopted as the weight mean of four results.

2.3 CIEMAT/NIST counting method

The CIEMAT/NIST method is based on statistical Free Parameter model of the scintillation photons distribution and their detection probabilities in counting systems consisting in two photomultipliers at 180° from each other [8]. An efficiency versus extinction curve is obtained from a set of ^{3}H standard samples in order to characterize the experimental condition. This experimental curve is compared with the theoretical curve of Free Parameter versus ^{3}H efficiency. Thus, a new Free Parameter versus Extinction curve is obtained [9]. Measurements of the radionuclide samples to be standardized are performed on the commercial equipment and the quench values obtained are interpolated in the Free Parameter Curve versus ^{3}H standard extinction curve. The value of the Free Parameter obtained is interpolated in the theoretical curve of Free Parameter versus radionuclide efficiency to find its activity (Bq).

The computational codes for theoretical calculation of the detection efficiency curve CIEMAT/NIST method depending on the Free Parameter and take into account the quenching from the Birks equation, the radionuclide decay schemes, the particles and photon and particle interactions in the scintillation cocktail and system measurement evaluated by use Monte Carlo simulation codes ([10,11]).

Figure 1 shows the representation of this process of semi-empirical standardization. The CIEMAT / NIST method, despite its dependence on the ^{3}H standard for the characterization of the experimental condition, has excellent application for the standardization of beta emitters, with great accuracy and minimized uncertainty. However, its application to the standardization of radionuclides that decay by electron capture and by internal conversion of electrons is of high complexity, since in these cases the theoretical curve of Free Parameter versus Efficiency offers more than one possibility of efficiency values.

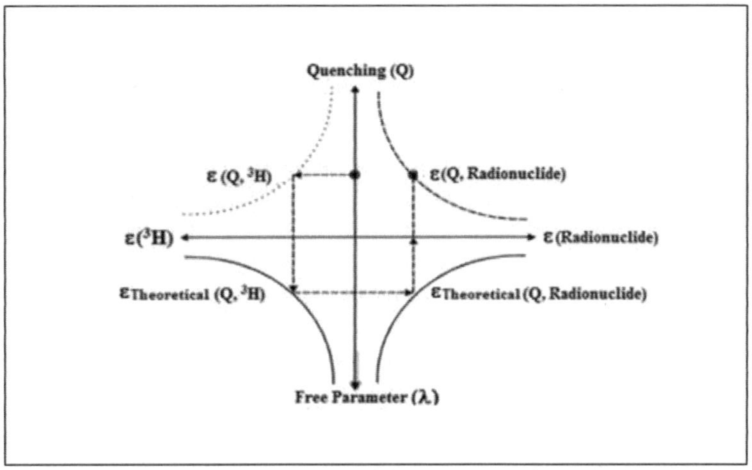

Figure 1. CIEMAT/NIST model: Diagram of standardization.

3. Results and discussion

3.1 Anticoincidence measurements

Twenty liquid scintillation sources were prepared from commercial scintillation cocktails manufactured by PerkinElmer, In c. (USA): ten from the brand OPTIPHASE Hisafe3 and 10 from Ultima Gold. The samples were prepared using 15 mL of cocktail scintillation plus 20 to 30 mg of ^{59}Fe solution contained in a glass vial with low potassium content. The beta particles are detected in the liquid scintillator and the pulses from the two coupled photomultipliers are processed in the corresponding electronic chain. For counting in the gamma channel an energy window encompassing the peaks of 1099.2 and 1291.6 keV was fitted. The beta efficiency variation, in order to obtain the extrapolation curve, was made by electronic discrimination using pulse height analyzer module in the beta channel and ranged from 0.80 to 0.95. The statistical Grubbs'test [12] was used to detect outliers in the data set of 20 results. The results of three of them were discarded for the final calculation of the activity because they presented inconsistent values. The final activity value adopted was the weighted mean of 17 results giving 468.73±1.59 kBq/g at reference date of June 1st, 2014, 00:00 UTC. The uncertainty refers to ~68% level of confidence (k=1).Relative uncertainty components for activity concentration by anticoincidence counting method are presented in Table 1.

Table 1. Relative uncertainty components for activity concentration by anticoincidence counting method.

Uncertainty components, in % of the activity concentration due to	Type	Standard relative uncertainty (%)
Extrapolation of efficiency curves	A	0.17
Weighing	B	0.05
Counting statistics	A	0.25
Lived-time	B	0.01
Background	B	0.15
Half-life	B	0.021
Timing	B	0.010
Combined relative uncertainty		0.34
Expanded relative uncertainty (k=2)		0.68

IOP Conf. Series: Journal of Physics: Conf. Series **1044** (2018) 012060 doi:10.1088/1742-6596/1044/1/012060

3.2 Coincidence measurements

Ten sources were prepared by dropping known masses of ^{59}Fe onto VYNS film previously gold coated on both sides. The solution masses ranged from 16 mg to 30 mg. For liquid scintillation counting five samples were prepared using 15 mL commercial OPTIPHASE HISAFE3 cocktail contained in a glass vial with low potassium content. The masses of ^{59}Fe added to each cocktail ranged from 20 to 30 mg. The gamma window condition in the PC-NaI set-up counting was done around the 1099 and 1291 keV. For PC-HPGe the gamma window was set around 1291 keV photopeak and for LS-NaI counting was around the two photopeaks simultaneously. The range of beta efficiency variation in PC counting was 0.90 to 0.65 using inactive carriers of $FeCl_3$ at various concentrations in source preparation. For liquid scintillation counting the efficiency variation, from 0.91 to 0.76, was carried out using electronic discrimination. The experimental results of the coincidence measurements were fitted using nonlinear regression Levenberg-Marquardt algorithm [7]. The final result reported was adopted as the weight mean of four results. Table 2 presents the activity results with respective standard uncertainties of the fitting procedures and the final result determined as the weighted mean of these results. The uncertainty budget is presented in Table 3.

Table 2. Results of the activity concentration by coincidence counting. The uncertainties refer to extrapolation procedure component of efficiency curves for each gamma window counting condition.

Method used	Activity (kBq/g)	Standard uncertaint y (kBq/g)	Relative standard uncertainty (%)
PC-NaI(Tl) (Gamma window: 1099 keV)	475.14	2.88	0.60
PC-NaI(Tl) (Gamma window: 1291 keV)	471.42	2.80	0.59
LS-NaI(Tl) (Gamma window: 1099+1291 keV)	473.12	1.94	0.41
PC-HPGe (Gamma window: 1291 keV)	475.51	3.39	0.70
Laboratory final result (weighted mean)	473.99	0.89	0.19

Table 3. Relative uncertainty components for activity concentration by coincidence counting method.

Uncertainty components, in % of the activity concentration due to	Type	Standard relative uncertainy (%)
Extrapolation of efficiency curves	A	0.19
Weighing	B	0.050
Dead-time	B	0.0061
Resolving time	B	0.016
Gandy effect	B	0.068
Background	B	0.48
Half-life	B	0.024
Counting statistics	A	0.18
Combined relative uncertainty		0.51
Expanded relative uncertainty (k=2)		1.0

The weighted mean was adopted as the final activity of coincidence measurements given 473.99 kBq/g with evaluated standard uncertainty of ±2.37 kBq/g at reference date. The consistency of the activity results was analyzed based on the use of Birge ratio criterion [13] where the ratio between external and internal uncertainties should be close to 1 or less. The Birge ratio calculated was 0.770 suggesting that the results are consistent.

3.3 CIEMAT/NIST measurements

The theoretical calculations of the ^{59}Fe detection efficiency were performed by computational code CN2003 [11]. A set of eight standard tracer tritium (^{3}H) samples in 10 mL of cocktail OPTIPHASE Hisafe3 for the commercial Wallac 1414 Liquid Scintillation Counter characterizing in terms of the quenching parameter by the introduction of the chemical quenching agent in increasing degree. The measurements of eight ^{59}Fe samples also in OPTIPHASE Hisafe 3 were performed under the same conditions to the tracer. In this condition the Free Parameter is common to the both radionuclides ones, allowing the correlation between their experimental and theoretical efficiencies, according the fig. 1. The final result of activity, adopted as the weighted mean of 8 results gave 469.33±1.80 kBq/g at reference date. The uncertainty refers to standard uncertainty (k=1). Relative uncertainty components for activity concentration by CIEMAT/NIST counting method are depicted in Table 4.

Table 4. Relative uncertainty components for activity concentration by CIEMAT/NIST counting method.

Component	Type	Relative standard uncertainty (%)
^{3}H activity	B	0.052
^{3}H weighing	B	0.0020
^{3}H Quenching (SQPE)	A	0.088
^{3}H statistic counts	A	0.0081
Nuclear and Atomic data	B	0.10
Photomulplier assimetry	B	0.23
kB value	B	0.10
^{59}Fe weighing	B	0.0011
^{59}Fe Quenching (SQPE)	A	0.023
^{59}Fe statistic counts	A	0.0070
^{59}Fe source activities	A	0.27
Combined relative uncertainty		0.40
Expanded relative uncertainty (k=2)		0.79

3.4 Calibration factors and half-life

The estimation of the half-life was performed by measuring the decay of ^{59}Fe for a period of 25 days in the ionization chamber. No gamma impurity was detected using calibrated HPGe spectrometer. The logarithm of the ion charge, which is the proportional to activity, measured in the ionization chamber as a function of the measurement time is a linear function and presented as $\ln(A) = -\lambda\Delta t + a$, where A is the ion charge, Δt is the time interval from the first measurements and λ is the decay constant of ^{59}Fe. Least square method using Levenberg-Marquardt algorithm [7] was performed to determine these parameters. The value of half-life determined is (44.547 ± 0.035) days which is 0.12% larger than the suggested half-life in this comparison (44.495 ± 0.008) days. The components of uncertainties evaluated come from the fitted λ value (0.0014%), chamber stability checked by ^{226}Ra sample measurements during the period (0.05%) and average charge measurements (0.06%) giving a total of 0.078%. The consistency of the results obtained by the three methods is shown in Figure 2.

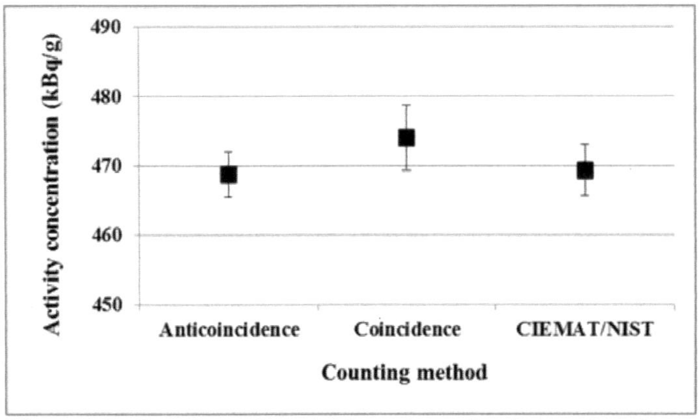

Figure 2. Results of activities determined by the three methods. The uncertainty bars refer to coverage factor k=2.

4. Conclusions

A ^{59}Fe solution was standardized by the live-timed 4πβ-γ anticoincidence with extended dead-time method. The validation of this result was carried out by two other methods: coincidence and CIEMAT/NIST. The results were equivalent within the uncertainties evaluated at 95% confidence level. The results of the coincidence method, using samples prepared in thin VYNS films and liquid scintillation cocktails showed statistically equivalent results. The value of the anticoincidence method was adopted as the reference activity in LNMRI because it presented the lowest uncertainty. The calibration factors for three IG11 model ionization chambers were determined increasing the number of calibrated radionuclides available for users of nuclear medicine. The half-life was determined which was statistically consistent at 95% level of confidence with the suggested value of comparison.

References

[1] Bryant J 1967 *IAEA Symposium on Standardization of Radionuclides* **SM-79**/211
[2] de Carlos J E and Granados C E 1973, *Nucl Instr. and Meth.* **112** 209
[3] Baerg A P, Munzenmayer K and Bowes,G C 1976 *Metrologia* **12** 77
[4] Bouchard J 2000 *Appl. Radiat. Isot.* **52** 441
[5] Bouchard J 2002 *Appl. Radiat. Isot.* **56** 269
[6] da Silva C J, Iwahara A, Poledna R, Bernardes E M O, de Prinzio M A R, Delgado J U and Lopes R T 2008 *Appl. Radiat. Isot.* **66** 886
[7] da Silva W C P 2011 LAB Fit Curve Fitting Software. Federal University of Campina Grande, Brazil
[8] da Cruz P A L, da Silva C J, Iwahara A, Loureiro J S., de Oliveira A E., Tauhata L and Lopes R T 2016 *IOP Publishing, Journal of Physics. Conference Series* **733** 012099
[9] Malonda A G, Coursey B M 1987 *Appl. Radiat. Isot.* **39** 165
[10] OECD/NEA, 2008. Penelope: A Code System for Monte Carlo Simulation of Electron and Photon Transport. Nuclear Agency Energy/Organization for Economic Co-operation and Development
[11] Gunther E 2003 Program CN2003: A program to calculate the LC efficiency of a nuclide vs. efficiency the tracer H-3 (CIEMAT/NIST). PTB/Germany
[12] Grubbs F E 1950 *Annals of Mathematical Statistics* **21(1)** 27
[13] Birge R T 1932 *Phys. Rev.* **40** 207

IOP Conf. Series: Journal of Physics: Conf. Series **1044** (2018) 012061 doi:10.1088/1742-6596/1044/1/012061

Strategies to overcome the crisis and leverage the legal control of measuring instruments in Brazil

A N Soratto, B A Rodrigues Filho, L L Nunes
National Institute of Metrology, Quality and Technology – Inmetro
Av. Nossa Senhora das Graças, 50, Xerém, Duque de Caxias, Rio de Janeiro, Brazil

E-mail: ansoratto@imetro.sc.gov.br

Abstract. The economic crisis in Brazil, has led to the reduction of the services regarding periodic verifications in legal metrology, however it has been an opportunity to incentive the initiatives focusing the efficiency of the legal metrological control in Brazil. The present study presents the impact of the reduction of the budget allocate to legal metrology and the initiatives carried out by the National Institute of Metrology, Quality and Technology – INMETRO, and the Brazilian Network of Metrology and Quality – RBMLQ-I, in order to leverage the efficiency the services provided in legal metrology to the commerce and industry. A survey with specialists working in the sector was conducted in order to identify strategies to improve the performance of the services in legal metrology. As a result, we present and discuss the main proposed strategies, such as partnership with either different regulatory bodies or private enterprises, fraud control and the development of information systems. The presented strategies aims to improve the legal control of measuring instruments more efficiently.

1. Introduction

Due to one of the biggest financial crisis in Brazil, the Federal Government has introduced several budget cuts to the governmental bodies during 2015 and 2016, including Inmetro, the National Metrology Institute (NMI), responsible for the legal metrological control in Brazil.

Consequently, this budget shortage has also impacted the Brazilian Network of Legal Metrology and Quality (RBMLQ-I), the branches of Inmetro, represented by 27 delegated bodies, covering the entire territory, conducting verifications and surveillance activities in legal metrology. In order to minimize the effects of the crisis, both Inmetro and the RBMLQ-I are developing and implementing initiatives optimizing the efficiency of the activities in legal metrology [1-6].

The present study evaluates the impact of the economic crisis in the results of the RBMLQ-I, discussing strategies to increase the number of periodic verifications of measuring instruments in the commerce and industry.

2. Materials and methods

In order to identify the more relevant measuring instruments to the present study, we applied a risk analysis approach, according to the Welmec 5.3 Guide [7].

The number of verifications from 2014 to 2016 for each class of measuring instrument is obtained using the Inmetro Integrated Management System (IMS). We analyzed the data regarding periodic verification since it represents the activity demanding more resources to be executed in the legal control of measuring instruments in Brazil.

We also conducted a survey to specialists from the RBMLQ and Inmetro, as a tool to support the strategies to increase the efficiency of the legal control of measuring instruments in Brazil.

3. RBMLQ-I results in 2014, 2015 and 2016

After the first semester of 2015, RBMLQ-I has suffered a significant reduction of infrastructure to support the activities of legal metrology, due to the budget shortage of the Federal Government, leading to a considerable impact of periodic verifications and inspections of regulated products in both industry and commerce in Brazil

Table 1 presents the results of verifications and inspections carried out by RBMLQ-I from 2014 to 2016 [8]. A risk analysis method was conducted in order to prior the measuring instruments to be studied: Non-automatic weighing instruments; fuel dispensers; and sphygmomanometers [7].

Table 1. RBMLQ periodic verifications in 2014-2016

Type of measuring instruments	2014	2015	2016
Non-automatic weighing instruments Class III e IV	1,338,327	1,074,783	889,232
Non-automatic weighing instruments Class I and II	46,496	39,814	32,963
Fuel dispensers	363,810	343,821	360,752
Sphygmomanometers	83,519	67,691	52,103

Table 1 shows a significant decrease of quantities regarding periodic verifications for Non-Automatic Weighing Instruments (NAWI) class III and IV, plummeting down 34% from 2014 to 2016. In the same period, sphygmomanometers periodic verification had a reduction of approximately 38%.

An Integrated Management System allowed a significant increase of periodic verification in the last five years [8], which allowed controlling and managing the performance of the parties working in subsequent verifications. Then, the current reduction in verifications is a critical scenario to legal metrology in Brazil.

Based on the perspective of a scarce budget for 2017, it is necessary to implement new strategies to enhance the operational activities of the legal control of measuring instruments permitting the execution of activities using limited resources.

4. Strategies to leverage the metrological control in Brazil

The proposed strategies in this section are based on the survey applied to specialists in legal metrology and management from the RBMLQ-I and Inmetro, and in successful initiatives already in use.

4.1 Strategy for the legal control of sphygmomanometers and medical scales

The number of verified instruments among medical devices that demand metrological verification represents only 5%.

Then, a partnership among Inmetro and the health regulatory authorities to require the Verification Certificate of both sphygmomanometers and medical scales in order to emit the authorization for a clinical facility would increase significantly the demand for verifications. However, an infrastructure improvement would be necessary to attend this demand. The use of private services would also be an option to attend it.

4.2 Strategy for a partnership with private stakeholders

As in several legal metrology models, a partnership to private entities to conduct verifications activities would be an alternative to overcome the decreasing numbers of the legal metrology model in Brazil. The participation of private stakeholders in state activities tend dismiss the demanded structure of public bodies and the usage of public resources, allowing the RBMLQ-I act in new areas and in quality surveillance of these bodies. Due to well-established requirements, the participation of private stakeholders may increase the productivity of metrological verifications, improving the technical quality of the services.

4.3 Strategy for re-distribute the workforce

The re-allocation of the workforce to finalistic areas, especially verification and inspection areas, would provide more efficacy to RBMLQ-I. Managers may incentive and offer the necessary conditions for an administrative worker to be engaged in technical activities. Recent studies showed that 45% of the workforce in RBMLQ-I are properly located in finalistic areas, when this percentage should be superior to 60% [9]. Then, managers may incentive and provide the demanded necessities to incentive the migration to finalistic areas. This strategy has showed that the bodies with majority of their workforce in finalistic areas are more productive and efficient regarding the services of periodic verification of measuring instruments.

4.4 Strategies to eliminate metrological frauds

An increasing number of metrological frauds has been noted in the last five years, mainly due to new technologies used in measurements, especially in fuel control and distribution [6]. The losses dud to metrological frauds, for example in the fuel market of US$ 300 million, considering 1% of fraudulent petro pumps and a volume deviation of 10% [3]. Some initiatives of Inmetro and partnerships among RBMLQ-I and regulatory bodies have allowed the identification of a set of technological frauds, permitting proper legal penalties to the offenders [6]. The fraud control is vital to the maintenance of fair competition and consumers' defense in the country.

4.5 Strategy for acting in more needy areas of the legal control

The System to Monitor the Coverage of Services – SIMCS has showed that several sectors of the economy are not been properly attended by legal metrology services. This information technology (IT) system provides support identifying the location of properties of measuring instruments, using information of several databases, according to the economic field of an enterprise. It provides tool to managers to plan and act more efficiently, in poorly explored areas in the legal metrology control. This strategy has the potential to increase significantly the efficiency of periodic verification services in the Brazilian market.

4.6 Strategy for online market surveillance

Several websites are nowadays putting inadequate measuring instruments in the market (scales, sphygmomanometers and others), such as not approved type ones. Consequently, managers working either to RBMLQ-I or Inmetro may direct efforts to online market surveillance to combat this practice protecting the market of inappropriate and not reliable devices. This strategy increases the reliability in the market demanding few resources.

4.7 Strategy for the legal control of chronotachographs

Regardless the instrument was not mention on the risk analysis, the periodic verification can be increased demanding few resources. The participation of other players as the transport regulation bodies can increase significantly the quantitative of verifications, demanding the Verification Certificate as a requirement to obtain the annual license for vehicles. In the last year, only 35% of the fled that demands for the instrument was periodically verified. This initiative would not increase the costs of verifications since they are conducted by private inspection bodies, although it would increase the reliability of measurements for chronotachographs.

IOP Conf. Series: Journal of Physics: Conf. Series **1044** (2018) 012061 doi:10.1088/1742-6596/1044/1/012061

5. Conclusions

The economic crisis in Brazil since 2015 has affected the services of legal metrology, especially the legal control of non-automatic weighing instruments and sphygmomanometers.

To confront this scenario, Inmetro and RBMLQ-I have reached for new strategies to increase the efficiency and effectiveness of metrological verifications, due to partnership to regulatory bodies, private stakeholders, fraud control and implementation of IT.

Consequently, these strategies aim to improve the control in legal metrology consuming fewer resources. Then, partners in several other countries, considering local necessities, may also use the presented strategies to improve their legal metrology systems.

References

[1] Rodrigues Filho B A and Soratto A 2017 Legal Metrology in Brazil *OIML Bulletin* **LVIII**, 1, January

[2] Soratto A N, Pohlmann Filho O, Paiva M R, Giordani R B and Bringhenti C 2014 Development of a system to increase the legal metrological control of measuring instruments in Brazil *International Journal of Metrology and Quality Engineering* **5**, n. 3, p. 5

[3] Rodrigues Filho B A and Gonçalves R F Measuring the economic impact of metrological frauds in trade metrology using an input-output model *Advances in Production Management Systems. Initiatives for a Sustainable World* p. 624

[4] Pohlmann Filho O et al 2014 Innovaciones de Inmetro en la gobernabilidad de la Red Brasilera de Metrología Legal y Calidad *9th International Symposium "Metrologia 2014" Proceedings*

[5] Rodrigues Filho B A, Soratto, A N and Gonçalves R F 2015 Information systems as a tool to improve legal metrology activities *Advances in Production Management Systems* **459** p. 121

[6] Leitão F O, Vasconcellos M T and Brandão P C R 2014 Contramedidas de hardware e software sobre el fraude de alta tecnologia al surtidor de combustible bajo al alcance de metrologia legal *9th International Symposium "Metrologia 2014" Proceedings*

[7] Welmec Guide 5.3 Risk Assessment Guide for Market Surveillance: Weigh and Measuring Instruments 2011 *European Cooperation in Legal Metrology*

[8] Relatório Consolidado do Plano de Trabalho da RBMLQ 2010 a 2014 *Sistema de Gestão Integrado* (Rio de janeiro: Inmetro)

[9] Levantamento da Distribuição da Força de Trabalho da Rede Brasileira de Metrologia Legal e Qualidade – Inmetro 2015 (Rio de Janeiro: Inmetro)

IOP Conf. Series: Journal of Physics: Conf. Series **1044** (2018) 012062 doi:10.1088/1742-6596/1044/1/012062

Surface plasmon resonance features of corrugated gold films: wavelength interrogation mode for exhaled gas detection

E P Rodrigues[1], **A M N Lima**[1], **L C Oliveira**[1,2], **T A T de Sousa**[1,3] **and H Neff**[1]

1 Graduate Program in Electrical Engineering (PPgEE), Universidade Federal de Campina Grande (UFCG), Campina Grande, PB, Brasil
2 Universidade Federal do Semi-árido (UFERSA)
3 Instituto Federal de Alagoas - Campus Palmeira dos Índios

E-mail: {eloise.rodrigues,leiva.oliveira,tiago.sousa}@ee.ufcg.edu.br, {helmut,amnlima}@dee.ufcg.edu.br

Abstract. The characterization of a surface plasmon resonance sensor operating in the grating mode has been explored theoretically and experimentally. Wavelength interrogation mode was exploited at specular reflection in the gas phase for thin gold films, deposited onto a corrugated polymer substrate by sputtering. Some tests using water vapor were performed and the achievable responsivity S_{λ}^{gm} at specular reflection was determined experimentally to be 87 nm/RIU and the limit of detection of 13.2×10^{-3} RIU was calculated. The good agreement between experimental and theoretical results confirms the correctness of the methodology and the feasibility of the proposed sensing system.

1. Introduction

Surface plasmon resonance (SPR) sensors are highly versatile optical tools and widely used for unlabeled real time monitoring of biochemical interactions in an aqueous environment [1, 3]. SPR sensors are applied to a variety of research disciplines that range from genomics, proteomics, drug discovery and development to medical diagnostics. In this study, the suitability of a specially designed SPR sensor for human gas exhaled sensing is explored.

The phenomenon SPR accounts for an exceptional surface sensitivity to changes of the refractive index [7], of the surrounding dielectric phase. Coupling to and resonant interaction of the SP-wave with an incoming photon, requires that the SP-wave vector k_{sp} (eq.1), matches to the wave vector of the incident light k_{ph}, corresponding to energy-momentum conservation. To augment k_{ph}, light propagation must occur within a dielectric material, where the refractive index $n > 1$. Under conditions, where the incident light beam propagates through vacuum or gas, with $n \approx 1$, the reflective surface must be spatially modulated through a corrugated surface profile, to establish appropriate momentum transfer. The former condition is known as the attenuated total reflection configuration, or Kretschmann set-up, the latter denoted as the grating mode (GM). The classical SPR sensor configuration, also known as the Kretschmann configuration, whose the incident light is reflected from the film and leaves the prism, utilizing attenuated total reflection (ATR) conditions. In the GM, light propagates at the opposite side, within gas or vacuum and is either specularly reflected at a corrugated metal film ($m = 0$), or diffracted into higher order ($m \geq 1$). A sketch of the geometry is outlined in Figure 1 who shows the type of grating studied, which has its profile similar to a sawtooth wave. The incident light beam, p-polarized, with a fixed angle of 40° and spanning several wavelengths characterizes the wavelength interrogation study (WIM).

$$k_{sp} = \frac{2\pi}{\lambda}\left(\frac{\varepsilon'_m \varepsilon_d}{\varepsilon'_m + \varepsilon_d}\right)^{\frac{1}{2}} \tag{1}$$

where λ is the wavelength of the incident light, ε'_m is the real part of the relative permittivity of the metal and ε_d is the permittivity of the dielectric.

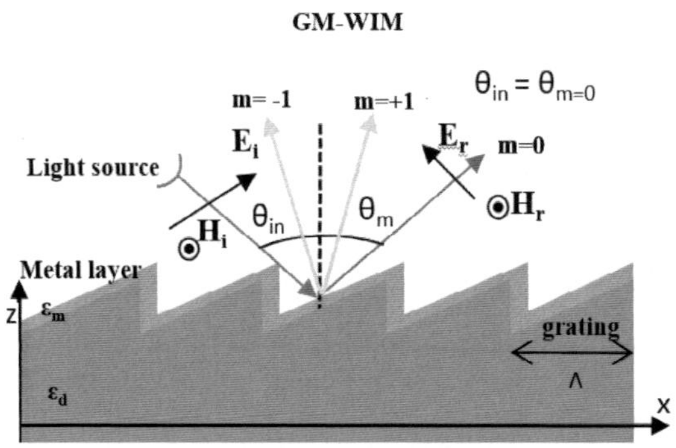

Figure 1. Geometry of the device grating. The thin metallic layer shows the periodic grating. P-polarized light has the electric field component parallel to the grooves of the grating. The reflectance conditions at the grating surface for GM-WIM condition is showed.

For the GM mode, the light beam, p-polarized, coming directly from the source strikes the surface of the metal grating. P-polarized means that the electric field is parallel to the plane of incidence, while S-polarized means that the electric field is perpendicular to this plane. In this case, a material with good optical properties will not be necessary, the surface plasmons wave (SPW) being produced via coupling in the grating. The grating boosts the incident wave through the grating moment vector, which is expressed by [5]:

$$k_{ph} = \kappa_x + m\kappa_g \tag{2}$$

where κ_x denotes the x component of the surface wave vector, m are orders of diffraction (positive or negative integers), $\kappa_g = 2\pi/\Lambda$ is the vector of the diffraction grating and Λ the periodicity. Once the resonance conditions are satisfied, vector of the wave of plasmons k_{sp} is coupled to the vector k_{ph} and both propagates in the same frequency, thus $k_{ph} = k_{sp}$.

In this work the construction of a SPR sensor system based on the periodic grating was approached. Simulations and practical experiments were performed in order to determine the sensitivity and limit of detection for the device built.

2. Methodology

An attempt is made for the first time to characterize the grating mode under spectrally broadband SPR excitation conditions. The used mode was Grating Mode-Wavelength Interrogation Mode (GM-WIM). The device was fabricated in an optical diffraction grating, composed of a polymer and a metallic thin film, and has been used how study tool. Figure 4, shows the grating profile; the device comprises a 50 nm thin sputter deposited gold film, exhibits a maximum grating amplitude of 50 nm, and a periodicity of nearly 500 nm.

Simulations of the scenario were performed in the COMSOL software as well as practical experiments, which validate the results obtained from this tool. The software tool allows to study this type of SPR simulation for different types gratings, since it is based on the direct solution of electromagnetic field equations. Due the waveform presented of the grating hasn't sinusoidal profile, shown in Figure 4, a Fourier series expansion was performed and thus the first harmonic was used for simulation, as shown in Figure 2. In the simulation the structure was built with only one period of the grating, however, with periodic configuration. The piece has a multilayer structure containing the substrate and the metal grating. The upper region of the grating consists of the dielectric, which used the air, since the incidence occurs directly on the metal surface. The incident light beam was configured with P polarization, whose electric field is parallel to the plane of incidence. The structure used for simulation was created following the parameters of the Table 1. Computation results for propagation of the electromagnetic wave in sinusoidal corrugated surface provides the ratio (Ip/Is) between electromagnetic fields of the reflected wave.

Table 1. Parameters used in the computational study.

Parameter	Value
Periodicity (Λ)	478 (nm)
Height of the structure	500 (nm)
Width of the structure	478 (nm)
Height of the metallic layer	50 (nm)
IR of air	1
IR of substrate (BK7)	1.52
Incident angle	$\theta_{in}(°)$
Wavelength	λ
Frequency	$3 \times 10^8/\lambda$

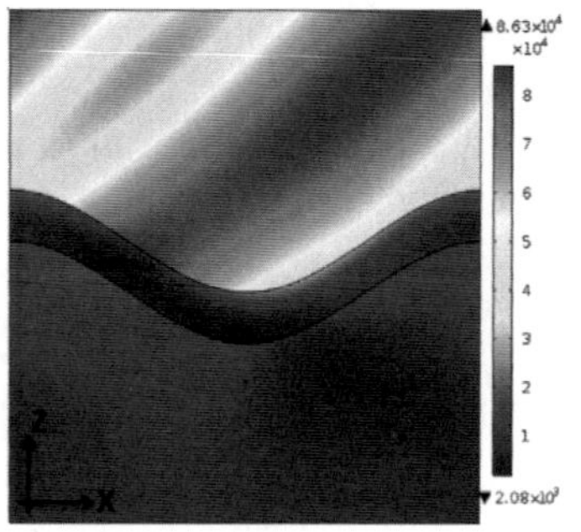

Figure 2. Simulated structure for GM-WIM conditions. The waveform of the grating is reproduced with the first harmonic resulting from the Fourier series expansion. The bluish region shows the deposited metal grid and the substrate. The variation of the electric field strength for a p-polarized incident light beam, given in V/m, is seen in the colored region.

The experimental device (Figure 3) was made in the laboratory with the deposition of a metallic layer of 50 nm, through the process of spputering. The metal chosen to constitute the metal grating was gold, which according to [9], due to its excellent characteristics for excitation of the plasmons $\varepsilon(\lambda)$ is the major metallic materials used in SPR applications.

Figure 3. Device used in the experiments. The device has two regions with similar gratings, which can be used for light incidence.

A photographic image of the optical set-up is shown in Figure 5. It comprises a non-polarized white light, a collimating lens, a polarizer and a compact optical spectrometer. The optical sourse used in the experiment is a stabilized tungsten-halogen light source providing a constant-intensity, 10 mW blackbody (blackbody spectrum spans both in the visible and near-infrared regions) radiation spectrum between 360 and 2600 nm. In both situations the incident angle of the light was of 40° and ambient temperature of 23°.The spectrometer captures the intensity of light reflected by the grating at two different times, being the reference curve with the light s polarized (Is) and the curve resulting from the excitation of the plasmos p polarized (Ip) light. The spectral position of the resonance is extracted from the reflectance ratio Ip/Is of the s- and p-polarized components.

The software used is a tool created to process the data obtained from the spectrometer, whose capture occurs through the USB connection. Once all components are correctly positioned, the software is initialized to plot the SPR curves and the sensorgram, making a pixel $\times$ λ relation. Some parameters such as curve smoothing, integration time and minimum detector algorithm can be changed from a graphical interface, to obtain a stable signal. The SPR curve is generated from the ratio between signals captured with different polarizations, and the sensorgram determines the change in the minimum of this curve as a function of the refractive index, with reference initially to the index of refraction of air.

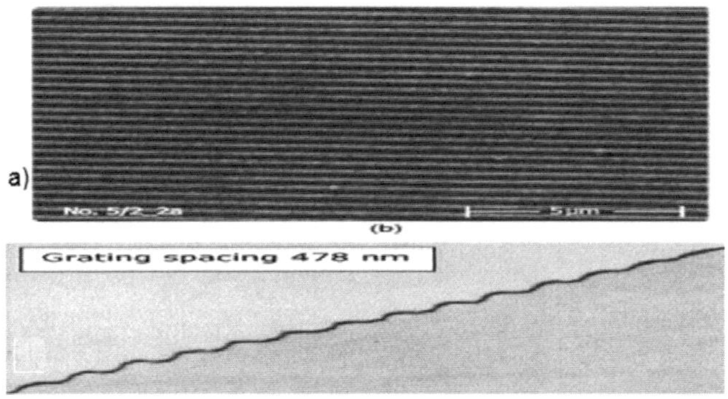

Figure 4. SEM and TEM images of the corrugated surface

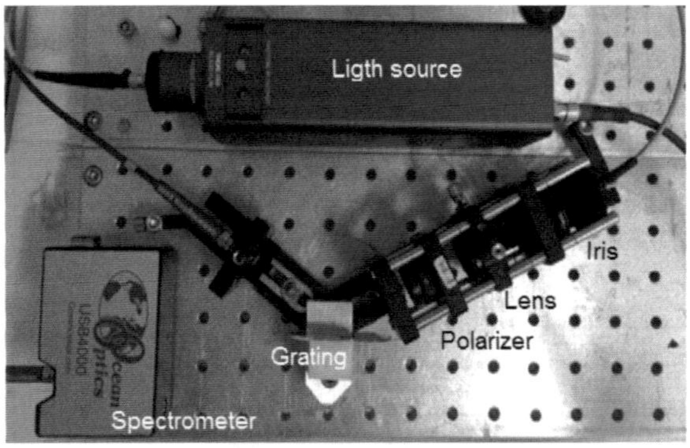

Figure 5. SEM and TEM images of the corrugated surface

IOP Conf. Series: Journal of Physics: Conf. Series **1044** (2018) 012062 doi:10.1088/1742-6596/1044/1/012062

3. Results

Experimentally recorded Ip/Is (p/s) intensity ratio as function of wavelength, insets show p and s-polarized spectra. Theoretically calculated p/s intensity ratio, as function of wavelength in for an incident angle of 40°. Figures 4 and 5 exhibit a grating resonance in the GM-WIM for $m = 0$ along with the associated responsivities given in the inset, numerically obtained values for $S_{\lambda}^{gm} = \frac{d\lambda_R}{dn_A}$ (λ_R and n_A denote the resonance wavelength and the refractive index of the analyte, respectively) are marginally smaller than those calculated.

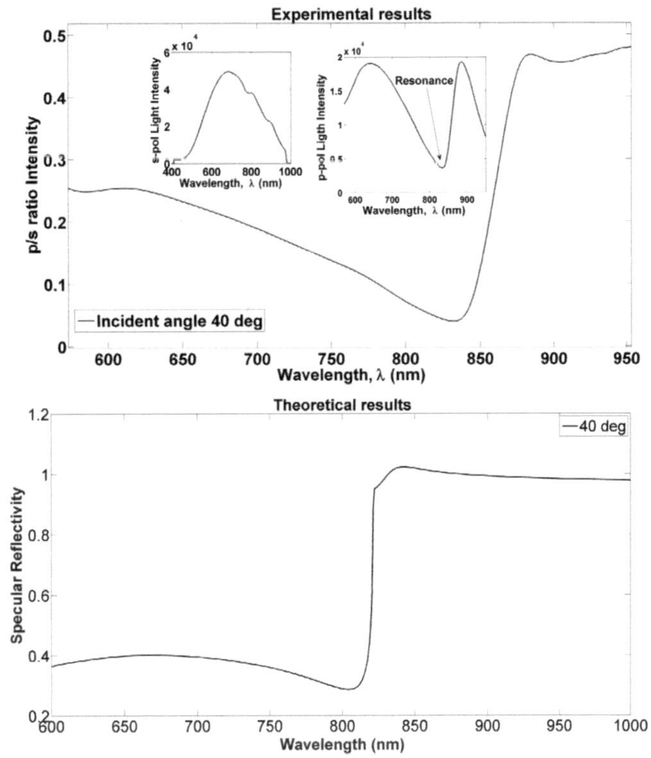

Figure 6. Experimental and calculated results as function of wavelength for an incident angle of 40°.

The temporal evolution and shift of the resonance minimum, generated upon exposure of the grating surface to a flash of condensing exhaled air, taken from a minimum hunt algorithm and processed in real time, is shown in the WIM-sensorgram of Figure 7. The flow of exhaled air at 38±2°C condenses on the Gold surface at 23°C to a nanometer thin, quickly desorbing multi-layer of water molecules. Its maximum thickness was independently verified by fast laser ellipsometry (Sentec GmbH) to 110±10 nm. The digitally processed SP-output signal is a change of the refractive index (RIU). This quantity is proportional to the film thickness [7]. To extract the SP-sensor responsivity S_{λ}^{gm} in the gas-phase, the RIU-value of a known protein film with known thickness and refractive index [8] is re-scaled to that of a condensed water film with known thickness on the gold layer, comprising a refractive index of 1.33.

The sensorgram, displaying the variation of the resonance wavelength as function of time to adsorption from the gas phase, is illustrated in Figure 7. At $t = 20$ s, $t = 48$ s and $t = 79$ s (see Figure 7), the grating surface was briefly exposed to exhaled water vapor. The data set illustrates the effect of rapid vapor condensation onto and desorption from its surface, comprising a rather short time span of 3-5 s. The maximum achievable wavelength change was 2.2 nm. After a fast

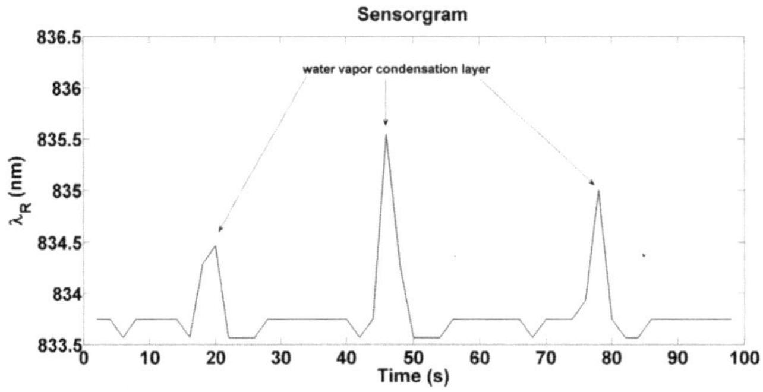

Figure 7. Experimentally obtained sensorgram for GM-WIM conditions.

initial vapor condensation, water molecules are desorbing at a similar rate [8]. The identical gas adsorption experiment has been performed on the same gold film, albeit analyzing the reflected beam by means of a laser ellipsometer, comprising a response time of 2 s. The responsivity S_{λ}^{gm} was estimated to 87 nm/RIU. The limit of detection as taken from the sensorgram is 13.2×10^{-3} RIU. It is important to point out that such small responsivity value results from the thin film of condensed exhaled air, which would increase by a factor 4-5 in case of a homogeneous adjacent analyte phase and thickness of several hundred nm.

Conclusions

This work shows the design of an SPR sensor system based on a periodic structure to detect humans gases exhaled. A minimal difference is found among the methods used, however, it can be justified by two factors: the deposition is not perfectly linear, as it is considered in the software, and the smoothing of the profile obtained through the Fourier series first harmonic. Therefore, it is possible to affirm that there are an agreement between experimental and theoretical results confirms the correctness of the methodology and the feasibility of the proposed sensing system.

References

[1] Abbas, A R; Linman, M.J. and Cheng, Q; *New trends in instrumental design for surface plasmon resonance-based biosensors.* Biosensors and Bioelectronics, 26 p. 1815: 2011.

[2] Tong, L; Wei, H; Zhan and S; Xu, H; *Recent advances in plasmonic sensors.* Sensors, 14 p. 79597973: 2014.

[3] Homola, J; *Surface plasmon resonance sensors for detection of chemical and biological species.* Chem. Rev., 108, p. 462, 2008.

[4] Otto, A; *Excitation of nonradiative surface plasma waves in silver by the method of frustrated total reflection.* Z. F. Physik, 216 p. 398: 1968.

[5] Fontana, E; *Theoretical and experimental study of the surface plasmon resonance effect on a recordable compact disk.* Applied Optics, 43 No. 1: 2004.

[6] Kretschmann, E and Raether, H; *Radiative Decay of Non Radiative Surface Plasmons Excited by Light.* Z. Naturforsch. A, 23 p.2135-2136: 1968.

[7] Thirstrup, C; Zong, W; Borre, M; Neff, H; Pedersen, H C and Holzhueter, G; *Diffractive optical coupling element for surface plasmon resonance sensors.* Sensors and Actuators B: Chemical, 100 n. 3, p. 298:2004.

[8] Neff, H; Laborde, H M and Lima, A M N; *Periodic protein adsorption at the gold/biotin aqueous solution interface: evidence of kinetics with time delay,* Scientific Reports, 6 2016.

[9] Oliveira, L C; de Sousa, T A T; Loureiro, F C C; Melcher, E U K; Lima, A M N and Neff, H; *Sensitivity of an optical sensor that operates at angular or wavelength interrogation.* I2MTC Proceedings,1 p. 14761481:2013.

IOP Conf. Series: Journal of Physics: Conf. Series **1044** (2018) 012063 doi:10.1088/1742-6596/1044/1/012063

Taking degrees of freedom from uncertainty into minimum weight estimate for analytical balances

F L Cacais[1,2], J U Delgado[1] and V M Loayza[2]
[1] Laboratório Nacional de Metrologia das Radiações Ionizantes / Instituto de Radioproteção e Dosimetria – LNMRI/IRD – Brazil
[2] Instituto Nacional de Metrologia, Qualidade e Tecnologia – Inmetro – Brazil

E-mail: facacais@gmail.com

Abstract. There is a great need to weigh very small mass samples in analytical balances due to cost reduction and sustainability requirements so weighing results should accomplish the accuracy requirements in selecting and quality checking tests. The minimum weight is an important balance performance requirement for small mass weighing and is defined as the load for which the relative expanded uncertainty is equal to the accuracy requirement. When "uncertainty" of uncertainty of repeatability is regarded in minimum weight determination, the lack of information about degrees of freedom can cause an underestimated value. Thus, here is shown a modified equation for minimum weight calculation which regards the uncertainty of repeatability uncertainty to take account of the degrees of freedom. Besides, it is developed a simple approximated equation for minimum weight calculation.

1. Introduction

Analytical electronic balances are widespread weighing instruments in fields of productive processes [1, 2]. Nowadays, there is a great need of weighing very small mass samples in analytical balances due to cost reduction and sustainability requirements [3] which impose stringent quality requirements to the process. Therefore, weighing results should accomplish the accuracy requirements stated according to the intended quality for the final processes results. The required weighing accuracies commonly are defined as a relative one specified as upper thresholds for relative expanded uncertainty, which relies on the balance properties such as repeatability, eccentricity, nonlinearity and sensitivity [4]. Thus, to achieve quality targets, also balance properties should be subject to performance requirements which should be assessed before selection of the weighing instrument and periodically checked when it is in use [5]. Measurement uncertainty evaluation before balance selection could be performed by assessing properties specification from manufacturer [6] and in use, should be performed taking account: calibration [7], testing of properties results [8] and others effects on weighing [9].

In small mass weighing, an important user's performance requirement is the minimum weight defined as the load for which the relative expanded uncertainty is equal to the accuracy requirement [10]. In minimum weight evaluation the uncertainty dominant component is repeatability [11] so the expanded uncertainty is calculated by the multiplication of repeatability uncertainty and the coverage factor obtained by the effective degrees of freedom (repeatability degrees of freedom) from Student's distribution [12]. By using this method to calculate minimum weight, one does not regard the measurement uncertainty as a measurement quantity with uncertainty and this assumption is mainly important when repeatability uncertainty is derived from a few repetitions (i.e., few degrees of freedom) as in the case of degrees of freedom obtained from weighing instruments regulations or from performance checks.

This work proposes to complement the minimum weight calculation by taking account of the uncertainty of measurement by regarding that the measurement variance (square of uncertainty) is a Chi-squared distributed variable. This is a routine assumption when one supposes that the weighing reading obeys a Gaussian distribution [13].

Moreover, this study shows a modified equation to calculate minimum weight taking account uncertainty to repeatability uncertainty for degrees of freedom in the range from 1 to 1000. This equation is obtained by regarding coverage probabilities of 95.45% and 95% for, respectively, the reading and uncertainty and will be able to help users in tasks of selection and checking weighing instruments.

2. Uncertainty of uncertainty in minimum weight.

The minimum weight is defined as the load $R_{\min}$ for which the relative expanded uncertainty $U(R_{\min})/R_{\min}$ is equal to the accuracy requirement Req, equation (1).

$$\frac{U\left(R_{\min}\right)}{R_{\min}} = Req \tag{1}$$

The expanded uncertainty $U(R_{\min})$ is defined as the multiplication of standard uncertainty u_c and the coverage factor $k_{p,\nu\text{eff}}$, regarding a coverage probability p and effective degrees of freedom ν_{eff} (2).

$$U\left(R_{\min}\right) = k_{p\nu_{\text{eff}}}\, u_c \tag{2}$$

The standard uncertainty may be written in linear form by the product of repeatability uncertainty S_{rp} and a multiplicative term λ, which encompass the other uncertainty components (3). The underlying hypothesis is that the repeatability is non null.

$$u_c = \lambda\, S_{\text{rp}} \tag{3}$$

For weighing uncertainties, degrees of freedom from repeatability ν_{rp} are the main contribution for effective degrees of freedom ν_{eff} [10], because the degrees of freedom of the systematic components are regarded infinite. So, the ν_{eff} calculation by Welch-Satterthwaite formula [12] may be simplified (4).

$$\nu_{\text{eff}} = \lambda^4\, \nu_{\text{rp}} \tag{4}$$

In small mass measurement the only two significative uncertainty components are repeatability and the balance resolution but repeatability uncertainty is the most meaningful [14]. Thus, in calculating the minimum weight, λ is very close to unit.

Thus, minimum weight $R_{\min}$ can be redefined from equation (1- 4) by (5).

$$R_{\min} = \frac{k_{p\nu_{\text{rp}}}\, S_{\text{rp}}}{Req} \tag{5}$$

In a former method to calculate minimum weight, uncertainty of repeatability is regarded as a non measurement quantity (i.e. zero uncertainty variable) what implies S_{rp} is the true value for repeatability uncertainty. On the other hand, S_{rp} value is estimated by the weighing readings which are a measurement quantity, thus S_{rp} also should be regarded an uncertainty variable.

By the assumption that the weighing readings are variables from a normal probability distribution and S_{rp} is estimated by the sample standard deviation the repeatability variance is Chi-squared distributed with ν_{rp} degrees of freedom. By this approach, the repeatability uncertainty is an estimate to weighing reading standard deviation σ, the true repeatability uncertainty value. Although the standard deviation of the weighing readings is not known, one can obtain its range of possible values from the Chi-squared confidence interval, calculated to some coverage probability α from the quantiles Q [13], equation (6).

$$\left(\frac{\nu_{rp}}{Q_{\left(\frac{\alpha}{2},\nu_{rp}\right)}}\right)^{\frac{1}{2}} S_{rp} \leq \sigma \leq \left(\frac{\nu_{rp}}{Q_{\left(1-\frac{\alpha}{2},\nu_{rp}\right)}}\right)^{\frac{1}{2}} S_{rp} \tag{6}$$

In equation (6), the multiplicative terms for the 95% standard confidence interval approaches to 1 as ν_{rp} increases. Thus, for few degrees of freedom the thresholds of the confidence interval deviate significantly from S_{rp}. Regarding that the minimum weight is defined as the true value of repeatability uncertainty, it is necessary to include de upper threshold of the confidence interval in equation (5) in order to assure that the true repeatability uncertainty will be accounted for (7).

$$R_{min} = k_{p,\nu_{rp}} \left(\frac{\nu_{rp}}{Q_{\left(1-\frac{\alpha}{2},\nu_{rp}\right)}}\right)^{\frac{1}{2}} \frac{S_{rp}}{Req} \tag{7}$$

Usually, specifications of manufacturer include the repeatability uncertainty but no information about ν_{rp} is provided. However, for countries where the weighing instruments are subject to type approval according to OIML, the manufacturer's repeatability uncertainty is achieved by ten repetitions [15] implying ν_{rp} value is nine.Thus, minimum weight can be calculated in selection of the balance for some weighing task. On routinely checking tests the value of degrees of freedom are not very different from this. Table 1 shows minimum weight values calculated for ν_{rp} equals nine with and without uncertainty of uncertainty term and also it is calculated for infinite ν_{rp}. In this calculation the taking value for repeatability uncertainty was 0.03 mg a common value specified by manufacturers for balances with maximum capacity of 200 g and resolution 0.01 mg. Moreover, the accuracy requirement *Req* value was taking 0.1% ("gold rule" [16]), the quantiles Q and $k_{p,\nu rp}$ were calculated, respectively for Chi-square and Student's distributions.

Table 1. Minimum weight values (mg).

Parameters: S_{rp} = 0.03 mg , *Req* = 0.1%, p = 95.45%, α = 95%		
R_{min} (ν_{rp} =9, Q = 1.8 k= 2.3)	R_{min} (ν_{rp} =9, k=2.3)	R_{min} (ν_{rp} = ∞, Q =1 k=2)
127	70	60

The minimum weight values show that to taking account uncertainty of uncertainty doubles the value in relation to the others. Indeed, the value of R_{min} could be reduced if was known ν_{rp} and S_{rp} from historical data [17], regarding there was not reproducibility problems [18]. These are the assumption when one use the uncertainty of repeatability from manufacturer's manual, but the information about degrees of freedom is not provided. In checking tests, users could use theirs estimate repeatability uncertainty historical data to reduce minimum weight.

3. Minimum weight approximated equation

A five terms approximated equation has been developed (8) which allows calculating minimum weight directly from degrees of freedom ν_{rp}, repeatability uncertainty S_{rp} and accuracy requirement *Req*. This equation is valid for ν_{rp} range from 1 to 1000, coverage probabilities of 95.45% and 95% for respectively, the reading and uncertainty. The approximation has a maximum absolute relative error of 2%.

$$R_{min} = \left(A_0 + A_1 e^{-\frac{\nu_{rp}}{t_1}} + A_2 e^{-\frac{\nu_{rp}}{t_2}} + A_3 e^{-\frac{\nu_{rp}}{t_3}} + A_4 e^{-\frac{\nu_{rp}}{t_4}} \right) \frac{S_{rp}}{Req} \quad (8)$$

Table 2 shows the values and standard deviation for parameters of equation (8). The approximated equation may have utility when one does not wish to use formulas for calculating quantiles specifying additional terms of probability, for example, when one is not very familiar with uncertainty calculations. Thus this formula could help users in tasks of selection and checking weighing instruments.

Table 2. Values of the parameters for R_{min} approximated equation.

Parameter	Value	Std. deviation
A_0	2.1102	0.0008
A_1	3.32	0.04
t_1	11.2	0.1
A_2	54.0	0.8
t_2	1.46	0.01
A_3	18237	146
t_3	0.2640	0.0006
A_4	0.519	0.006
t_4	144	2

4. Conclusion

The minimum weight is an important performance requirement for measuring small mass sample which should be assessed in selection of analytical balances and in quality checking tests. By regarding uncertainty of repeatability uncertainty, the minimum weight dependence on degrees of freedom of repeatability become explicit and shows the importance of knowing the degrees of freedom for the assessments. In consequence, the lack of information about degrees of freedom of repeatability uncertainty can cause an underestimated minimum weight value in selection and checking tests on analytical balances, therefore the final process cost reduction and sustainability requirements may not be achieved.

Acknowledgments

One of the authors wishes to thank IRD/CNEN and Pronametro/Inmetro for the financial scholarship sponsor.

References

[1] Nater R *et al* 2009 *Dictionary of Weighing Terms - A Guide to the Terminology of Weighing* (Berlin: Springer)

[2] USP 2013 Weighing on an Analytical Balance *General Chapter 1251*: *USP36 – NF31* (Rockville, MD: US Pharmacopeia)
[3] Reichmuth A 2007 Weighing Small Samples on Laboratory Balances *Transverse disciplines in metrology: Proc. Int. Metrology Conf.* (Hoboken: Wiley) pp. 641–56
[4] Fritsch K 2013 GWP® - The Science-Based Global Standard for Efficient Lifecycle Management of Weighing Instruments *NCSLI Measure* **8** 3
[5] Chan C C *et al* 2010 *Practical Approaches to Method Validation and Essential Instrument Qualification* (Hoboken: Wiley)
[6] Reichmuth A 2004 The Uncertainty of Weighing Data Obtained with Electronic Analytical Balances *Microchim. Acta* **148** 3
[7] González A G *et al* 2005 Estimation of the uncertainty of mass measurements from in-house calibrated analytical balances *Accred Qual Assur* **10** pp. 386–91
[8] González A G and Herrador M A 2007 The assessment of electronic balances for accuracy of mass measurements in the analytical laboratory *Accred Qual Assur* **12** pp. 21–9
[9] Malengo A 2014 Buoyancy effects and correlations in calibration and use of electronic balances *Metrologia* **51** pp. 441–51
[10] EURAMET 2015 *Guidelines on the calibration of non-automatic weighing instruments*: *Version 4.0 (11/2015)* (Braunschweig: EURAMET)
[11] Fritsch K 2015 Calibration of Weighing Instruments: Measurement Uncertainty, Minimum Weight and the Safe Weighing Range *NCSLI Measure* **10** 3
[12] JCGM 2008 Evaluation of measurement data - Guide to the expression of uncertainty in measurement (GUM): *JCGM 100:2008* (Sevres: BIPM)
[13] Roussas G 2003 *Introduction to Probability and Statistical Inference* (Burlington: Academic Press)
[14] Reichmuth A 2001 Weighing Accuracy *Estimating Measurement Bias & Uncertainty of a Weighing(V 2.1)* (Greifensee: METTLER TOLEDO)
[15] OIML R-76-1 2006 Non-automatic Weighing Instruments Part 1: Metrological Requirements - Tests (Paris: BIML)
[16] Kowalski T 2001 *Calibration in the Pharmaceutical Laboratory* (Boca Raton, F: CRC Press)
[17] NIST/SEMATECH 2003 *e-Handbook of Statistical Methods* (Gaithersburg, MD: NIST)
[18] Clark J P and Shull H 2001 Reproducibility: a Major Source of Uncertainty in Weighing: Report in conjunction with Contract No.DE-AC09-96SR18500 with the US Department of Energy

IOP Conf. Series: Journal of Physics: Conf. Series **1044** (2018) 012064 doi:10.1088/1742-6596/1044/1/012064

Test equating for measuring system progress in longitudinal surveys of student academic achievement

C Nguyen
P Griffin
M Wu
Melbourne Graduate School of Education
University of Melbourne, Vic 3010 Australia
Email: cuc@unimelb.edu.au

Abstract. In 2001 and 2007 Vietnam conducted its first and second surveys of student achievement at primary school level in two key subject areas: literacy and mathematics. In order to track the progress of grade 5 students at the system level, the 2001 and 2007 tests were linked using common items. This paper presents the process used to link the two literacy tests. Several issues arose in the equating process of the two tests. This paper discusses these issues and the implications for assessment design and analysis to enable reliable and valid measures of student progress in longitudinal large-scale studies.

1. General description of research questions, objectives and theoretical framework

1.1. Introduction
Monitoring student achievement over time provides a means of tracking system progress over time and assessing possible associations between educational policies and student achievement over the designated period of time. Comparison of student achievement between two points in time is important for evaluating the likely impact of pedagogical interventions on student achievement. There are, however, significant challenges in ensuring that the comparisons are valid, given changes to curriculum and testing procedures. This paper reports on the techniques used to equate the tests of literacy achievement from the Vietnam national surveys of student achievement conducted in 2001 and 2007 in order to place the 2007 cohort on the same scale as the 2001 cohort. The 2001 test has been calibrated, item parameters have been published and the scale has been interpreted [1]. The purpose of this study is to link the 2007 test items to the continuum established in 2001 in order to place the 2007 sample on the same scale as the 2001 sample.

1.2. Framework and research questions
Internationally, there are a number of studies that monitor change in student performance over time, such as the Trends in International Mathematics and Science Study, the Program for International Student Achievement, the United States National Assessment of Educational Progress and the Australian National Testing Program. Each of these studies is characterised by specific approaches to testing, linking and equating. To establish the equivalence of tests over a period of years, many studies link tests across time using a subset of common items. In order to make definitive statements about changes in outcomes, however, the relative difficulty of the items used in the tests should be the same.

To monitor test difficulty, the tests need to be equated, or placed on the same measurement scale, so that they can be directly compared.

The use of item response models, and specifically the Rasch (1960) model, provides one way of achieving this. Item response modelling provides estimates of the difficulty of the items and the abilities of the test-takers on the same measurement scale. This process is called calibration, and is a prerequisite for equating. Three assumptions are required by the Rasch model [2]. First, the construct or variable being measured has a single dimension – that is, only one trait or dimension governs the responses of persons to the items. This refers to the fit of the items when adding a new set of items to the existing set of items. Second, person and item measures can be placed on the same scale. The difference between item difficulties should be stable and free from sample of people who are tested [3].Therefore the common items should not demonstrate differential item functioning (DIF) in relation to test times or versions. And in order to meet these requirements, one further assumption is needed. Each item must be independent of all others, so that each can be used as a pointer or indicator of position on the variable being measured. If this is not done and tests are designed in a way that creates local item dependency, as in assessment instruments that include passages of text and also items embedded in those passages, increased linkage errors can result [4, 5]. Monseur & Berezner (2007) pointed out the need for calculating and including equating errors when reporting the trends of educational achievement over time. Failing to recognise linking errors leads to an underestimation of standard errors and thus increases the Type I error rate, thereby resulting in reporting of significant changes in achievement when in fact the changes are not significant [4]. Large-scale surveys of student achievement often describe a continuum of student abilities and report student achievement in terms of level of proficiency. Therefore the descriptions of the derived scales after equating should be compared to the original scale descriptions as a way to examine if the underlying proficiency continuum was the same after equating. It should also be noted that equating usually requires large samples and depends on the equating method and design [6]. The location of test items in a test may impact item difficulties. The same items when moved from the beginning of a test to the end of a test may change their difficulties [7, 8]. It has also been argued that the link items should be spread out to cover different level of a scale continuum. In addition, the number of items must meet a minimum requirement [9].

The research questions to be addressed in this paper are:

- To what extent do the common items of the two tests, 2001 and 2007 Grade 5 reading tests (to be used for calibration), meet the requirement for 'good link items' under the Rasch model?
- What is the best method for equating the 2001 and 2007 literacy tests?

2. Methodology

2.1. The data and the sample

The data were sourced from two national studies of student achievement in Vietnam. The first study, in 2001, assessed both students and teachers in mathematics and reading. The second study, in 2007, assessed students only. The multistage sampling process that governed the collection of the achievement sample data is described in detail in the first Vietnam study report on the 2001 survey [1]. The 2001 sample include 72,666 Grade 5 students while the 2007 sample included 59,601 Grade 5 students. The data therefore, meet the requirement of large samples for equating design [6].

The two literacy tests were developed using the same test blueprint [1] which meets the assumption and requirement of the Rasch model for test equating purpose [9]. The 2001 literacy test included 60 items while the 2007 literacy test included 40 items. There were 12 common items between the two tests which was considered small compared to the minimum number of items required suggested by Wu et al. (2016) [8]..The 12 items were embedded in the three passages. This is not an ideal situation as it may create an issue of local item dependency which in turn can lead to increased linkage errors [4,5]. The locations of items were not the same across the two tests which may impact on the item difficulties and therefore may lead to item differential functioning (DIF). Given the data characteristics, great attention has been given to identifying a suitable method that enhances reliability and validity of the test equating process for this common item design.

2.2. The analysis

Before conducting test equating, the first step was to examine the extent to which the common items met the requirements of good link items. In accordance with the literature review presented in the previous section, the following criteria were checked:

- The extent to which the items fit the Rasch model: It is required that when the two tests were calibrated separately the link items should have good fits, indicating that the link items are in agreement with other items in measuring a single construct of students' literacy competency. It is also required that the fit measures of the common items, when test equating is conducted, are good.
- The extent to which the link items cover a range of student abilities
- The extent to which the links items behave consistently across time: This means that there is a relative stability of difficulties of the link items across different calibrations.
- Equating error: As the equating error should be included when describing the progress trend from 2001 to 2007, it is expected that the equating error should be relatively small.
- The extent to which the proficiency scale underlying the 2007 test be interpreted in the same way as the 2001 test.
- The extent to which the descriptions of the proficiency levels established in 2001 match the content of the items clustered in the those levels when applying the cut scores used in 2001 to the 2007 continuum.

As a result of the first step, items which did not meet these requirements would normally be removed and a set comprising only link items that had stable difficulties and met the above criteria would be retained. The situation in this study is problematic, as the link items do not meet the requirements of good link items. However, as the number of link items in this study was 12 and did not meet the required number, [8] removing items is not a good solution.

3. Results

3.1. Evaluation of the link items

In order to check if the common items fit well with other items to measure a single domain, the Weighted Mean Squared Residuals (MNSQ) and Weighted t-statistics of the common items in the 2001 and 2007 test calibrations were examined. The MNSQ of the common items in two calibrations were relatively close to 1 which shows that the common items had a good fit. The Weighted t-statistics, however, were statistically significant. When sample sizes are large, t-statistics tend to be significant [8]. In this study the 2001 and 2007 sample sizes were 60000 and 72000 respectively. Therefore the significance of the t-statistics in this case can be ignored. As a result, it can be considered that the common items fit well with other items to measure a single domain.

The difficulties the common items obtained from the separate calibrations of the two tests are presented in Table 1. Column 1 presents the reading passages which are associated with items. Column 2 presents the position of the link items in the 2011 test while column 3 presents the position of the link items in the 2007 test. Column 4 presents the item difficulties of the link items when calibrating 2001 test items alone. Column 5 presents the item difficulties of the link items when calibrating 2007 test items alone. Column 6 presents the adjusted item difficulties of the link items in 2007 onto the 2001 item scale using "a shift" G_0107 which was calculated using the following formula: G_0107=

$\sum_{i}^{k}(d\_i01 - d\_i07)/k$ where k is the number of link items between 2001 and 2007, and d_i01 and d_i07 are the estimated difficulties of the link item i in each test [10]. It can be seen from column 6 that the 2007 item difficulties has adjusted to have the same mean with the 2001 item difficulties but still keep their own standard deviation. Column 7 presents the difference in difficulties of the link items between 2001 and 2007 by subtracting the adjusted 2007 item difficulties from the 2001 item difficulties.

Table 1. The difficulties the common items: free calibration and transformed using shift mean.

Reading passage	Item position in 2001 test	Item position in 2007 test	2001 logit from free calibration	2007 logit from free calibration	2007 adjusted logit using shift mean	Difference of 2001 and 2007 item difficulties (d_i01-d_i07adj)
(1)	(2)	(3)	(4)	(5)	(6)	(7)
A	48	9	1.47	0.57	0.88	0.59
	49	10	-0.1	-0.88	-0.57	0.47
	50	11	0.21	-0.63	-0.32	0.53
	51	12	0.41	-0.27	0.04	0.37
B	6	18	-1.54	-1.38	-1.06	-0.48
	7	19	-0.86	-0.75	-0.43	-0.43
	8	20	-0.26	-0.65	-0.34	0.08
	9	21	0.37	0.20	0.51	-0.14
	10	22	-0.17	0.07	0.38	-0.55
C	28	36	-1.18	-0.84	-0.53	-0.65
	29	37	-0.61	-0.88	-0.56	-0.05
	30	38	0.8	0.23	0.54	0.26
Mean			-0.12	-0.43	-0.12	0.00
SD			0.85	0.58	0.58	0.45

It can be seen from Table 1 that:

- The positions of the common test items in the tests changed between 2001 and 2007. Item positions in the test were found to be associated with a difference in item difficulty. The items become more difficult if they are moved from the beginning of a test to the end of a test. On the other hand items become easier if moved from the end of a test to the beginning of a test. This finding is consistent with what Kingston & Dorans (1984) and Wu et al. (2016) pointed out.
- The set of common items covers a relatively wide range of student abilities.
- The standard division of the item difficulties in 2001 and 2007 are different, providing evidence that item difficulties were not stable across time.
- There are 8 items whose differences in difficulties ranged from 0.37 to 0.59 logit scores. This means that the difficulties of these items are different across two tests and are not qualified as link items for test equating purposes.

There are also other methods of transforming 2007 scale into the same scale of 2001 test calibration: Transform to have the same Mean and SD, using a shift; and Transform to have the same Mean and SD, using regression. Each of these methods identified different subsets of items which were considered as unstable across the two tests.

3.2. Calibration using 12 common items

Given the situation that there was a small number of common items, test equating was conducted with all 12 common items. Four approaches to estimating the link items difficulties were used to equate the two tests:

- A1-Anchoring: the 2001 item difficulties were used to calibrate the 2007 items and estimate 2007 student ability.
- A2-Transform 2007 item difficulties to the same mean and standard deviation of the 2001 item difficulties using a shift mean: In the first stage, the 2007 and 2001 tests were calibrated separately. In the second stage, the item difficulties of 2007 common test items were transformed to have the same mean and standard deviation as the 2001 common items. Other items in the 2007 test were also transformed using the equation for transforming the common items.
- A3-Shift item mean score and no change to standard deviation: In the first stage, the 2007 and 2001 tests were calibrated separately. In the second stage, the item difficulties of 2007 common test items were transformed to have the same mean as the 2001 common items. Other items in the 2007 test were also shifted using the same shift applied for the common items.
- A4-Transform item difficulties to the same mean and standard deviation using regression: In the first stage, the 2007 and 2001 tests were calibrated separately. In the second stage, the item difficulties of the 2007 common test items were transformed to have the same mean and standard deviation as the 2001 test using regression. Other items in the 2007 test were also transformed using the same equation.

The population means which were calculated using the four test equating methods are presented in Table 2.

Table 2. Student ability estimates using different methods – 12 link items.

	Minimum	Maximum	Mean	Std. Deviation
A1_Anchor	29.93	861.99	523.66	98.66
A2_Transform to have the same Mean and SD, using a shift (shift and scale method)	29.02	899.89	542.46	108.11
A3_Shift to the same mean(shift method)	33.84	855.06	518.77	97.24
A4_ Transform to have the same Mean and SD, using regression (scale method)	32.10	879.23	532.05	102.88

It should be noted that the student logit scores have been transformed into the 2001 scale with Mean of 500 and standard deviation of 100.

The linking error was calculated using the following formula

$$\text{Se\_link} = \sqrt{\frac{Var\_parameterdiference}{n}}$$

[4].

where n is the number of common items.

The equating error was also transformed onto the scale of 2001 reading with a mean of 500 and a standard deviation of 100. As a result, the equating error was 10.3 points, which is 0.1 Standard Deviations (SD) of the student estimates. Here is a summary of the findings from test equating:

- The fit statistics of the common items obtained from the equating exercises were not good, showing that as a result of test equating, the common items may be measuring something different from the other items.
- Depending on the methods of test equating, the gain in literacy scores varied and ranged from 0.19 SD to 0.42 SD
- Item ordering and location along a literacy continuum was changed dramatically depending on the equating methods, which put the test validity and continuum interpretation under threat.

The test equating using 12 items did not lead to a reliable and valid measures of item difficulties and student abilities. There is a need to try to eliminate the bad link items.

3.3. Identifying poor links and calibration with a smaller number of better link items

The number of items was too small to consider removing poor link items. However, given the situation, an attempt to identify poor link items to remove was made. It was found that, depending on methods of transforming the 2007 common items onto the same sale of 2001 common items, the items flagged as the most unstable items changed. There should be a range of combination of items demonstrating unstable difficulties across the two tests. Technically, all these combinations should be tried as optional and the results should be compared before an optimal solution is found. For this project, it was then decided to combine the expert judgment and psychometrics analysis to come up with a set of four items for test equating with acknowledgement of limitations. A range of test equating was conducted using only four link items; the population mean scores were relatively stable. The test equating error when using the four link items of was smaller compared that when using 12 items. The interpretation of the continuum as a result of test equating was compatible to the continuum description established in 2001.

4. Discussion and recommendations

The analysis demonstrates that the design of the 2007 test was not well prepared for equating with the previous test, though an attempt to use the same test blue print and to have common items was made. A strong association found between the change in item positions in the test and the change in item difficulties reinforces the notion that the common items should be placed in similar sections in different tests to avoid item difficulty change between tests [8]. There are multiple methods of identifying items which demonstrate instability in difficulties between the two tests. Each method indicated a different subset of items for removing. Each subset of link items used for test equating and each method lead to different population estimates. While a careful evaluation using the criteria presented above was conducted, the solution used to equate the two tests has a number of disadvantages due to an inappropriate design.

References

[1] World Bank. 2004 *Vietnam: reading and mathematics assessment study*. Volume 2. Hanoi: World Bank
[2] Rasch G 1960 *Probabilistic models for some intelligence and attainment tests*. (Copenhagen: Danmarks Paedagogiske Institut).
[3] Wright, B. D. & Masters G. N. 1982 *Rating scale analysis. (*Chicago: Mesa Press).
[4] Monseur, C., & Berezner, A. (2007). The computation of equating errors in international surveys in education. *Journal of Applied Measurement*, 8(3), 323–325.

[5] Monseur, C., Sibberns, H., & Hastedt, D. (2008). Linking errors in trend estimation for international surveys in education. *IERI Monograph Series: Issues and Methodologies in Large-Scale Assessments,* 1, 113–122.

[6] Kolen, M. J., & Brennan, R. L. (2014). *Test equating, Scaling and Linking. Methods and Practices.* Third edition. (New York: Springer-Verlag).

[7] Kingston, Neal M.; Dorans, Neil J.. (1984). *Item location effects and their implications for IRT equating and adaptive testing.* Retrieved from the University of Minnesota Digital Conservancy, http://hdl.handle.net/11299/101880.

[8] Wu, M. Tam H. P. & Jen T. 2016 *Educational Measurment for applied researchers (*Singapore: Springer)

[9] Dorans, N. J., Moses, T., & Eignor, D. *(*2010*). Principles and practices of test score equating (ETS Research Report No. RR-10-29). (Princeton, NJ: ETS).*

[10] Wright, B. D., & Stone, M. H. (1979). *Best test design.* (Chicago: MESA Press)

The coordinate measuring machines, essential tools for quality control of dimensional and geometrical specifications of technical components, in the context of the industry 4.0

F. Ferreira and H. Guerra

CATIM - Technological Centre for the Metal Working Industry, Porto, Portugal.

fernando.ferreira@catim.pt
helder.guerra@catim.pt

Abstract. The industry, as part of the economy, focuses its activity on the production of technical parts which are highly mechanized using sophisticated and automated machines. Over the years, the ability to innovate has proved to be an important factor in industrial development, especially in technological development in line with different requirements of the industry and the efficiency of production processes, reflecting in the quality improvements of the technical components produced. For control of dimensional and geometrical requirements of the technical components, companies use the 3D coordinate measuring machines. These sophisticated measuring machines follow the effectiveness of the latest production systems according to the requirements of industry 4.0, using information technology that promotes quick and easy inspections with large data acquisition, which facilitates the interaction of the entire production process. This paper aims to give an overview of the importance of the measuring processes, using coordinate measuring machines, the large data acquisition of the dimensional and geometrical characteristics, how to export this data to the production machines in order to correct the machine parameters and to obtain technical components in accordance with the technical requirements.

1. Introduction

The globalization process that the world has been experiencing, has contributed to introduce in the world economy a troubled and complex period which has led many countries to face very difficult economic conditions.

Nowadays the industry based on modern and smart production systems has brought new improvements in production management, elimination of waste and increase of the investment in research and development processes.

However, the production processes in industrial environments continue to be affected by several factors. The production processes of individualized products in industrial environments are exposed to several factors that (in a more or less significant way) affect the expected specifications, forcing constant adjustments in the parts and in the production control means. These factors must be considered to not compromise the quality of the final product.

The inevitable introduction of digital technologies into the industry and the need to produce sophisticated technical components has helped innovation processes to reduce development time, introduce new products in the market and improve the management of the short product life cycles.

Industry 4.0, based on new concepts of smart factories where all the machines are connected in a network, allows performing operations in real time, taking as a starting point the data obtained and processed instantly.

In order to control the production process, the support of the metrology is fundamental to ensure the technical specifications of the products.

The metrology, widely recognized as one of the most important areas of knowledge to support innovation processes, presents itself as one of the sciences that has the greatest applicability in scientific and industrial environments, maintaining itself as a strong pillar for scientific and technological development, contributing to the competitiveness of the companies in the world economy, and contributing largely to their sustainability.

2. The support of the 3D metrology to the production processes

Coordinate measuring machines are machines that allow to locate point coordinates on three-dimensional objects. They allow integration of both dimensions and orthogonal relationships.

Normally, these machines are used to eliminate difficult and time-consuming measurements with traditional single-axis devices such as calipers, micrometers and height gages.

Complex parts can be measured quantitatively, and data can be stored for later or immediate use.

The more complex the part to be controlled and measured, the more useful the coordinate measuring machine becomes.

The coordinate measuring machines have taken a lead role in many industries such as aerospace, automotive, electronics, plastics, and semiconductor, and are an equipment that is perfectly in line with the new trends of the industry 4.0.

"In the context of the Industry 4.0 concepts, the importance of measurement technology and its digital integration into production processes becomes increasing significant. Measurement technology plays an essential role in capturing the real world."[1]

The measurement technologies are nowadays required to be more fast, accurate, reliable and flexible. In fact, the coordinate measuring machines are measuring instruments that integrate the mentioned characteristics and can carry out fast measurements.

The integration of coordinate measuring machines into production processes allows for faster measurement results and facilitates a faster delivery of measurement results. The reduction of lead time occur because the displacement time of the parts between the production areas and the control laboratories are eliminated, since robots are used to place the parts on the coordinate measuring machine (figure 2) and the measurement processes are already optimized.

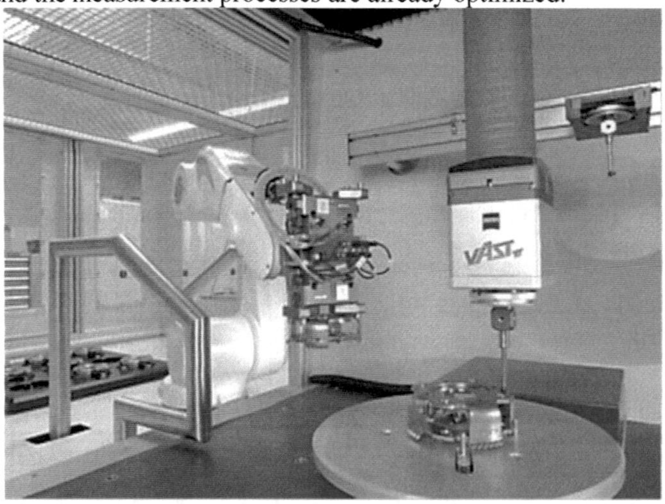

Figure 1. Example for faster metrology due to the automated integration of a coordinate measuring machine into material flow by robot loading. [2]

Through the accredited calibration of the coordinate measuring machines, the accuracy of the measurements as well as their traceability are ensured.

This type of metrology equipment is very reliable, presents measurement uncertainties that cover the accuracy of most of the produced parts and is very flexible as it allows the use of different measurement techniques (figure 2).

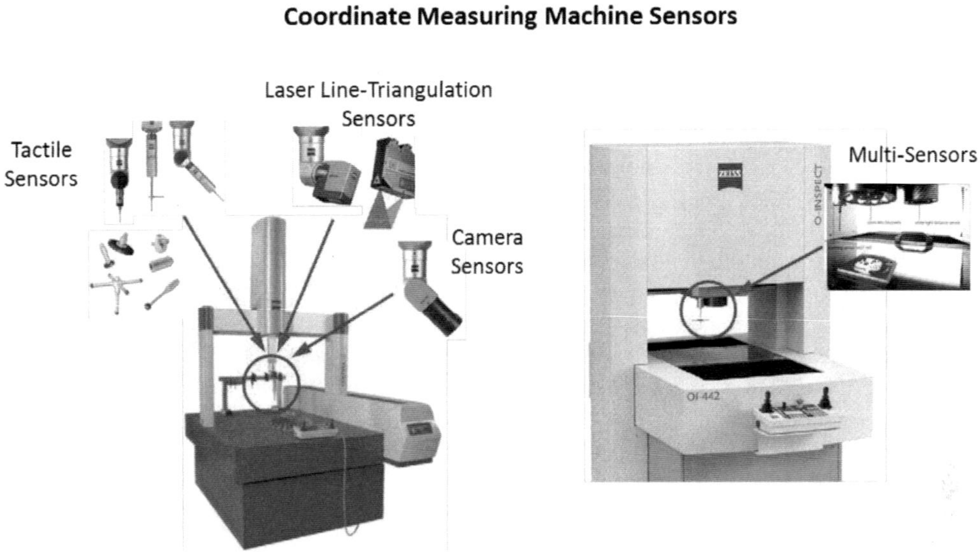

Figure 2. Examples of multi-sensor implementations in coordinate measuring machines.

In the field of dimensional and geometrical metrology, the coordinate measuring machines in industrial environments are an important resource for the quality systems, since they monitor manufacturing processes, contribute to reduce errors during the manufacturing process and contribute to the inspection of technical specifications, which ensure the quality of the final product.

Considering the need to establish increasingly tight controls on the parts produced in order to control the whole process, the coordinate measuring machines are the means used that support the productive processes, in particular the automated processes.

Once project and technical drawing whit all dimensional and geometrical specification of a part is completed, it is sent to the production and using automatic production machine to obtained the real part, which is subsequently sent to dimensional and geometric control in coordinate measuring machines from which a measurement report is obtained that allows validating the part (figure 3).

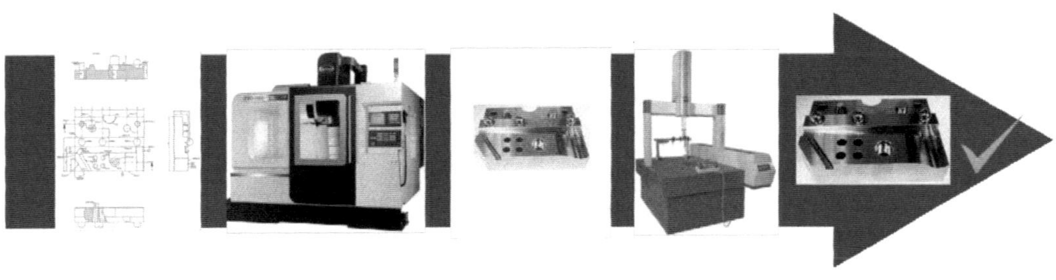

Figure 3. Production and control process.

The control of the technical components, using coordinate measuring machines (refined measuring instruments that work with contact and non-contact probes systems), has become widespread in the most diverse areas of the industry, with special relevance in the demanding automotive, space and aeronautical industries, following the rhythm with the production of automated machines.

3. The data acquisition and the need to establish communication between coordinate measuring machines and production machines in order to correct failures and control the process in an automatic way

Measurement processes, as we understand them today, work offline and in laboratories, being almost exclusively dedicated to the evaluation of the technical characteristics of the manufactured components.

The existing interaction with the coordinate measuring machines and the automatic production machines is still very small and passes through human performance most of the times.

However, the coordinate measuring machines that track the effectiveness of the latest production systems according the requirements of the industry 4.0, the new business paradigm, use information technology that promotes quick and easy inspections with the acquisition of large volume of data, which will facilitate the interaction of the entire production process efficiently, ensuring compliance.

In an expected model of a smart production system all resources must be connected (figure 4). In this model, all dimensional and geometrical specifications of the technical drawing of a part are sending to automatic production machine and the cycle begins.

The technical specifications are inserting in to the software of the automatic production machine and the part is produced. The produced part is placed in to the coordinate measuring machine, especially by automation means (robots), then the measurement process occurs and the measurement report is generated regarding the state of the part, the analysis and processing of this data are directly available in production allowing automatic adjustments in the automatic production machine, if necessary. All generated data is stored on servers for future evaluation and consultation. These data are accessible to all stakeholders.

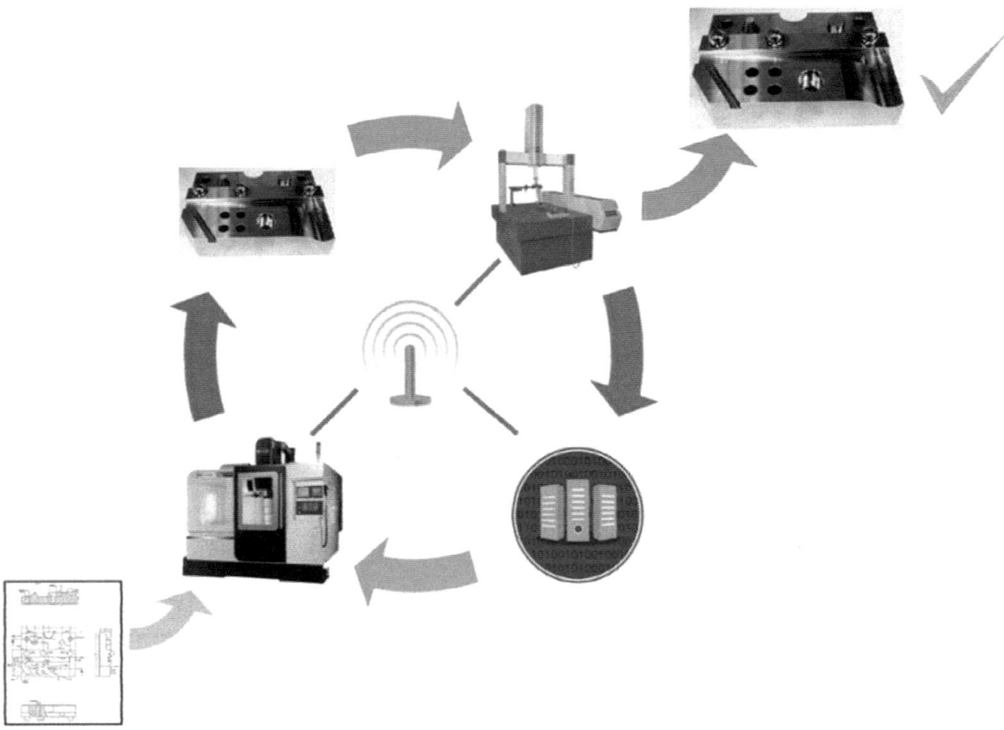

Figure 4. Production and control process in accordance with industry 4.0.

One of the great challenges for the industry 4.0 concept is dealing with the amount of data acquired in a dimensional and geometrical analysis performed on the products, i.e. how to file and process all the information acquired by the measuring instruments, contemplating the associated uncertainties so that this information be disseminated on the network for that all stakeholders have access in real time.

"Measurement technology and the information thereby gained is rather the pacemaker in the concept of industry 4.0, and this is made clear in particular by its role in the cyber-physical production system of linking together the "cyber-world" and the "real world" (Figure 5)"[1]

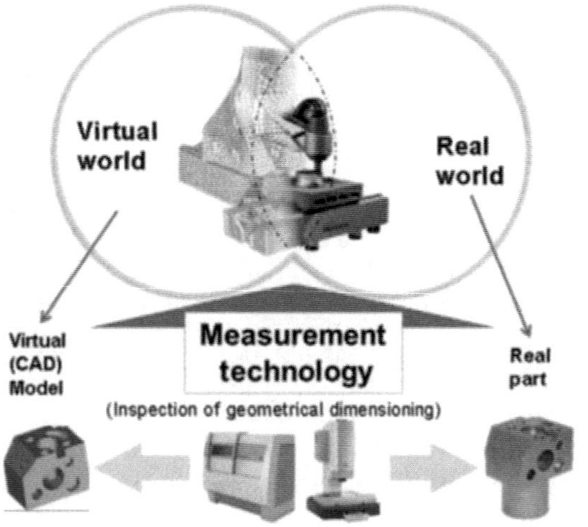

Figure 5. Measurement technology for linking the virtual and real worlds, with an example from coordinate metrology for comparing the nominal (CAD model) and actual product shape.[1]

4. Added value for companies

In the industry 4.0 context, manufacturing intelligence provides that all the information that comes from different sources allows connecting different management systems of quality control of the company with the manufacturing process, therefore improving the analysis and the product.

For the industry 4.0 paradigm, a real-time and smart metrology with online communication between the coordinate measuring machine and the production machine is absolutely necessary.

This connection allows direct measurement in production to become a reality, which means that the costs of measure and produce tend to fall and this trend will lead to the fundamental characteristics of products being controlled 100%, which generates data used in statistical controls.

Since test results and statistical controls are available to all stakeholders in the manufacturing process and to the customer as well, decisions are no longer centralized, which means that change decisions are made by the customers themselves.

It is expected that with the implementation of industry 4.0 concepts in companies, these can have significant increases in productivity and efficiency should be significantly improved.[3]

With increased efficiency and productivity, companies will have a return on defects reduction, optimization of the costs and, above all, the full satisfaction of customers.

5. Conclusions

The coordinate measuring machines are in fact an essential tool for quality control of dimensional and geometrical specifications of technical components in the context of the industry 4.0. This equipment allows to control the production process, measure the production parts, save the large volume data, control the autonomous machines, send large volume data to do simulations and provides the integration of all systems.

This process allows to control the stability and the good performance of the industrial production, facilitates the interaction of the entire production process, optimizing the existing resources in a company and improving the efficiency and sustainability.

References

[1] Dietrich I, Jürgen B, Michael H, Karin K, Eberhard M, Martin P, Robert S, Jochen S and Klaus-Dieter S 2016 *Challenges and trends in manufacturing measurement technology – the "Industrie 4.0" concept*
[2] Dietrich I, Alessandro G and Jürgen B 2014 *CHALLENGES AND TRENDS IN MANUFACTURING METROLOGY – VDI/VDE ROADMAP*
[3] Albert A, Bartosz G, Tobias P, Viktoriia B and Tobias S 2016 *Procedure for defining the system of objectives in the initial phase of an industry 4.0 project focusing on intelligent quality control systems*
[4] Lasi H, Fettke P, Kemper H, Feld T and Hoffmann M 2014 *Bus. Inf. Syst. Eng.* **6** 239-42
[5] Wright I 2017 *Metrology and Quality Assurance in Industry 4.0*
[6] Toni V 2017 *Metrología 4.0 para la Industria 4.0*
[7] David B 2017 *Indústria 4.0 – Aplicação a Sistemas de Manutenção*

IOP Conf. Series: Journal of Physics: Conf. Series **1044** (2018) 012066 doi:10.1088/1742-6596/1044/1/012066

The evaluation of the expert competence in field of metrology difference methods

O Velychko[1] and T Gordiyenko[2]
[1]State Enterprise "Ukrmetrteststandard", 4, Metrologichna Str., 03143, Kyiv, Ukraine
[2]Odesa State Academia of Technical Regulation and Quality, 15, Kowalska Str., 65020, Odesa, Ukraine

E-mail: velychko@ukrcsm.kiev.ua

Abstract. The practical results of evaluation of the expert's competence for a few field of measurement for difference methods are presented. Comparison of results of evaluation of expert competence difference methods is made. Results of evaluated concordance of expert's data for a few fields of measurement are shown. Special software and universal software are used for processing received expert's data.

1. Introduction

The group expert evaluation is widely used in various fields [1-5]. They are intended to resolve problematic issues concerning certain activities to find solutions. For this is expedient to consider the opinions of qualified experts with special skills or knowledge in particular field [6-9]. Considering the practical competence of each expert is involved for group expert evaluation taking into account their objective professional data allows increasing the reliability and accuracy of such group expert evaluation.

The Rasch model or analysis is used to measure latent traits like attitude or ability. This model is a lot simpler and intuitive than many other models, and created from actual data. People can also be placed on the same measurement scale [10,11]. In [12] has been one attempt to link metrological and psychometric (Rasch) characterisation of MAN as a measurement instrument. In [13] a gentle introduction to Rasch measurement models for metrologist is presented. Establishing a link of the got results at the expert's evaluation by means of Rasch model with expert's competence requires further researches with an aim from comparableness with the got results by other evaluation methods.

Authentic knowledge of the real status of measurements of certain physical quantity is very important. The real status of measurements should be considered in two main approaches: traditional approach (calibration, verification and testing of measuring instruments for definition of the metrological characteristics, metrological expertise of methodologies of the measurements, etc.) [14,15], and not traditional approach (expert's evaluation of status of measurements for improvement of metrological work in concrete field, etc.) [16,17]. Second approach can be one of the useful means solution of noted issue. An account the practical competence of each expert for evaluation of status of measurements allows to increase confidence level of expert's evaluation.

2. Practical evaluation of expert competence in field of metrology

In [18] there suggested methodology of evaluation of expert's competence that belong to the sphere of comparative evaluation of level of expert competence in various fields of activities. For implementation of suggested methodology, corresponding criteria are set for the numerical score of expert competence of certain field.

Within the framework of realization of expert's evaluation of the real status of measurements on the specially worked out criteria for evaluation of expert competence: E1 – education; E2 – total work experience; E3 – experience in field of metrology; E4 – experience of expert work in field of metrology; E5 – work status. The quantitative descriptions of expert's competence were appraised by means of special software ("Competence UD 1.1" and "Competence AHP 1.1" [19,20]) for ten fields of measurements.

The view of these software windows for 21 experts for measurement of time and frequency are shown in Figure 1 with evaluated results, for example. 6 experts ("Competence UD 1.1") and 7 experts ("Competence AHP 1.1") from 21 evaluated experts (29 % and 33 % accordingly) have a level below than set (least competent experts) by evaluation with using Pareto principle (80/20) and Laurence curve.

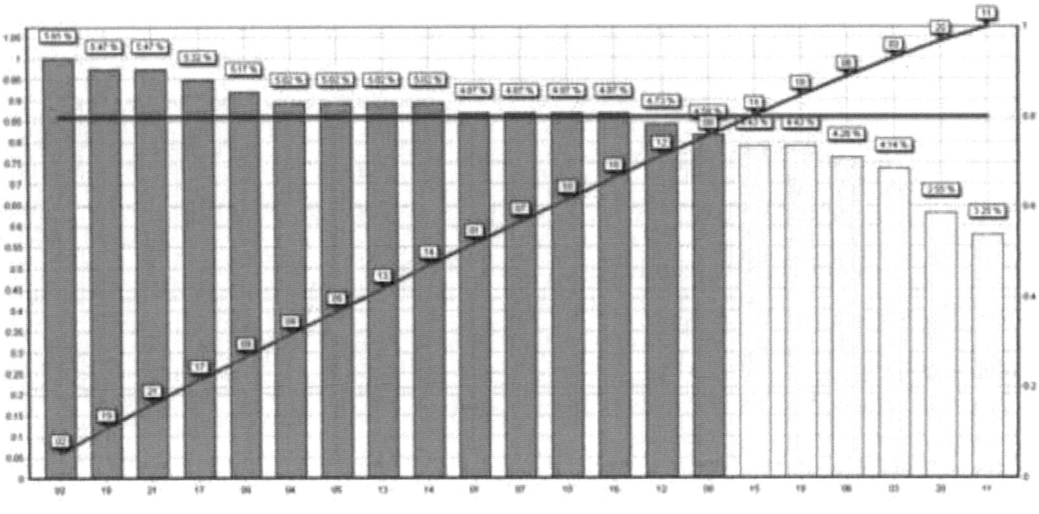

a

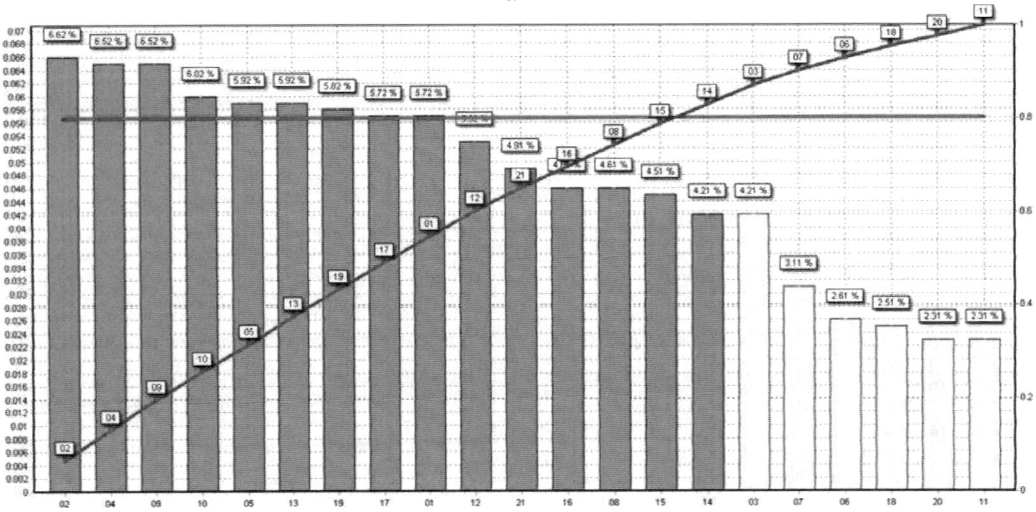

b

Figure 1. Results of evaluation of expert's competence for measurement of time and frequency (*a* – software "Competence UD 1.1"; *b* – software "Competence AHP 1.1").

3. Comparison of results of evaluation of expert competence difference methods

Values of used competence coefficients k_k are in range from 0 – minimum to 1 –maximum. Competence coefficients (k_{UD} – for method evaluation with taking into account uncertainty data (UD), k_{AHP} – for method evaluation which based on Analytic Hierarchy Process (AHP) [15,18]) for all experts as average k_k (k_{UD} or k_{AHP}) and for difference methods (method UD and method AHP) for ten field of measurement are given in Table 1 (AC is alternating current, DC is direct current, U is voltage, k is transformation coefficient).

Table 1. Competence coefficients and variations of expert's competence coefficient for UD and AHP methods for all field of measurements.

	Field of measurement	Number of experts	k_{UD}	k_{AHP}	R_{UD}	R_{AHP}
1	Electrical power	26	0.40	0.15	0.60	0.85
2	Electrical capacity	14	0.64	0.33	0.36	0.67
3	Electrical inductance	14	0.61	0.17	0.39	0.83
4	Phase shift angle	11	0.54	0.19	0.46	0.81
5	Time and frequency	21	0.58	0.30	0.42	0.70
6	Voltage AC	12	0.56	0.20	0.44	0.80
7	High-voltage DC	14	0.54	0.18	0.46	0.82
8	High-voltage AC (U)	15	0.54	0.18	0.46	0.82
9	High-voltage AC (k)	16	0.54	0.17	0.46	0.83
10	High AC current	15	0.54	0.20	0.46	0.80
		Average:	0.54	0.20	0.45	0.79

The variation of competence coefficient for experts was calculated on a formula:

$$R = k_{k\max} - k_{k\min} = 1 - k_{k\min}, \qquad (1)$$

where: $k_{k\max}$ and $k_{k\min}$ are maximal and minimal competence coefficient for experts accordingly. The got variations of competence coefficient R for measurement of time and frequency amount such values: R_{UD} is 0.42 and R_{AHP} is 0.70.

Results of competence evaluation of 21 experts' of difference methods (method UD and method AHP) for measurement of time and frequency are shown on Figure 2, for example.

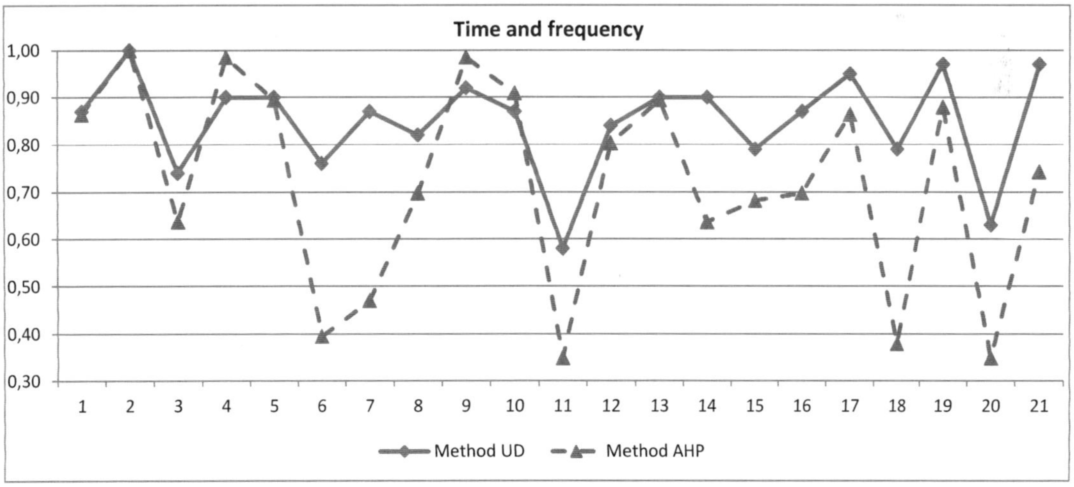

Figure 2. Results of evaluation of the expert's competence of difference methods for time and frequency measurement.

The correlation of the evaluation results, which got UD and AHP methods, is obvious. Results got for method AHP have a more great range of values of competence coefficient k_{AHP}.

The evaluated average grades taking into account the expert's competence for difference methods (method UD and method AHP) and for ten fields of measurement (see Table 1) are presented on Figure 3.

The correlation of the evaluation results got two methods is obvious, but values of competence coefficient for method AHP (k_{UD} from 0.40 to 0.64, average for ten field of measurement is 0.54) are smaller that values of competence coefficient for method UD (k_{AHP} from 0.12 to 0.33, average for ten field of measurement is 0.20).

For the analysis of concordance of the obtained data were used Kendall's coeficient of concordance and Pearson's χ^2 criteria. Results of evaluated concordance of expert's data for ten fields of measurement are presented on Figure 4.

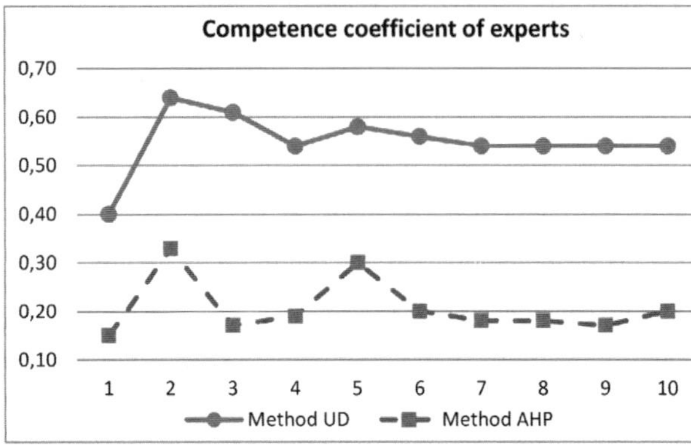

Figure 3. Evaluated average grades taking into account the experts competence for UD and AHP methods.

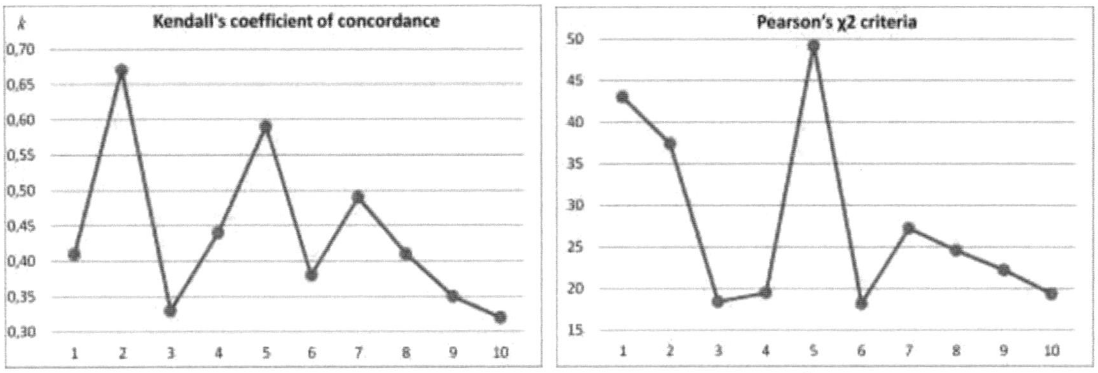

a *b*

Figure 4. Evaluated concordance of experts data
(*a* – Kendall's coeficient of concordance; *b* – Pearson's χ^2 criteria).

The Kendall's coeficient of concordance was calculated on a formula:

$$W = \frac{12S}{M^2\left(N^3 - N\right) - M\sum_{i=1}^{M} T_i}, \tag{2}$$

where:

$$T_i = \sum_{q-1}^{Q}\left(t_q^3 - t_q\right); \tag{3}$$

S is a sum of squares of deviations from middle value;
T_i is common number of the same ranks for i-th expert for all criteria of evaluation of the expert competence;
t_q is number of the same ranks for i-th expert for all criteria of evaluation of the expert competence;
Q is number group of the same ranks for i-th expert for all criteria of evaluation of the expert competence;
q is the same ranks for i-th expert for all criteria of evaluation of the expert competence.

The value of Pearson's χ^2 criteria was calculated on a formula:

$$\chi^2 = \frac{S}{MN(N+1) - \frac{1}{N-1}\sum_{i=1}^{M} T_i}. \quad (4)$$

Values of Kendall's coeficient of concordance changes in limits from 0.32 to 0.68 that corresponds to the "low" (from 0.20 to 0.36), "middle" (from 0.37 to 0.64) and "high" (from 0.64 to 0.80) levels of concordance on a Harrington scale [21,22].

At application of Pearson's χ^2 criteria was checked up for a condition: $\chi^2 > \chi^2_{T(0.05;M-1)}$, where M is number of criteria that are executed for all fields of measurement (from 18.68 to 54.39).

Kendall's coeficient of concordance and Pearson's χ^2 criteria for all field of measurement are given in Table 2.

Table 2. Kendall's coeficient of concordance and Pearson's χ^2 criteria for all field of measurement.

	Field of measurement	Number of experts	W	Level of concordance	χ^2	$\chi^2_{T(0.05;M-1)}$
1	Electrical power	26	0.41	middle	54.39	37.65
2	Electrical capacity	14	0.67	high	22.81	22.36
3	Electrical inductance	14	0.33	low	27.03	22.36
4	Phase shift angle	11	0.44	middle	18.68	18.31
5	Time and frequency	21	0.59	middle	38.36	31.41
6	Voltage AC	12	0.38	middle	24.67	19.68
7	High-voltage DC	14	0.49	middle	33.20	22.36
8	High-voltage AC (U)	15	0.41	middle	34.32	23.69
9	High-voltage AC (k)	16	0.35	low	33.53	25.00
10	High AC current	15	0.32	low	32.22	23.69

4 Conclusion

Different methods of evaluation of expert's competence are expedient to apply as useful instrument for the comparative estimation of expert's competence on the basis of their objective data on the set criteria. This allows more reasonably carrying out the selection of the most competent experts for forming of group from the evaluation of certain problem questions in certain fields of activity. The group expert evaluation with involved experts in field of metrology can be a useful tool to establish the real status for specific measurements. Special and universal software can be used for mathematical processing of obtained expert data. It is expedient to conduct an analysis of concordance of the obtained data with using Kendall's coeficient of concordance and Pearson's χ^2 criteria.

5. References

[1] Saaty T L 1992 *The Hierarchon: A Dictionary of Hierarchies* (Pittsburgh, Pennsylvania: RWS Publications)
[2] Litvak B G 1996 *Expert estimations and making decision* (Moscow: Patent, in Russian)
[3] Drake P R 1998 *Intern. J. of Engineering Educ.* **14 (3)** 191–196
[4] Navneet Bhushan 2004 *Strategic Decision Making: Applying the Analytic Hierarchy Process.* (London: Springer-Verlag)
[5] Grabovetskii B E 2010 *Methods of expert estimations: theory, methodology, directions of the use* (Vinnytsa: VNTU, in Ukrainian)
[6] Chernyshova T Y 2009 *Managements, computing engineering and informatics* **1 (19)** Part 1 168–173 (in Russian)
[7] Kolpakova T A 2011 *Radioelectronics, informatics, management* **1** 40–43 (in Russian)
[8] Kalinina I O, Gozhii O P and Musenko G O 2012 *Scientific works of Bleak Sea State University. Computer technology* **191** 116–123 (in Ukrainian)
[9] Pendrill L 2014 *NCSLI Measure J. Meas. Sci.* **9** No. 4 24–35

[10] Slinde J A and Linn R L 1979 *Appl. Psych. Meas.* **3** No. 4 437–452
[11] Bond T G and Fox C M 2001 *Applying the Rasch model: fundamental measurement in the human sciences* (Mahwah: Lawrence Erlbaum Associates)
[12] Pendrill L and Fisher W P, Jr. 2015 *Measurement* **71** 46-55
[13] Mari L and Wilson M 2013 *Journal of Physics: Conf. Series*. IOP Publication **459** 012002
[14] Velichko O N 1999 *Meas. Techn.* **42** 1109–1115
[15] Velichko O N 2010 *Meas. Techn.* **53** 721–726
[16] Velychko O and Gordiyenko T 2015 Evaluation of competence of the experts in field of metrology and instrumentations. *XXI IMEKO World Congress "Measurement in research and industry"* (Prague, Czech Republic)
[17] Velychko O and Karpenko S 2016 *Journal of Physics: Conf. Series*. IOP Publication **772** 012043
[18] Velychko O M, Gordiyenko T B and Kolomiets L V. 2015 *Metallurgical and Mining Industry* **2** 262–271
[19] Velychko O M, Karpenko S R, Gordiyenko T B and Kolomiets L V. 2015 *Metallurgical and Mining Industry* **7** 401–407
[20] Velychko O M, Gordiyenko T B and Kolomiets L V. 2015 *Metallurgical and Mining Industry* **7** 396–401
[21] Harrington E C 1965 *Industrial Quality Control* **21** No. 10 494–498
[22] Postnikov V M and Spiridonov S B, 2013 *Sci. and Educ.* **8** 395–412 (in Russian)

The features of evaluation technical experts' competence in the field of metrology

O Velychko, S Karpenko and I Karpenko
State Enterprise "Ukrmetrteststandard", 4, Metrologichna Str., 03143, Kyiv, Ukraine

E-mail: velychko@ukrcsm.kiev.ua

Abstract. In the paper the improved method was proposed for evaluation of technical experts' competence in metrology. Comparative analysis of the proposed method which is laid the basis for the software algorithm "Competence TE 1.0" and Borda Count method is applied to the electric power measurements was done for group technical experts. Experimental evaluation of accuracy proposed approach and traditional Borda Count method of voting is done in the field of electricity measurement.

1. Introduction

Expert evaluations are carried out in different areas of life in modern society for solutions of specific problems, taking into account the views of the technical experts (TE). Their essence consists in averaging the opinions of TE on the issues to be considered in various ways. Obtaining reliable estimates should be based on the opinion of qualified TE having special skills or knowledge in a particular field of activity. To do this is necessary correctly carry out the selection of TE who is involved in research and scientific and technical expertise of specific issues [1-4].

For the selection of TE, it is advisable to develop certain criteria for assessing the level of competence of TE. Accounting for the competence of TE should increase the reliability of the TE evaluation.

The most common TE methods include the ranking method, the direct estimation method, the pair comparison method [5-6]. Also widely used methods based on voting theory, such as the simple majority rule, the Borda Count method of voting (BCM), the Kemeny rule [7-9].

In order to objectively evaluation of the quality of independent work in the field of metrology is developed special software "Competence TE 1.0", designed to assess the competence of TE. In the framework of the software "Competence TE 1.0" founded an improved method, that takes into account the uncertainty of the data (UD), analysis of data consistency with the use Kendall coefficient of concordance (KCC) and analysis of consistency data using Pearson correlation coefficient (PCC).

The actual task is the development of software for both the processing of the results of the group expert evaluation itself and for the processing of objective data on TE involved in such an assessment. To implement these tasks, it is necessary to develop special software tools based on various algorithms.

2. Proposed improved method

Appropriate criteria are set for scoring expertise of TE and industry specific calculated to implement the improved method UD [10]: the average score $\overline{x}_i$ for each i-th TE from M TE for all criteria UD N;

the relative average $\bar{x}_{ri}$ for each i-th TE from M TE; the normalized average $\bar{x}_{ni}$ for each i-th TE from M TE; the overall average $\bar{X}$ for all M TE for all evaluation criteria is calculated according to the formula:

$$\bar{X} = \sum_{i=1}^{M} \bar{x}_i \, ; \qquad (1)$$

the relative overall average $\bar{X}_r$ for all M TE is calculated according to:

$$\bar{X}_r = \sum_{i=1}^{M} \bar{x}_{ri} = 1 \, ; \qquad (2)$$

the average normalized score $\bar{X}_n$ for all M TE is calculated according to:

$$\bar{X}_n = \sum_{i=1}^{M} \bar{x}_{ni} \Big/ M \, . \qquad (3)$$

Later parameters characterizing the consistency of the data of the TE are calculated: KCC taking into account the associated ranks. The received values of KCC are analyzed and concluded on the degree of consistency according to the data. If it is necessary, examines the score of correction values for certain criteria UD.

The resulting value of PCC χ^2 for the level of confidence 0.05 is compared with a critical value $\chi^2 > \chi^2_{\mathrm{T}(0{,}05;M-1)}$. In case if the resulting criterion value less tabulated critical value, than are needed consideration of score correction values for specific criteria (Figure 1). TE are ranked in descending the order $\bar{x}_{ni}$.

Selection of TE is based on the Pareto principle with display results of UD on a special chart (Pareto diagram).

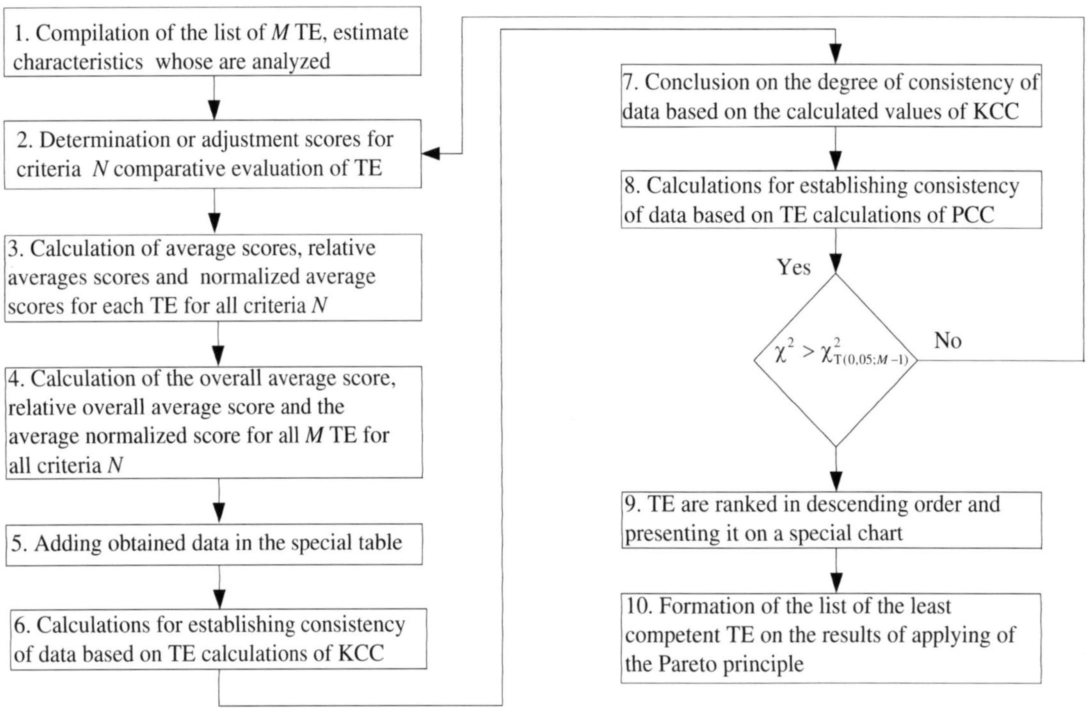

Figure 1. Algorithm of the uncertainty of the data method.

3. The results of evaluation of technical expert competence

3.1 Processing data by Borda Count method

One of the main problem when processing TE information is choosing the best alternative (or candidate). One of the way to determine the winner and rank objects by the degree of manifestation of a certain feature are to assign points to each candidate based on its rank on each ballot. Finding a consensus relationship is a task of voting, in which the set E is a set of candidates, and λ is the set of voters. By methods of the theory of voting, based on the constructed rankings, one can find the only consensus relation, i.e. take such a decision, which will be approved by all participants in the process [10-11].

One of them the BCM also gives a ranking of the candidates. The method is that each TE is assigned a number of points depending on its place in the ranking: 0 points for the last position, 1 point for the penultimate position, etc. to $n - 1$ points for the first position. Points of each TE are summed across all rankings, and its position in ranking of consensus is determined by the received sum. If the total sum of the ranks are denoted as S_i for every alternative a_i, then the best alternative is that which for $S_{ik} \geq S_{i1} \geq S_{i2} \geq \ldots S_{in}$.

For calculation points was used $(n \times n)$ electoral matrix $S = [s_{ij}]$, where:

$$s_{ij} = \sum_{k=1}^{m} b_{ij}^{k}, \quad i, j = 1,2,..,n \tag{4}$$

$b_{ij} = 1$ if $E_i \succ E_j$; $b_{ij} = 0.5$ if $E_i \sim E_j$; $b_{ij} = 0$, if $E_i \prec E_j$.

The sum of points earned for each TE on the ranking BCM in Table 1, is the sum of the corresponding line items of electoral matrix (Figure 2).

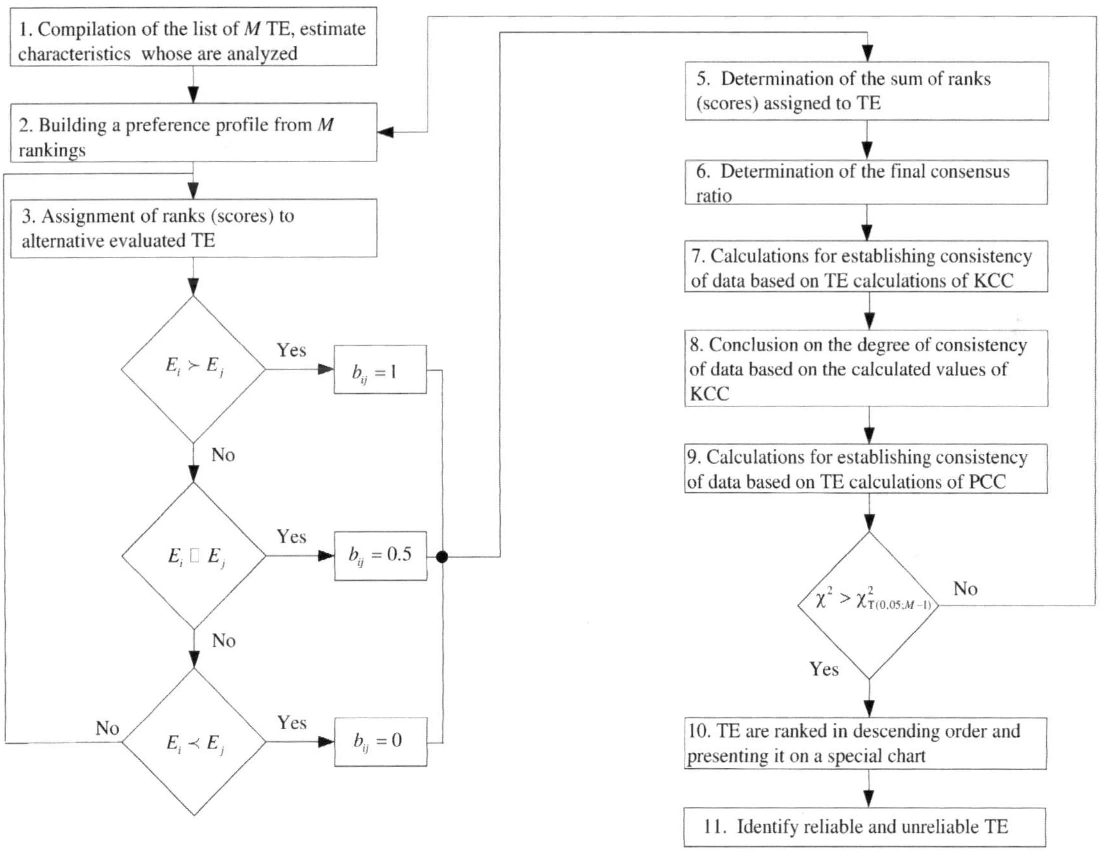

Figure 2. Algorithm of Borda Count method.

Apply BCM for evaluation TE competence. For example, the BCM was used to assess the problem of TE competence.

Preference profile is based on questionnaires developed for the 26 TE:

λ₁: $E_{24} \succ E_1 \sim E_2 \sim E_3 \sim E_4 \sim E_5 \sim E_6 \sim E_7 \sim E_8 \sim E_9 \sim E_{10} \sim E_{11} \sim E_{12} \sim E_{13} \sim E_{14} \sim E_{15} \sim E_{16} \sim E_{17} \sim E_{18} \sim E_{19} \sim E_{20} \sim E_{21} \sim E_{22} \sim E_{23} \sim E_{25} \sim E_{26}$;

λ₂: $E_3 \sim E_7 \sim E_9 \sim E_{19} \sim E_{21} \sim E_{24} \succ E_{10} \sim E_{11} \sim E_{12} \sim E_{13} \sim E_{14} \sim E_{17} \succ E_{18} \sim E_{26} \succ E_6 \sim E_{15} \sim E_{22} \succ E_{16} \sim E_{25} \succ E_{20} \succ E_1 \sim E_2 \sim E_4 \sim E_5 \sim E_8 \sim E_{23}$;

λ₃: $E_3 \sim E_6 \sim E_7 \sim E_9 \sim E_{10} \sim E_{11} \sim E_{12} \sim E_{13} \sim E_{17} \sim E_{18} \sim E_{21} \succ E_{15} \sim E_{19} \sim E_{24} \succ E_8 \sim E_{14} \sim E_{16} \sim E_{22} \sim E_{26} \succ E_{23} \succ E_1 \sim E_2 \sim E_{25} \succ E_4 \sim E_{20} \succ E_5$;

λ₄: $E_3 \sim E_6 \sim E_7 \sim E_{10} \sim E_{11} \sim E_{12} \sim E_{13} \sim E_{14} \sim E_{15} \sim E_{17} \sim E_{18} \sim E_{19} \sim E_{21} \sim E_{22} \sim E_{24} \sim E_{26} \succ E_8 \sim E_9 \sim E_{16} \sim E_{23} \succ E_{25} \succ E_1 \sim E_2 \succ E_4 \sim E_5 \sim E_{20}$;

λ₅: $E_{20} \succ E_{24} \succ E_2 \sim E_4 \sim E_9 \sim E_{10} \sim E_{12} \sim E_{15} \sim E_{18} \sim E_{21} \sim E_{22} \sim E_{23} \sim E_{25} \succ E_1 \sim E_3 \sim E_6 \sim E_{13} \sim E_{14} \sim E_{17} \sim E_{19} \succ E_8 \succ E_7 \sim E_{11} \succ E_5 \sim E_{16} \sim E_{26}$.

Table 1. Competence coefficients for technical experts by using two methods.

Technical expert, E_i	E_1	E_2	E_3	E_4	E_5	E_6	E_7	E_8	E_9	E_{10}	E_{11}	E_{12}	E_{13}
Total rank, S_i	31	40	84	35	17	69	76	36	78	84	70	84	75
Technical expert, E_i	E_{14}	E_{15}	E_{16}	E_{17}	E_{18}	E_{19}	E_{20}	E_{21}	E_{22}	E_{23}	E_{24}	E_{25}	E_{26}
Total rank, S_i	65	70	39	75	81	75	45	90	67	45	100	46	50

The sum of points earned for each TE on the ranking BCM, is the sum of the corresponding line items of electoral matrix. The final ranking of consensus in accordance with the BCM is:

$$\beta_{BM} = \{E_{24} \succ E_{21} \succ E_3 \sim E_{10} \sim E_{12} \sim E_{10} \sim E_{12} \succ E_{18} \succ E_9 \succ E_7 \succ E_{13} \sim E_{17} \sim E_{19} \succ E_{11} \sim E_{15} \succ E_6 \succ E_{22} \succ E_{14} \succ E_{26} \succ E_{25} \succ E_{20} \sim E_{23} \succ E_2 \succ E_{16} \succ E_8 \succ E_4 \succ E_1 \succ E_5\}.$$

3.2 Comparison results of evaluation of technical expert's competence by two methods

For TE competence evaluation involved a group of 26 TE in field of electric power measurements whose competence were evaluated by two methods.

Calculated competence coefficients k_{UD} and k_{BM} are presented in Table 2.

Table 2. Competence coefficients for technical experts by using two methods.

Technical expert	E_1	E_2	E_3	E_4	E_5	E_6	E_7	E_8	E_9	E_{10}	E_{11}	E_{12}	E_{13}
k_{UD}	0.58	0.60	0.88	0.56	0.40	0.86	0.81	0.57	0.88	0.91	0.81	0.91	0.88
k_{BM}	0.31	0.40	0.84	0.35	0.17	0.69	0.76	0.36	0.78	0.84	0.70	0.84	0.75
Technical expert	E_{14}	E_{15}	E_{16}	E_{17}	E_{18}	E_{19}	E_{20}	E_{21}	E_{22}	E_{23}	E_{24}	E_{25}	E_{26}
k_{UD}	0.84	0.86	0.65	0.88	0.81	0.86	0.60	0.91	0.84	0.67	1.00	0.70	0.70
k_{BM}	0.65	0.70	0.39	0.75	0.81	0.75	0.45	0.90	0.67	0.45	1.00	0.46	0.50

Statistical data for 26 TE in field of metrology are shown in Figure 3 for two methods.

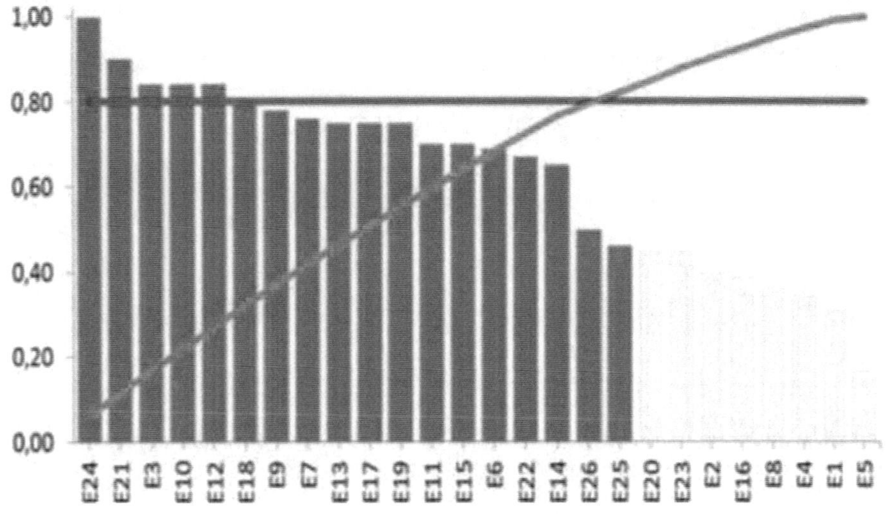

a) Borda Count method.

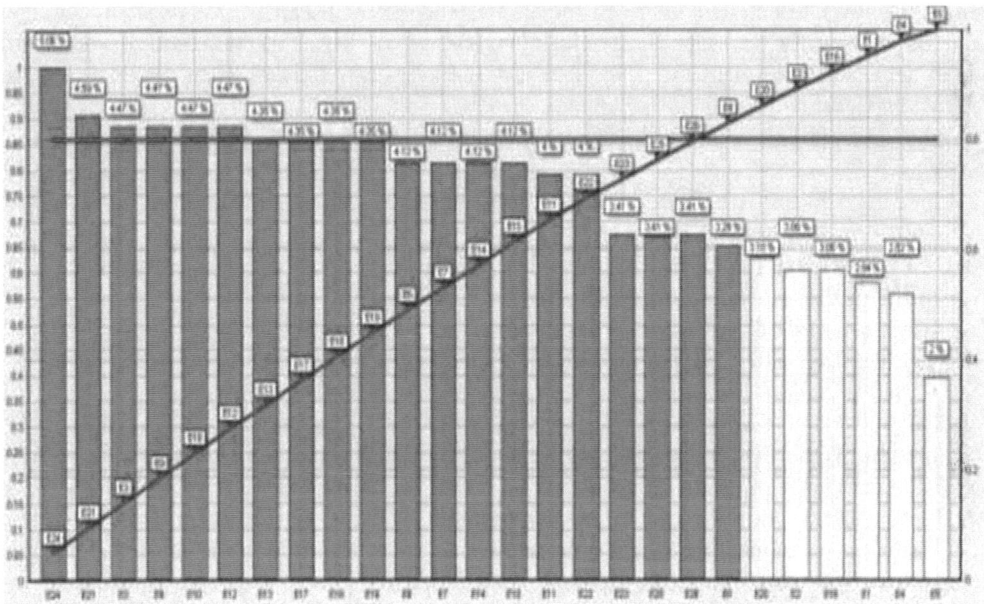

b) method of uncertainty data.

Figure 3. Results of evaluation of technical experts' competence

TE number E_{20}, E_{23}, E_2, E_{16}, E_8, E_4, E_1, E_5 were rated as the least competent and TE number E_{24} as the most competent by BCM;

TE number E_8, E_{20}, E_2, E_{16}, E_1, E_4, E_5 were rated as the least competent and TE number E_{24} is also as the most competent with the help of software "Competence TE 1.0".

Both methods have determined that TE numbers E_1, E_2, E_4, E_5, E_8, E_{16}, E_{20} are not competent. And TE number E_{24} is the most competent. TE number E_5 is the least competent. Competence of TE number E_{23} needs further analysis.

For visualization of the conclusions of the above construct a correlation curve of distribution of competence coefficients for the two methods is presented (Figure 4).

IOP Conf. Series: Journal of Physics: Conf. Series **1044** (2018) 012067 doi:10.1088/1742-6596/1044/1/012067

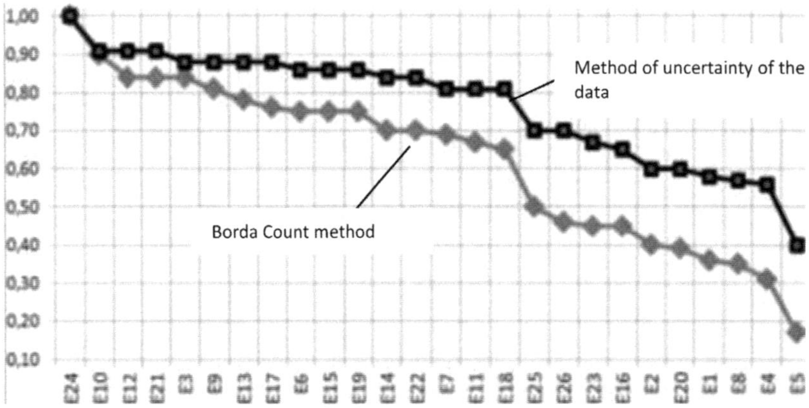

Figure 4. Results of evaluation of the technical expert' of competence by two methods for electric power measurements.

Conclusion

The proposed improved method evaluation of TE competence in the field of electric power measurement allows effectively assesses the competence of the TE in any field of activity. This method is the basis of the developed software "Competence TE 1.0". The comparative analysis made it possible to demonstrate its efficiency. The data ranking helps to efficiently identify, among a group of TE the most and least competent.

References

[1] Grabovetskii B E 2010 Methods of expert estimations: theory, methodology, directions of the use (Vinnytsa, VNTU) **171**

[2] Kalinina I O, Gozhii O P and Musenko G O 2012 Taking into account of competence of experts in the methods of multi-criterion analysis in the tasks of rational choice *Scientific works of Bleak Sea State University. Computer technology* **191** 116–23

[3] Kolpakova T A 2011 Determination of competence of experts at the acceptance of group decisions *Radioelectronics, informatics, management* **1** 40–43

[4] Velychko O and Gordiyenko T 2015 Evaluation of competence of the experts in field of metrology and instrumentations *Proc. 21st IMEKO World Cong. (Prague, Czech Republic, 30 August – 4 September 2015)*

[5] Velychko O M, Gordiyenko T B and Kolomiets L V 2015 Hardware-software complexes for evaluation of competence level of experts *Metallurgical and Mining Industry* **7** 396–401

[6] Velychko O M, Karpenko S R, Gordiyenko T B and Kolomiets L V 2015 Special software for experts evaluation *Metallurgical and Mining Industry* **7** 401–07

[7] Parhami B 1994 Voting algorithms *IEEE Trans. Reliab.* **43** 617–29

[8] Muravyov S V 2014 Aggregation of preferences as a method of solving problems in metrology and measurement technique *Measurement Techniques* **57** 132–38

[9] Muravyov S V, Khudonogova L I and Marinushkina I A 2016 Representation of interval data by weak orders yields robustness of the data fusion outcomes *J. Phys.: Conf. Ser.* **772** 012064

[10] Velychko O M, Gordiyenko T B and Kolomiets L V 2015 Methodologies of expert's competence evaluation and group expert evaluation *Metallurgical and Mining Industry* **2** 262–71

[11] Muravyov S V and Marinushkina I A 2013 Intransitivity in multiple solutions of Kemeny Ranking Problem *J. Phys.: Conf. Ser.* **459** 012006

[12] Berg D E, Norine S, Su F E, Thomas R and Wollan P 2010 Voting in agreeable societies *Amer. Math. Monthly* **117** 27–39

IOP Conf. Series: Journal of Physics: Conf. Series **1044** (2018) 012068 doi:10.1088/1742-6596/1044/1/012068

The impact of time-of-use electricity tariffs for Brazilian residential consumers using smart meter real data

F S Azevedo[1] and R F Calili[1]
[1]Pontifical Catholic University of Rio de Janeiro - PUC-Rio, Posgraduate Programme in Metrology, Marquês de São Vicente Street, 225 - Gávea, 22451-900, Rio de Janeiro - RJ, BRAZIL

Email: flavia.silveira.azevedo@gmail.com

Abstract - In 2011, the Brazilian Electricity Regulatory Agency (ANEEL) approved a new tariff structure for distribution services, the White Tariff, a time-varying tariff modality for retail consumers connected to low-voltage distribution grids. The aim of this paper is to demonstrate existing distortions in White Tariffs established by ANEEL for two Brazilian distribution utilities: Ampla and Coelce. Furthermore, through the analysis of the consumption profiles of residential customers located in these two companies, this paper reveals that those distortions can provide differentiated signals for similar consumers' profile and also distinct revenue impacts for both distribution utilities. The real data came from electronic meters with mass memory. These measurements were carried out through specific devices that provide not only the total consumption but also its breakdown considering the relevant electric equipment existing in each analyzed household. As a result, taking into account the studied samples, 66.67% of Ampla's clients would benefit from the new tariff compared to 98.33% of Coelce's, which would lead to loss revenue of 1.96% for Ampla and 7.73% for Coelce.

1. Introduction

The implementation of Smart Grid can be considered the main instrument of the current modernization in the electric power sector. Smart Grids are able to monitor and manage, using advanced digital meter technology. In this way, it enables active actions among them, sending real time information as consumption, and quality of service.

Smart meters, a technological evolution of electronic meters, are essential smart grid components. Besides providing additional information for utilities and consumers, it presents the possibility to offer different tariff options, such as hourly rates, for example, as the advanced meters can record customers consumption in real time [1].

With the information provided by the smart meter and an appropriate tariff policy, it will be possible to promote change in the electricity consumption behavior - the ultimate concept of demand response, since the consumer can manage their energy use in order to obtain the lowest cost.

This paper aims to show the application impacts of time-of-use tariffs for Brazilian residential consumers using smart meter real data.

2. Time-varying Tariff

Tariff is an effective tool for modifying agent consumption patterns, so an optimized tariff structure is essential to ensure a more efficient system operation. The use of electricity varies over time and the capacity of the equipment which makes up the electrical system is determined by the maximum use of electrical energy per unit of time, even if it occurs only during certain hours of the day.

The peak load demand, which occurs when there is a high concurrent use, imposes risk and cost to the electrical system because it occurs only a few hours of the year [2]. In this way, a consumer that increases its consumption in periods of high load on distribution systems contributes to the system expansion needs. In addition, peak load is usually supplied by fossil fuels and/or energy stored in hydroelectric reservoirs, resulting in increase of emissions and environmental impacts.

An alternative to hire additional infrastructure to maintain a secure balance between generation and demand of electricity is the implementation of Demand Response mechanisms, which can be defined as a change in the electricity utilization by end users in relation to their usual consumption, in response to changes in energy prices that encourage the reduction of use in periods of high energy costs [2]. These mechanisms are feasible only with smart grid and smart meter implementation.

The application of time-varying tariffs that signal to customers the different costs of energy supply contributes, therefore, to a more optimized use of the electric system infrastructure, as it encourages consumers to use the energy outside the peak hours. Among the benefits this type of tariff are increase of generation capacity and network, reduction of the overall costs of the electric system, fairer price in retail market since there is more correspondence between the costs each consumer imputes to the system and the value at which energy is charged. As a result, there are the following consequences: customer energy bill reductions, improvement of customer service (automatic billing, less theft of energy, quality of energy etc.), favoring the expansion of renewable and intermittent distributed generation (such as solar and wind), and possibility to introduce electric vehicles in the market. Besides it also involves possible greenhouse gases (GHG) emissions reductions [3].

Aiming at sending time-varying economic signals to retail electricity consumers and optimizing the use of the distribution network, the Brazilian Electricity Regulatory Agency (ANEEL) approved the creation of a time-varying tariff modality for retail consumers connected to low-voltage distribution grids, the White Tariff, with R$/kWh tariffs that vary according to the time of day [4].

In Brazil, low voltage consumers are charged via supply tariffs, basing exclusively on energy consumption, in R$/kWh. The White Tariff will be an alternative to the existing Conventional Tariff, a flat tariff (no differentiation over different periods within a day) that is currently applied to all consumers [4]. This new tariff modality is conditioned to the substitution of the current lectromechanical metering by smart meter, which the first model was approved by the Brazilian National Institute of Metrology, Quality and Technology (INMETRO) in 2016.

3. Metering

In most of the countries, the classic electromechanical meter is still the main way of measuring energy consumption in residences and small businesses. This traditional electricity meters show consumption in kWh, record consumption cumulatively and are read manually. In some places, monthly the supplier sends a meter reader to each residence in order to read the meter, in others, monthly billing is based on a consumption estimate and a correction, if necessary, is done when the reading is done, therefore with a delay. Tariffs are based on a unit (kWh) rate, independent of time of use.

Electronic meters can have multiple functionalities. Interval meters are electronic meters that have the ability to record electricity consumption on frequent intervals, usually 15 or 30 minute intervals, which are able innovative pricing schemes to retail electricity customers, as more complex time-varying rates, for example. Smart meters can be equipped with communications technology that allows the supplier read the meter remotely.

Communications technology improves functionalities allowing greater interaction between the players in the chain. Two-way communications systems enable innovative services for the customer and options for the supplier as remote connection and disconnection, loss of supply detection and communication to the supplier, ability to interface with load control technology.

The most recent evolution of the smart seter is based on the introduction of bidirectional capabilities from two different points of view: communication and also energy. Therefore, these meters also have the capability to record electricity that is imported from the grid as well as electricity that is exported to the grid, allowing for the measurement of energy production from micro-generators (photovoltaic, for example).

The European Parliament, in the 2012/27/EC directive, defined a smart metering system as "an electronic system that can measure energy consumption, providing more information than a conventional meter, and can transmit and receive data using a form of electronic communication".

The benefits of smart metering can be divided into two main categories:

(i) Operational benefits: The avoided cost of meter reading is one of the most significant operational benefits. Non-technical loss reduction, as losses due to theft, can also be an important benefit as well as better outage detection that can reduce field visits. It can be possible to reduce technical losses with more complete and frequent data from nodes and networks. Furthemore, accurate billing also improve the efficiency of metering services reducing the frequency of customer complaints and savings in re-issuing bills. Moreover, utilities can use the big variety of data generated from measurements to make a variety os forecasts using predictive analysis, which can enabling them to take proactive action [5].

(ii) Demand response benefits: Smart meters are able to offer customers more choice of payments, to provide additional consumption information and to facilitate micro-generation installation. The smart meters can act as a platform for automated forms of demand response by connecting with smart appliances, such as the smart thermostat, to control loads directly. Direct load control and time-varying prices can move consumption from peak to off-peak periods.

However, demand response impacts of smart metering depend on the tariffs offered by suppliers, the number of customers that opt to the new tariffs and/or load control and the availability of customer to respond to new tariffs and change use habits, contributing to energy efficiency of the energy system. In addition, smart meters may remove the cost to customers of installing additional metering equipment to link the microgeneration units to the grid.

Metering service consists of several activities that do not necessarily have to be carried out by a single party: (i) meter provision (supplying metering equipment); (ii) meter operation (installation, operation and maintenance); and (iii) meter reading and data processing. Traditionally, meters have been owned and activities have been undertaken by network operators but in some countries there is competition in metering [5]. And, in the last decade, most countries have committed to adopt changes in technology to enable them to meet targets on sustainability and ensure security of energy supply [6].

In the European Union, the 2006 Energy Services Directive (2006/32/EC) requires Member States to "ensure that, in so far as it is technically possible, financially reasonable and proportionate in relation to the potential energy savings, final customers for electricity, natural gas, district heating and/or cooling and domestic hot water are provided with competitively priced individual meters that accurately reflect the final customer's actual energy consumption and that provide information on actual time of use" [7]. Also, smart metering is one of the most relevant tool in promoting more responsive demand in the context of improving security of supply and reducing CO2 emissions.

It is anticipated that almost 800 million smart meters would have been installed globally by 2020. While Europe leads the penetration rate – growing from 15% in 2010 to 85% in 2020. When the rollout is complete, China would see the largest install base with over 435 million meters followed by the United States with 135 million meters [6].

Although the deployment of smart meters is happening worldwide, utilities are focused on different benefits from them. For North American utilities, for example, the priorities are support outage management and increasing grid reliability. In contrast, European utilities are much more focused on consumer-related capabilities and meeting regulatory mandates [8], including reduction of greenhouse gases (GHG). According to the European Commission, member states have committed to deploy 200 million smart meters by 2020 (Electric Directive 2009/72/EC). This implies that more than 70% of end-users will be covered by smart srid technology. In 2013, the total number of installed smart meters was estimated at 61 million, mainly as the result of large rollouts in Italy, Sweden, Finland, and Denmark. [9]

Starting in 2001, Italy has been the first case of deployment of smart metering systems on large scale in Europe: in 2016 more than 35 million customers have their smart meter installed and working. Now Italy is preparing a new upgrade in the technology of smart meters which will enable greater efficiency in the supply chain and allow customers to benefit of new innovative services. [10]

Advanced meters are the predominant metering technology installed and operational throughout the United States. According to the Energy Information Administration (EIA) 2014 data, 58.5 million advanced meters were operational nationwide out of a total of 144.3 million meters, indicating a 40.6 percent penetration rate [12]. Deployments are projected to reach 70 million smart meters by the end of 2016 and 90 million by 2020. More than 30 electric companies in the United States have fully deployed smart meters. [11]

China has become the world's largest market for smart electricity meters because of several initiatives by the Chinese national government. The installed base of smart meters in China is expected to grow from 139 million units in 2012 to 377 million units by 2020, reaching 74% market penetration. [6]

4. Methodology

The data presented in this paper come from a Research and Development Project (R&D) lead by Pontifical Catholic University of Rio de Janeiro (PUC-Rio) with costumers from two distribution utilities in Brazil: Ampla and Coelce.

Regarding the measurement, two types of electronic meters were used: one that measured the total consumption of homes called SAGA 2000 and the other, Powersave meter, specifically

IOP Conf. Series: Journal of Physics: Conf. Series **1044** (2018) 012068 doi:10.1088/1742-6596/1044/1/012068

designed for the project, consists of intelligent sockets individually installed in household appliances. Both meters have an important property of mass storage in all measurements done.

A sample of residential customers of each utility was extracted, using information provided by them. It was measured the appliances that had a greater representation in the overall household consumption (using Powersave meter). The other devices, including the lighting system, "other uses", were obtained by the difference between the total load curve and the sum of the load curve of the meter, being used 240 measurements (120 of each electricity distribution utility). The meters remained in the homes for nine consecutive days between the months of July and November of 2012 and recorded the energy consumption in intervals of 15 minutes, storing and recording them in their mass memory. These measurements provided information for the daily load curve and monthly electricity consumption.

5. Results

It was observed, in the current situation and considering the analyzed samples, the revenue of Ampla, which has the highest tariff, is 23.27% above that of Coelce, although consumption is only 8.15% higher. Considering a compulsory migration for the White Tariff by all consumers studied, 66.67% of consumers would benefit in Ampla area and 98.33% in Coelce, simply from adopting the new tariff without changing their consumption behaviors and, therefore, without bringing any benefit to the electric system. At the same time the revenues of both distribution utilities would be reduced: Ampla in 1.96% and Coelce in 7.73%. Taking into account only the benefited consumers, as the White Tariff indeed consists of a voluntary, opt-in time-varying tariff modality, Ampla's loss would increase to 2.78% and that of Coelce, 7.81%, increasing the financial imbalance of the distribution companies.

By simulating one company´s market using the other´s tariff, it is possible to conclude that, in addition to the tariff distinction, there is a differentiation, albeit minor, in the way the energy is used throughout the day. When the load curves of the measured equipment consumptions are analyzed, it is observed that the equipment has different weights in the electrical system peak:

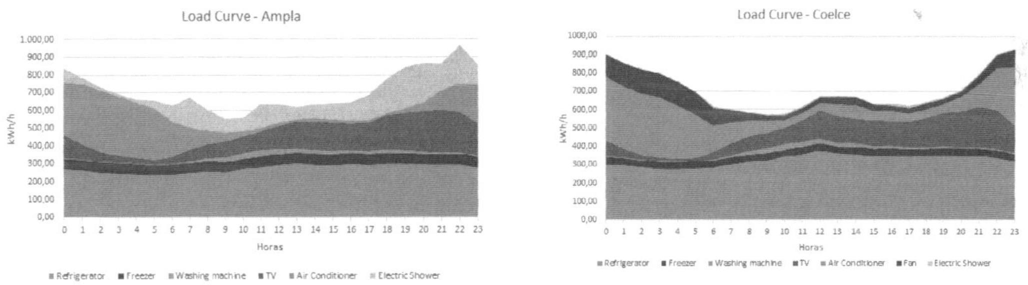

Figure 1: Load Curve registered in Ampla and Colce

6. Conclusion

The expansion of deployment of smart meters is taking place worldwide as one of the most relevant tool in promoting more responsive demand in the context of improving security of supply and reducing CO2 emissions.

Following this tendency, Brazilian Electricity Regulatory Agency (ANEEL) approved the creation of a time-varying tariff modality for retail consumers connected to low-voltage distribution grids, the White Tariff.

However, the study concludes that the migration to the White Tariff, as it is established, would not bring immediate benefits to the electric system, since many consumers would benefit without any change of electricity consumption behavior.

In addition, it was found that the White Tariff did not provide the same incentives to the two distribution companies, compromising the equality criterion of benefits and duties. Therefore, in the way the White Tariff is defined, it would bring greater losses to one company than the other.

Time-varying tariff main objective is to benefit the system with greater security and reliability and also postpone expansion investments. Therefore, it is necessary to consider if a two-part tariff would be more objective in order to send economic signals to retail electricity consumers, since the White Tariff reflects only the energy consumption not taking in account the demand.

7. Acknowledgment

The authors would like to thank the R&D program of the Brazilian Electricity Regulatory Agency (ANEEL) for the financial support (R&D project PDPD-0383-9004/2009).

8. References

[1] Wolak F A 2011 *Do residential customers respond to hourly prices? Evidence from a dynamic pricing experiment*, American Economic Review.

[2] Gyamfi S, Krumdieck S and Urmee T 2013 *Residential peak electricity demand response - Highlights of some behavioral issues*, Renewable and Sustainable Reviews.

[3] Faruqui, A, Hledik R and Palmer J 2012 *Time-Varying and Dynamic Rate Design*, Report for Regulatory Assistance Project.

[4] Ferreira R S, Barroso L A, Lino P R, Valenzuela P and Carvalho M M 2013 *Time-of-use tariffs in Brazil: Design and Implementation Issues*, Innovative Smart Grid Technologies Latin America.

[5] Haney A B, Jamasb T, Pollitt M G 2009 *Smart Metering and Electricity Demand:Technology, Economics and International Experience*, Cambridge University Press

[6] *Smart Meter Security Survey*, 2016, Report by The Global Smart Grid Federation

[7] European Union (2006). *Directive 2006/32/EC of the European Parliament and of the Council of 5 April 2006 on energy end-use efficiency and energy services and repealing* Council Directive 93/76/EEC. Official Journal of the European Union.

[8] *Realizing the Full Potential of Smart Metering.* Accenture's Digitally Enabled Grid Program.

[9] Pérez N, Hernández L, de la Vega D, Angulo I 2016 *State of the Art and Trends Review of Smart Metering in Electricity Grids*, Applied Sciences

[10] Bettenzoli E, Cirillo D, De Min M, Lo Schiavo L, Pitì A 2017 *The Italian Case on Smart Meters in the Electricity Market: a New Wave of Evolution is Ready to Come*, The International Confederation of Energy Regulators Chronicle

[11] Electric Company Smart Meter Deployments: *Foundation for a Smart Grid* 2016 Institute for Electric Innovation Report

[12] *Assessment of Demand Response and Advanced Metering* 2016 Staff Report Federal Energy Regulatory Commission

IOP Conf. Series: Journal of Physics: Conf. Series **1044** (2018) 012069 doi:10.1088/1742-6596/1044/1/012069

The problem of patient-centred outcome measurement in psychiatry: why metrology hasn't mattered and why it should

S P Barbic[1], S J Cano[2] and S Mathias[3]

[1] Faculty of Medicine, Department of Occupational Science and Occupational Therapy, the University of British Columbia, Vancouver, BC V6T 2A1, Canada
[2] Modus Outcomes, Spirella Building, Letchworth Garden City, SG6 4ET, UK
[3] Faculty of Medicine, Department of Psychiatry, the University of British Columbia, Vancouver, BC, V6T 2A1

E-mail: skye.barbic@ubc.ca

Abstract. In psychiatry, there is a call for clinicians to use patient-centred outcome data routinely at the point of care to help tailor treatment plans to meet patient preferences and needs. Given that many decisions in psychiatry are constructed from patient narratives, it is critical that the conceptual, empirical, and measurement structure underlying patient reported outcome measures is robust and patient-centred. Here, we argue for the systematic accruement of patient-centred data in psychiatry to meaningfully enhance the treatment of mental disorders. Specifically, we suggest three crucial considerations for system transformation: (1) the engagement of international patient research partners to conceptualize and prioritize outcomes; (2) the application of modern test theory to develop and evaluate patient-centred outcome measures; and (3) funding allocation accountable to evidence-based services prioritized by patients.

1. Introduction

The treatment of mental disorders is a global health challenge, costing one *trillion* dollars each year [1]. One in four people will experience a mental disorder at some point in their lives and 450 million people currently suffer from mental health challenges [1-3]. As a result, mental illness is the leading cause of ill-health and disability worldwide [2]. To reduce the impact of mental illness, transformative change and service redesign are necessary. In this paper, we propose three crucial considerations for system transformation: (1) the engagement of international patient research partners to conceptualize and prioritize outcomes; (2) the application of modern test theory and metrological standards to develop and evaluate patient-centred outcome measures; and (3) the alignment of funding towards outcome driven services prioritized by patients.

2. The engagement of patient research partners to conceptualize and prioritize outcomes

Patient-oriented research and clinical practice are described collectively as a continuum of research and care that engages patients as partners, focuses on patient-identified priorities, and improves outcomes through the identification of treatment effects; mapping these effects to pathology and guiding clinical care [4]. The "patient engagement" movement was developed to enhance researcher and clinician understanding of the effects of disease and treatment on patients' daily lives. Rheumatology, oncology, and plastic surgery are examples of fields in medicine that have successfully emphasized patient-partners in research in the last decade to improve endpoint outcome measurement [5, 6]. Through their iterative, inclusive consensus processes, outcomes important to patients such as participation, function, and fatigue have become endorsed as targets for treatment and clinical trials in both fields [7]. Systematic approaches to defining patient priorities in psychiatry have yet to be developed. However, psychiatry is not alone. In a recent editorial published in the *British Medical Journal*, Coulter (2017) notes that the behaviour of measuring what matters to patients is "surprisingly rare" in medicine, with a mere 11% of patient-reported outcome measures actually asking patients which outcomes are worth measuring [8].

In response to this trend, national initiatives, such as Canada's Strategy for Patient Oriented Research [4], the Patient Centred Outcome Research Institute [9] in the United States, and the European Patients' Academy on Therapeutic Innovation [10] have been developed to incorporate the perspectives of patients in research and help individuals feel empowered about their healthcare choices. As the demand increases for accountability of clinical mental health services to be person- and family-centred, engagement of patients at all levels of clinical research design are necessary. Yet patient engagement in isolation may not be the solution. Unfortunately, many of the outcomes in psychiatry are not directly observable. This presents systematic challenges for capturing them and characterizing treatment effectiveness. Unlike other fields of medicine, outcomes in psychiatry are predominantly latent and cannot be quantified with imaging or analytical data obtained from tests such as bloodwork. Therefore, practical reasons exist to incorporate patient-reported data to justify clinical decisions, care pathways, and research endeavors. If treatment selection or clinical efficacy and effectiveness trials are to emphasize patient-prioritized outcomes, a robust practical approach to measuring these types of outcomes in psychiatry is needed.

3. The problem with health outcomes measurement in psychiatry

Current approaches to measuring clinical outcomes in psychiatry (i.e., depression, anxiety, neurocognitive deficits) and patient-centred outcomes such as quality of life and functioning, often utilize multiple item scales or questionnaires. With multiple item scales, scores from each item are combined to give a total score. This practice may seem clinically sensible; combining multiple items reduces random error, therefore enhances scale reliability [11]. With large enough sample sizes, estimates to support the psychometric strength of the rating scale can be achieved [11]. Most clinicians invest significant faith in a rating scale that has been deemed 'reliable and valid'. However, understanding the clinical tangibility of the total scores is necessary. For example, what does a score of 15 mean on an instrument purporting to measure depression, which ranges from 0-60? How does that same score of 15 compare to the total scores generated by dozens of other existing depression scales? Interpretation of these scores has been a longstanding challenge in psychiatry. Communicating results to patients, families, and other health professionals can often be difficult and abstract. This limits the potential to systematically convey information about treatment impact and quality assurance.

Another concern for psychiatry and health outcomes measurement is that increasing pressure exists to record patient-reported data electronically in real time. A recent article in the *New England Journal of Medicine* suggests that real time collection of person-reported data can improve patient quality of life, enhance patient-clinician communication, reduce emergency department utilization, and lengthen survival [12]. However, one of the biggest challenges in psychiatry is how to collectively identify patient outcomes that are important to stakeholders, let alone quantify these outcomes robustly in real time. It is unlikely that meaningful quantification of outcomes in psychiatry can occur if key stakeholders do not speak the same language or share the same conceptualization and targets for care. We emphasize that establishing a common measurement language with interpretable units is crucial for linking patient-centred research, clinical, and policy activities. The realization of these units in practice can provide significant value added to current approaches to psychiatric care. This will be in line with a fundamental application of metrology to social sciences that has been well documented for decades:

> *"Such adoption* [of the units] *immediately imposes the duty to obtain the various standards that is, the national prototypes by which units can be translated into practice for application in other fields"* [13] (p.51).

4. Solutions to health outcomes measurement in psychiatry

Fortunately, psychometric approaches, such as Rasch Measurement Theory (RMT) [14], exist as solutions to this problem. RMT has unique properties that are invaluable to fields where outcomes are mostly latent and inferred from the patient [15]. These include references for traceability and means of evaluating measurement uncertainty for both patients and items. As well, RMT methods embrace the opportunity for key stakeholders to work together to hypothesize and experimentally test the composition of a variable of interest and how it can be mapped hierarchically along a clinical continuum. Should the items in a rating scale be unidimensional and represent a harmonized hierarchy, total scores are interpretable and can therefore directly inform treatment. The implications for this are profound. Meaningful scale scores allow patients, health practitioners, administrators, and policy-makers to work collaboratively towards common goals and be accountable to each other. Specifically, the hierarchy of items can inform where a patient is at on the continuum and provide indication for where they are going in either direction along the health continuum). This approach to measurement provides clinicians and patients with a compass and interpretable language, pointing the way forward for innovation and evidence-based care. As a result, the application of metrological standards to psychiatry has significant potential to ensuring greater quality, accountability, and accessibility of what Canada's Strategy for Patient Oriented Research (SPOR) calls:

"the right care at the right time for the right patient"[4].

In the light of pressure for demonstrating outcomes and performance standards in the field, psychiatry is ideally placed to consider the application of metrology for its future. Based on well documented standards form the late 1960s [13], we propose the following broad aims for measurement standards to be considered by psychiatry:

- The development of a scheme to produce outcome measures that have been approved by all stakeholders for which the instrument has been developed, specifically patients.
- The development of the organization necessary for translating and practicing metrological accuracies derived from clinical outcome measures developed in the field.
- The specifications for accuracies required for various contexts of use in mental health care.
- The training of psychiatry and mental health professions in applied metrology to equip clinicians with the skills necessary to select outcome measures that are fit for purpose for their context of use and clinical service intentions.

5. Applied patient-centred outcome funding allocation

The routine use of robust patient-centred outcomes in psychiatry also provides an opportunity to help drive how health care is funded [16-18]. Globally, benchmarking standards are ensuing to promote high performing health care systems that achieve better access, improved quality, and greater efficiency of care. The United Kingdom, Netherlands, Finland are examples of nations that have witness significant changes in health practice and outcomes which can be attributed to a range of policy initiatives and incentives to support and reward high quality outcome driven activity. As an example, the Netherlands have built a nationwide infrastructure to collect of patient-reported data to organize mental health care. Annually, 50% of 1,000,000 treatments performed yearly in the Dutch mental health care have complete outcome data on patients that can be evaluated [19] . This data includes assessment of symptomology and functioning in patients at regular intervals. Feedback is immediately given to clinicians to ensure that progress towards treatment goals is monitored, clinicians are motivated to continue the chosen treatment or investigate alternative treatments in a shared-decision making capacity with patients. The strength of the Dutch approach has been to work collaboratively with patients, clinicians and stakeholders to select outcome domains and measures. However, increasingly the quality of the measures in the field that are commonly used has been critiqued [8]. Opportunity exists to organize data from this benchmarking exemplar for various sections in mental health to iteratively test the extent to which the measures meaningfully capture the outcomes they purport to measure. This would create a common language of performance between patients, clinicians, researchers and key stakeholders. As well, this would allow for the comparison of performance metrics reported by clinical services to evidence-based guidelines, in turn informing funding decisions. This is critical because without a measurement and benchmarking system there is limited ways to:

- Provide a sound basis for communication of treatment decisions and effectiveness between stakeholders;
- Use a commonly accepted measurement approach that can inform health services and treatments, compare individuals over time, and compare systems or services.
- Provide a measurement basis which will permit and encourage the interchange of meaningful scientific and clinical information.
- Development for a standard for how services are funding based on performance and quality of care.

Through benchmarking, there is the potential to learn how well psychiatric interventions perform and align to outcomes that are meaningful to the patient. Benchmarking has been used rigorously by other areas of social science such as Education [20] to monitor progress of stakeholders. Under the assumption that outcomes are derived from rigorous patient-centred conceptual and measurement models, benchmarking can inform the allocation of funding towards the areas of greatest need. This ensures decision makers and psychiatrists are equipped with the evidence needed to develop plans for how to make improvements or adapt best practices to optimize the mental health and wellbeing of patients.

6. Application of patient-centred outcome measurement in transformative change and service redesign: an exemplar in psychiatric program innovation and evaluation

In September 2014, a proposal entitled *Transforming Access to Health and Social Services for Transition-Aged Youth (12-25)* was submitted to the Select Standing Committee (SSC) on Children and Youth in British Columbia, Canada. The proposal called for the creation of a network of health and social service centres across Canada that would provide youth- and family-centred services to young adults with mental health disorders. The proposal resulted in the establishment of the Canadian "Foundry" and funding support for eleven sites. With the goal of implementing timely, evidence-based, and youth and family-centred services across the province, the Foundry team recognized early in 2016 that there was a significant gap in the literature regarding the mental health needs of these key stakeholders. Not surprisingly, the team also identified a lack of measurement tools fit for purpose to measure mental health outcomes for youth.

Early in project conception, the Foundry prioritized (1) patient and family engagement, (2) measurement rigor, and (3) allocation of funding to patient priorities. The model of care addresses a critical research-to-practice continuum gap by:

- Developing a patient-reported data system that enables real time data to be collected to inform services and immediate psychometric validation of the outcomes measures using Rasch measurement methods. The goal of this approach is to produce estimates for items and total scores that are fit for purpose to measure the mental health needs of youth receiving Foundry services, thereby allowing for comparisons between patients, services, and clinical sites.

- Using patient reported data to develop a real time evolving profiles of the mental health needs of youth from both the perspectives of patients and their families.
- Allocating funding to services that reflect the immediate needs of youth and the services needed to support them along the full range of the health continuum.

In this novel initiative, patient-reported information is gathered daily about patient-outcome performance and provides real-time evidence for clinical decision makers and researchers to advance care based on a rigorous understanding of measurement and how it can be used to enhance the health of this population. Psychometric summaries, using Rasch measurement theory, in parallel with regular patient, family, and stakeholders advisory council meetings, ensures quality assurance standards and funding are in place to support patient needs and priorities. The value in this approach lies in the identification and presentation of clinical data that can be successfully applied to clinical decision making and system design.

7. Conclusion
In 2017, many countries are presented with ideal political climates that prioritize patient-oriented and outcome-driven care in medicine. Accurate measurement can increase the quality and impact of patient-oriented research and clinical care. Yet, measurement control in psychiatry poses many challenges as the measurement objects are often not directly observable. Fortunately, promising approaches to address these challenges now exist and are being readily adopted by other areas in the social sciences. It is therefore an excellent time to mobilize metrology infrastructures and intelligence to tackle challenges in measuring patient-reported outcomes in psychiatry. Similar to Canada's Foundry initiative or the Netherlands' use of benchmarking standards, collaboration efforts between clinicians, patients, policy makers, and metrologists has potential to lead to improved quality of care for individuals living with mental illness globally. Reliable, comparable and efficient use of diagnostic and therapeutic methods across all areas of psychiatry can improve health care, limit costs and foster excellence in patient-centred research and innovation.

References

[1] Chisholm D *et al* 2016 *Lancet Psychiatry* **3** 415-24

[2] World Health Organization 2013 *Mental Health Action Plan 2013-2030* (Geneva, Switzerland)

[3] Patel V et al 2016 *Lancet* **387** 1672-1685

[4] Canadian Institutes for Health Research 2016 *Strategy for Patient-Oriented Research* (Ottawa, Canada)

[5] Kirwan J R *et al* 2017 *Value in Health* **20** 481-486.

[6] Cohen W A 2016 *J Plast Reconstr Aesthet Surg* **69** 149-162

[7] Bartlett S J *et al* 2015 *J Rheumatol* **42** 2185-9.

[8] Coulter A 2017 *BMJ* **356** j816.

[9] Selby J V *et al* 2015 *JAMA* **314** 2235-6

[10] Amato A *et al* 2017 *Annali dell'Istituto superiore di sanita* **53** 104-7

[11] Cano S and Hobart J 2011 *Pat. Pref. Adherence* **5** 279-90

[12] Basch E 2017 *N. Engl. J. Med.* **376** 105-8

IOP Conf. Series: Journal of Physics: Conf. Series **1044** (2018) 012069 doi:10.1088/1742-6596/1044/1/012069

[13] White C E 1968 *Measurement standards as a valuable* **13** 51-8

[14] Rasch G 1980 *Probabilistic models for some intelligence and attainment tests* (Copenhagen, Demark)

[15] Barbic S and Cano S 2016 *BJPsych Bull* **40** 243-4

[16] Reeve B 2013 *Qual Life Res* **22** 1889-905

[17] Ahmed S *et al* 2012 *Med Care* **50** 1060-70

[18] Kirwan J R *et al* 2014 *J Rheumatol* **41** 1011-5

[19] de Beurs E *et al* 2016 *Clinical psychology & psychotherapy* **23** 308-18

[20] Tasopoulou K and Tsiotras, G 2017 *Benchmarking: An International Journal* **24** 617-34

IOP Conf. Series: Journal of Physics: Conf. Series **1044** (2018) 012070 doi:10.1088/1742-6596/1044/1/012070

The unreasonable effectiveness of theory based instrument calibration in the natural sciences: What can the social sciences learn?

A J Stenner[1,4], M H Stone[2] and W P Fisher Jr[3]

[1] Chief Scientist, MetaMetrics, Inc., Durham, North Carolina, USA
[2] Professor Emeritus, Department of Psychology, Aurora University, Aurora, Illinois, USA
[3] Research Associate, BEAR Center, Graduate School of Education, University of California, Berkeley, California, USA

E-mail: jstenner@lexile.com

Abstract. In his classic paper entitled "The Unreasonable Effectiveness of Mathematics in the Natural Sciences," Eugene Wigner addresses the question of why the language of Mathematics should prove so remarkably effective in the physical [natural] sciences. He marvels that "the enormous usefulness of mathematics in the natural sciences is something bordering on the mysterious and that there is no rational explanation for it." We have been similarly struck by the outsized benefits that theory based instrument calibrations convey on the natural sciences, in contrast with the almost universal practice in the social sciences of using data to calibrate instrumentation.

1. Introduction

Why is mathematics so remarkably effective in the natural sciences, and what might the social sciences have to learn from the way it is used in those sciences? Is the effectiveness of mathematics in the natural sciences truly "unreasonable," as Wigner [1] put it? Previous research on Maxwell's foundational contributions shows the effectiveness of mathematical model-based reasoning to be rooted in everyday thinking [2], and not in any special capacities associated with scientists or their objects of study. Rasch's [3] adoption of Maxwell's method of analogy [2] set the stage for extending the effectiveness of mathematics into the social sciences [4,5].

In our ongoing explorations of the ways the natural sciences and social sciences invoke, define, and engage in measurement, we have identified a number of differences that are not as epistemologically necessary or predetermined as is popularly imagined [6-14]. We have, to some benefit, contrasted human temperature thermometry (Nextemp™ thermometers; see Appendix A) with the testing of

[4] To whom all correspondence should be addressed.

mathematical ability and the measurement of English language reading ability. Although cataloging these differences has been useful, we now believe they are all traceable to a common cause. Physical science measurement virtually without exception takes place in the context of well-developed substantive theory, experimental evidence, and instruments calibrated to uniform unit standards. In the natural sciences, theories are not just compelling stories about the relationships between measurement outcomes (such as the count of cavities turning black on a Nextemp thermometer), unit standards (degrees Celsius), and measurement mechanisms (a chemical specification equation). They are instead sufficiently elaborated and precise in their specifications that they can be used to calibrate instrumentation.

In contrast, throughout the behavioral and social sciences, instrument calibration depends on data, is typically devoid of theory, and is not traceable to a unit standard. We hypothesize that most of the observed differences between behavioral and physical science measurement are traceable to these foundational differences. The absence of theory is the primary determinant of the need for data-based calibration and the lack of efficient methods for defining units and traceability to them. Further, we offer an example of a theory-referenced reading and text measurement system in the educational sciences that exhibits key theoretical, experimental, and instrumentation features analogous to those of human thermometry. Finally, we review the affordances shared by human thermometry and reading measurement [15].

2. A reading example

A consensus unit and systems for ensuring traceability to it are typical of most natural science measurement. Sometimes, as in temperature measurement, the unification process is not fully completed, but for the vast majority of natural science attributes (referred to as constructs in psychometrics), a unification process has resulted in diverse instrument makers sharing a unit of measure even when the measurement mechanisms vary from manufacturer to manufacturer. Mercury in glass tube thermometers for human temperature measurement differ substantively with Nextemp technology, but produce comparable results. Though the measurement mechanisms are drastically different they both report out in either Fahrenheit or Celsius units. In the case of Nextemp thermometry, a chemical specification equation calibrates the instrument in ^{0}C or ^{0}F. The chemical specification equation derived from experimental evidence enforces the unit, which is embodied in the instrument to a known degree of uncertainty. Similarly, in reading measurement, a text complexity specification equation enforces the unit and ensures that 100L of difference between two readers, two texts or a reader/text encounter is invariant over any of 100+ English reading tests that, at present, employ the unit [14].

Strictly parallel instruments are typical in the natural sciences. Such instruments share a common correspondence table that links a measurement outcome (count of cavities turning black on a Nextemp thermometer) to a ^{0}C or ^{0}F. The ability to manufacture essentially identical instruments in large quantities is a hallmark of natural science measurement. The specification equation is the recipe for manufacturing and calibrating clones of an instrument. The social sciences borrow the concept and talk about 'parallel' instruments or 'alternate forms' and advertise that say, form A and B produce exchangeable measures. But without a specification equation it is impossible to manufacture copies or clones that share the same correspondence table. The reading measurement specification equation can be used to build strictly parallel clones of any reading test (see Appendix B). No such capability exists, for example, for mathematics, and this is so precisely because, at present, there exists no specification equation for mathematical ability that can calibrate mathematics test items (see Appendix C). Different mathematics tests are empirically linked to a common scale through large scale, expensive field studies typically involving thousands of students.

Typical Rasch model applications in the social sciences are singly prescriptive. The major prescription that data must meet is non-intersecting item characteristic curves (ICCs), which relate the probability of a correct response to the difference between person ability and item difficulty. The data are used to estimate person and item parameters with no a priori constraints on the item parameters. Mathematics ability measurement is achieved in this way, as is typical of much social science measurement. Because there is no strong substantive theory for 'mathematical ability,' there is no specification equation and, thus, no potential for theoretically calibrating items/instruments. Instrument calibrations depend on sample data and a property of the Rasch model: when data fit the model *differences* between persons and *differences* between items are independent of items and persons, respectively.

Contrast this singly prescriptive measurement framework with the doubly prescriptive models underlying Nextemp human thermometry and the theoretical framework for reading. In both these cases strong substantive theory coupled with either a Guttman model or a causal Rasch model requires not just data fit to the model but also data fit to the theory specified item/instrument calibrations. For Nextemp a chemical specification equation is used as a recipe for the chemical compound that fills each cavity. By precisely varying the amount of additive the difference between any two adjacent cavities in sensitivity to the green component of light is precisely .2 degrees Fahrenheit. The chemical specification equation enforces this common unit difference for each of the 44 adjacent cavity differences across the 9° F operating range for the instrument.

When data fit a doubly prescriptive Rasch model absolute person measures (not merely differences) are independent of items and instruments and are independent of person sample precisely because no person data figures in the instrument calibration process. Theory calibrated Rasch models are, thus, doubly prescriptive: prescriptive as to Rasch model requirements and prescriptive as to the substantive theory i.e. item/instrument calibrations. Person misfit to a doubly prescriptive model signals that the measurement mechanism that transmits variation in the attribute to the measurement outcome (often a count) is not working as intended for that individual. Frequent failures of theoretical invariance forces reexamination of the substantive theory, the measurement mechanism and instrument calibration procedures. Theoretical invariance can be tested within person over time (e.g. reading ability growth trajectories) and when intra individual theoretical invariance holds across persons then inter-individual theoretical invariance necessarily holds i.e. the attribute is homologous [16-20].

Molenaar [18-20] shows that inferences moving in the reverse direction, inferring from inter-individual factor structures something about intra-individual factor structures, is fraught with complications. The fact that so much of social and psychological measurement is based upon factor analysis of inter-individual variation prompted Molenaar [18,20] to call for a Kuhnian revolution, a paradigm shift in the concepts and methods of measurement in psychology. This paper is intended as another in a series of contributions to this revolution [5-15].

3. Conclusion

Unification of measurement refers to a 200-year-old process whereby dozens if not hundreds of distinct scales for measuring a common attribute are, sometimes quickly and more often slowly, reduced to one, two or three exchangeable units of measure. The history of temperature measurement is a paradigmatic case [21,22] that parallels many contemporary measurement movements in the social and behavioral sciences. Typically, an attribute (construct) captures the imagination of a community of scholars and engineers and different tests, instruments, mechanisms, and scales are proposed for measuring the attribute, and each is uniquely named. Once there is consensus that the selfsame attribute is being measured across these various devices small scale linking studies are undertaken to build conversion tables to re express one unit in one or more other units. More advanced linking studies reduce the link to an equation ^{0}F = ^{0}C * 9/5 +32 making for quick and easy conversions. Since at this stage there is often not much to elevate one scale about the competition the market place takes over and '*unification*', with all its time and cost savings eventually prevails. Sometimes unification is swift and decisive but more often, particularly in the social sciences, metrology is poorly understood and unification plods along.

A useful case study of unification in the social sciences is the longstanding network of reading measures that has linked 100+ English language reading tests across the world, 250,000 book measures and 200 million article measures. The unification process is 27 years old and is accelerating but is far from complete. This effort drew inspiration and strategies from the history of the unification of temperature [13,20,21].

Rather than using factor analysis of inter individual data to define an attribute structure and then asking if this structure obtains when examining intra individual data we suggest the use of substantive theory (in the form of specification / calibration equations) to establish the universality of attribute structure and measurement mechanism at the individual level. Once this is accomplished there is no puzzle about whether between person differences have the same structure as within person differences – of course they do. So, what this analysis reveals is that it is problematic to study between person variation at one point in time to glimpse truths about within person structures over time [8,19]. But the surprise is that if we start with within-person theory-referenced measurement, where in the extreme no two persons have any items in common over 5 years of measurement, then we would not stop for a moment to puzzle about the validity of the claim that at the end of year 1 Jane was higher than Bob but at the end of year 5 Bob was higher than Jane (i.e. a claim about inter-individual variation.) This is yet another benefit of theory based instrument calibration.

Several key features distinguishing physical science and behavioral science measurement systems can be traced to the absence of substantive theory sufficiently developed that said theory can be used to calibrate measurement instruments. Once such a calibration/specification equation is available most of these distinguishing features can quickly and easily imported into the behavioral sciences.

References

[1] Wigner E P 1960 *Comm. Pure Appl. Math.* **13** 1-14

[2] Nersessian N J 2002 *Essays in the History and Philosophy of Science and Mathematics* ed. D Malament (Lasalle, Illinois: Open Court) 129-66

[3] Rasch G 1960 *Probabilistic models for some intelligence and attainment tests* (Reprint, Chicago: University of Chicago Press, 1980) (Copenhagen, Denmark: Danmarks Paedogogiske Institut)

[4] Fisher W P Jr and Stenner A J 2013 *Pacific Rim Objective Measurement Symposium 2012 Conference Proceedings* eds Q Zhang and H Yang (Berlin, Germany: Springer-Verlag) 1-11

[5] Fisher W P Jr 2010 *J Phys.: Conf. Series* **238** 012016

[6] Stenner A J, Smith M 1982 *Perceptual and Motor Skills* **55** 415-26

[7] Stenner A J, Smith M, Burdick D S 1983. *J Ed Meas* **20** 305-16

[8] Williamson G L, Fitzgerald J, Stenner A J 2013 *Educational Researcher* **42** 59-69

[9] Stenner A J, Burdick H, Sanford E E, and Burdick D S 2006 *J Appl Meas* **7** 307-22
[10] Burdick D S, Stone M H and Stenner A J 2006 *Rasch Meas. Trans.* **20** 1059-60
[11] Stenner A J, Stone M 2010 *J App Meas* **11** 244-52
[12] Fisher W P Jr 2009 *Measurement* **42** 1278-87
[13] Stenner A J, Fisher W P Jr, Stone M H and Burdick D S 2013 *Frontiers Psychol.: Quantitat. Psychol. and Meas.* **4** doi: 10.3389/fpsyg.2013.00536
[14] Fisher W P Jr and Stenner A J 2016 *Measurement* **92** 489-96
[15] Cano S, Vosk T, Pendrill L and Stenner A J 2016 *J. Phys.: Conf. Series* **772** 012025
[16] Borsboom D and Dolan C V 2007 *Measurement* **5** 236-63
[17] Borsboom D, Kievit R A, Cervone D and Hood S B 2009 eds J Valsiner, P C M Molenaar, M C D P Lyra and N Chaudhary *Dynamic Process Methodology in the Social and Developmental Sciences* (New York: Springer) 67-98
[18] Molenaar P C M 2004 *Measurement* **2** 201-18
[19] Hamaker E L, Nesselroade J R, Molenaar P C M 2007 *J. Res. Personality* **41** 295-315
[20] Molenaar P C M and Newell K M *Individual pathways of change* (Washington, DC: American Psychological Association)
[21] Chang H 2004 *Inventing temperature* (New York: Oxford University Press)
[22] Sherry D 2011 *Studies Hist. Philos. Sci.* **42** 509-24
[23] Medical Indicators Inc 2006 *NexTemp* http://medicalindicators.com/nextemp-2/
[24] Simpson M A, Kosh A, Elmore J, Bickel L, Stenner A J, Fisher W P Jr., et al. 2015 April 15-19 *Family group mathematics item generation theory: Large scale implementation* Chair M A Simpson (Chicago: National Council on Measurement in Education)
[25] Borsboom D, Cramer A O, Kievit R A, Scholten A Z and Franic S 2009 *The concept of validity* ed R Lissitz (Charlotte, NC Information Age Publishing) 135-70
[26] Borsboom D 2005 *Measuring the mind* (Cambridge, England: Cambridge University Press)

Appendix A. The NexTemp Thermometer

"The NexTemp Thermometer is a thin, flexible, paddle-shaped plastic strip containing multiple cavities. In the Fahrenheit version, the 45 cavities are arranged in a double matrix at the functioning end of the unit. The columns are spaced 0.2°F intervals covering the range of 96°F to 104.8°F….Each cavity contains a chemical composition comprised of three cholesteric liquid crystal compounds and a varying concentration of a soluble additive. These chemical compositions have discrete and repeatable change-of-state temperatures consistent with an empirically established formula to produce a series of change-of-state temperatures consistent with the indicated temperature points on the device. The chemicals are fully encapsulated by a clear polymeric film, which allows observation of the physical change but prevents any user contact with the chemicals. When the thermometer is placed in an environment within its measure range, such as 98.6°F (37.0°C), the chemicals in all of the cavities up to and including 98.6°F (37.0°C) change from a liquid crystal to an isotropic clear liquid state. This change of state is accompanied by an optical change that is easily viewed by a user. The green component of white light is reflected from the liquid crystal state but is transmitted through the isotropic liquid state and absorbed by the black background. As a result, those cavities containing compositions with threshold temperatures up to and including 98.6°F (37.0°C) appear black, whereas those with transition temperatures of 98.6°F (37.0°C) and higher continue to appear green" [23]. Thus, the observed outcome is a count of cavities turned black. The measurement mechanism is an encased chemical compound that includes a varying soluble agent that changes optical properties according to changes in temperature. Amount of soluble agent can be traded off for change in human temperature to hold number of black cavities constant.

Appendix B. Edsphere

The Edsphere™ technology for measuring English language reading ability employs computer-generated, four-option, multiple choice cloze items built on the fly for any prose text. Counts correct on these items are converted into reading measures in the standard unit via an applicable Rasch model. Individual cloze items are one-off creations and disposable; an item is used only once. The cloze and foil selection protocol ensures that the correct answer (cloze) and incorrect answers (foils) match the vocabulary demands of the target text. The text complexity measure and the expected spread of the cloze items are given by a proprietary text theory and associated equations. Thus, the observed outcome is a count of correct answers. The measurement mechanism is a text with a specified complexity and an item generation protocol consistent with that text complexity measure. The text complexity measure can be traded off for a change in reading ability to hold constant the number of items answered correctly.

Appendix C. Mathematics ability measurement

Mathematics ability measurement consists of a common supplemental metric that locates students relative to a taxonomy of mathematical skills, concepts, and applications. In order to develop the framework, several tasks were undertaken: (1) develop a structure of mathematics that spans the developmental continuum from first grade content through Algebra I, Geometry, and Algebra II content, (2) develop a bank of items that have been field tested, (3) develop the unit scale (multiplier and anchor point) based on the calibrations of the field-test items, (4) validate the measurement of mathematics ability as defined by the framework, and (5) link extant tests of mathematical ability to the scale. The process of scale unification for mathematics ability is well underway [24].

At present the attribute "mathematical ability" is unspecified; i.e. there is no specification equation and associated analyzer that can be used to locate 'math text' on the scale. Rather, data intensive methods are employed to calibrate instrumentation and human intensive qualitative analysis is employed to locate math text (e.g. a chapter on adding fractions with uncommon denominators) on the scale. The vast majority of social science attributes are similarly unspecified. By contrasting Nextemp thermometry, theory-based reading measurement, and data-based mathematics ability measurement we hope to illuminate the chasm of difference between instrumentation that employs strong substantive theory and that which does not.

For the vast majority of measurement systems, it is the case "that the difference between any two points for one individual is qualitatively the same as a corresponding difference between two individuals at one time point" [25]; that is, the attribute is homologous. The same cannot be said for many measurement systems used in the social sciences. We propose that the routine adoption of theory based instrument calibrations will pave the way for homologous attributes in the social sciences, thus, assuring that the attribute on which I differ from myself over time is the same attribute on which I differ from my brother [26].

Thinking the future: development of metrology education in Brazil

L S Alves and J M Granjeiro
Postgraduate Program in Biotechnology – Inmetro,
Av. Nossa Senhora das Graças, 50, Xerém, Duque de Caxias/RJ - Brazil

E-mail: lsalves@inmetro.gov.br

Abstract. Brazilian reference documents to organize actions in metrology have chapters dedicated to the education which is described as lacking fundamental metrological concepts in many qualification areas and with incipient activities of specific teaching on the theme. Such documents highlight the challenge to strengthen the knowledge of Metrology. The objective of this study was to make a survey of educational activities described on websites of institutions participating in the quality infrastructure of BIPM member states in South America. This information allowed identifying actions to develop the metrology education in Brazil. The research was classified as exploratory with the qualitative approach. Three categories were set: "Qualification" (the presence-based training courses were the highest occurrence); "Materials" (Audiovisual materials were the most published); and "Others". Argentina was the country with the highest amount and diversity of activities. Some proposed actions were the translation of six materials; disclosure and distribution of 15 materials; extinction of the technical course in Metrology; organization of improvement/post-high school courses; feasibility of specialization qualification level; courses models of INTI's (undergraduate and doctorate levels); and events for teacher training. Moreover, this study offers subsidies to set the strategic guidelines to consolidate and develop a metrological culture and education in Brazil.

1. Metrology education in documents from three Brazilian entities

The document "Physics for Brazil: Thinking about the Future", published by the Brazilian Physical Society in 2005 [1, p. 140-141], emphasizes that Metrology is one of the multidisciplinary challenges to further develop physics and that the major Brazilian need in such area is to attract professionals with high scientific and technologic qualification. According to the document, the metrology area is associated with the demands of a society that intends to become fully industrialized, with a qualitative step for reaching the capacity to conceive and perform experiments, with a common and standardized language, with procedures that ensure the reliability in measurement results, with the safety and wellbeing and with the technical barriers to trading.

In the document entitled "Consolidation of the recommendations from the 4th National Conference of Science, Technology and Innovation towards the Sustainable Development" [2, p. 57] one of the recommendations is the support to qualification programs to tackle metrology challenges in new areas, such as biotechnology, nanotechnology, weather changes and renewable energies.

The document entitled "Strategic Guidelines for the Brazilian Metrology" [3-5] is published every five years since 2003 and always dedicates one of its chapters to education in metrology. The importance of metrology education and its challenges are described in four statements:

- Metrology is essential to increase the quality of products and processes with the consequent increase of competitiveness of industries and of the country's industrial development.
- There is a blatant lack of fundamental metrology concepts in many qualification areas.
- Nowadays new qualification areas such as health, safety and environment depend on metrology knowledge areas in the utilization of sophisticated high-tech equipment and in the employment of technical regulations and standards.
- Metrology teaching activities are incipient.

Following this context, the objective of this study was to make a survey of educational activities described on websites of congener institutions participating in the quality infrastructure of *Bureau International des Poids et Mesures* (BIPM) member states in South America [6]. Reflecting about these activities and comparing the performance of institutions allowed identifying actions to develop the metrology education in Brazil, such as initiatives to reproduce, key institutions to create partnerships and emerging topics to develop new activities.

The research was classified as exploratory with qualitative approach, developed in order to provide an overall vision on a certain fact and is fundamental when the chosen theme is poorly explored [7].

2. Survey of educational activities

Educational activities may be developed in three "teaching ways" [8] or "education modes" [9]: Formal Education, Non-formal Education and Informal Education.

Formal Education occurs in official teaching institutions recognized by the legal system of the country, as well as the Brazilian Ministry of Education (MEC). It is the educational system that is both hierarchical and chronologically structured, with all its parts – bureaucratic and curricular – interconnected and mutually dependent [8 to 11].

The Non-formal Education is any educational activity organized out of the established formal system, with focus on serving specific users and with evident learning goals. Some of the characteristics that make it different from the Formal Education are: less hierarchy, less bureaucracy, variable duration and independent activity without the need of a sequential progression system, of an official teaching institution and of legal recognition. It is also known as non-school or extra-school education [8 to 11].

Informal Education is the process to acquire values and knowledge and to develop attitudes and skills occurring during the whole life thanks to the daily interaction experience with the environment – family, work, and leisure [8 and 10]. The experience of informal education is not intended to reach a specific public, which prevents learning goals from being evidenced and measured, thus making impossible to issue certificates attesting the development of competences.

According to this classification of educational activities, three categories were set for the research: Qualification (Formal Education and Non-formal Education); Materials (Informal Education); Others: activities diverging from the classification of the other two categories.

Member states of the BIPM located in South America are Argentina, Brazil, Chile, Colombia, Uruguay and Venezuela. [6] With the exception of Venezuela, with a click on the name of each one of the countries, one of the information categories is "Quality Infrastructure", showing the list of all participating institutions. From the research on the websites of those institutions, all educational activities with their information available online were accounted and described.

The research identified educational activities in 24 institutions in five countries:

- Qualification: 1168 presence-based and 69 distance-based training courses; 3 technical courses, 34 presence-based and 1 distance-based improvement/post-high school courses; 82 undergraduate courses, 98 specialization courses, 76 master's degree course and 40 doctorate degree courses;
- Materials: 276 articles, 9 magazines (207 ed.), 20 bulletins (339 ed.), 19 reports, 12 manuals, 54 books, 68 folders, 2 Newsletter (152 ed.), 46 booklets, 4 Strategic Plans, 6 Institutional Memories (37 ed.); 819 videos, 334 audios and 10533 pictures (audiovisual); 2 Posters, 1 CD Rom; 2 Didactic Sequences;

- Others: Seminaries and Lectures, Awards and Contests; Qualification Programs; Newsletter Subscription; Disclosure Space or Museum; Virtual Museum; Mobile Museum; Projects and Special Websites; Software; Games; Guided Visits; Publishing House; Radio Programs; Virtual Campus; Qualification Centers; Planning Course; "*De Acuerdo* Magazine" (interinstitutional).

Among the activities classified as "Qualification", the presence-based training courses were the activities of highest occurrence. Audiovisual materials (videos, pictures and audios) were the most published ones. Argentina was the country with the highest amount and diversity of educational activities.

3. The quality infrastructure of BIPM member states in South America

3.1. Argentina

Quality infrastructure is formed by six institutions: *Instituto Nacional de Tecnología Industrial* (INTI), *Comisión Nacional de Energía Atómica* (CNEA), *Ministerio de Economía y Producción* (MECON), *Sistema Métrico Legal Argentino* (SIMELA), *Organismo Argentino de Acreditación* (OAA), *Instituto Argentino de Normalización* (IRAM).

Link to SIMELA on the BIPM's website directs to the website of the *Dirección Nacional de Defensa del Consumidor*. Probably there is an error in the name of the institution on BIPM's website, as no institution associated with the name *Sistema Métrico Legal Argentino* (SIMELA) was found. Link to the website of the *Ministerio de Economía y Producción (MECON)* directs to a website that references two ministries: *Ministerio de Hacienda* and *Ministerio de Finanzas*.

3.2 Brazil

Quality infrastructure is formed by six institutions: National Institute of Metrology, Quality and Technology (INMETRO); National Metrology Laboratory of Ionizing Radiations (LNMRI/IRD); National Observatory/Time Service Division (ON/DSHO); General Accreditation Coordination (CGCRE/INMETRO); Brazilian Association of Quality Control (ABCQ) and Brazilian Association of Technical Standards (ABNT).

Link to the website of the General Accreditation Coordination (CGCRE/INMETRO) on the BIPM's website directed to a page in English that is not available. The research was made on the Portuguese page of CGCRE/INMETRO and it had no information on educational activities.

3.3 Chile

Quality infrastructure is formed by ten institutions: *Instituto Nacional de Normalización* (INN), *Instituto Nacional de Salud Publica* (ISP), *Instituto de Estudios, Medicción y Certificación de Calidad* (CESMEC LTDA), *Calibraciones Industriales S.A* (CISA), *Corporación Nacional del Cobre* (CODELCO), *Laboratorio Custodio de los Patrones Nacionales de la Magnitud Longitud. Nodo de la Red Nacional de Metrología* (DICTUC S.A)., *Empresa Nacional de Aeronáutica* (ENAER), *Instituto de Investigaciones y Control* (IDIC), *Universidad de Concepción* (UDEC), *Red Nacional de Metrología de Chile* (RNM).

The websites of the *Instituto de Estudios, Medicción y Certificación de Calidad* (CESMEC LTDA) and of the *Empresa Nacional de Aeronáutica* (ENAER) had no information on educational activities.

Link to ENAER's website on the BIPM's webpage did not correctly direct to the website and search for the name of the company using Google® search tool identified another address. Link to the website of *Instituto de Investigaciones y Control* (IDIC) on the BIPM's webpage directed to an inactive website of the *Laboratório de Força* belonging to Red Nacional de Metrología de Chile (RNM). On the BIPM's website, the link to *Universidad de Concepción* (UDEC) directed to the website of the *Laboratorio Custodio de los Patrones Nacionales de las Magnitudes Eléctricas* belonging to UDEC. In cases where the institutions of the quality infrastructure listed on BIPM's website are laboratories of bigger institutions, the educational activities were searched in the websites of such institutions.

3.4 Colombia

Quality infrastructure is formed by two institutions: *Instituto Nacional de Metrología da Colombia* (INM) and *Instituto Colombiano de Normas Técnicas y Certificación* (ICONTEC).

3.5 Uruguay

Quality infrastructure is formed by four institutions: *Laboratorio Tecnológico del Uruguay* (LATU), *Instituto Uruguayo de Normas Tecnicas* (UNIT), *La Administración Nacional de Usinas y Trasmisiones Eléctricas* (UTE) and *Laboratorios Tecnogestión do Ministerio de Industria, Energía y Minería* (MIEM – LSMRI).

Considering that the institution *Laboratorios Tecnogestión* is part of the *Ministerio de Industria, Energía y Minería* (MIEM – LSMRI), educational activities were searched on the website of the ministry.

4. Results from survey related to strategic guidelines to consolidate and develop a metrological education and culture in Brazil - years 2013-2017

The results from survey were related to five strategic guidelines (i, ii, vii, viii, ix) concern Formal Education that are in force [5, p.53]:

- i. Intensify programs to insert metrology contents in disciplines of higher education and professionalizing courses;
- ii. Foster and stimulate the production and publication of literature, including didactic books, thesis, studies and researches in the scope of metrology;
- vii. Consolidate and expand professional technical teaching programs and Inmetro's graduate programs, thus expanding the offer to society of programs related to Metrology, Quality and Technology;
- viii. Support and encourage implementation of high school level professionalizing technical courses and schools in all regions of Brazil in accordance with the *Pronatec* Program with focus on areas such as metrology, quality and compliance evaluation;
- ix. Consolidate the Program of Technologic Residence in Legal Metrology, which includes teacher qualification actions and the holding of specialization courses in Legal Metrology with the purpose of qualifying professionals to work in the State Institutes of Weights and Measurements (IPEM).

In order to insert metrology contents in already established subjects (Guideline i), there must be materials published on the subject (Guideline ii) to be then used as either didactic or support material, as well as teacher qualification actions (Guideline ix) to disclose materials and encourage their adoption.

The survey demonstrated 15 materials for education in metrology and quality, nine in Portuguese:

- Bulletin: "*Miradas 2016. Diseño Industrial. Información teórica, de actualidad y casos prácticos, vinculadas con las distintas áreas del diseño*" (INTI);
- Manual: "*Introducción a las Tecnologías de Gestión", "Introducción a las Tecnologías de Gestión. Manual Módulo 2" and "Introducción a las Tecnologías de Gestión. Manual Módulo 3.*" (INTI)
- Book: "*Sistema Internacional de Unidades: SI. — 1ª Edição Brasileira da 8a edição do BIPM", "Sistema Internacional de Unidades Suplemento 2014: Atualizações para a 1ª Edição Brasileira (2012) referente a 8ª Edição do BIPM (2006) da Brochura SI", "Vocabulário Internacional de Metrologia: Conceitos fundamentais e gerais e termos associados (VIM 2012)", "Avaliação de dados de medição –Uma introdução ao "Guia para a expressão de a expressão de incerteza de medição e a documentos correlatos –INTROGUM 2009"; "Avaliação de dados de medição - Guia para a expressão de incerteza incerteza de medição –*

GUM 2008", "Avaliação da Conformidade. 6ª edição", "Medida, Normalização e Qualidade - Aspectos da história da metrologia no Brasil" (INMETRO).

- Folder: "*A tomar medidas - La Metrología es la Ciencia de las Mediciones*" (INTI); "*Normalización" and "Certificación" (IRAM); "Certificação de Produtos" and "Certificação de Sistemas*" (ABNT).
- Magazine: "*De Acuerdo*" (identified on websites of RNM, INM and LATU).

Actions to qualify teachers appeared in institutions INTI, CNEA and INMETRO.

The offering of technical courses and graduate programs (Guidelines vii and viii) occurs in eight institutions. Three of them also offer undergraduate programs. Courses related to education in metrology and quality are:

- one Technical course: Metrology (INMETRO)
- 26 Improvement / Post-High school: "*Asistentes en calidad con orientación en Metrología", "Calidad en la gestión integral de los Procesos", "Calidad y seguridad de sistemas de información*" (INTI); "Fundaments on radioprotection and metrology" (LMRI/IRD); *"Gestión de la calidad para el sector educativo", "Sistema de gestión de calidad e inocuidad en la industria de alimentos", "Gestión de calidad MECI-GP 1000","Sistemas de gestión de calidad NTC ISO 9001: 2015", "Sistemas de gestión integrados HSEQ. NTC ISO 9001:2015, NTC ISO 14001: 2015 y NTC OHSAS: 2007", "Sistema de gestión ambiental SGA. NTC ISO 14001: 2015", "Sistemas de gestión seguridad industrial y salud ocupacional", "Sistemas de gestión seguridad industrial y salud ocupacional. Decreto 1072 2015", "Básico sistema de gestión de calidad para dispositivos médicos - ISO 13485", "Acreditación IPS", "Gestión de la seguridad del paciente", "Gestión de la seguridad del paciente con énfasis en habilitación", "Acreditación para entes territoriales", "Responsabilidad médico legal", "Gestión de la calidad modelo ISO 9001-virtual" (ICONTEC); "Técnico en Gestión de la Calidad UNIT-ISO 9000", "Supervisor (Asistente) en Gestión de la Calidad UNIT-ISO 9000", "Especialista en Gestión de la Calidad UNIT-ISO 9000", "Supervisor (Asistente) en Gestión de la Seguridad y la Salud Ocupacional UNIT-OHSAS 18000", "Especialista en Gestión de la Seguridad y la Salud Ocupacional UNIT-OHSAS 18000", "Especialista en Gestión Ambiental UNIT-ISO 14000", "Especialista UNIT en Recursos Humanos para Sistemas de Gestión"* (UNIT)
- two Undergraduate programs:"*Ingeniería Industrial con orientación en eficiencia y calidad industrial*" (INTI); *"Diploma Superior en Sistemas UNIT de Gestión"* (UNIT)
- six Specialization: "*Calidad industrial en alimentos", "Calidad Industrial", "Tecnología y Gestión de la Seguridad contra Incendio*" (INTI); *"Innovación y Sistemas de Gestión", "Implementación y Auditoría Interna de Sistemas Integrados de Gestión","Gestión de Calidad y Acreditación de Laboratorios Clínicos".* (UDEC)
- four Masters: "*Calidad Industrial*" (INTI); "Metrology and Quality" (INMETRO), "Radioprotection and Dosimetry" (LMRI/IRD); *"Gestión Integrada de Medio Ambiente, Riesgos Laborales y Responsabilidad Social Empresarial"* (UDEC)
- two Doctorate: "*Calidad y Innovación Industrial*" (INTI); "Radioprotection and Dosimetry" (LMRI/IRD)

5 Conclusion: basis for a project on metrology education in Brazil

Even though it is necessary to reach deeper knowledge on all materials and qualifications existing in the institutions, the survey enabled the identification of some actions capable of being used as basis for a project focused on metrology education in Brazil. The actions are:

- Extinction of the technical course in Metrology of INMETRO and the organization of improvement/post-high school courses: the only occurrence of a technical course in Metrology (INMETRO) may indicate that this educational level is not adequate to a specialized qualification. The amount of courses in improvement/post-high school level shows this level better for a specialized metrology education.

- Undergraduate level by means of a course model of INTI's "*Ingeniería Industrial con orientación en eficiencia y calidad industrial*" in two modes - a new qualification in Engineering and discipline ideas to integrate the already existing qualification curricula.
- Improvement specialization courses: graduate courses offered by quality infrastructure institutions in Brazil fall under the masters and doctorate levels. The six specialization courses identified at INTI and UDEC demonstrate the feasibility of such qualification level.
- Research about qualification events to capacitate professors and the doctorate in "*Calidad y Innovación Industrial*" from INTI: they concern the initial and final steps of a formal education path in metrology and quality. However, there was not much information on the websites.
- Translation for six materials into Portuguese and disclosure and distribution of all 15 materials identified as hard and soft copies.

INTI is a key institution to create partnerships and knowing better his initiatives can be the first step to establish the interchange between congener institutions and to launch Brazilian education in metrology and quality.

Moreover, this study offers subsidies to set the strategic guidelines to consolidate and develop a metrological culture and education in Brazil for the next five years - period between 2018 and 2022.

6. References

[1] Chaves A; Shellard R C (ed) 2005 Physics for Brazil: Thinking about the Future. <http://www.sbfisica.org.br/v1/arquivos_diversos/publicacoes/FisicaBrasil_Dez05.pdf>

[2] BRAZIL. Ministry of Science and Technology. Centro de Gestão e Estudos Estratégicos 2010 *Consolidation of the recommendations from the 4th National Conference of Science, Technology and Innovation towards the Sustainable Development.* <http://www.cgee.org.br/atividades/redirect.php?idProduto=6819>

[3] BRAZIL. Ministry of Industry, Foreign Trade and Services. Conselho Nacional de Metrologia, Normalização e Qualidade Industrial. Comitê Brasileiro de Metrologia. 2003 *Strategic Guidelines for the Brazilian Metrology 2003-2007* <http://www.inmetro.gov.br /metcientifica/dirEstrategica/diretrizesEstrategicas_11_02_2003.pdf>

[4] BRAZIL. Ministry of Industry, Foreign Trade and Services. Conselho Nacional de Metrologia, Normalização e Qualidade Industrial. Comitê Brasileiro de Metrologia. 2008 *Strategic Guidelines for the Brazilian Metrology 2008-2012* <http://www.inmetro.gov.br /metcientifica/dirEstrategica/diretrizesEstrategicas.pdf>

[5] BRAZIL. Ministry of Industry, Foreign Trade and Services. Conselho Nacional de Metrologia, Normalização e Qualidade Industrial. Comitê Brasileiro de Metrologia. 2013 *Strategic Guidelines for the Brazilian Metrology 2013-2017* <http://www.inmetro.gov.br /legislacao/resc/pdf/RESC000246.pdf>

[6] Bureau International des Poids et Mesures (BIPM). *Member States.* <http://www.bipm.org/en/about-us/member-states/>

[7] Gil A C 1999 *Métodos e Técnicas da Pesquisa Social* **5 ed**. (São Paulo, Atlas)

[8] Bianconi M L, Caruso F 2005 *Cienc. Cult.* Educação não-formal. 57, n. 4 <http:// cienciaecultura.bvs.br/scielo.php?pid=s0009-67252005000400013&script=sci_arttext>

[9] Pastor Homs M I 2001 *Revista Española de Pedagogia* Orígenes y evolución del concepto de educación no formal **220** <https://dialnet.unirioja.es/descarga/articulo/23701.pdf>

[10] Coombs P H, Prosser R C, Ahmed M 1973 *International Council for Educational Development for UNICEF.* New Paths to Learning for Rural Children and Youth. *Apud* [7]

[11] Gadotti M 2005 *Institut international des droits de l'enfant. Droit à l'éducation: solution à tous les problèmes ou problème sans solution?* A questão da educação formal/não-formal. <http://www.virtual.ufc.br/solar/aula_link/llpt/A_a_H/estrutura_politica_gestao_organizacional/aula_01/imagens/01/Educacao_Formal_Nao_Formal_2005.pdf>

IOP Conf. Series: Journal of Physics: Conf. Series **1044** (2018) 012072 doi:10.1088/1742-6596/1044/1/012072

Uncertainty evaluation for the dynamic measurement of deflections with a falling-weight-type impulse load device

L Martins[1], A Ribeiro[1], J Alves e Sousa[2], A Freire[1], S Fontul[1], F Batista[1] and A Maia[1]

[1] LNEC – National Laboratory for Civil Engineering, Lisbon, Portugal
[2] IPQ – Portuguese Institute for Quality, Caparica, Portugal

Corresponding author's e-mail address: lfmartins@lnec.pt

Abstract. The paper describes the uncertainty evaluation related to the dynamic measurement of deflections using a falling-weight-type impulse load device. This equipment is used in road and airfield pavements in-situ testing, contributing for the assessment of their performance analysis and characterization. This study aims at the determination of the deflection measurement uncertainty, following the GUM approach, where quantities are considered constant in time, and its comparison with the results obtained from recently developed approaches dedicated to time-varying measurement quantities.

1. Introduction

Testing activities related to road and airfield pavements have an important role in assuring the safety of persons and vehicles in transportation networks, which is a major concern to all the entities related to their design, construction, maintenance and management. In this context, several standard test methods are available, contributing for knowledge improvement of the tested pavements, namely, their structural performance and characterization.

Deflection measurement with a falling-weight-type impulse load device [1-2], often referred as Falling Weight Deflectometer (FWD), is one of these standard test methods, in which an impulse load is applied to the pavement surface and its vertical deflection response is measured at points spaced radially outward from the load axis. The impulse load is generated by a weight dropped on a buffer system and is transmitted through a plate resting on the pavement surface. In general, the test apparatus is mounted in a vehicle or on a trailer towed by a vehicle, as shown in Figure 1.

Figure 1. Airfield pavement testing using the FWD.

In addition to the measurement of influence quantities such as air and pavement temperature, the two main quantities measured in the FWD test – deflection and force – constitute time-varying quantities with standard requirements, namely, for the measurement accuracy. In order to comply with this requirement, it is crucial to establish SI traceability through the calibration of the individual measurement chains and to perform a rigorous measurement uncertainty evaluation, which is particularly relevant when flexible pavements are evaluated by FWD, considering that temperature has a relevant influence on pavement behavior and on its dynamic response under the applied load.

The main focus of GUM framework [3,4] is on quantities that are constant in time, considered suitable for a wide range of measurement applications. Measurement accuracy requirements are often more demanding and often time-dependent, thus creating challenges to metrology research, to traceability and to measurement uncertainty evaluation for dynamic quantities [5-9].

The main objective of this paper is, at a first stage, to perform a measurement uncertainty analysis based on the current GUM method and, at a second stage, to compare the results with those produced by recent developed approaches dedicated to the evaluation of dynamic measurement uncertainty, in this case, applied to deflection measurement with a FWD. The measurement process will be briefly described in a dedicated section, as well as the theoretical background on measurement uncertainty evaluation for both time-constant and time-variable quantities. Results obtained from both approaches will be present and compared in order to draw conclusions about the most suitable measurement uncertainty evaluation, taking into account the standard accuracy requirement related to the studied pavement test.

2. Deflection measurements with a falling-weight-type impulse load device

The FWD is the most widely used equipment in Europe and USA for pavement structural evaluation, as it performs non-destructive tests and allows the evaluation of distinct layers that constitute the pavement structure [10]. The use of FWD is even more important nowadays for rehabilitation purposes of the existing pavements, in order to adopt proper reinforcement solutions according to the real condition of the in service pavement.

In the studied standard test method, the FWD is placed over the desired test location, followed by the lowering of a loading plate (with a 300 mm or a 450 mm diameter) and of the deflection sensors (typically nine geophones are used, radially distributed outward from the load axis) to the pavement surface, which must be as clean as possible of loose rock particles and debris. Both the loading plate and the deflection sensors must rest on a firm and stable surface in order to assure suitable test repeatability.

Afterwards, the FWD weight is raised to the height that, when dropped, will apply the required pulse force to the pavement, with an approximate shape of a half-sine wave and a peak force between 45 kN and 250 kN, depending on the characteristics of pavement to be tested (typically, lower dynamic loads for road pavements and higher loads for airfields). When the weight is dropped on the pavement surface, the deflection sensors in the FWD measure the resulting vertical movement whereas a load cell mounted over the load plate measures the corresponding applied pulse force. At least three loading sequences are performed, one to ensure a proper contact of the plate on the pavement surface and the other two in order to evaluate the test repeatability. The main metrological requirements for the mentioned measuring chains are: (i) 1 μm resolution, repeatability of 2 μm and 2% accuracy for the deflection measurement; (ii) 0,2 kN resolution and 2 % accuracy for the force measurement.

3. Experimental data

The experimental data used in the measurement uncertainty evaluation presented in this paper was obtained by LNEC from real pavement testing with a FWD (manufactured by Carl Bro, model PRI 2100). SI traceability of the measuring chains is provided by an accredited calibration laboratory.

In this case, deflection measurements with LNEC's FWD are supported by the use of nine geophones, which typically have a non-linear frequency response in the low frequency range (approximately comprised between 0,2 Hz and 7 Hz). This fact justifies the addition of a dedicated filter to each of the nine geophones, with a natural frequency equal to the geophone's cut-off frequency, in order obtain a linear frequency response in a frequency interval between 0,2 Hz and 300 Hz. The knowledge of the geophones cut-off frequencies (Table 1) is critical for filter compensation (symmetrical to the uncompensated geophone frequency response) and consequently for the deflection measurement accuracy (for example, a 0,5 Hz deviation in the geophone cut-off frequency of the compensation filter can result in a 5% deviation in the deflection peak value).

Table 1. Cut-off frequencies of LNEC's FWD geophones.

Geophone identification	Cut-off frequency / Hz	Geophone identification	Cut-off frequency / Hz
1	4,435	6	4,423
2	4,546	7	4,833
3	4,695	8	4,473
4	4,742	9	4,882
5	4,594	---	---

The calibration of LNEC's FWD geophones is carried out on a controlled dynamic shaker, which includes a SI traceable displacement measurement standard. The adopted metrological procedure includes three sequential tests: (i) the quantification of the linear relation (defined by the slope and offset parameters) between the geophone output digital signal and the displacement reference value, based on the peak values synchronization. This assumes a maximum permissible deviation of 0,1% between the geophone and the standard displacement measurements (Table 2 indicates the parameters estimates for the studied FWD, based on its last calibration certificate); (ii) the analysis of the geophone frequency response, in order to assess the linearization effect resulting from the applied compensation filter; (iii) the linearity test, in which calibration deviations are evaluated for an extended range of dynamic pulses in terms of rise time and amplitude.

Table 2. LNEC's FWD geophone parameterization.

Geophone identification	Slope / µm	Offset	Geophone identification	Slope / µm	Offset
1	11,348	13,666	6	11,326	11,327
2	11,280	-2,424	7	10,994	-0,681
3	10,864	2,7945	8	11,335	-15,6025
4	10,896	-1,121	9	10,663	16,8795
5	11,216	5,916	---	---	---

The evaluation of the dynamic measurement uncertainty, described in the following section, was supported by experimental data obtained in a road test performed in November 2016 by LNEC's Transport Infrastructures Division (Transportation Department), using a load circular plate with a 150 mm diameter and a low-pass filter with a cut-off frequency equal to 150 Hz. Table 3 presents an example of the experimental deflection peak values in one drop, while Figure 2 presents the graphical representation of the corresponding deflection dynamic record.

Table 3. Experimental deflection peak estimates values - drop no. 1, 2016/11.

Geophone identification	1	2	3	4	5	6	7	8	9
Radial offset, /cm	0	30	45	60	90	120	150	180	210
Deflection peak value, /µm	60	33	20	12	6	5	4	3	2

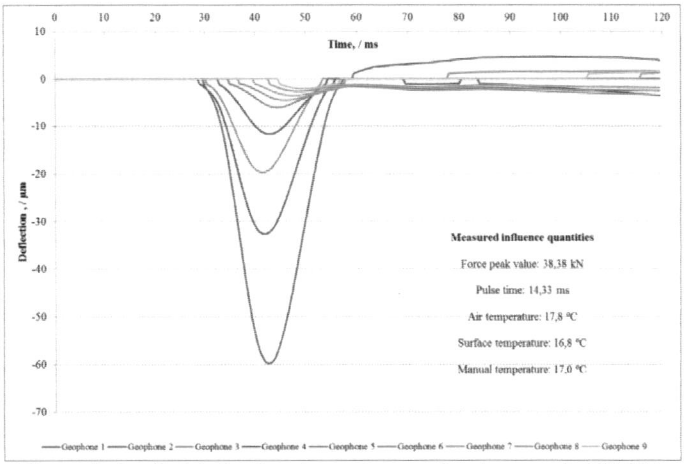

Figure 2. Studied deflection dynamic record – drop no. 1, 2016/11.

4. Evaluation of measurement uncertainty

4.1 Time-constant approach

This measurement uncertainty evaluation approach follows the conventional guidelines of the GUM [3], applied to the deflection measurement in the FWD test. In this case, the following uncertainty components where identified (quantification details are presented in Table 4):

- calibration of the FWD's geophones – this uncertainty component corresponds to the calibration uncertainty of the reference displacement transducer (LVDT) related to the shaker used in the geophone calibration experimental apparatus;
- instrumental drift – based on the historical record of calibration certificates related to LNEC's FWD, the geophones instrumental drift was calculated and the obtained maximum value (for a time period of 36 months) was used in the quantification of this measurement uncertainty;
- short-term repeatability of the deflection measurement – this uncertainty component was quantified based on available information regarding short-term repeatability tests performed after the FWD calibration; an average experimental standard deviation was determined taking into account the nine geophones of LNEC's FWD;
- relative deviations between geophones – following the FWD calibration and in addition to the short-term repeatability test, another test is performed – the stacking tower test – where all the FWD's geophones are installed in a dedicated structure in order to simultaneously measure the same deflection, which allows determining relative deviations between geophones; this uncertainty component was quantified based on the maximum experimental standard deviation of the geophones relative deviations, observed in LNEC's FWD calibration records;
- linearity – this uncertainty component was quantified based on available information of LNEC's FWD manufacturer, regarding the maximum linearity deviation obtained from experimental tests with similar equipment;
- systematic deviation of the deflection peak value – corresponds to the typical deviation (according to the LNEC's FWD manufacturer) between deflection peak values measured by the geophone and the reference measurement standard in the FWD calibration process, considering the use of a compensation filter in agreement with each geophone cut-off frequency;
- rounding-off of deflection peak value – the output file of LNEC's FWD provides the geophones deflection peak values estimates rounded to µm, as mentioned in the test standard [1].

Table 4. Deflection measurement uncertainty budget.

Standard uncertainty component $u(x_i)$	**Uncertainty source**	**Probability distribution**	**Relative standard uncertainty $u_r(x_i)$**	**Degrees of freedom**
$u_{cal}(d)$	Calibration	Gaussian	0,007 5%	50
$u_{drf}(d)$	Instrumental drift	Uniform	0,003 6% / √3 = 0,002 1%	50
$u_{rep}(d)$	Short-term repeatability	Gaussian	0,50 %	17
$u_{dev}(d)$	Relative deviations	Gaussian	0,32 %	8
$u_{lin}(d)$	Linearity	Uniform	0,87% / √3 = 0,50%	50
$u_{sys}(d)$	Systematic deviation	Uniform	0,18% / √3 = 0,10%	50
$u_{rnd}(d)$	Rounding-off	Uniform	0,023% / √3 = 0,013%	50
Combined relative standard uncertainty, $u_{c\,r}(d)$ = 0,78%				
Effective degrees of freedom, ν_{eff} = 60				
Expansion factor, k = 2,00				
95% expanded relative uncertainty, $U_{95\%}(d)$ = 1,6%				

The obtained 1,6% expanded measurement uncertainty value complies with the deflection measurement accuracy requirement (2%) mentioned in the FWD test standard [1]. The main uncertainty contributions are related to the short-term repeatability and the linearity components.

4.2 Time-varying approach

In recent years, the theoretical background regarding the evaluation of dynamic measurement uncertainty was established by several authors [5-9] and is available for use in both calibration and testing applications, as it is the case of the deflection dynamic measurement with a FWD studied in this paper.

In the analysis of a dynamic measurement (shown in Figure 3) [9], the measurement system is assumed to be linear and time-invariant and with a known dynamic behavior, where the measurand is a continuous function $x(t)$. The performed n discrete-time observations $y[n]$ are considered as the result of an analogue-to-digital conversion (ADC) of the system output signal, affected by noise disturbance, $\varepsilon[n]$. In this approach, a linear estimation in the discrete-time domain, $x[n]$, is performed by digital filtering with an appropriate deconvolution filter, and is used later to make an inference of the continuous-time measurand $x(t)$. This approach and the corresponding measurement uncertainty evaluation were implemented in the PyDynamic software package for Python 3.x, developed and validated by PTB and NPL [11].

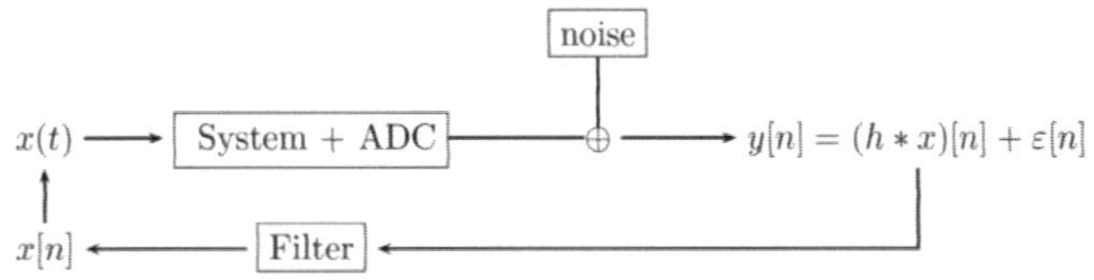

Figure 3. Analysis of a dynamic measurement.

The following input information was used for the study of the dynamic measurement uncertainty:

- the time record of the deflection estimates obtained from geophone no. 1 (shown in Figure 2), where a deflection peak value of 60 μm was obtained in a measuring interval of 120 ms with a sampling frequency of 40 kHz;
- the measurement uncertainty contribution of 0,78% (see Table 4, in the previous section 4.1) for the deflection output signal;

- a low-pass filter for deconvolution with a cut-off frequency equal to 150 Hz, since the deconvolution process is an ill-posed inverse problem, requiring regularization in order to provide reasonable estimates;
- the geophone's nominal frequency response (amplitude and phase estimates and uncertainties), according to the manufacturer's specifications [12]; it should be notice that correlation between different frequencies as well as between amplitude and phase values is assumed to be negligible.

The sequential calculation procedure included several steps of uncertainty propagation [5]: (i) through the frequency domain by the Direct Fourier Transform; (ii) to the real and imaginary parts of the geophone frequency response; (iii) through the inverse system; (iv) through the low-pass filter; (v) and back to the time domain. A 95% expanded relative uncertainty of 0,25% was obtained from the time-variant approach for the deflection peak value estimate of 60 μm, which is quite lower than the 1,6% dispersion value obtained by the time-constant value.

5. Conclusions

The performed uncertainty evaluation showed significant differences between the dispersion values obtained from the time-constant and time-varying approaches. However, both complying with the accuracy requirement mentioned in the FWD test standard (lower than 2%).

Moreover, further improvements are still required in the implemented time-varying approach in order to validate the measurement uncertainty obtained from experimental data, namely, the use of the real geophone frequency response (instead of the nominal response given by the geophone manufacturer) and deformation reference data, in order to study the occurrence of systematic deviations induced by the regularization process in the time-varying approach.

References

[1] ASTM 2009 *ASTM D4694-09 Standard Test Method for Deflections with a Falling-Weight-Type Impulse Load Device*
[2] ASTM 2003 *ASTM D4695-03 Standard Guide for General Pavement Deflection Measurements*
[3] JCGM100 2008 *Evaluation of measurement data. Guide to the expression of uncertainty in measurement* (GUM 1995 with minor corrections) 1st edition
[4] JCGM101 2008 *Evaluation of measurement data. Supplement 1 to the "Guide to the expression of uncertainty in measurement". Propagation of distributions using a Monte Carlo method* 1st edition
[5] Eichstädt S and Wilkens V, 2016 GUM2DFT – a software tool for uncertainty evaluation of transient signals in the frequency domain *Meas. Sci. Technol.* **27** 5 055001 (12pp)
[6] Eichstädt S, Wilkens, A, Dienstfrey, A, Hale, P, Hughes, B and Jarvis C, 2016 On challenges in uncertainty evaluation for time-dependent measurements *Metrologia* **53** S125-35
[7] Eichstädt S, Link A, Harris P M and Elster C, 2012 Efficient implementation of a Monte Carlo method for uncertainty evaluation in dynamic measurements *Metrologia* **49** 3 401-10
[8] Eichstädt S, Elster C, Esward T J and Hessling J P 2010 Deconvolution filters for the analysis of dynamic measurement processes: a tutorial *Metrologia* **47** 5 522-33
[9] Eichstädt S 2012 *Analysis of Dynamic Measurements. Evaluation of dynamic measurement uncertainty* Informationstechnik PTB-IT-16 (Braunschweig: Physikalisch-Technische Bundesanstalt)
[10] Fontul S 2004 *Stuctural Evaluation of Flexible Pavements Using Non-Destructive Tests* (Coimbra: FCTUC)
[11] PyDynamic, https://github.com/eichstaedtPTB/PyDynamic (2017-05-11)
[12] Input/Output 1999, SM-6 Geophone / Sensor Nederland BV, Specification ref. 121026C

Validation of a high resistance calibration method based on a binary voltage divider

V Cabral, L Ribeiro, J A Sousa

IPQ – Portuguese Institute for Quality, Caparica, Portugal

Corresponding author's e-mail address: vcabral@ipq.pt

Abstract. A method to measure standard resistors in the range 10 kΩ to 1 TΩ, based in automated bridge using a binary voltage divider is described, tested and validated by comparison with other method based in a Modified Wheatstone Bridge. Evaluation of measurement uncertainty is also validated.

1. Introduction

The aim of this work is to describe a binary voltage divider (BVD) [1][2] bridge system implemented in IPQ, the National Measurement Institute (NMI) of Portugal, and the validation process of its performance through a comparison of its results with a validated Modified Wheatstone Bridge (MWB) method, including an approach to identify the relevant uncertainties sources of the BVD bridge, and the results of the two methods will be compared in order to get an experimental evidence of the uncertainty to the BVD setup.

The validation procedure includes the overall process of uncertainty evaluation, under the GUM uncertainty framework and supplements [3][4]. The validation of the measurement uncertainty evaluation is particularly important in situations where the mathematical model is based on ratios, as it has been shown elsewhere [7]. The GUM conventional approach will be validated using a Monte Carlo method (MCM) [4] in order to propagate the distributions through the functional relationships referred above. The influence of the choice of PDF for the input quantities, based on the experimental data available, will be studied.

2. The BVD Bridge

The simplified diagram of the BVD bridge setup is represented in Figure 1, whereas the MWB is shown in Figure 2 (to be developed in chapter 3). The digital voltmeter (DVM) and the voltage source are independent equipment connected to the acquired commercial bridge from "*Measurement International*" (MI 6000B).

The bridge works in an automated way to find the balance between the output voltage of the BVD and the test voltages, V_1 to V_4, using the DVM as null detector. The balance voltages condition detected by the DVM corresponds to the determination of the ratios r_i by the BVD. The ratio, R, between the two resistors R_x and R_s calculated by the bridge software, can be expressed in terms of the measured quantities by:

$$R = \frac{R_x}{R_s} = \frac{\frac{V_1}{E} - \frac{V_2}{E}}{\frac{V_3}{E} - \frac{V_4}{E}} = \frac{r_1 - r_2}{r_3 - r_4} \quad (1)$$

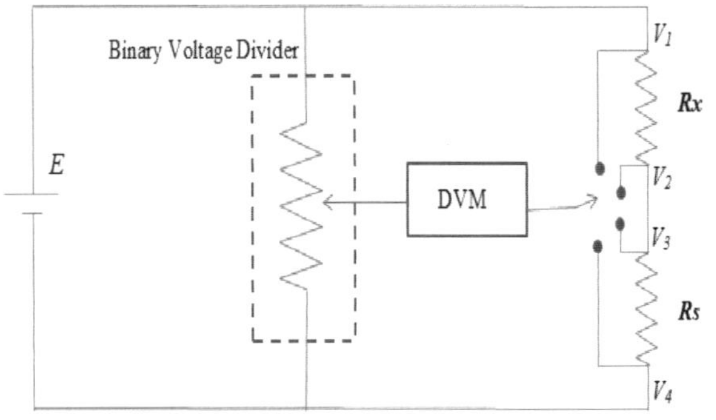

Figure 1. Simplified diagram of the BVD bridge.

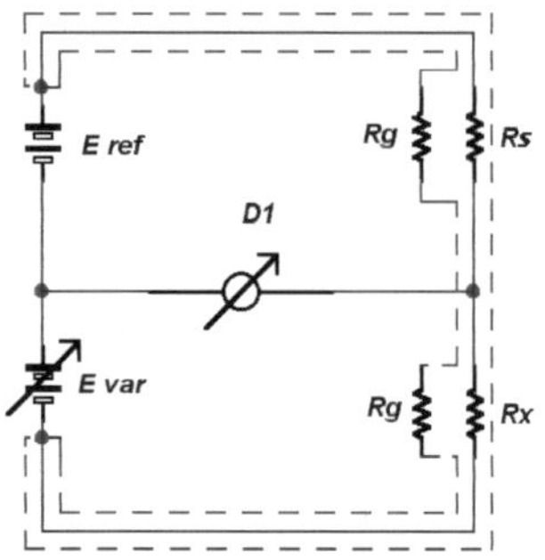

Figure 2. Simplified diagram of the MWB

Therefore, the accuracy of the system depends, first of all, on the three independent instruments used. From the measurement equation, the result does not depend on the absolute value of the source voltage, but on its stability along the time interval of the measurements. The DVM is used as null detector and therefore the resolution, zero offset stability and linearity in the mV range are the relevant DVM parameters to be taken into account for the accuracy of the system. The BVD is based on a 13-bit resolution resistive divider made up of resistors with low temperature coefficients. Its auto-calibration feature assures the determination of corrections factors for each bit, with an extended resolution enabled by the resolution of the DVM. Besides the instrumentation, the thermoelectric and leakage effects are also relevant influence quantities. The current leakage effect has significant impact in the final ratio results, due to the higher resistance values involved, as described in [5].

2.1. The bridge MI6000B measurement model

The accompanied software calculates each voltage ratio, V_i , by the following relation:

$$V_i = (r_i + c_i) \times E + (V_{i,b} - V_{i,off}) \tag{2}$$

where:

- r_i ratio determined by the binary divider for each of the balanced voltages V_i .
- c_i correction determined by the bridge auto-calibration feature, to each stage of the BVD.
- E voltage applied to the resistors by the voltage source and measured with the DVM.
- $V_{i,b}$ residual voltage measured, in balance condition, between the BVD output and the resistor terminal. This measurement strategy allows an extended resolution of the 13-bit binary divider, limited by the quantity $E/2^{13}$, to the resolution of the DVM.
- $V_{i,off}$ offset voltage of the circuit measured by the DVM, shortcutting the inputs.

After sequential measurements of E followed by the four V_i values, the software calculates ratio R:

$$R = \frac{R_x}{R_s} = \frac{V_1 - V_2}{V_3 - V_4} \quad (3); \qquad R_j = \frac{R_d + R_i^{-1}}{2} \quad (4); \qquad R_m = \frac{\sum_{j=1}^{n} \frac{R_{d,j} + R_{i,j+1}}{2}}{n} \quad (5)$$

It is also available an "interchange" option that allows that each R is the mean of two sequential measurements, R_j, with the change of the resistors in the circuit to remove the thermoelectric effect error of the circuit, where R_d is the direct ratio R_x/R_s and R_i is the inverse ratio R_s/R_x . Finally, it is possible to choose a number (n +1) of sequential ratio measurements to produce a mean value R_m.

2.2. An approach to evaluate the measurement uncertainty

In this section, a simplified approach will be presented to evaluate the measurement uncertainty taking as a starting point the known functional relationship implemented by the bridge software. It will not take into account the correlation between the input quantities based on voltage measurements and of the relation (5). The experimental standard deviation of the measurements is estimated from the repeatability of the output quantity, R. The reproducibility is taken from R relates to different sequential cycles. The relevant uncertainty about the residual leakage currents in the system is formulated as an input quantity in the model equation, but since there is no immediate way to estimate this source of uncertainty, it will be a target to estimate indirectly by the results comparison obtained with the two distinct methods addressed in the work. Equation (5) may be rewritten as:

$$R = R_m + \delta R_{rep} + \delta R_{repro} + \delta R_{leak} \quad (6)$$

The evidence of correction factors, δ, to each input quantity in expression (2) above results in following expression:

$$V_i = (r_i + c_i + \delta c_i) \times (E + \delta E) + (V_{i,b} + \delta V_{i,b} - V_{i,off} - \delta V_{i,off}) \quad (7)$$

The terms δ above are estimated as a zero-expected value.

δc_i	error correction in the c_i determination including stability along calibrations.
δE	drift in the E value during each measurement cycle: V_1 to V_4.
$\delta V_{i,b}$	error correction in the $V_{i,b}$ due to linearity and finite resolution of the DVM.
$\delta V_{i,off}$	error correction in the $V_{i,off}$ due to the finite resolution of the DVM.
δR_{rep}	auxiliary quantity to represent the global repeatability of the setup (one cycle).
δR_{reprod}	auxiliary quantity to represent the global reproducibility of the setup (sequential cycles).
δR_{leak}	error due to the imperfect correction of leakage current effects of the system.

It is known that bridge software measures and corrects the leakage effect, using a "null virtual configuration of the circuit", and offers the availability of "an isolated guard source that follows the voltage being measured and keeps the guard circuit at the same potential. The resulting leakage current is then practically zero", but how such feature is implemented is not addressed by the supplier. [5]

The variance of R, associated with the GUM uncertainty framework [3], is expressed in a compact way, as a function of its covariance and partial derivatives by the matrix relation:

$$u^2(R) = d^T COV_R d \quad (8)$$

This simplified approach assumes the linearized model entailed by the GUM approach, i.e., linear model, Gaussian input quantities and Central limit theorem, which requires validation for any model not fulfilling those conditions. With most input quantities having uniform distribution in the ratio models, a validation with a Monte Carlo method [4] will be attempted.

2.2.1. Uncertainty of the V_i terms

The uncertainty of V_i can be estimated based on the measurement model of equation (7) (for simplicity we will omit the subscript i):

$$u(V)^2 = E^2\, u(\delta c)^2 + (r + c)^2 u(\delta E)^2 + u(\delta V_b)^2 + u(\delta V_{off})^2 \quad (9)$$

From the experience in working with the binary divider, the accuracy for each binary factor correction determination, by the software bridge, is taken as less than a few parts in 10^{-8}. From the observation of the factor correction values change, between periodic auto-calibrations made by the bridge, we can also assume a few parts in 10^{-8} to its stability. To each bit activated in the determination of r_i ratios, we will consider a value of $5\text{x}10^{-8}$ to the δc interval limit. The uncertainty related to the Voltage Source, E, is due to the its drift during each measurement cycle of V_1 to V_4 and was confirmed experimentally to the used source as being limited to 10^{-7}. Moreover, the range of V values measured by the DVM depends on the resolution of the binary divider applied to the E value (for E equal to 100 V, the quantity V could reach 12 mV); for the DVM used the linearity of the 100 mV range was verified to be less than 1 ppm and its resolution is 10 nV. Finally, the uncertainty of δV_{off} is affected by the finite resolution of the DVM and could be estimated, for a resolution of 10 nV.

Table 1. Uncertainty values associated with quantity V in equation (11)

Source	$u(\delta c)$	$u(\delta E)$	$u(\delta V_b)$	$u(\delta V_{off})$
Value	$\frac{\sqrt{n}\times 5\times 10^{-8}}{\sqrt{3}}$	$\frac{10^{-7}\times E}{\sqrt{3}}$	$\sqrt{\left(\frac{1\times 10^{-6}\times V}{\sqrt{3}}\right)^2+\left(\frac{5\times 10^{-9}}{\sqrt{3}}\right)^2}$	$\frac{5\times 10^{-9}}{\sqrt{3}}$

where n is the number of bits activated to determine each r_i

2.3. Tests to 100 MΩ : 10 MΩ ratio

Experimental measurements at 100 V were made to explore the system behavior in different connection configurations of the bridge, the DVM (Agilent 3458A) and the two Guildline 9330 resistors, with nominal values of 10 MΩ as R_s and 100 MΩ as R_x. Each measurement followed the configuration proposed in the bridge manual [1] for this range of resistance: 30 repeated measurements to each V_i; 6 (n+1) sequential measurements for a mean R value determination and with the interchange of the resistors as formulated in equation (5). The uncertainty associated with the leakage current correction in its value is one of the open questions in this paper, but an estimation of its value will be attempted based on the results obtained with another experimental method taken as reference and comparing the results of both methods

2.3.1. Time stability, repeatability of measurements and guarding

From several series of continuous measurements, it was established the need for long periods of stabilization. Figure 3 shows that several hours are needed to obtain an experimental standard-deviation, $s(R)$, of about 0.13 ppm. Figure 4, show two different configurations: active guard ON: a standard deviation of 0.34 ppm was observed; active guard OFF: 1.37 ppm. The difference in the mean values is 0.10 ppm less than the standard deviation, which shows that, for this particular range, the leakage current in the DVM connection does not play a relevant role as systematic effect but increase the noise in the measurement results. For measurements taken with active guard ON an interval due to reproducibility, $(R_{máx} - R_{min})$ of 1.1 ppm was observed. This means that reproducibility is about one order of magnitude larger than the repeatability and should be taken into account in the configuration of a measurement setup to produce meaningful results

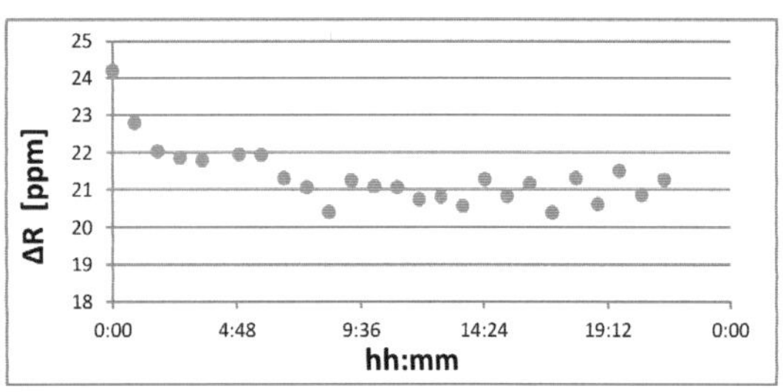

Figure 3. Ratio measurements, stabilization period.

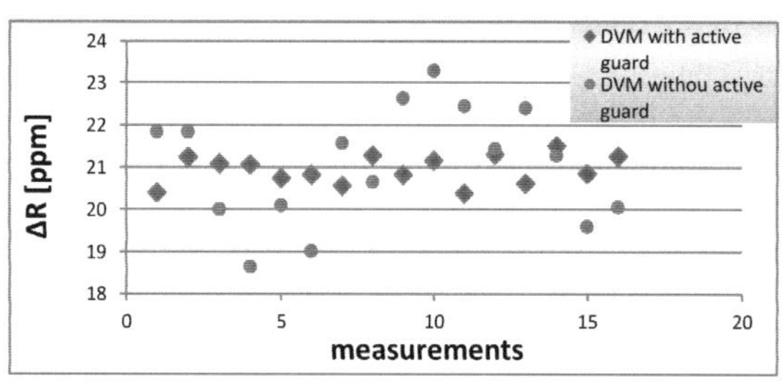

Figure 4. Ratio measurements, after stabilization.

To observe the relevance of the active guard, a measurement was made floating the screen of the connections (between the resistors and the bridge). A change in the mean value of the ratio from 20.9 ppm to 31.9 ppm was observed. It can be concluded that the leakage currents corrections plays an essential role in this setup and particular attention must be paid to the way how guard is implemented.

2.3.2. Results for the ratio 100 MΩ : 100 MΩ and uncertainties

With the input values indicated in Table 1, the results are illustrated in Figure 5 and Table 2 for the GUM (line) and MCM (bar) approaches. It can be seen that assuming a Gaussian distributions for the measurand, despite the fact that all input quantities are taken as having a uniform distribution, except for the repeatability, produces a results that is significantly higher than the value obtained by propagating the true uniform distributions. Vertical lines represent 95 % confidence limits. This is an important result to alert those that think that only complicated models can produce erroneous results when using the GUM. In this case the GUM significantly overestimates the measurement uncertainty.

Table 2. Uncertainty values with GUM and MCM methods

Quantity	R_d	$u(R_d)$	R_i	$u(R_i)$	R_m	$U(R_m)$
GUM	10.000235	0.8 ppm	0.099998	1.7 ppm	10.000209	2.0 ppm
MCM	10.000234	0.5 ppm	0.099998	1.0 ppm	10.000209	1.2 ppm

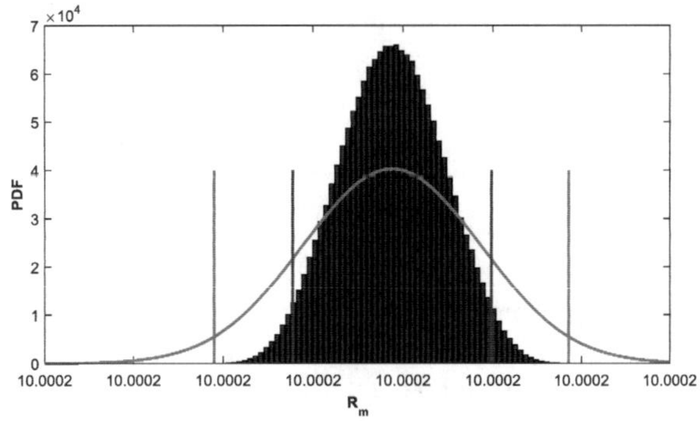

Figure 5. Uniform and Gaussian distributions for V_i.

3. The Modified Wheatstone Bridge

The simplified diagram of the MWB setup implemented in IPQ [6] is represented in Figure 2. This measurement method evaluates the ratio of two high value standard resistors (R_s and R_x) based on a well-known ratio of two dc voltages. The guard resistors R_g are chosen to have a value 10 times the value of R_s and R_x, respectively.

The equilibrium condition, measured by an electrometer is described by the expression (10). The ratio between the two voltages is equal to the ratio of the two resistors, and one can say that the uncertainty with which the ratios are determined is the same. In a first measurement, there is an imbalance situation translated by a current δI_1 (11). To cancel this current, assuming that the two branches of the bridge are practically independent of each other, we can simply change the E_{var} voltage of a value corresponding to (12).

$$R = \frac{E_{var}}{E_{ref}} = \frac{R_x}{R_s} \quad (10); \qquad E_{var1} = \frac{R_x \times E_{ref}}{R_s} + \delta I_1 \times R_x \quad (11); \qquad \delta E_1 = \delta I_1 \times R_x \quad (12)$$

With this new voltage value, a second measurement is carried out, obtaining a current already very close to zero. In the implemented procedure, a third measurement is performed to test the linearity. For this, a 20% increase over the previously calculated δE is introduced. Zero compensation is applied to each set of three measurements by subtracting the mean of the two consecutive zero readings. From equations above and for each of determinations, we can assume:

$$\frac{E_{var}}{E_{ref}} = R + \delta I \times R_x \quad (13)$$

The theoretical value at which δI is cancelled, R, is calculated by a linear adjustment of the three points previously determined ($\frac{E_{var}}{E_{ref}}$, δI) and corresponds to the intercept of the line fit. The mean square error value of this adjustment gives us a measure of the uncertainty of this value. The other uncertainty sources are due to the accuracy of the voltage ratio determination and to the electrometer accuracy in the δI measurements. This entire process is then repeated by inverting the polarity of the connections with a low noise switch so as to override thermal effects in the measuring circuit. The mean of the two theoretical R values is then considered to best describe the ratio of the resistors under evaluation, R_m. Neglecting differences in the resistors thermal behaviours (as they were kept under a bath controlled temperature within ± 0.01 °C), considering a global repeatability and reproducibility estimated for n sequential measurements of R_m, δR_{rep}, and assuming the correction error due to current leakage effects on the system, δR_{leak}, a final equation for the output quantity can be formulated:

$$R = R_m + \delta R_{rep} + \delta R_{leak} \quad (14)$$

4. Methods comparison and conclusions

Measurements were made to determine the 10:1 ratio resistors allowing comparing the results of both systems setups presented, with the following final results:

Table 3. Ratio results and Uncertainty from different methods

System	Ratio measured ppm	Expanded Uncertainty ppm	Uncertainty method approach
BVD	20.9	2.0	GUM
		1.2	MCM
MWB	17.1	3.7	GUM

An agreement of about 3.8 ppm has been determined that is in accordance with the combined uncertainty of the comparison 4.2 ppm for the GUM method approach and 3.9 for the MCM method approach. As the uncertainty values presented for each system setups (BVD and MWM) don't take into account any specific contribution due to leakage currents, we can conclude that, in those particular setups, the automated corrections guaranteed by the BVD software seems to be effective. For the resistors type used, it was seen that leakage currents could reach 11 ppm in the difference ratios. The use of the guarded voltage in DVM and resistors connections has a determinant influence in the correction of leakage currents and in the noise measurements.

The simplified uncertainty approach here presented, based in the model equation of the measurements implemented by the BDV Bridge, conducts to values that are in accordance with other proven method (MWD). The uncertainty assumed to the factor correction values of the BVD have a dominant weight in the combined uncertainty resulting from the model equation derived from the way as the bridge effectively works.

Further measurements will be needed to consolidate this preliminary result and extended to higher values of resistance where the effects of leakage currents will be more relevant.

References

[1] Automatic High Resistance Ratio Bridge – MODEL 6000B – Operator Manual; Measurements International; 6000B.OM Rev10.

[2] Adrew F Dunn, "Measurement of Resistances Ratios in The Range to 100 Megaohms", IEEE Transactions on Instruments Measurements, Vol. 40, n. 2, ISSN 0018-9456, pp 278-280.

[3] *Evaluation of measurement data. Guide to the expression of uncertainty in measurement*, JCGM 100:2008 (GUM 1995 with minor corrections), 1st edition, September 2008.

[4] *Evaluation of measurement data. Supplement 1 to the "Guide to the expression of uncertainty in measurement". Propagation of distributions using a Monte Carlo method*, JCGM 101:2008, 1st edition, 2008.

[5] Roman Honing, "Pratical Aspects of High Resistance Measurements", Cal Lab – The international Journal of Calibration, jan-mar 2010.

[6] L. Ribeiro, M. Nunes, M., "Software Application For Semi-Automatic Measurement Of High Value Standard Resistors"; *in proc* of Trends In Electrical Measurement And Instrumentation; ISBN 972-98115-4-7; (11th IMEKO TC-4 Symposium) 208/211, Instituto Superior Técnico - Lisboa; September 13-14, 2001.

[7] J. Alves e Sousa, A. Silva Ribeiro, "The choice of method to the evaluation of measurement uncertainty in metrology", *IMEKO XIX World Congress*, 9-11 September, 2009, Lisbon.

[8] Duane Brown, Measurements International Resistance Measurements Systems w/Sub PPM.